Autonomous Marine Vehicle Operations—2nd Edition

Autonomous Marine Vehicle Operations—2nd Edition

Guest Editors

Xiao Liang
Rubo Zhang
Xingru Qu

MDPI

Basel • Beijing • Wuhan • Barcelona • Belgrade • Novi Sad • Cluj • Manchester

Guest Editors

Xiao Liang
Naval Architecture and Ocean Engineering College
Dalian Maritime University
Dalian
China

Rubo Zhang
College of Mechanical and Electronic Engineering
Dalian Minzu University
Dalian
China

Xingru Qu
College of Mechanical and Electronic Engineering
Dalian Minzu University
Dalian
China

Editorial Office
MDPI AG
Grosspeteranlage 5
4052 Basel, Switzerland

This is a reprint of the Special Issue, published open access by the journal *Journal of Marine Science and Engineering* (ISSN 2077-1312), freely accessible at: https://www.mdpi.com/journal/jmse/special_issues/6939G7X6D6.

For citation purposes, cite each article independently as indicated on the article page online and as indicated below:

Lastname, A.A.; Lastname, B.B. Article Title. *Journal Name* **Year**, *Volume Number*, Page Range.

ISBN 978-3-7258-5019-8 (Hbk)
ISBN 978-3-7258-5020-4 (PDF)
https://doi.org/10.3390/books978-3-7258-5020-4

Contents

About the Editors

Xiao Liang

Xiao Liang received his B.S. degree in Naval Architecture and Ocean Engineering from Harbin Engineering University, China, in 2003, his M.S. degree in Fluid Mechanics from Harbin Engineering University in 2006, and his Ph.D. in Design and Construction of Naval Architecture and Ocean Structure in 2009. He is currently a Professor with the School of Naval Architecture and Ocean Engineering, Dalian Maritime University, China. He has authored and co-authored more than 80 papers in scientific journals. His current research interests include decision making and advanced control, unmanned technology and swarm intelligence in maritime applications.

Rubo Zhang

Rubo Zhang received his B.S. degree in Automatic Control of Aircraft and his M.S. degree in Navigation Control of Aircraft from Harbin Shipbuilding Engineering Institute, Harbin, China, in 1984 and in 1987, respectively. He received his Ph.D. in Control Theory and Control Engineering from Harbin Engineering University, Harbin, China, in 1999. He is currently a Professor with the College of Mechanical and Electronic Engineering, Dalian Minzu University, Dalian, China. He has published over 100 papers in journals and conferences. His research interests include intelligent robot hardware and software architecture, multi-robot technology, and autonomous decision-making technology in complex environments.

Xingru Qu

Xingru Qu received his B.S. degree in Naval Architecture and Ocean Engineering from Shandong Jiaotong University, Jinan, China, in 2015, and his M.S. and Ph.D. degrees in Design and Construction of Naval Architecture and Ocean Structure from Dalian Maritime University, Dalian, China, in 2022. He is currently a Lecturer with the College of Mechanical and Electronic Engineering, Dalian Minzu University, Dalian, China. He has published over 20 papers in journals and conferences. His research interests include guidance and control as well as the game of autonomous marine vehicles.

Journal of
Marine Science and Engineering

MDPI

Editorial

Autonomous Marine Vehicle Operations—2nd Edition

Xiao Liang [1,*], Rubo Zhang [2,*] and Xingru Qu [2,*]

[1] School of Naval Architecture and Ocean Engineering, Dalian Maritime University, Dalian 116026, China
[2] College of Mechanical and Electronic Engineering, Dalian Minzu University, Dalian 116600, China
* Correspondence: liangxiao@dlmu.edu.cn (X.L.); zhangrubo@dlnu.edu.cn (R.Z.); qxr@dlnu.edu.cn (X.Q.)

1. Introduction

In recent years, the field of autonomous marine vehicles has undergone remarkable advancements, with unmanned surface vehicles (USVs) and unmanned underwater vehicles (UUVs) demonstrating transformative potential for oceanographic exploration and marine applications [1,2]. However, practical deployment of individual vehicles confronts substantial technical barriers arising from system nonlinearities, tightly coupled dynamics, multiple inputs and outputs, parametric uncertainties, and complex constraints. These operation complexities become exponentially amplified in networked swarm systems, where high-dimensional state spaces and dynamically evolving network topologies introduce unprecedented challenges in autonomous coordination. Current research frameworks systematically address these issues through three interconnected domains: perception, decision-making, and control [3,4]. On this background, the newly launched Special Issue "Autonomous Marine Vehicle Operations—2nd Edition" is founded, including 15 selected hot topics.

In the field of perception for USVs and UUVs, current research emphasizes the acquisition and fusion of multi-source information under complex marine environments. This primarily involves the cooperative perception of multimodal sensors, including sonar, radar, and optical/infrared cameras, coupled with deep learning algorithms for target detection and tracking [5,6]. Existing studies widely adopt spatiotemporal alignment of multi-sensor data and feature fusion methods enhanced by attention mechanisms to improve robustness under adverse weather conditions. However, significant challenges remain, such as interference from underwater optical scattering, limited accuracy in small target identification, and asynchronous multi-sensor data integration. These issues are particularly pronounced in low-visibility scenarios, such as nighttime or foggy conditions, where the false detection rate of current methods exceeds practical application requirements.

Decision-making research focuses on intelligent planning and collaborative decision in dynamic environments, encompassing single-vehicle path planning, multi-vehicle task allocation, and human–machine collaborative decision [7,8]. Popular approaches include improved A* algorithms, deep reinforcement learning, and game-theoretic models, addressing challenges such as trajectory optimization under wind, wave, and current disturbances; collision avoidance with sudden obstacles; and dynamic multi-objective scheduling [9,10]. While significant progress has been achieved in simulation environments, real-world marine applications still face bottlenecks such as cumulative errors in environmental modeling, insufficient decision-making real-time performance, and communication reliability in multi-vehicle systems. To address these challenges, some scholars have proposed digital twin-driven virtual–physical hybrid decision validation methods, which enhance system adaptability by constructing high-fidelity digital replicas of marine environments [11]. However, mechanisms for data acquisition and model updates remain to be further refined.

The core focus of control systems lies in motion control and actuator optimization under complex disturbances, with key advancements targeting wave–current disturbance suppression, actuator fault tolerance, and energy consumption optimization [12,13]. Current methodologies are dominated by adaptive control, model-predictive control, and sliding-mode control, often combined with Kalman filtering for state estimation and reinforcement learning for parameter self-tuning [14]. While these control algorithms achieve high precision in speed and heading control, challenges such as actuator saturation and instability in power distribution persist under extreme sea conditions. Emerging research trends highlight the deep coupling of "perception-decision-control" frameworks, with end-to-end learning approaches being introduced to enhance the whole system performance [15]. However, foundational issues, including model interpretability and safety verification, remain critical barriers to further progress.

2. An Overview of Published Articles

This Special Issue concentrates on autonomous marine vehicle operations in complex environments, containing 15 published articles. Main contributions are delineated below.

To address the path planning of USVs in dynamic environments, contribution 1 proposes a multi-agent proximal policy optimization-based planning approach. This method incorporates an action critic network and reward function aligned with collision avoidance rules while leveraging an improved YOLO model to enhance target detection accuracy. The proposed framework enables USVs to execute rational evasive maneuvers in scenarios such as head-on encounters, crossing situations, and overtaking.

To overcome the limitations of traditional formation control methods, including lagging formation adjustments, oscillatory obstacle avoidance paths, and poor adaptability to dynamic environments, contribution 2 introduces a novel formation control approach integrating virtual structures and artificial potential fields. By employing a dynamic velocity regulation mechanism, the proposed method achieves adaptive formation adjustments and robust obstacle avoidance in dynamic scenarios.

Recognizing the inadequacy of traditional integrated navigation in meeting the demands for miniaturization and swarm operations, contribution 3 combines the lightweight Doppler Velocity Log and Ultra Short Baseline to propose a cost-effective, high-precision underwater navigation method. This approach dynamically corrects sensor installation biases, mitigating the impact on velocity and position measurements, and demonstrates the feasibility and superiority through experimental validation.

Contribution 4 addresses the challenge of maritime collision avoidance by leveraging expert knowledge to quantify the fuzzy terminology (e.g., "very large ship") in the COLREGs as membership functions. These functions are integrated into a collision avoidance decision system, effectively enhancing autonomous vehicles' ability to assess collision risks.

Considering the large parameter size and high computational complexity of existing target detection models, contribution 5 proposes a model compression method combining pruning and knowledge distillation. By constructing inter-layer dependency graphs to automatically identify redundant parameter groups, and employing a hybrid strategy of output and feature distillation, the proposed method minimizes detection accuracy loss while significantly reducing model size.

To address the limitations of underwater fusion localization accuracy, contribution 6 introduces a Lie group-based dual-coordinate representation method. The proposed approach constructs a state-space model encompassing rotation matrices, positions, velocities, and sensor biases, while employing a Lie algebra disturbance model to handle inertial bias. This method avoids the accuracy loss typically associated with linearization in traditional extended Kalman filtering.

Tackling the rationality and compliance of autonomous collision avoidance in complex marine environments, contribution 7 presents an NSGA-II algorithm enhanced with good point sets. A dual-objective fitness function based on collision risk and path cost is established, with membership functions quantifying the influence of various parameters on collision risk. By explicitly assigning responsibilities under different scenarios based on regulatory rules, the approach achieves a balance between safety and efficiency.

To address the inefficiency and data scarcity of sonar image acquisition, contribution 8 develops a diffusion model-based image generation method for specific categories. A nonlinear gain enhancement algorithm compensates for echo intensity at greater distances, while category information is incorporated into the model. By training a noise prediction model, the method generates high-quality and diverse sonar images.

Contribution 9 proposes a stereo image-based localization technique for UUVs to address high-precision underwater-positioning challenges. By capturing multi-view images along pre-planned trajectories, a 3D geometric model containing point features, line segments, and angular features is generated. Using feature detection, epipolar constraints, and correlation analysis, the method matches model features with trajectory data, enabling object recognition and pose estimation.

To address formation control challenges under time-varying communication delays and input saturation constraints, contribution 10 proposes a distributed prescribed performance control method based on state prediction. Real-time predictors estimate the states of neighboring nodes, while nonlinear mapping functions confine formation errors within predefined dynamic boundaries, thereby improving both transient and steady-state performance.

To mitigate the issues of differential explosion and chattering in sliding-mode control, contribution 11 introduces a bio-inspired sliding-mode control method. Integral sliding mode surfaces are employed to reduce steady-state tracking errors, while hyperbolic tangent functions replace the traditional sign function to suppress oscillations. A dynamic attenuation mechanism smooths the virtual velocity of underactuated USVs, ensuring uniformly ultimately bounded system errors.

Addressing the challenges of low resolution, high noise, and multipath reflections in underwater sonar images, contribution 12 develops a Gaussian clustering and maximum clique matching-based registration method. Local geometric features are described using Gaussian blocks, while a distance ratio-based adaptive mismatch rejection strategy is introduced. Temporal and non-temporal constraints are employed to suppress error accumulation, significantly enhancing the robustness of sonar image feature matching.

In addressing the angular velocity measurement errors in underwater navigation, contribution 13 introduces the Underwater Gyros Denoising Net (UGDN) based on deep learning. This network integrates dilated convolutions and memory modules to extract spatiotemporal features from Inertial Measurement Unit sequences, dynamically compensating for gyroscope angular velocity measurement errors. The UGDN optimizes calibration parameters through end-to-end learning, significantly enhancing the accuracy of underwater navigation systems.

To address the decoupling of underwater positioning and communication systems, contribution 14 proposes a low-cost positioning algorithm based on acoustic communication. This method utilizes acoustic communication to simultaneously achieve positioning and data transmission. By compensating for Doppler frequency shifts, it enhances speed estimation accuracy and constructs a lightweight Extended Kalman Filter framework. This framework optimizes the state covariance matrix to suppress the accumulation of inertial navigation errors, thereby improving the reliability of underwater positioning.

In response to the need for underwater natural gas pipeline inspection, contribution 15 develops an autonomous pipeline tracking and damage detection method based on UUVs. This approach employs convolutional neural networks to process pipeline images captured by onboard cameras, utilizing feature extraction and support vector machine classifiers to detect damages. By integrating navigation data, the system dynamically calibrates the position of the damage relative to the starting point, enhancing the efficiency and accuracy of underwater pipeline inspections.

3. Conclusions

The 15 contributions in this Special Issue collectively address critical technical bottlenecks in autonomous marine vehicle operations through innovative methodologies spanning navigation, control, perception, and collaborative systems. Solutions to dynamic path planning are exemplified by multi-agent proximal policy optimization with action critic networks (Contribution 1) and collision avoidance frameworks integrating COLREGs-compliant risk quantification (Contribution 7). Navigation advancements feature lightweight Doppler-ultrashort baseline fusion for swarm operations (Contribution 3) and Lie group-based dual-coordinate localization (Contribution 6), while acoustic communication-driven positioning achieves simultaneous data transmission and inertial error suppression (Contribution 14). Control innovations include bio-inspired sliding-mode designs mitigating chattering through hyperbolic tangent functions (Contribution 11) and distributed prescribed performance strategies addressing time-varying delays via state prediction (Contribution 10). In perception, diffusion-model-based sonar synthesis overcomes data scarcity through nonlinear gain compensation (Contribution 8), complemented by Gaussian-clustering registration enhancing feature-matching robustness (Contribution 12). Cross-domain integration is evident in digital twin-validated formation control (Contribution 2), stereo vision-driven UUV localization (Contribution 9), and CNN-powered pipeline damage detection (Contribution 15). Computational efficiency gains are achieved via dependency graph-guided model compression (Contribution 5), whereas deep learning-enhanced gyro-denoising networks (Contribution 13) demonstrate the potential of data-driven sensor calibration. These works systematically advance marine autonomy through three synergistic paradigms: hybrid physics–AI architectures (e.g., reward-aligned reinforcement learning and Lie algebra-augmented filtering), resource-aware optimization (e.g., pruning-distillation compression and acoustic-Doppler fusion), and regulatory-compliant intelligence (e.g., membership function-based COLREGs interpretation). The current research works shown in this Special Issue should not only be considered as the results of investigation accomplished by the respective scholars, but as a starting point, encouraging readers to continue with new studies.

List of Contributions

1. Wen, N.; Long, Y.; Zhang, R.; Liu, G.; Wan, W.; Jiao, D. COLREGs-Based Path Planning for USVs Using the Deep Reinforcement Learning Strategy. *J. Mar. Sci. Eng.* **2023**, *11*, 2334. https://doi.org/10.3390/jmse11122334.
2. Liu, G.; Wen, N.; Long, F.; Zhang, R. A Formation Control and Obstacle Avoidance Method for Multiple Unmanned Surface Vehicles. *J. Mar. Sci. Eng.* **2023**, *11*, 2346. https://doi.org/10.3390/jmse11122346.
3. Liu, J.; Yu, T.; Wu, C.; Zhou, C.; Lu, D.; Zeng, Q. A Low-Cost and High-Precision Underwater Integrated Navigation System. *J. Mar. Sci. Eng.* **2024**, *12*, 200. https://doi.org/10.3390/jmse12020200.

4. Kristić, M.; Žuškin, S. Quantification of Expert Knowledge in Describing COLREGs Linguistic Variables. *J. Mar. Sci. Eng.* **2024**, *12*, 849. https://doi.org/10.3390/jmse12060849.
5. Cheng, C.; Hou, X.; Wang, C.; Wen, X.; Liu, W.; Zhang, F. A Pruning and Distillation Based Compression Method for Sonar Image Detection Models. *J. Mar. Sci. Eng.* **2024**, *12*, 1033. https://doi.org/10.3390/jmse12061033.
6. Wang, C.; Cheng, C.; Cao, C.; Guo, X.; Pan, G.; Zhang, F. An Invariant Filtering Method Based on Frame Transformed for Underwater INS/DVL/PS Navigation. *J. Mar. Sci. Eng.* **2024**, *12*, 1178. https://doi.org/10.3390/jmse12071178.
7. Liang, Z.; Li, F.; Zhou, S. An Improved NSGA-II Algorithm for MASS Autonomous Collision Avoidance under COLREGs. *J. Mar. Sci. Eng.* **2024**, *12*, 1224. https://doi.org/10.3390/jmse12071224.
8. Zhang, F.; Hou, X.; Wang, Z.; Cheng, C.; Tan, T. Side-Scan Sonar Image Generator Based on Diffusion Models for Autonomous Underwater Vehicles. *J. Mar. Sci. Eng.* **2024**, *12*, 1457. https://doi.org/10.3390/jmse12081457.
9. Bobkov, V.; Kudryashov, A. A Method for Recognition and Coordinate Reference of Autonomous Underwater Vehicles to Inspected Objects of Industrial Subsea Structures Using Stereo Images. *J. Mar. Sci. Eng.* **2024**, *12*, 1514. https://doi.org/10.3390/jmse12091514.
10. Zhang, H.; Jiang, Y.; Gao, R.; Li, H.; Li, A. Prescribed Performance Formation Tracking Control for Underactuated AUVs under Time-Varying Communication Delays. *J. Mar. Sci. Eng.* **2024**, *12*, 1533. https://doi.org/10.3390/jmse12091533.
11. Dong, Z.; Tan, F.; Yu, M.; Xiong, Y.; Li, Z. A Bio-Inspired Sliding Mode Method for Autonomous Cooperative Formation Control of Underactuated USVs with Ocean Environment Disturbances. *J. Mar. Sci. Eng.* **2024**, *12*, 1607. https://doi.org/10.3390/jmse12091607.
12. Zhuang, L.; Chen, X.; Lu, W.; Yan, Y. Graph Matching for Underwater Simultaneous Localization and Mapping Using Multibeam Sonar Imaging. *J. Mar. Sci. Eng.* **2024**, *12*, 1859. https://doi.org/10.3390/jmse12101859.
13. Cao, C.; Wang, C.; Zhao, S.; Tan, T.; Zhao, L.; Zhang, F. Underwater Gyros Denoising Net (UGDN): A Learning-Based Gyros Denoising Method for Underwater Navigation. *J. Mar. Sci. Eng.* **2024**, *12*, 1874. https://doi.org/10.3390/jmse12101874.
14. Garin, R.; Bouvet, P.-J.; Tomasi, B.; Forjonel, P.; Vanwynsberghe, C. A Low-Cost Communication-Based Autonomous Underwater Vehicle Positioning System. *J. Mar. Sci. Eng.* **2024**, *12*, 1964. https://doi.org/10.3390/jmse12111964.
15. Kartal, S.K.; Cantekin, R.F. Autonomous Underwater Pipe Damage Detection Positioning and Pipe Line Tracking Experiment with Unmanned Underwater Vehicle. *J. Mar. Sci. Eng.* **2024**, *12*, 2002. https://doi.org/10.3390/jmse12112002.

Acknowledgments: As Guest Editors of the Special Issue "Autonomous Marine Vehicle Operations—2nd Edition", we would like to express vast gratitude toward all authors whose valuable works were published.

Conflicts of Interest: The authors declare no conflicts of interest.

References

1. Wang, Y.; Wang, C.; Han, Q.; Wang, X. Networked and Deep Reinforcement Learning-Based Control for Autonomous Marine Vehicles: A Survey. *IEEE Trans. Syst. Man Cybern.-Syst.* **2025**, *55*, 1. [CrossRef]
2. Restreop, E.; Matous, J.; Pettersen, K.Y. Tracking Control of Cooperative Marine Vehicles Under Hard and Soft Constraints. *IEEE Trans. Control Netw. Syst.* **2024**, *11*, 4. [CrossRef]

3. Xu, H.; Zhang, X.; He, J.; Yu, Y.; Cheng, Y. Real-Time Volumetric Perception for Unmanned Surface Vehicles Through Fusion of Radar and Camera. *IEEE Trans. Instrum. Meas.* **2024**, *73*, 1–12. [CrossRef]
4. Liu, J.; Chen, H.; Xie, S.; Peng, Y.; Zhang, D.; Pu, H. Trajectory planning for unmanned surface vehicles in multi-ship encounter situations. *Ocean Eng.* **2023**, *285*, 2. [CrossRef]
5. Wang, B.; Jiang, P.; Gao, J.; Huo, W.; Yang, Z.; Liao, Y. A lightweight few-shot marine object detection network for unmanned surface vehicles. *Ocean Eng.* **2023**, *277*, 114329. [CrossRef]
6. Wu, Z.; Wu, Z.; Chen, X.; Lu, Y.; Yu, J. Self-Supervised Underwater Image Generation for Underwater Domain Pre-Training. *IEEE Trans. Instrum. Meas.* **2024**, *73*, 1–14. [CrossRef]
7. Cui, Z.; Guan, W.; Zhang, X. Collision avoidance decision-making strategy for multiple USVs based on Deep Reinforcement Learning algorithm. *Ocean. Eng.* **2024**, *308*, 118323. [CrossRef]
8. Luo, J.; Zhang, Y.; Zhuang, J.; Su, Y. Intelligent Task Allocation and Planning for Unmanned Surface Vehicle (USV) Using Self-Attention Mechanism and Locking Sweeping Method. *J. Mar. Sci. Eng.* **2024**, *12*, 179. [CrossRef]
9. Yin, S.; Xu, N.; Shi, Z.; Xiang, Z. Collaborative path planning of multi-unmanned surface vehicles via multi-stage constrained multi-objective optimization. *Adv. Eng. Inform.* **2025**, *65*, 103115. [CrossRef]
10. Gao, K.; Gao, M.; Zhou, M.; Ma, Z. Artificial intelligence algorithms in unmanned surface vessel task assignment and path planning: A survey. *Swarm Evol. Comput.* **2024**, *86*, 101505. [CrossRef]
11. Liu, G. Control Strategies for Digital Twin Systems. *IEEE-CAA J. Autom. Sin.* **2024**, *11*, 1. [CrossRef]
12. Li, W.; Ge, Y.; Guan, Z.; Gao, H.; Feng, H. NMPC-based UAV-USV cooperative tracking and landing. *J. Frankl. Inst.* **2023**, *360*, 11. [CrossRef]
13. Qu, X.; Liang, X.; Hou, Y. Fuzzy State Observer Based Cooperative Path-Following Control of Autonomous Underwater Vehicles with Unknown Dynamics and Ocean Disturbances. *Int. J. Fuzzy Syst.* **2021**, *23*, 6. [CrossRef]
14. Yuan, W.; Rui, X. Deep reinforcement learning-based controller for dynamic positioning of an unmanned surface vehicle. *Comput. Electr. Eng.* **2023**, *110*, 108858. [CrossRef]
15. Qu, X.; Jiang, Y.; Zhang, R.; Long, F. A Deep Reinforcement Learning-Based Path-Following Control Scheme for an Uncertain Under-Actuated Autonomous Marine Vehicle. *J. Mar. Sci. Eng.* **2023**, *11*, 1762. [CrossRef]

Journal of
Marine Science and Engineering

MDPI

Article

COLREGs-Based Path Planning for USVs Using the Deep Reinforcement Learning Strategy

Naifeng Wen, Yundong Long, Rubo Zhang *, Guanqun Liu *, Wenjie Wan and Dian Jiao

College of Mechanical and Electronic Engineering, Dalian Minzu University, Dalian 116600, China; wennaifeng@dlnu.edu.cn (N.W.); l2905975877@outlook.com (Y.L.); 19841141350@163.com (W.W.); jd0120666@163.com (D.J.)

* Correspondence: zhangrubo@dlnu.edu.cn (R.Z.); liuguanqun@dlnu.edu.cn (G.L.)

Abstract: This research introduces a two-stage deep reinforcement learning approach for the cooperative path planning of unmanned surface vehicles (USVs). The method is designed to address cooperative collision-avoidance path planning while adhering to the International Regulations for Preventing Collisions at Sea (COLREGs) and considering the collision-avoidance problem within the USV fleet and between USVs and target ships (TSs). To achieve this, the study presents a dual COLREGs-compliant action-selection strategy to effectively manage the vessel-avoidance problem. Firstly, we construct a COLREGs-compliant action-evaluation network that utilizes a deep learning network trained on pre-recorded TS avoidance trajectories by USVs in compliance with COLREGs. Then, the COLREGs-compliant reward-function-based action-selection network is proposed by considering various TS encountering scenarios. Consequently, the results of the two networks are fused to select actions for cooperative path-planning processes. The path-planning model is established using the multi-agent proximal policy optimization (MAPPO) method. The action space, observation space, and reward function are tailored for the policy network. Additionally, a TS detection method is introduced to detect the motion intentions of TSs. The study conducted Monte Carlo simulations to demonstrate the strong performance of the planning method. Furthermore, experiments focusing on COLREGs-based TS avoidance were carried out to validate the feasibility of the approach. The proposed TS detection model exhibited robust performance within the defined task.

Keywords: COLREGs; USV cooperative path planning; multi-agent proximal policy optimization; deep learning; target detection

1. Introduction

The key to the unmanned surface vehicle (USV) cooperative path-planning problem is the adaptive selection of the best collision-avoidance action [1]. In the procedure, the subproblems below are predominantly considered, i.e., cooperatively avoiding static obstacles, avoiding collision within the USV fleet and between USVs and target ships (TSs) [2]. Several challenges are posed to the planning process. High-dimensional action and state spaces present a major difficulty to the efficiency of the online planner that is required to respond to various scenarios in time [3,4]. Meanwhile, the dynamic TS avoidance problem is complicated since it can greatly increase the dimension of the planning space [5,6].

The development of the International Regulations for Preventing Collisions at Sea (COLREGs)-compliant navigation has been in two areas: the complexity of ship encounter scenarios and the evolution in methodologies.

Studies [7–12] have contributed to the advancement in research on ship collision avoidance by presenting systematic approaches and providing insights into the interpretation of COLREGs rules. Research described in [7] offers valuable insights into ship collision avoidance based on COLREG 72, which can be useful for Officers of the Navigational Watch (OONW), both onboard and remotely, as well as for autonomous systems. However, the research is supported by examples drawn from various works that, despite their significant influence in current literature, may not be from the correct standpoint. In [8], Kim and Park provide insights into COLREGs sailing rules based on the perspectives of navigators and researchers. In [9], the recent progress in COLREGs-compliant navigation of USVs from traditional to learning-based approaches is reviewed in depth.

In [10], Yim and Park presented a systematic approach to model evasive action aimed at preventing collisions in a give-way situation at the minimum-distance moment. The researchers established a conceptual framework for such evasive action and identified COLREGs-compliant maneuvers through a simulation based on ship-handling scenarios. In [11], Kim and Park proposed a method for determining the appropriate timing and necessary actions to ensure ship collision avoidance in accordance with COLREGs rules, using Bayesian-regularized artificial neural networks (BRANNs). In [12], Hagen et al. expressed the COLREGs rules mathematically, providing insights into their interpretation through the selection of parameters and weights.

There are mainly four classes of conventional methods that make a great effort to accommodate COLREGs rules in their path-planning modules [9].

Rule-based methods involve hand-crafted designs and focus on simple ship-encounter scenarios and, hence, cannot be easily extended to more complex ship-encounter scenarios. Hybrid methods, such as A*-variants and rapidly exploring random-tree variants, suffer from the complexity of the multi-TS-avoiding problem and the high dimension of the multi-USV planning space. Reactive methods, such as the artificial potential field and velocity obstacle methods, have difficulties in uncertain TS course prediction. Optimization-based methods are also hard to conduct in complex TS encounter scenarios.

Additionally, traditional research usually focuses on simple 1–1 ship-encounter scenarios dictated by rules 13–16 in Part B of the COLREGs. However, the USV path-planning method becomes more challenging when considering multi-TS encounter scenarios. When planning those algorithms in simulated or real scenarios, simplified and basic assumptions of COLREGs are far from its real complexity [7,8].

Therefore, traditional methods are not able to fully use the seamanship of experienced mariners to solve complex situations, and they can hardly be considered powerful nonlinear approximators of the optimal value and policy functions.

Different from traditional methods, as a paradigm in the field of machine learning (ML) to achieve multi-agent collaboration, the deep reinforcement learning (DRL) model mainly studies the synchronous learning and evolution of agent strategies that are used to plan the coordinated movement of formation in real time [13]. By interacting with the environment, it continuously optimizes the agent's action strategy. The value function of different action strategies in the current state is estimated, and high-return actions are executed to avoid performing low-return or punitive actions. The deep network module of DRL is utilized to fit the motion model of USVs, enabling smooth control and avoiding falling into the local optimal solution [13]. DRL is well suited for situations where the optimal decision-making strategy is not known beforehand and for dealing with non-stationary environments where the underlying dynamics may change over time [13]. These characteristics make it an effective and powerful tool for our task.

DRL methods can be divided into three categories [14]. The value-based methods, such as deep Q-network variants, estimate the optimal values of all different states and then derive the optimal policy using the estimated values [15,16]. The policy-based methods, such as trust-region policy optimization (TRPO) [17] and proximal policy optimization (PPO) [18], optimize the policy directly without maintaining the value functions. Actor–critic methods, such as the deep deterministic policy gradient (DDPG) [19], twin-delayed DDPG (TD3) [20], and soft actor–critic (SAC) [21], can be viewed as a combination of the above two methods, and they maintain an explicit representation of both the policy (the actor) and the value estimates (the critic).

The vanilla policy gradient (VPG) method directly optimizes the policy by utilizing the gradient of the expected reward, which can lead to slow convergence [17]. In response to this challenge, TRPO ensures that each update to the policy parameters remains within a trusted region to prevent large parameter updates that could potentially degrade the performance of the policy [17]. Additionally, TRPO exhibits the capability of effectively handling high-dimensional state spaces. However, it is relatively complicated, and it is not compatible with architectures that include noise (such as dropout) or parameter sharing (between the policy and value function, or with auxiliary tasks), and it has poor data efficiency. To address this problem, the PPO method emerged. PPO converts the constraint term of TRPO into a penalty term to decrease the complexity of constrained optimization problems, and it uses only first-order optimization [18]. Multi-agent PPO (MAPPO) is a version for the multi-agent partially observable Markov's decision-making process [22]. The multi-agent DDPG (MADDPG) method is another state-of-the-art method besides MAPPO. In MADDPG, each agent takes the other agents as part of the environment, and agents in the same region cooperate with each other to determine the optimal coordinated action [19]. However, it is hard to achieve stability due to the complexity of the hyperparameters.

In [23], a multi-USV automatic collision-avoidance method was employed based on a double deep Q network (DDQN) with prioritized experience replay. In [24], the PPO algorithm and a hand-crafted reward function are used to encourage the USV to comply with the COLREGs rules.

Sawada et al. [25] proposed a collision-avoidance method based on PPO, and it uses a grid sensor to quantize obstacle zones by target and a convolutional neural network (CNN) and long–short-term memory (LSTM) network to control the rudder angle. Xu et al. [26] proposed a COLREGs intelligent collision-avoidance (CICA) algorithm that tracks the current network weight to update the target network weight, which improves stability when learning the optimal strategy.

The study in [27] employed a collision-avoidance framework that divides all encounter scenarios into seven types according to the avoidance constraints of the COLREGs for different encountered scenes.

In [28], the COLREGs and ship maneuverability were considered in the reward for achieving multi-ship automatic collision avoidance, and the optimal reciprocal collision-avoidance (ORCA) algorithm was used to detect and reduce the risk of collision.

In the work by [6], the action space and reward function were improved by incorporating the reciprocal velocity obstacle (RVO) scheme. Gate-recurrent unit-based networks were utilized to directly map the state of varying the number of surrounding obstacles to the corresponding actions.

In [5], the collaborative path-planning problem was modelled as a decentralized partially observable Markov decision-making process and used to devise an observation model, a reward function, and an action space suitable for the MAPPO algorithm for multi-target search tasks. In the research described in [29], a system was constructed to switch between path-following and collision-avoidance modes in real time, and the collision hazard was perceived through encounter identification and risk calculation.

In [30], the Q learning method was applied for optimizing the state action pairs to obtain the initial strategy, then PPO was used to fine-tune the strategy. In [31], agents were trained using a mixture of observations from different training environments and linearity constraints were imposed on both the observation interpolations and the supervision (e.g., associated reward) interpolations. In the study described in [32], multiple targets were simultaneously optimized to enhance the PPO.

The comparative methods are outlined in Table 1. The majority of current DRL methods focus on generating COLREGs-compliant paths using a COLREGs-based reward function. However, the high degree of randomness in early-stage action selection can lead to unpredictable strategy gradient updates, making it challenging to achieve model convergence.

Table 1. Introduction to comparative methods.

	Method	Advantage	Limitation
Path replanning methods	Rule-based methods	COLREGs rules integrated into path replanning	relies on hand-crafted design; hard to extend to complex ship-encounter scenarios
	Hybrid methods: A star	fast; COLREGs-compliance incorporated into path replanning	hard to extend to more complex ship-encounter scenarios
	Reactive methods: velocity obstacle	fast; COLREGs compliance enforced by integrating forbidden zones	accurate TS course information required
	Optimization-based methods	optimal; COLREGs rules naturally formulated as constraints	relatively high computational burden
DRL methods	Value-based methods: Q-learning, DQN	optimal policy derived from estimates of the optimal values of all different states	overestimation bias; high dimensionality; hard to strike a balance between exploration and exploitation
	Policy-based methods: MAPPO, TRPO	directly optimizes the policy without maintaining the value functions	careful design of reward functions required
	Actor–critic methods: MADDPG, MATD3,	explicit representation of both the policy and the value estimates	computationally expensive; hyperparameter sensitivity

Moreover, in the case of USVs, it is crucial for their paths to be both feasible and optimal, while ensuring that multiple USVs can maintain their formation and reach their respective goals [33,34]. Additionally, being COLREGs-compliant does not guarantee an ideal evasive behavior as it may result in overly conservative or inattentive responses to unexpected TSs.

By leveraging a large dataset of pre-recorded USV trajectory data and employing powerful DRL-based methods, it is possible to derive promising solutions that not only ensure that USV navigation adheres to COLREGs rules but that it also replicates the good seamanship exhibited by experienced mariners. Furthermore, path planning in dynamic environments poses a complex problem due to the need to plan multiple paths simultaneously while ensuring collision avoidance within the USV fleet and between USVs and TSs. Therefore, it is crucial that the planner is efficient. MAPPO offers high efficiency in learning, fast convergence, and improved stability, making it an ideal choice as the basic path planner for our task.

Subsequently, this research seeks to propose a two-stage multi-agent reinforcement learning (MARL) scheme based on the MAPPO algorithm, incorporating a centralized training and a decentralized execution strategy. The following innovations are presented:

1. We introduce a COLREGs-compliant action-evaluation module to compute action probabilities that align with COLREGs regulations when encountering multiple TSs. The module parameters are learned from a dataset of pre-recorded USV trajectories. By fusing the probability vector and the candidate action vector from the actor network, we select an action that is the most feasible for the encountered situation. Our reward function incorporates both COLREGs and seamanship considerations, providing a dual heuristic approach to guide the selection of COLREGs-compliant actions.

2. We propose a policy network that can handle multiple aggregation goals, obstacles, and dynamic TSs. To achieve this, we have defined the action space, observation space, and reward function for the policy network. Additionally, we have designed actor and critic networks.
3. A TS motion-detection network is constructed to provide guidance for the decision-making process of the MARL model.

2. Planning Problem Definition

2.1. USV Motion Model

Figure 1 shows the Earth-fixed inertial frame {i} and the body-fixed frame {b}. The positive direction of the X-axis of frame {b} coincides with the USV heading direction, and the origin is located at the barycenter of the USV.

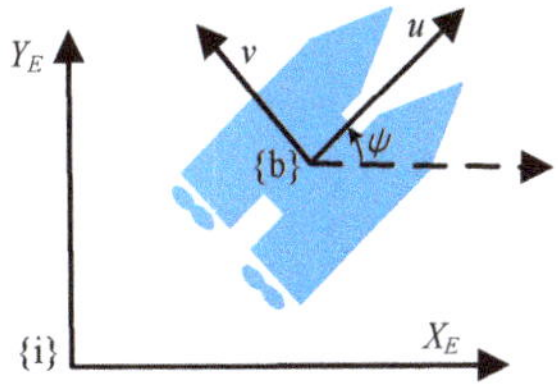

Figure 1. Schematic of the USV motion model.

The kinematic model is as follows:

$$\begin{cases} \dot{x} = u \cdot \cos\psi - v \cdot \sin\psi \\ \dot{y} = u \cdot \sin\psi + v \cdot \cos\psi \\ \dot{\psi} = r \end{cases} \tag{1}$$

To guarantee the feasibility of the resulting path, we give zero mask values to the actions that violate motion constraints; for example, the maximum velocity and acceleration, which are defined by the performance limitations of the USV.

2.2. Cooperation Problem Description

The cooperation is defined as a loosely coupled one whereby USVs choose the aggregation targets freely in terms of the minimum aggregation cost and collision-avoidance actions. The objective is illustrated as follows:

$$\begin{array}{l} \operatorname{argmin}\left\{ \sum\limits_{i=1}^{N} \sum\limits_{j=1}^{M} s_{ij} \right\} \quad s.t.|p_m - p_t| > r_o \text{ and} \\ \min\limits_{r}(|p_{iM} - q_i| < d_{agg}) \end{array} \tag{2}$$

where N represents the number of USVs and the variable M is the number of waypoints on the path of the ith USV, s_{ij} is the cost of the ith USV reaching the jth waypoint on its path, and s_{ij} considers the turning angle, velocity, and the distance travelled. If the distance between the end-point of a path and the corresponding USV's aggregation point is smaller than d_{agg}, then the USV will have achieved the aggregation target. The planned paths must be collision-free, meaning that the minimum distance between USVs as well as that between USVs and TSs must exceed the safety radius r_o. Here, p_m represents the position of a USV, and p_t indicates the position of a sailing obstruction.

2.3. COLREGs Rules for Collision Avoidance

A USV may encounter head-on, crossing, and overtaking situations, as shown in Figure 2.

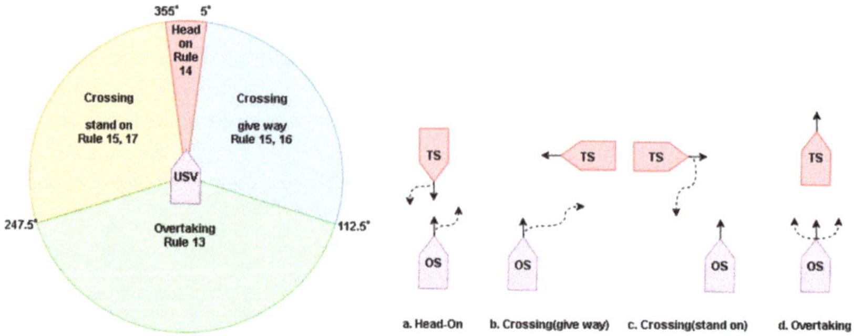

Figure 2. Illustrations of the COLREGs.

Referring to studies [7–12], the applicable COLREGs rules for this research are as follows:

- Rule 14: head-on. When two power-driven vessels are meeting on reciprocal or nearly reciprocal courses such that this involves the risk of a collision, each shall alter its course to starboard so that each shall pass on the port side of the other.
- Rule 15: crossing situation. The USV has the option to either stand on or give way to the TS. When two power-driven vessels are crossing such that this involves the risk of a collision, the vessel that has the other on its own starboard side shall keep out of the way and shall, if the circumstances of the case permit, avoid crossing ahead of the other vessel.
- Rule 16: action by the give-way vessel. Every vessel, which is directed by these rules to keep out of the way of another vessel shall, so far as possible, take early and substantial action to keep well clear of another vessel.
- Rule 17: action by the stand on vessel. (i) Where one of two vessels is to keep out of the way, the other shall keep her course and speed. (ii) The latter vessel may, however, take action to avoid collision by its maneuver alone, as soon as it becomes apparent to it that the vessel required to keep out of the way is not taking appropriate actions in compliance with these rules. When, from any cause, the vessel required to keep her course and speed finds itself so close that collision cannot be avoided by the action of the give-way vessel alone, it shall take such action as will best aid to avoid collision.
- Rule 13: overtaking. The COLREGs states that "any vessel overtaking any other shall keep out of the way of the vessel being overtaken". The above description stipulates that it is the responsibility of the overtaking ship to avoid a collision, but there is no clarity as to what action said ship should take to avoid a collision. Therefore, in the overtaking situation, we do not define a specific reward function but directly use the reward functions in the base layer to evaluate the avoidance actions of the USV.

Most of the recent research has focused on addressing more complex scenarios, such as areas with restricted visibility that have obstructions and busy narrow channels governed by a traffic separation scheme. These scenarios involve interactions with vessels that may not comply with the COLREGs. For these scenarios, rules 2(b), 8, and 17 should be considered [9].

Rule 2(b): responsibility. Under special circumstances, a departure from the rules may be made to avoid immediate danger.

Rule 8: actions to avoid collision. Actions shall be made with ample time. If there is sufficient sea-room, alteration of course alone may be the most effective. Reduce speed, stop, or reverse if necessary. Action by a ship is required if there is a risk of collision, and when the ship has right-of-way.

Rule 17: actions by the stand-on vessel. Where one of two vessels is to keep out of the way, the other shall keep her course and speed. The latter vessel may, however, take action to avoid collision by her maneuver alone, as soon as it becomes apparent to her that the vessel required to keep out of the way is not taking appropriate actions in compliance with these rules.

In Figure 2, the yellow area represents the situation where the ship is crossing from the port side of other ships, the blue area represents the starboard crossing situation, the red area represents the head-on situation, and the green area represents the overtaking situation in which the USV is being overtaken. Figure 2a–d illustrate the actions that each vessel should take according to the COLREGs.

Figure 3 shows a typical collision-avoidance situation when a TS crosses the USV path [2,35], where v_i denotes the USV velocity, v_o represents the velocity of the TS, v_{io} is the relative velocity of the USV with respect to the TS, r_o is the safe radius from the USV to the TS, $p_i = (x_i, y_i)$ denotes the position of the USV, $p_o = (x_o, y_o)$ denotes the position of the TS, and the turning direction is represented by $n_{\text{io}\perp}$, which is perpendicular to the line $p_i p_o$.

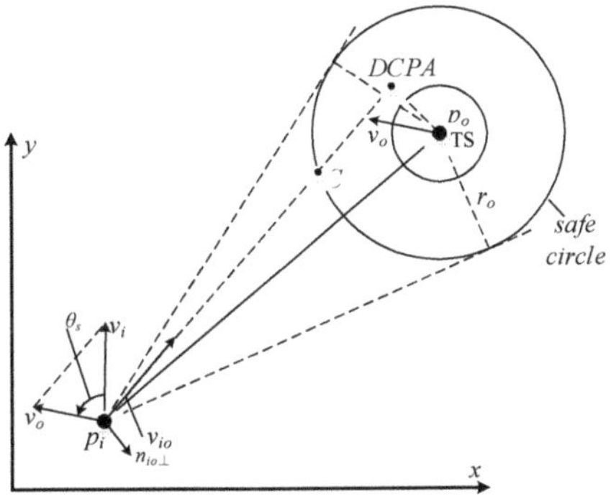

Figure 3. TS avoidance situation evaluation according to the COLREGs.

We observe that the direction of collision avoidance for the give-way vessel, as specified by the COLREGs, is primarily determined by the absolute angle (θ_s, $0<|\theta_s|<\pi$) from the motion direction of the USV to that of the TS in the crossing situation. Therefore, the collision-avoidance direction for the give-way vessel corresponds to the steering direction of the USV that increases the magnitude of $|\theta_s|$ [35].

Since encounters are dynamic situations, continuous monitoring is required. We assume that USVs remain vigilant for other vessels. If another vessel (including TSs or other USVs) does not comply with the COLREGs, the USV should take collision-avoidance actions in time.

2.4. Collision Detection

The collision prediction method is the collision-cone method [2,35], which determines the collision area according to the velocity relationship between the USVs and between USVs and TSs. The collision-cone method operates as follows. Taking the TS avoidance problem as an example, when a TS enters the visible range of the USV ($d_o \leq 600$), the collision navigational angle range is computed according to the relative velocity of the USV with respect to the TS.

$$\begin{cases} v_\theta = v_i \sin(\alpha - \theta) - v_o \sin(\beta - \theta) \\ v_r = v_i \cos(\alpha - \theta) - v_o \cos(\beta - \theta) \end{cases} \tag{3}$$

As illustrated in Figure 4, we denote v_θ to be the component of v_{io} in the direction perpendicular to the line $p_i p_o$, and we denote C to be the intersection of the expected USV path with the safe circle of the TS, if the USV and the TS keep their velocities. We also define v_r to be the component of v_{io} in the direction of $p_i p_o$. If $v_\theta = 0$, then the relative trajectory of the USV to the obstacle is right on the line of $p_i p_o$. If $v_r > 0$, then the projection of v_{io} on $p_i p_o$ is positive, and the USV has the tendency to get close to the obstacle; otherwise, the

USV is more likely to leave the TS alone. Figure 4 shows that collision probably occurs at the point C. If $v_r > 0$, and the direction of v_{io} is within the domain from p_iA to p_iB, which are the tangents of the circular bounding box of the TS from the USV location, then a collision possibly occurs. We denote $(v_\theta)_{pA}$ and $(v_\theta)_{pB}$ to be the components of the v_{io} in the direction perpendicular to the line p_iA and p_iB, respectively. If any one of the conditions $v_\theta = 0$ and $v_r > 0$, $(v_\theta)_{pA} \cdot (v_\theta)_{pB} \leq 0$, or $v_r > 0$, is met, then the USV will probably collide with the obstacle. The practical style of the second condition is ${d_o}^2 {v_\theta}^2 \geq {r_o}^2 \left({v_r}^2 + {v_\theta}^2\right)$, where d_o is the distance between the USV and the obstacle. The collision detection algorithm is as follows:

$$\begin{cases} v_\theta = 0 \; and \; v_r > 0 \\ v_r > 0 \; and \; d_{io}^2 \cdot v_\theta^2 \geq r_o^2 \cdot (v_r^2 + v_\theta^2) \end{cases} \tag{4}$$

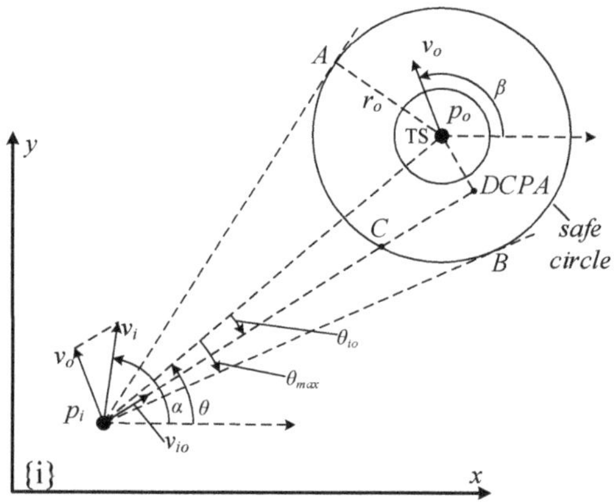

Figure 4. Illustration of the collision detection via the collision-cone method.

3. MAPPO Algorithm Design

3.1. Observation Space Design

Each USV can observe the environment information, including the information of TSs within its sensing range, and they share their observations to construct the global observation through communication [5]. The observation consists of the positions and navigation states of both USVs and TSs, as well as the environmental information, such as the positions and shapes of obstacles. The observation vector has a fixed dimension, and any missing values are filled with zeros.

The inner formation observational features are employed to describe the state of the USVs within the communication range. For USV_i, the observational feature of the neighboring USV_j is set to $o_{ij} = \{l, v_x, v_y, x_i, y_i, x_j, y_j, \Delta v_x, \Delta v_y, e_i\}$, where l is the location of the estimated aggregation target that is available to all USVs, and v_x and v_y represent the velocity components of USV_i in the x and y directions, respectively. To avoid complex coupling between velocities, orthogonal components are used to express the actions of the USV. In this context, x_i and y_i represent the position of the first vessel (denoted as USV_i), while x_j and y_j represent the position of another vessel (denoted as USV_j). The variable e_i represents the TSs and obstacle features observed by USV_i. The variables Δv_x and Δv_y represent the differences between the velocities of USV_i and USV_j, respectively. Specifically, this equation considers three USVs, four obstacles, and three TSs in the vicinity of a USV.

3.2. USV-State Feature Design

The USV-state feature is represented as $S_i = \{ name, done, s_c, r_c, x, y, v_x, v_y, c_{ij}, n_c\}$, where *name* represents the name of the USV state, *done* indicates whether the USV has reached an aggregation goal, s_c represents the collision state of the USV, r_c represents the optional actions determined by the current USV state, and x and y represent the USV position.

To address the restricted visibility problem in narrow channels, practical communication is simulated by incorporating observations from USVs with added noises into a global observation during the model training process. The USV can take communication actions to broadcast information to other USVs, and c_{ij} represents the communication utterance transferred from USV_j to USV_i, while n_c represents the communication noise. Noise is added to the communication utterance of a USV to simulate the practical environment.

3.3. Reward Function Design

The reward calculation combines guidance rewards and sparse rewards. During task execution, each USV receives a guiding reward at each time step, which aims to guide the USVs to chase the aggregation targets and to form a formation. The average reward of all USVs is calculated as the collective reward of the USV formation. The reward function, R, consists of guiding rewards, COLREGs-compliant rewards, collision rewards, and time-consumption rewards weighted accordingly.

To guide the USVs to approach the aggregation targets during the early stage of training, the USVs receive negative rewards based on their distances to the targets at each time step, until the USVs reach the target.

$$r_t = \begin{cases} 0, & d_t \leq d_{agg} \\ -w_{ta} \cdot (d_t - d_{agg}), & d_t > d_{agg} \end{cases} \tag{5}$$

where w_{ta} is the coefficient that limits the range of reward changes; if the distance d_t of a USV to its nearest target is less than d_{agg}, the USV is considered to have reached the aggregation point. We set $w_{ta} = 1$, $d_{agg} = 20$, and all the hyperparameters are set empirically.

The guiding reward is the estimation of USVs reaching the targets, while the time-consuming rewards reflect the actual time consumption from the start points. A USV will receive a negative reward of Δt (-0.1) if it takes one additional step.

The control on turnings is more difficult than that on a linear path; thus, a USV will receive a negative reward if it needs to perform a steering maneuver. We assign a reward of $R_s = -0.6$ to turning points when the trajectory turning angle exceeds 45 degrees.

The obstacle-avoidance reward consists of two parts, which are the static obstacle-avoidance reward and the dynamic obstacle-avoidance reward. The static obstacle-avoidance reward is calculated using the following formula.

$$r_{ca} = -R_c \cdot e^{-k_c d_o} \tag{6}$$

where d_o is the minimum distance between a USV and its obstacles; k_c is the weight, which we set to be 1; and R_c represents the reward amplitude. If $d_o > 600$, then $R_c = 0$; when $300 \leq d_o \leq 600$, $R_c = 1.5$; when $100 \leq d_o < 300$, $R_c = 2$; and if $d_o < 100$, then $R_c = 10$. The reward is a negative value that decreases exponentially along with the decrease in the distance between the USV and the obstacles.

The dynamic avoidance reward considers the USVs avoiding sailing obstructions (including other USVs and TSs) in terms of the COLREGs. If a USV action enters the collision areas of other USVs or TSs and the minimum distance, d_o, from the USV_i to other sailing obstructions is smaller than 600, then a negative reward, denoted as r_i, is added to the total reward. As shown in Formulas (7) and (8), the reward consists of two parts: the COLREGs-compliant reward, which is based on adherence to the rules, and the obstacle-avoidance reward, which focuses on avoiding vessels regardless of the rules.

Once the encountering scenario is determined by the COLREGs, the calculation of the COLREGs-compliant reward will remain unchanged. During the dynamic obstacle-avoidance procedure, the USV continuously monitors other vessels. If another vessel that is responsible for giving way according to the COLREGs does not comply with the rules, the USV takes collision-avoidance actions based on their distance, regardless of the specific rules.

$$r_i = -R_{rule} \cdot e^{-k_c d_o} \tag{7}$$

The dynamic obstacle avoidance constant, R_{rule}, is computed as:

$$R_{rule} = \alpha \cdot R_c + R_{col} \tag{8}$$

where the weight, α, is set to be 0.5.

The head-on reward is shown in Formula (9). According to the COLREGs, the USV should turn to the starboard side in a head-on scenario; if the USV turns left, a penalty will be added to the reward. The rewards for the give-way and stand-on situations, shown in Formulas (10) and (11), are similar to the head-on reward. There is no guidance on the action the USV should take to avoid a collision in the overtaking situation. Therefore, we do not define a specific reward function for this scenario but, instead, directly use the distance to evaluate the avoidance actions of the USV. Finally, the head-on reward, give-way reward, and stand-on reward are combined to form the COLREGs-compliant reward, as shown in Formula (12).

$$R_{h\_on} = \begin{cases} 0, & if\ turns\ to\ the\ starboard\ side \\ -R_{port}, & if\ turns\ to\ the\ port\ side \end{cases} \tag{9}$$

Give-way reward:

$$R_{g\_way} = \begin{cases} 0, & if\ turns\ to\ the\ starboard\ side \\ -R_{port}, & if\ turns\ to\ the\ port\ sidet \end{cases} \tag{10}$$

Stand-on reward:

$$R_{s\_on} = \begin{cases} 0, & if\ turns\ to\ the\ starboard\ side \\ -R_{port}, & if\ turns\ to\ the\ port\ side \end{cases} \tag{11}$$

The aggregate:

$$R_{col} = R_{h\_on} + R_{g\_way} + R_{s\_on} \tag{12}$$

3.4. Action Space Design

To ensure the planning efficiency, we use the discrete action space that distributes uniformly by 45 degrees around a USV. As shown in Figure 5, the action probabilities are determined by the actor network; the appropriate actions in green are selected, while actions in red are suppressed.

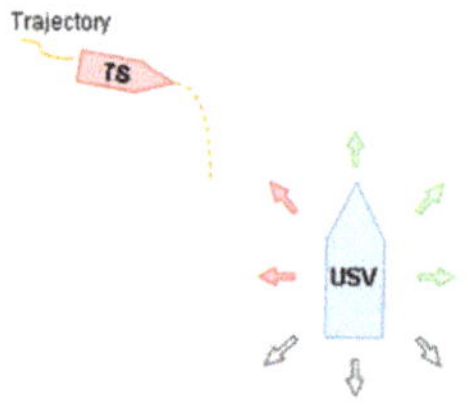

Figure 5. Illustration of the action space.

3.5. Policy Network Design

Figure 6 shows the policy network in this research. The environment considers that USVs sail confronting static obstacles and TSs. An observation block is constructed for each USV to map the USV state feature, S, to the observation feature, O, by interacting with the environment.

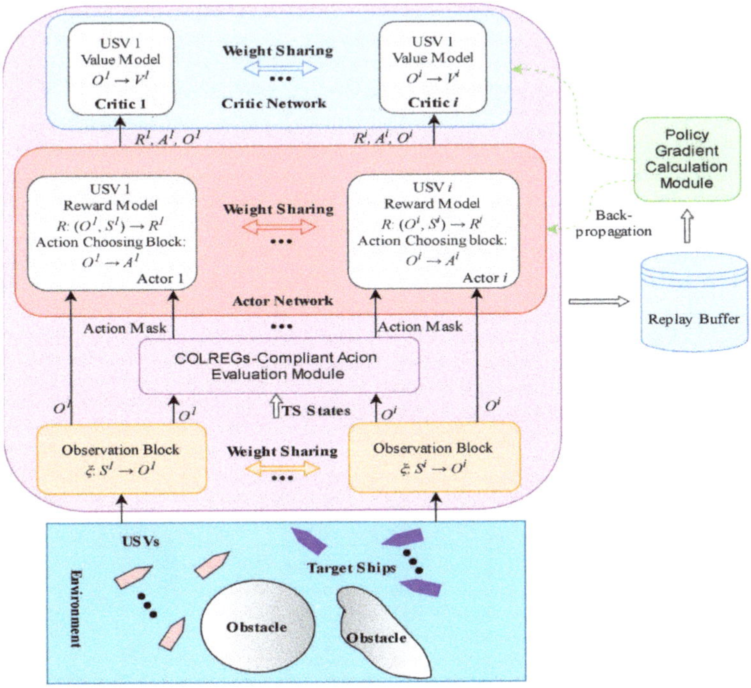

Figure 6. Illustration of the policy network.

The COLREGs-compliant action-evaluation module is used to distinguish the actions satisfying the COLREGs from those violating the rules. The MLP-based actor network is responsible for calculating the reward, R, and mapping the observation feature into the optimal action, A. The critic network, which assesses the chosen action, is also established using the MLP network, and it outputs the value V.

To optimize policies, the network alternates between sampling data from the policy and performing optimization on the sampled data. The policies represented by the S, O, A, R, V vectors are saved in the replay buffer. We use the mini-batch method. After collecting data for multiple epochs, the data is transferred from the buffer to the policy gradient calculation module for the optimization of the actor and critic networks. By sharing weights among networks, the planner can learn from the cooperative action-selection experiences. When the policy network is used for path planning in a real environment, the actor networks plan actions based on the observation features without critic networks.

The loss function of the network adopts the clip strategy [18], with $\varepsilon = 0.1$, as follows.

$$L^{CLIP}(\theta) = \frac{1}{NM}\sum_{i=1}^{N}\sum_{k=1}^{M}\min\left\{r_{\theta,i}^{(k)}A_i^{(k)}, clip(r_{\theta,i}^{(k)}, 1-\varepsilon, 1+\varepsilon)A_i^{(k)}\right\} \quad (13)$$

3.6. COLREGs-Compliant Action-Evaluation Network Design

The COLREGs-compliant action-evaluation module is built based on the MLP network. We used a pre-training strategy, training the COLREGs-compliant action-evaluation network separately from the other parts of the policy network. The COLREGs-compliant action-evaluation network is pre-trained using pre-recorded USV trajectory data, which includes the state vector and global observation vector, as well as the USV actions for avoiding the TSs. The action labels align with our action space. The final established dataset consists of 5000 data items, with an almost equal number of avoidance behaviors for each category (head-on, crossing and give-way, crossing and stand-on, and overtaking), ensuring a balanced dataset.

The COLREGs-compliant action-evaluation network consists of three fully connected (FC) layers, with parameter sizes of 18×64, 64×64, and 64×7 for the respective layers. The Tanh activation function is applied after the first two FC layers, and the softmax function is used following the last FC layer to obtain the COLREGs-compliant probabilities of actions. During the pre-training procedure, the CrossEntropyLoss function is utilized. The remaining parameters, including the optimizer and the learning rate, are set to be identical to those of the policy network. The pre-trained network is utilized directly without applying any parameter updates via the policy gradient algorithm.

The probability vector of actions compliant with the COLREGs will be multiplied by the dot product with the probability vector of chosen actions to select the final action with the highest result.

In Figure 7, the illustration depicts that the gym environment supplies the state vectors of the entities to the path-planning model. Additionally, the TS motion-detection network offers TS motion information, which includes courses and bearings, to the planner. Subsequently, the state vector and the observation vector are concatenated to form the input vector for both the actor network and the COLREGs-compliant action-evaluation network.

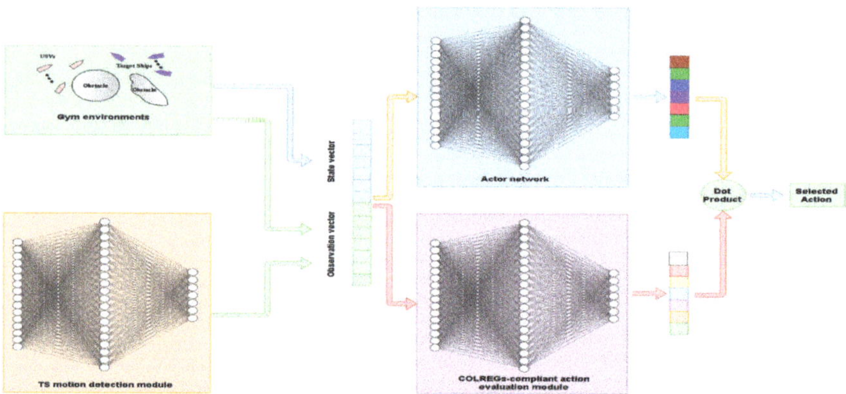

Figure 7. COLREGs-compliant action selection.

By integrating the prediction vectors from both the actor and action-evaluation networks, the policy network is capable of making well-informed decisions concerning the USV's path-finding and collision-avoidance actions while following the COLREGs.

4. TS Motion-Detection Network Design

This research recognizes the TS intention by the time series of RGB images. The bow, stern, and whole body of TSs on consecutive frames are detected. TSs are detected at different scales since they are sometimes relatively small in the overall picture. Subsequently, the motions of the TSs can be identified using the inter-frame difference scheme.

Practical TS images are relatively rare; thus, we adopt the pre-train and fine-tune paradigm to train the network. Firstly, the network is pre-trained using the SeaShip dataset [36]. The dataset is designed to train and evaluate ship object-detection algorithms. The dataset consists of 31,455 images covering six common ship types (ore carriers, bulk carriers, general cargo ships, container ships, fishing vessels, and passenger ships). All images are from approximately 10,080 real-world video footage captured by surveillance cameras in a deployed coastline video surveillance system. The data has been carefully selected to cover all possible imaging variations, such as different proportions, hull components, lighting, viewpoints, backgrounds, and occlusion.

After our TS detection model is pre-trained on the SeaShip dataset, the model is finetuned using the real-scene images to generalize it to our task.

The TS detection model is shown in Figure 8. The network is constructed based on the YOLO model [37]. The network utilizes the spatial pyramid pooling-based feature (SPPF) network to capture spatial features of images. The cross-stage partial (CSP) network is employed to enhance the CNN block and to reduce the computational burden [37]; CSPi_j refers to the *i*th CSP module utilizing *j* residual components. The CBS module comprises a convolutional network, a batch normalization network, and SiLU activation. It is utilized for feature extraction.

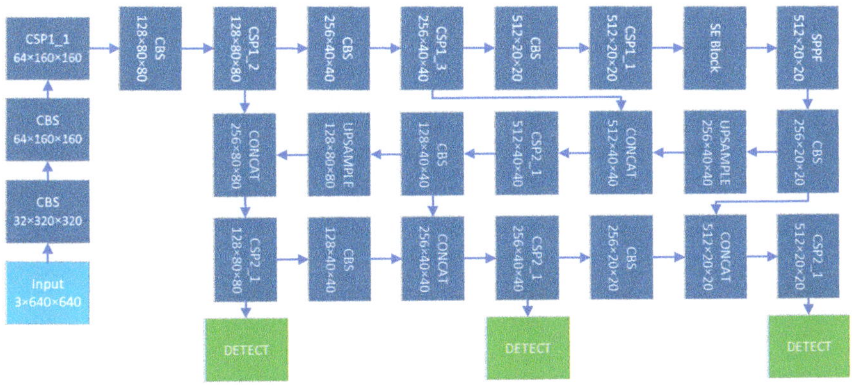

Figure 8. Illustration of the TS detection network. The dark blue boxes indicate the body of the model. The light blue box represents the model input block, while the green boxes represent the detection block.

The channel attention mechanism, namely squeezing and excitation (SE), is combined with the YOLO network to boost the feature-extraction ability, as shown in Formula (14) and Figure 9.

$$SE(x) = FC(x) \cdot Sigmoid(FC(Relu(FC(AvePool(x))))) \tag{14}$$

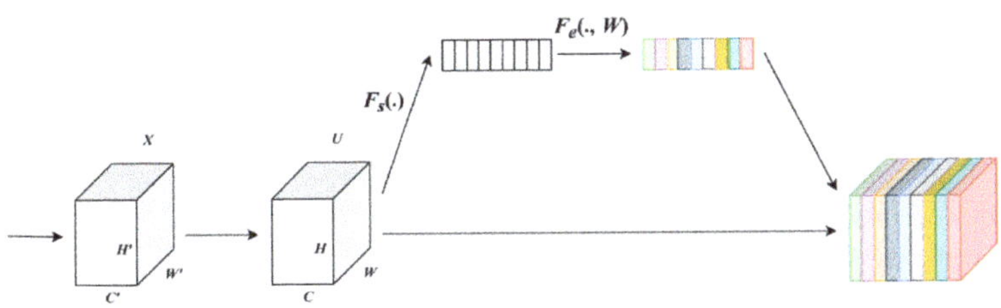

Figure 9. Illustration of the squeezing and excitation mechanism.

5. Experiments

5.1. Experimental Setting

Our MAPPO-based path-planning program runs on a computer equipped with an Intel(R) Core(TM) i7-7820HQ CPU @ 2.90 GHz and 16 GB of memory. Our planning model is compared with the MADDPG model in experiments. Due to the complexity of the MADDPG algorithm, we utilize a more powerful computer to run the program. The configuration is as follows: the operation system is Ubuntu 20.04.6 LTS, the CPU is Intel(R) Xeon(R) Silver 4214 @ 2.20 GHz with 48 cores and 2 threads per core, and the computer has 2 memories, each with a capacity of 8192 MB.

5.2. The Path-Planning Model Training Pipeline

The MAPPO-based path-planning model is trained using a multi-threaded approach, where each thread establishes a virtual environment to fully select data points by importance.

Observations are captured first, then step actions for a USV are captured in terms of rewards from the actor network until an episode is reached. The step rewards of the episode or a path are summed up and then normalized. The normalization is used to eliminate the impact of different scales of metrics.

The maximum number of steps per episode is 25, which means that the planner generates a local path of no more than 25 steps in the local planning time domain. The critic network aims to assess the cooperative action-selection strategy of the actor network by solving the value function. The obtained values are also normalized.

We adopt the mini-batch scheme, and the batch size is set to be 32 episodes. After collecting a batch of data, the actor and critic networks are trained three times, and the average metrics are used to compute the strategy gradient.

The learning rate is set to be 5×10^{-4}. The discount factor when calculating the rewards is set to be 0.99. The policy entropy coefficient is set to be 0.01. The epsilon value of the Adam optimization is set to be 1×10^{-5}. The simulation time step is 0.1 s. We use the learning-rate-decay method and gradient clip tricks when updating the network parameters. The orthogonal parameter initialization scheme is applied to the MLP layers of both the actor and critic networks to keep the weight matrixes orthogonal, thereby reducing the problem of the gradient vanishing and exploding.

The MLP blocks of the actor network consists of three fully connected (FC) networks, and the parameters are 18×64, 64×64, and 64×7 for the respective layers. The Tanh activation function is applied after the first two FC layers. Subsequently, the softmax function is used following the last FC layer to obtain the probabilities of choosing actions.

The MLP blocks of the critic network consist of three FC networks, with parameters of 54×64, 64×64, and 64×1 for the respective layers. The Tanh activation function is applied after the first two FC layers. The value can be obtained from the last FC layer.

5.3. Path-Planning Results

We configure Monte Carlo simulation environments for training the planner network whereas the aggregation targets for USVs and obstacles are randomly located, and the linear TS routes are also randomly generated. The planning objective of the USVs is to reach the aggregation targets while avoiding TSs and obstacles.

We trained our model for 1×10^6 iterations. Figure 10 shows the results of the Monte Carlo experiment for USV cooperation. The black-filled circles represent obstacles, the colored circles represent the aggregation positions of USVs, stars indicate the estimated aggregation targets, the triangles indicate the start points of the USVs, and the colored lines depict the obstacle-avoidance paths. Different colors are used to distinguish the relevant elements of different USVs. Since the planner needs to be trained based on the current model in a new environment that is significantly different from the known one, we train the planner for 1×10^6 iterations for efficiency.

We have observed that our planner can efficiently generate obstacle-avoidance paths in less than 1 s, and these paths can effectively avoid collisions while reaching the aggregation targets to maintain the formation. The success rate of achieving multiple objectives is higher than 95%. These observations testify to the efficiency and effectiveness of our method.

We train our planner for a total of 1×10^6 and 6×10^6 iterations. We evaluate the network and record metric values per 5000 training iterations, and the evaluation reward curves are shown in Figure 11. The action-space dimension is 7 and the observation-space dimension is 24. Figures 12 and 13 depict the training-reward curves for the MADDPG and multi-agent TD3 (MATD3) algorithms. MATD3 improves upon the original DDPG algorithm by incorporating additional techniques, such as using twin critic networks and delayed policy updates, to improve the coordination and learning in complex environments. Due to the complexity of the training process and the long duration required for each iteration, we trained both the MADDPG and MATD3 algorithms for 1×10^6 iterations.

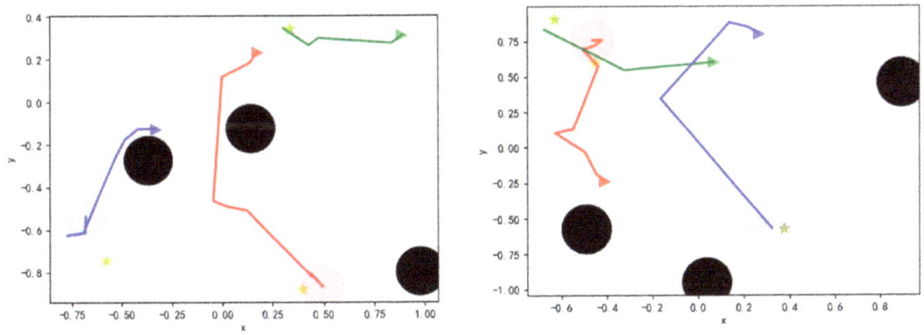

Figure 10. Monte Carlo experiment results.

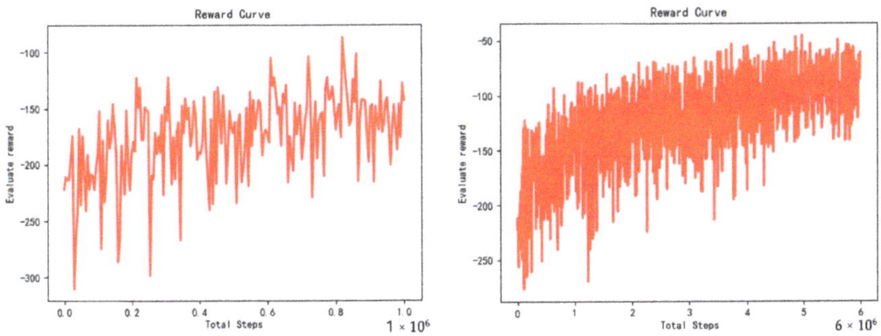

(**a**) (**b**)

Figure 11. Reward curves of the MAPPO-based planner. (**a**) Reward curve of training for 1×10^6 iterations; (**b**) Reward curve of training for 6×10^6 iterations.

Figure 12. Reward curve of MADDPG.

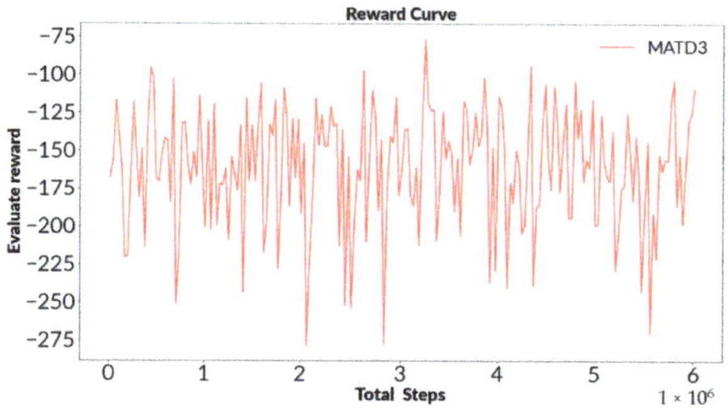

Figure 13. Reward curve of MATD3.

We observed that the reward curve of our method converges as the training iterations approach 1×10^6. The oscillation is relatively low, indicating that the method has relatively high stability. Figure 11b shows the training-reward curve for 6×10^6 training iterations. The rewards decrease as the training iterations increase, indicating that the method scales well with a large number of training iterations.

The curve became steady as the iteration time approached 5×10^6, which probably indicated that the model was sufficiently trained. Since training efficiency is also an important metric for the online planner model, we benchmarked our method against the state-of-the-art MADDPG model and the MATD3 model after training for 1×10^6 iterations. The oscillations in the reward curves of MADDPG and MATD3 are significant, and the decreases in rewards are not obvious. The observation implies that MADDPG and MATD3 may require more training iterations than our method.

Meanwhile, the duration of training for our MAPPO-based method with 6×10^6 iterations was less than 5 h, whereas the training durations for MADDPG and MATD3 with only 1×10^6 iterations exceeded 10 h. The above observations demonstrate that our method has a much higher training efficiency than MADDPG and MATD3 due to the low complexity of MAPPO.

The results of the Monte Carlo simulations show that our MAPPO algorithm outperforms both the MADDPG and MAPPO algorithms in terms of the average path length and the steering angle. Specifically, the average path length of our MAPPO algorithm is 0.489, which is lower than that of the MADDPG and MAPPO algorithms (0.683 and 0.582, respectively). This indicates that our algorithm is able to plan more optimal paths. Moreover, the average steering angle of our MAPPO algorithm is 2.14 rad, which is lower than that of the MADDPG and MATD3 algorithms (2.35 and 2.23, respectively). This suggests that our algorithm is also more efficient in steering the vehicle. Overall, these observations provide evidence that our method is capable of planning more optimal paths than the MADDPG and MATD3 models after being trained for 1×10^6 iterations in the simulation environment of the Monte Carlo simulations. This result highlights the potential of our algorithm to improve the performance of autonomous driving systems.

5.4. COLREGs-Based Collision-Avoidance Experiments

Figure 14 depicts the simulation results of USVs avoiding TSs, with randomly generated positions for both the USVs and the obstacles. The TS paths are also generated randomly. The obstacles are represented by black-filled triangles, rectangles, and circles. The USVs plan their paths to reach the aggregation targets, considering both low-cost paths and collision avoidance among the USVs and TSs. The red curves represent the paths of the three USVs, while the blue curves depict the paths of the TSs. The start points of the TS paths are indicated by blue triangles.

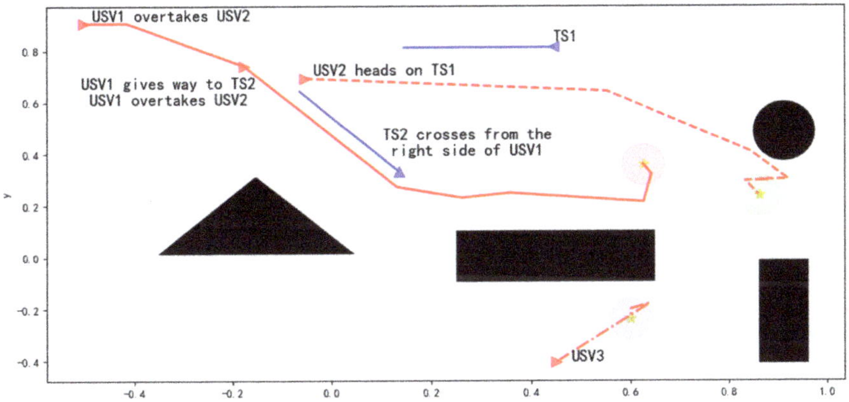

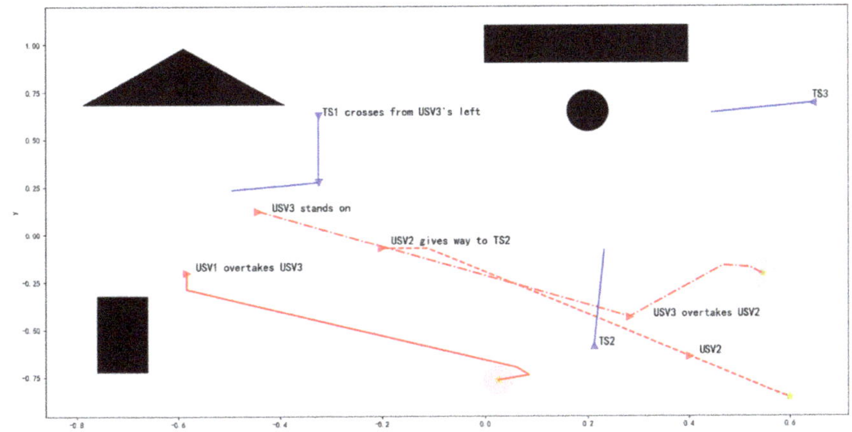

Figure 14. Simulations of USVs avoiding TSs in accordance with the COLREGs.

Figure 14 illustrates two representative scenarios of avoiding collisions in accordance with the COLREGs. USVs typically prioritize the most significant collision risks in the nearby area and subsequent collision-avoidance issues, taking actions according to the COLREGs to avoid a collision. The upper sub-figure illustrates that the first USV not only overtakes the second USV by turning to the starboard side in accordance with the COLREGs but also gives way to the TS crossing from the right by turning to the starboard side. Similarly, the second USV follows the COLREGs by turning to its starboard side when heading toward a TS.

The lower sub-figure illustrates that the first USV overtakes the third USV by maneuvering to its starboard side. When a TS crosses from the port side of the third USV, the USV stands on its course if it can pass the TS without risking a collision, in accordance with the COLREGs. In the subsequent segments of the voyage, the third USV turns to its port side to overtake the second USV. Simultaneously, the second USV adjusts its course to the starboard side to give way to the TS.

Figure 14 demonstrates that our path-planning method is capable of effectively planning multiple USV paths to achieve their targets while ensuring collision avoidance among USVs and TSs in accordance with the COLREGs regulations.

5.5. TS Detection Results

In the TS detection experiments, we detect the head, stern, and the entire body (i.e., all) of the TS. The corresponding PR curves are shown in Figure 15. When any part of the TS is detected, the motion intention of the TS can be captured by the detection results in consecutive frames using the inter-frame difference scheme. The blue curve indicates the mean average precision (mAP) curve when the intersection-over-union (IoU) threshold is 0.5. The metric values may imply that the model achieves a certain balance between recall and accuracy, and it demonstrates a relatively high overall performance. In the evaluation of the TS detection network on the SeaShips dataset, the 10-fold cross-validation method was employed. This method involves dividing the dataset into 10 subsets or folds of approximately equal size. For each fold, 70% of the images were used as the training set, while the remaining 30% of images were used as the test set. It is important to note that the training set and the test set did not have any overlapping images.

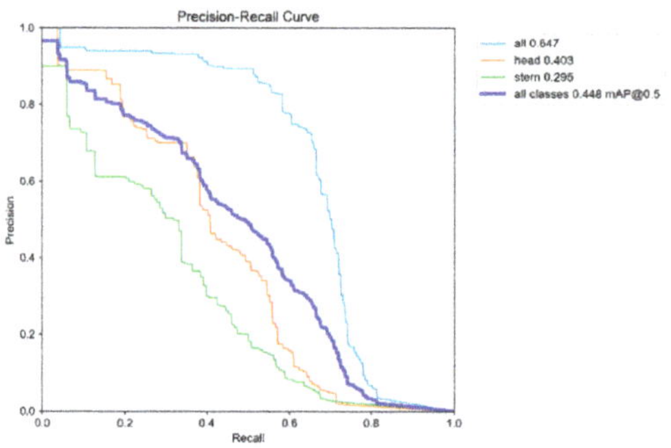

Figure 15. PR and mAP curves of the TS detection model.

Figure 16 presents the TS detection result curves from different perspectives. The training loss and validation (val) loss curves demonstrate that the model converges quickly. The precision, recall, and mAP curves imply high performance of the model.

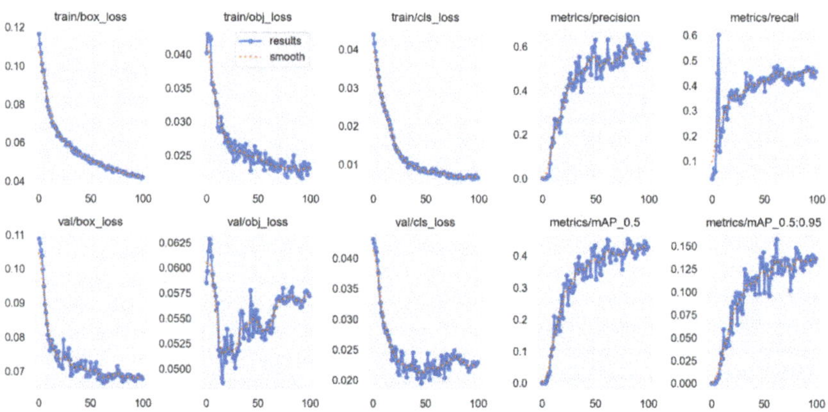

Figure 16. Result curves of the TS detection algorithm.

Figure 17 shows the result of the model detecting two TSs, and the model performed well in our TS detection task. The effectiveness of our model may benefit from the following factors. Since we use the time serial images of TSs, the detection of any part of a TS in consecutive frames is enough to recognize the motion of the TS. Meanwhile, the detections are conducted several times per second.

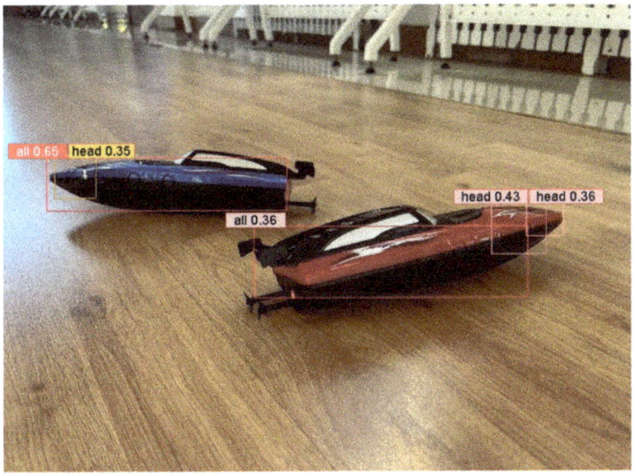

Figure 17. Illustration of TS detection.

6. Conclusions

This research proposes a two-stage path-planning method for multiple USVs based on the COLREGs. The method combines a cooperation module, a COLREGs-compliant action-evaluation module, and a TS detection module. The cooperative path-planning model for collision avoidance among USVs is constructed based on the MAPPO strategy, which utilizes a policy network capable of handling multiple aggregation goals, obstacles, and TSs. To achieve this, we define the action space, observation space, and reward function for the policy network, and design actor and critic networks.

- Monte Carlo experimental results confirm the effectiveness and efficiency of our path-planning method for formation aggregation and collision avoidance. We conducted these experiments by randomly specifying the positions of the USVs and obstacles. This approach allowed us to evaluate the performance of our method in diverse scenarios and validate its robustness.
- We benchmarked the simulation results against the MADDPG and MATD3 methods to validate the efficiency and the optimization performance of our approach.

After training the COLREGs-compliant action-evaluation calculation module using TS-avoiding trajectories, violations of USV actions that go against the COLREGs can be recognized and suppressed. This judgment is then used as heuristics for the actor network. Our reward function considers both COLREGs and seamanship principles.

- We conducted further experiments to test the feasibility of our collision-avoidance scheme based on the COLREGs within the USV fleet, as well as between USVs and TSs. We were able to confirm the practicality and effectiveness of our method in a realistic scenario.

The TS detection network is constructed based on the YOLO network and the squeeze-and-excitation scheme.

- Our proposed TS detection model performs well in our specific environment.

Primarily, we evaluate the algorithm through simulations conducted in gym environments. Additionally, we have conducted experiments on a semi-physical simulation platform where our algorithm acts as a local path planner, guiding the navigation of the system. The cooperative path-planning module is executed on individual ship-borne computers (PC-104) installed on each USV member. VxWorks 6.6 is utilized as the operating system, with Workbench 3.0 as the chosen development package.

In our future experiments, we aim to further advance our research by translating the algorithm into physical USVs. This will allow us to conduct practical implementation and comprehensive testing of the algorithm's capabilities.

Author Contributions: Conceptualization, N.W., R.Z. and G.L.; methodology, N.W., G.L., R.Z. and Y.L.; software, N.W., W.W. and Y.L.; data curation, Y.L., W.W. and D.J.; writing—original draft preparation, N.W.; writing—review and editing, N.W. and D.J.; visualization, N.W., Y.L. and W.W. All authors have read and agreed to the published version of the manuscript.

Funding: This work was supported by grants from the Natural Science Foundation of Liaoning Province (No. 2020-MS-135), the National Natural Science Foundation of China (No. 62272067), and the Key Laboratory of Intelligent Perception and Advanced Control of State Ethnic Affairs Commission (No. MD-IPAC-2019103).

Data Availability Statement: The core code is available at https://github.com/WenNaifeng/Reinforcement-Learning-MAPPO.

Conflicts of Interest: The authors declare no conflict of interest.

References

1. Campbell, S.; Naeem, W.; Irwin, G.W. A review on improving the autonomy of unmanned surface vehicles through intelligent collision avoidance manoeuvres. *Annu. Rev. Control* **2012**, *36*, 267–283. [CrossRef]
2. Chakravarthy, A.; Ghose, D. Obstacle Avoidance in a Dynamic Environment: A Collision Cone Approach. *IEEE Trans. Syst. Man Cybern. A Syst. Hum.* **1998**, *28*, 562–574. [CrossRef]
3. Liang, X.; Qu, X.; Wang, N.; Li, Y.; Zhang, R. Swarm control with collision avoidance for multiple underactuated surface vehicles. *Ocean Eng.* **2019**, *191*, 106516. [CrossRef]
4. Liang, X.; Qu, X.; Wang, N.; Li, Y.; Zhang, R. A Novel Distributed and Self-Organized Swarm Control Framework for Underactuated Unmanned Marine Vehicles. *IEEE Access* **2019**, *7*, 112703–112712. [CrossRef]
5. Xia, J.; Luo, Y.; Liu, Z.; Zhang, Y.; Shi, H.; Liu, Z. Cooperative multi-target hunting by unmanned surface vehicles based on multi-agent reinforcement learning. *Def. Technol.* **2022**, *29*, 80–94. [CrossRef]
6. Xue, D.; Wu, D.; Yamashita, A.S.; Li, Z. Proximal policy optimization with reciprocal velocity obstacle based collision avoidance path planning for multi-unmanned surface vehicles. *Ocean Eng.* **2023**, *273*, 114005. [CrossRef]
7. Maza, J.A.G.; Argüelles, R.P. COLREGs and their application in collision avoidance algorithms: A critical analysis. *Ocean Eng.* **2022**, *261*, 112029. [CrossRef]
8. Kim, J.K.; Park, D.J. Understanding of sailing rule based on COLREGs: Comparison of navigator survey and automated collision-avoidance algorithm. *Mar. Policy* **2024**, *159*, 105894. [CrossRef]
9. Hu, L.; Hu, H.; Naeem, W.; Wang, Z. A review on COLREGs-compliant navigation of autonomous surface vehicles: From traditional to learning-based approaches. *J. Autom. Intell.* **2022**, *1*, 100003. [CrossRef]
10. Yim, J.B.; Park, D.J. Modeling evasive action to be implemented at the minimum distance for collision avoidance in a give-way situation. *Ocean Eng.* **2022**, *263*, 112210. [CrossRef]
11. Kim, J.K.; Park, D.J. Determining the Proper Times and Sufficient Actions for the Collision Avoidance of Navigator-Centered Ships in the Open Sea Using Artificial Neural Networks. *J. Mar. Sci. Eng.* **2023**, *11*, 1384. [CrossRef]
12. Hagen, I.B.; Vassbotn, O.; Skogvold, M.; Johansen, T.A.; Brekke, E.F. Safety and COLREG evaluation for marine collision avoidance algorithms. *Ocean Eng.* **2023**, *288*, 115991. [CrossRef]
13. Yang, Y.; Wang, J. An overview of multi-agent reinforcement learning from game theoretical perspective. *arXiv* **2020**, arXiv:2011.00583.
14. Wang, X.; Wang, S.; Liang, X.; Zhao, D.; Huang, J.; Xu, X.; Dai, B.; Miao, Q. Deep reinforcement learning: A survey. *IEEE Trans. Neural Netw. Learn. Syst.* **2022**. [CrossRef] [PubMed]
15. Heiberg, A.; Larsen, T.N.; Meyer, E.; Rasheed, A.; San, O.; Varagnolo, D. Risk-based implementation of COLREGs for autonomous surface vehicles using deep reinforcement learning. *Neural Netw. Off. J. Int. Neural Netw. Soc.* **2022**, *152*, 17–33. [CrossRef] [PubMed]
16. Li, L.; Wu, D.; Huang, Y.; Yuan, Z. A path planning strategy unified with a COLREGS collision avoidance function based on deep reinforcement learning and artificial potential field. *Appl. Ocean Res.* **2021**, *113*, 102759. [CrossRef]
17. Schulman, J.; Levine, S.; Abbeel, P.; Jordan, M.; Moritz, P. Trust region policy optimization. In Proceedings of the International Conference on Machine Learning, Lille, France, 7–9 July 2015; PMLR: Lille, France, 2015; pp. 1889–1897.
18. Schulman, J.; Wolski, F.; Dhariwal, P.; Radford, A.; Klimov, O. Proximal Policy Optimization Algorithms. *arXiv* **2017**, arXiv:1707.06347.
19. Lowe, R.; Wu, Y.I.; Tamar, A.; Harb, J.; Pieter Abbeel, O.; Mordatch, I. *Multi-Agent Actor-Critic for Mixed Cooperative-Competitive Environments*; NeurIPS: Long Beach, CA, USA, 2017; pp. 1–12. [CrossRef]

20. Fujimoto, S.; Hoof, H.; Meger, D. Addressing function approximation error in actor-critic methods. In Proceedings of the International Conference on Continuous Control with Deep Reinforcement Machine Learning, Stockholm, Sweden, 10–15 July 2018; PMLR: Lille, France, 2018; pp. 1587–1596.
21. Lillicrap, T.P.; Hunt, J.J.; Pritzel, A.; Heess, N.; Erez, T.; Tassa, Y.; Silver, D.; Wierstra, D. Continuous control with deep reinforcement learning. *arXiv* **2015**, arXiv:1509.02971.
22. Yu, C.; Velu, A.; Vinitsky, E.; Gao, J.; Wang, Y.; Bayen, A.; Wu, Y. The surprising effectiveness of ppo in cooperative multi-agent games. *Adv. Neural Inf. Process. Syst.* **2022**, *35*, 24611–24624.
23. Zhai, P.; Zhang, Y.; Shaobo, W. Intelligent ship collision avoidance algorithm based on DDQN with prioritized experience replay under COLREGs. *J. Mar. Sci. Eng.* **2022**, *10*, 585. [CrossRef]
24. Meyer, E.; Heiberg, A.; Rasheed, A.; San, O. COLREG-compliant collision avoidance for unmanned surface vehicle using deep reinforcement learning. *IEEE Access* **2020**, *8*, 165344–165364. [CrossRef]
25. Sawada, R.; Sato, K.; Majima, T. Automatic ship collision avoidance using deep reinforcement learning with LSTM in continuous action spaces. *J. Mar. Sci. Technol.* **2021**, *26*, 509–524. [CrossRef]
26. Xu, X.; Lu, Y.; Liu, X.; Zhang, W. Intelligent collision avoidance algorithms for USVs via deep reinforcement learning under COLREGs. *Ocean Eng.* **2020**, *217*, 107704. [CrossRef]
27. Wang, W.; Huang, L.; Liu, K.; Wu, X.; Wang, J. A COLREGs-Compliant Collision Avoidance Decision Approach Based on Deep Reinforcement Learning. *J. Mar. Sci. Eng.* **2022**, *10*, 944. [CrossRef]
28. Wei, G.; Kuo, W. COLREGs-compliant multi-ship collision avoidance based on multi-agent reinforcement learning technique. *J. Mar. Sci. Eng.* **2022**, *10*, 1431. [CrossRef]
29. Rongcai, Z.; Hongwei, X.; Kexin, Y. Autonomous collision avoidance system in a multi-ship environment based on proximal policy optimization method. *Ocean Eng.* **2023**, *272*, 113779. [CrossRef]
30. Skrynnik, A.; Yakovleva, A.; Davydov, V.; Yakovlev, K.; Panov, A.I. Hybrid policy learning for multi-agent pathfinding. *IEEE Access* **2021**, *9*, 126034–126047. [CrossRef]
31. Wang, K.; Kang, B.; Shao, J.; Feng, J. Improving generalization in reinforcement learning with mixture regularization. *Adv. Neural Inf. Process. Syst.* **2020**, *33*, 7968–7978.
32. Khoi, N.D.H.; Van, C.P.; Tran, H.V.; Truong, C.D. Multi-Objective Exploration for Proximal Policy Optimization. In Proceedings of the 2020 Applying New Technology in Green Buildings (ATiGB), Da Nang, Vietnam, 12–13 March 2021; IEEE: Piscataway, NJ, USA, 2021; pp. 105–109.
33. Tam, C.K.; Richard, B. Collision risk assessment for ships. *J. Mar. Sci. Technol.* **2010**, *15*, 257–270. [CrossRef]
34. Statheros, T.; Howells, G.; Maier, K.M. Autonomous ship collision avoidance navigation concepts, technologies and techniques. *J. Navig.* **2008**, *61*, 129–142. [CrossRef]
35. Wen, N.; Zhao, L.; Zhang, R.B.; Wang, S.; Liu, G.; Wu, J.; Wang, L. Online paths planning method for unmanned surface vehicles based on rapidly exploring random tree and a cooperative potential field. *Int. J. Adv. Robot. Syst.* **2022**, *19*, 1–22. [CrossRef]
36. Shao, Z.; Wu, W.; Wang, Z.; Du, W.; Li, C. SeaShips: A Large-Scale Precisely Annotated Dataset for Ship Detection. *IEEE Trans. Multimed.* **2018**, *20*, 2593–2604. [CrossRef]
37. Kim, J.H.; Kim, N.; Park, Y.W.; Won, C.S. Object detection and classification based on YOLO-V5 with improved maritime dataset. *J. Mar. Sci. Eng.* **2022**, *10*, 377. [CrossRef]

Journal of
Marine Science and Engineering

MDPI

Article

A Formation Control and Obstacle Avoidance Method for Multiple Unmanned Surface Vehicles

Guanqun Liu, Naifeng Wen, Feifei Long and Rubo Zhang *

College of Mechanical and Electrical Engineering, Dalian Minzu University, Dalian 116600, China; liuguanqun@dlnu.edu.cn (G.L.); wennaifeng@dlnu.edu.cn (N.W.); dqpilf@126.com (F.L.)
* Correspondence: zhangrubo@dlnu.edu.cn

Abstract: This study introduces a method for formation control and obstacle avoidance for multiple unmanned surface vehicles (USVs) by combining an artificial potential field with the virtual structure method. The approach involves a leader–follower formation structure, where the leader autonomously avoids collisions using an artificial potential field based on the target's position as a reference. It also determines the ideal position of each follower in the formation based on its own position, heading angle, and the formation structure. To effectively avoid obstacles and maintain formation, the follower selects the position of the target or its ideal position as a reference during movement, depending on whether it is being repelled by obstacles. Additionally, this paper modifies the attractive force model of the traditional artificial potential field method to restrict the maximum magnitude of the attractive force when encountering repulsive forces, thus expediting departure from obstacle areas. The dynamic characteristics of USVs are taken into account by constraining the maximum linear speed and angular speed. Formation stability is ensured by maintaining a constant speed for the leader, while the linear speed of the follower varies based on the distance to the reference object during movement. Simulation experiments demonstrated that this method can effectively avoid obstacles and maintain formation.

Keywords: unmanned surface vehicle; obstacle avoidance; formation maintenance; artificial potential field; virtual structure

1. Introduction

USVs can be deployed in hazardous situations to perform specialized tasks, thereby minimizing potential casualties. They can also be utilized for monotonous and repetitive tasks, effectively reducing the workload on humans. In extreme marine environments, the capabilities of a single USV may not be sufficient. Therefore, it becomes necessary to integrate multiple USVs into a cluster, to tackle complex tasks through information integration. A cluster of USVs have stronger perception ability, larger operating range, and stronger risk resistance compared to a single USV [1]. Multiple USVs can be effectively deployed in complex and hazardous water areas where manned vessels cannot be dispatched or where individual USVs may not be able to complete tasks. These areas could include high-risk environments or areas contaminated by nuclear, biological, or chemical agents [2]. Multiple-USV collaborative formation operations are capable of performing a wide range of tasks, including coastal patrols and maritime rescue, without endangering the safety of naval personnel.

A cluster of USVs can be regarded as a multi-agent system, which refers to a system composed of multiple interacting agents that can jointly complete certain tasks [3]. In a multiple-USV system, each USV has a certain range of perception ability and can independently perform relatively simple tasks. When faced with complex tasks, they need to collaborate in order to complete the task. This collaboration typically requires information sharing and task allocation. For example, in a USV formation collision avoidance mission, each USV in the formation provides information support for optimized path planning and formation maintenance by sharing their respective perceived information. Usually, the information will be gathered by the USV designated as the leader. The leader makes collision avoidance decisions and updates the formation based on global information, and sends these decisions and formation information to the members of the formation [4–6].

In multi-agent systems, the collaborative formation is a fundamental issue as it plays a decisive role in efficiently executing specific tasks [7–9]. Formation collision avoidance aims to guide a group of USVs to form specific geometric configurations and control the motion of the formation. In the past few decades, collaborative formation research has mainly focused on unmanned aerial vehicles and intelligent robots, with relatively few studies on collaborative formation of USVs. With the expansion of the application range of USVs, research on the collaboration of multiple USVs has received much attention. Researchers have gradually deepened their research on formation control, continuously improving the performance of formation control algorithms, making multiple-USV formations suitable for various marine environments. The main problems for collaboration among multiple USVs include collaborative collision avoidance, formation, intersection, etc. [10]. Basic formation control approaches include behavior-based methods, virtual structure methods, artificial potential field methods, graph theory methods, and leader–follower methods [11].

The complex and dynamic marine environment, along with the presence of moving obstacles, presents significant challenges in designing formation control strategies. Factors such as the interference of wind, waves, and water flow, particularly the resistance of water, further complicate the problem. Many scholars have considered the effects of these interferences in the research of multiple-USV collaborative formations [12]. With the development of artificial intelligence, biomimetic optimization algorithms and artificial intelligence methods based on learning and gaming have made significant progress [4,13–15]. As a type of ship, USVs must comply with maritime traffic regulations in practical applications. The Convention on International Regulations for Preventing Collisions at Sea (COLREGs) stipulates the navigation rules that all ships must comply with at sea to prevent collisions. To expand the application range of USVs, scholars have introduced COLREGs into the research of USV technology. Kim [16] proposed a distributed local search algorithm to achieve multiple-ship formations and a distributed random search algorithm to reduce the amount of information transmission among multiple ships, thereby adjusting the formation faster. Ma [17] proposed quantitative standards for a wide range of complex collision avoidance situations involving a large number of USVs under COLREGs.

While researchers have conducted extensive research about USV formations and obtained promising results, the pursuit of simple and effective methods has always been a primary goal. Given the challenges of obstacle avoidance and formation maintenance, this paper uses an artificial potential field for collision avoidance and the virtual structure method for formation maintenance. The formation control adopts the leader–follower method. The leader independently avoids collisions based on the artificial potential field, using the position of the target as attractor, and generates the ideal position of each follower in the formation based on its own position, heading angle, and the formation structure. The follower selects the position of the target or its ideal position as the attractor during movement based on whether it is being repelled by obstacles or not. In order to quickly leave the obstacle area, this paper modified the attractive force model of the traditional artificial potential field (APF). This includes a limitation on the maximum value of the attractive force when it is influenced by the repulsive force, limiting the maximum value of the attractive force when subjected to the repulsive force. Taking into account the dynamic characteristics of USVs, this paper limits the maximum linear speed and maximum angular speed of USVs. For maintaining the stability of the formation, the leader moves at a constant speed, and the linear speed of the follower varies during the movement process, which may be greater than the speed of the leader, but must be smaller than the maximum linear speed.

The rest of this article is organized as follows: the APF method is described in Section 2; The formation control methods are described in Section 3; Section 4 presents the USV formation control and obstacle avoidance method; The simulation results and analysis are presented in Section 5; and Section 6 contains the conclusion.

2. Artificial Potential Filed

The APF method for path planning is a virtual force method proposed by Khatib [18]. The basic idea is to design the motion of a robot in its surrounding environment as an abstract artificial potential field. The target point generates "gravity" or "attractive force" on the mobile robot, and obstacles generate "repulsive force" on the mobile robot. Finally, the motion of the mobile robot is controlled by finding the resultant force. The path planned using an APF is generally relatively smooth and safe.

The goal point in the environment produces a gravitational potential field, guiding the USV toward it. Each obstacle in the environment produces a repulsive field, which in turn exerts a repulsive force on the USV, pushing the USV away from the obstacle. All the attractive forces produced by the goal and the repulsive forces produced by the obstacles generate a combined force, whose direction is the moving direction of the USV. A diagram of force analysis for an APF is depicted in Figure 1. F_{att} is the attractive force generated by the goal, F_{rep} is the repulsive force generated by the obstacle, and F is the combined force.

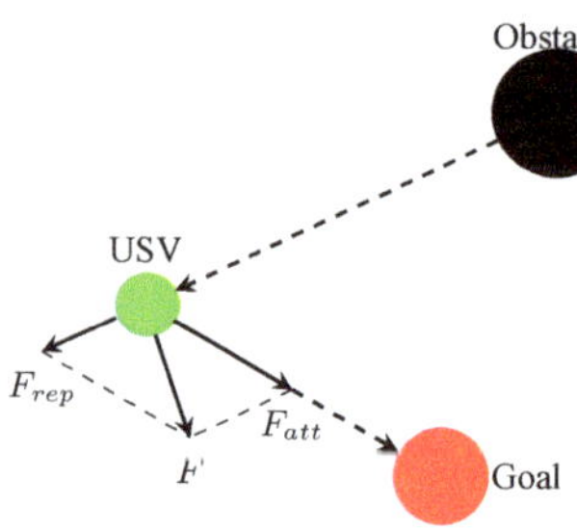

Figure 1. Diagram of the force analysis of an APF.

The potential function of the goal can be represented through

$$U_{att}(\mathbf{p}, \mathbf{p}_G) = \frac{1}{2}k_{att}\rho^2(\mathbf{p}, \mathbf{p}_G), \tag{1}$$

where $U_{att}(\mathbf{p})$ is the potential field at the position $\mathbf{p}$, the goal is at the position $\mathbf{p}_G$, $k_{att}(> 0)$ is the attractive gain, and $\rho(\mathbf{p}, \mathbf{p}_G) = \|\mathbf{p} - \mathbf{p}_G\|$ is the shortest distance between $\mathbf{p}$ and $\mathbf{p}_G$.

As the USV approaches the goal position, the potential field intensity generated by Equation (1) decreases, and the field intensity becomes 0 when it reaches the goal position. The potential fields generated with two different attractive gains are shown in Figure 2.

The force $\mathbf{F}_{att}$ generated by the goal potential field is the gradient of Equation (1),

$$\mathbf{F}_{att}(\mathbf{p}, \mathbf{p}_G) = -\nabla U_{att}(\mathbf{p}, \mathbf{p}_G) = k_{att}(\mathbf{p}_G - \mathbf{p}), \tag{2}$$

where $\mathbf{F}_{att}$ is a vector whose direction is from $\mathbf{p}$ to $\mathbf{p}_G$, ∇ is the gradient operator. The variation in the magnitude of $\mathbf{F}_{att}$ with the distance d between the goal and the USV is illustrated in Figure 3. From Equation (2) and Figure 3, it can be observed that the force has a linear relationship with the distance when the potential field function is described using Equation (2).

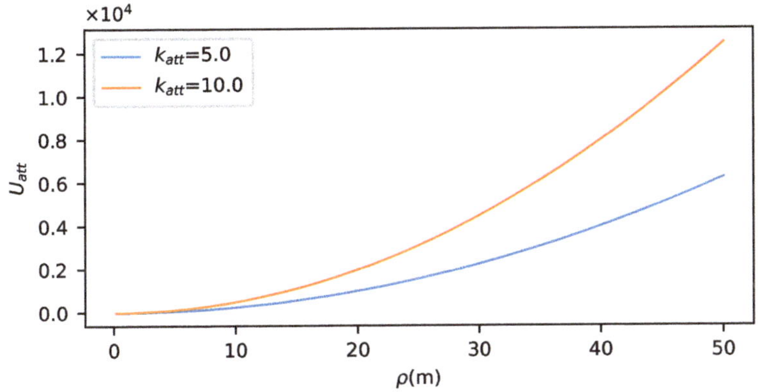

Figure 2. The potential field generated by the goal.

The potential function of the obstacle can be represented through

$$U_{rep}(\mathbf{p}, \mathbf{p}_O) = \begin{cases} \frac{1}{2}k_{rep}\left(\frac{1}{\rho(\mathbf{p}-\mathbf{p}_O)} - \frac{1}{\rho_0}\right)^2, & \rho(\mathbf{p}-\mathbf{p}_O) \leq \rho_0, \\ 0, & \rho(\mathbf{p}-\mathbf{p}_O) > \rho_0, \end{cases} \quad (3)$$

where U_{rep} is the potential field of the USV at the position $\mathbf{p}$, the obstacle is at the position $\mathbf{p}_O$, ρ_0 limits the influence range of the potential field, and $k_{rep}(> 0)$ is the repulsive gain. As the USV approaches the obstacle, the field intensity generated by Equation (3) increases rapidly.

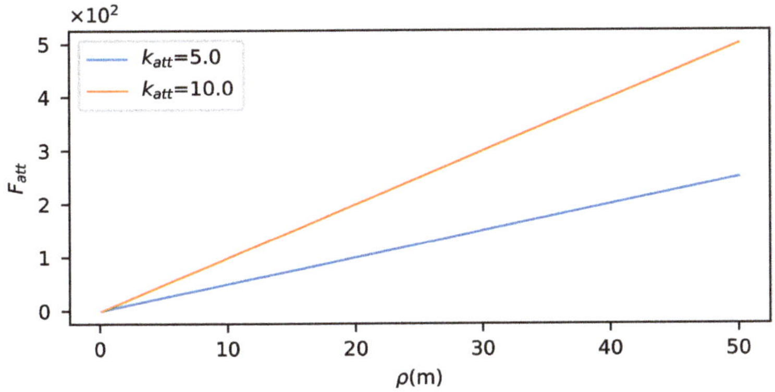

Figure 3. The force generated by the goal.

The intensity approaches infinity when $\rho(\mathbf{p}-\mathbf{p}_O)$ approaches 0. When $\rho(\mathbf{p}-\mathbf{p}_O) > \rho_0$, the field intensity is 0. The potential fields generated by the two different repulsive gains and $\rho = 1.0$ m are shown in Figure 4.

From Figure 4, it can be observed that the potential field of the obstacle decays rapidly with increasing distance, and the repulsive gains only affect the intensity of the repulsive field within a very close range of the obstacle.

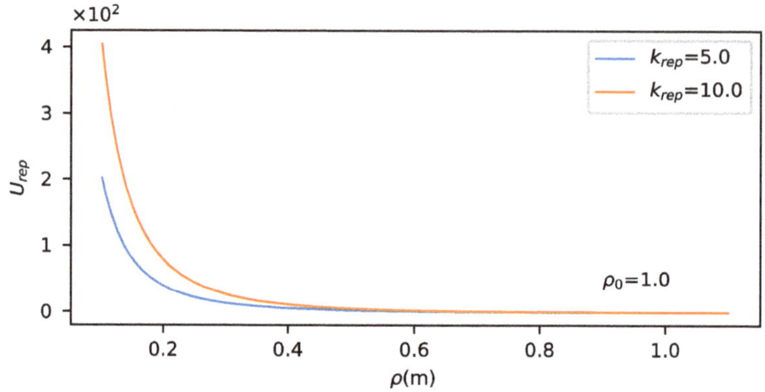

Figure 4. The potential field generated by the obstacle.

The repulsive force $\mathbf{F}_{rep}(\mathbf{p})$ generated by the obstacle is the gradient of Equation (3),

$$\mathbf{F}_{rep}(\mathbf{p},\mathbf{p}_O) = -\nabla U_{rep}(\mathbf{p},\mathbf{p}_O) = \begin{cases} k_{rep}\left(\frac{1}{\rho(\mathbf{p}_O-\mathbf{p})} - \frac{1}{\rho_0}\right)\frac{1}{\rho^2(\mathbf{p}_O-\mathbf{p})}\nabla\rho(\mathbf{p}_O-\mathbf{p}), & \rho(\mathbf{p}-\mathbf{p}_O) \le \rho_0, \\ 0, & \rho(\mathbf{p}-\mathbf{p}_O) > \rho_0. \end{cases} \quad (4)$$

where $\mathbf{F}_{rep}$ is a vector whose direction is from $\mathbf{p}_O$ to $\mathbf{p}$. The variation in the magnitude of $\mathbf{F}_{rep}$ with the distance d between the obstacle and the USV is depicted in Figure 5.

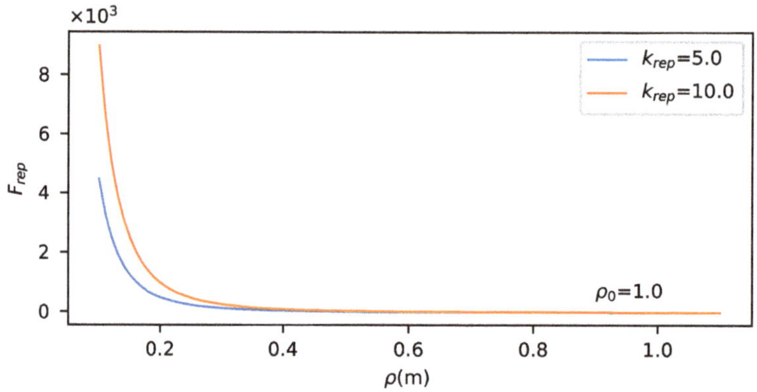

Figure 5. The force generated by the obstacle.

When the potential field is generated by one goal and multiple obstacles, the combined force $\mathbf{F}(\mathbf{p})$ received by the USV is

$$\mathbf{F}(\mathbf{p}) = \mathbf{F}_{att}(\mathbf{p},\mathbf{p}_G) + \sum_{i=0}^{M-1} \mathbf{F}_{rep}(\mathbf{p},\mathbf{p}_{Oi}), \quad (5)$$

where M is the number of obstacles in the potential field. The direction $\mathbf{F}(\mathbf{p})$ is the moving direction of the USV.

3. Formation Control Method

3.1. Behavior-Based Method

The behavior-based method controls the behavior of the entire formation by weighting the expected behaviors, including avoiding obstacles, maintaining the desired formation, and reaching the target point. In reality, for an USV fleet in formation, it is often necessary to simultaneously achieve multiple control objectives, such as formation maintenance and target search.Formation maintenance means that the USV fleet maintains a certain relative positional relationship when the USV fleet is moving. Target search means that the members of the formation search for the target area through the collaborative movement of the members and find the target object or location. These two control objectives have different impacts on the behavior of the formation members, so the behavior of the formation needs be adjusted accordingly. The behavior-based method compromises multiple control objectives and weights each control behavior to enable the formation members to coordinate and complete various tasks [19,20].

3.2. Virtual Structure Method

The virtual structure method utilizes the position and attitude information of each target point to form the formation. These target points usually correspond to the points on a rigid body. Compared to other formation methods, the control concept of this method is easier to understand and more convenient to implement. However, centralized control is required and the flexibility of the formation is insufficient, which also poses significant problems in obstacle avoidance [21,22].

3.3. Graph Theory Method

The graph theory method uses a control graph to define the shape of USV formations, where each node represents an USV and the directed edges between nodes represent the relative relationship between two USVs. The distributed nature of graph theory determines that it can be used for large-scale USV formations, making it easy to handle the addition or deletion of USV nodes in occasional queues. Using graph theory to describe the formation and the relationship between USVs, facilitating the main exchange between different formations. One disadvantage is that it is mainly limited to simulation research, and the implementation is relatively complex [23,24].

3.4. Leader–Follower Method

The basic idea of the leader–follower method is that, in a group composed of multiple USVs, one USV is designated as the leader and the other USVs are its followers. The followers follow the position and orientation of the leader USV at a certain distance interval. This method can also be extended on the basis of the above description, which can, not only specify one leader, but also specify multiple leaders. However, in the control of group formation, the leader is often designated as one. According to the relative positional relationship between leaders and followers, different network topologies can be formed, which means forming different formations. In a system that employs this method to control the formation of USVs, collaboration is achieved by sharing knowledge such as the state of the leader. The advantage of the leader–follower method is that simply giving the leader's behavior or trajectory can control the behavior of the entire USV population. The main drawback is that there is no clear formation feedback in the system. For example, if the USV leader advances too quickly, the following USV may not be able to track in time. Another drawback is that if the leader fails, the entire formation cannot be maintained [24–28].

4. USV Formation Control and Obstacle Avoidance Method

In this paper, we utilize a leader–follower and virtual structure method for formation control, and use an APF for obstacle avoidance. For simplicity, we simplify the dynamic description of the USVs by limiting the maximum linear speed and maximum angular speed. The leader is moving at a constant speed. To address the issue of limited flexibility in maintaining formation using virtual structure methods and to enhance the stability of the formation, the linear speed of the follower can be adjusted based on the distance between the follower and the object being tracked, which may be greater than the speed of the leader but must be smaller than the maximum linear speed.

The USV leader is influenced by the attractive force generated by the task target, as well as the repulsive force generated by all obstacles in the environment and other USVs in the fleet. The USV leader calculates its position and the heading angle in the next moment based on the combined direction of the attractive and repulsive force, combined with the linear speed and maximum angular speed constraints of the formation movement.

After determining the position and the heading angle of the leader in the next moment, the leader uses the virtual structure method to calculate the ideal position of each follower based on the position and the heading angle of the leader itself. The followers are attracted by the forces generated by their respective ideal positions or the target position, depending on whether they are repelled by obstacles or not, as well as the repulsive forces generated by all obstacles and the other USVs in the environment. The followers are guided by the direction of the combined forces generated by the attractive and repulsive forces. The followers calculate their heading angles in the next moment by combining the maximum linear speed and the maximum angular speed constraints during the follower's movement.

In order to quickly leave the obstacle area, this paper modifies the attractive force model of the traditional APF. We restrict the maximum magnitude of the attractive force when influenced by the repulsive force, to modify the direction of the combined force.

There are three points to note: first, the speed of the USV formation is the same as the speed of the leader, but the speed of the follower can be greater than the formation speed to meet the needs of maintaining the formation; second, the follower may be pulled by the target position or the ideal position, depending on whether they are repelled by obstacles or not; third, when a USV is subject to a repulsive force, the magnitude of its attractive force is restricted by that of the repulsive force. This restriction can impact the heading angle of the USV.

Below, we elaborate on the principles of the method proposed in this article.

4.1. Formation Representation

This article uses the body coordinate system of the leader to represent the formation. The heading direction of the leader is taken as the positive direction of the x-axis, and the y-axis coordinate direction of the leader is determined using the right hand rule. The origin O of the xy coordinate system can be specified as needed.

We represent the position of the follower i in the formation with (α_i, l_i). α_i is the azimuth in the leader body coordinate system, and l_i is the distance between the follow i and the origin, as shown in Figure 6. The coordinate (x_i, y_i) of the follower i in the leader body coordinate system can obtained through

$$\begin{cases} x_i &= l_i \cos(\alpha_i) \\ y_i &= l_i \sin(\alpha_i) \end{cases} \tag{6}$$

When the positions of all followers are specified, the formation of the USVs is determined.

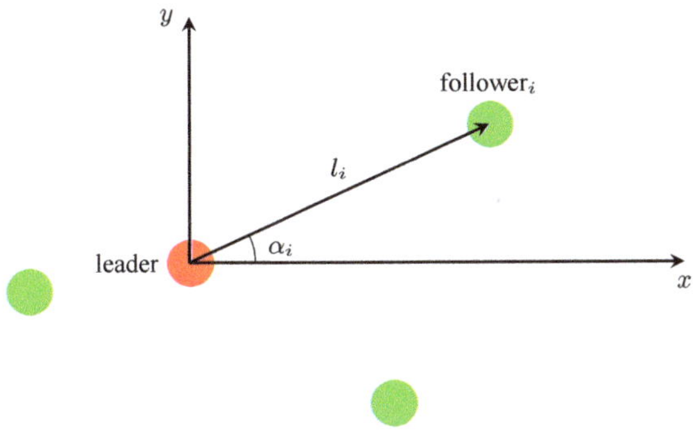

Figure 6. Formation representation in the leader body coordinate system.The red solid circle represents the leader, and the green solid circles represent the followers.

4.2. Obstacle Avoidance and Formation Update

Assume that the USV formation contains N USVs, which consist of 1 leader and $N-1$ followers, and there are M obstacles in the task scenario. We number the USVs, the id $i(0 \le i \le N-1)$ represents the i_{th} USV, and the id of the leader is 0. We also number the obstacles, the id $j(0 \le j \le M-1)$ represents the j_{th} obstacle. Assume that $\mathbf{p}_{Ii}$ represents the ideal position of the USV i in the world coordinate system, and $\mathbf{p}_{Ai}$ represents the actual position of the USV i in the world coordinate system, and $\mathbf{p}_{Oj}$ represents the position of the obstacle j in the world coordinate system.

4.2.1. The Unit Vector of the Combined Force on the Leader

The repulsive force on the leader is

$$\mathbf{F}_{rep}(\mathbf{p}_{A0}) = \sum_{i=0}^{M-1} \mathbf{F}_{rep}(\mathbf{p}_{A0}, \mathbf{p}_{Oi}) + \sum_{i=1}^{N-1} \mathbf{F}_{rep}(\mathbf{p}_{A0}, \mathbf{p}_{Ai}). \tag{7}$$

In order to quickly leave the obstacle area, we limit the maximum of the attractive force when subjected to the repulsive force to modify the direction of the combined force. We adjust the magnitude of the attractive force using

$$\mathbf{F}_{att}(\mathbf{p}_{A0}, \mathbf{p}_G) = \begin{cases} \frac{a\|\mathbf{F}_{rep}(\mathbf{p}_{A0}, \mathbf{p}_G)\|}{\|\mathbf{F}_{att}(\mathbf{p}_{A0})\|}\mathbf{F}_{att}(\mathbf{p}_{A0}, \mathbf{p}_G), & \|\mathbf{F}_{att}(\mathbf{p}_{A0}, \mathbf{p}_G)\| > \|\mathbf{F}_{rep}(\mathbf{p}_{A0})\|, \\ \mathbf{F}_{att}(\mathbf{p}_{A0}, \mathbf{p}_G) = \mathbf{F}_{att}(\mathbf{p}_{A0}, \mathbf{p}_G), & \|\mathbf{F}_{att}(\mathbf{p}_{A0}, \mathbf{p}_G)\| \le \|\mathbf{F}_{rep}(\mathbf{p}_{A0})\|, \end{cases} \tag{8}$$

where $a(0 < a < 1.0)$ is the weighting coefficient that can modify the direction of the combined force.

Now, we can calculate the unit vector of the combined force on the leader. From Equation (5), the combined force on the leader is

$$\mathbf{F}(\mathbf{p}_{A0}) = \mathbf{F}_{att}(\mathbf{p}_{A0}, \mathbf{p}_G) + \mathbf{F}_{rep}(\mathbf{p}_{A0}). \tag{9}$$

Thus, the unit vector of $\mathbf{F}(\mathbf{p}_{A0})$ is

$$\mathbf{F}_{unit}(\mathbf{p}_{A0}) = \frac{\mathbf{F}(\mathbf{p}_{A0})}{\|\mathbf{F}(\mathbf{p}_{A0})\|}. \tag{10}$$

4.2.2. The Next Moment Position and the Heading Angle of the Leader

Assume that θ_{yaw} represents the heading angle of the USV formation, which is also the heading angle of the leader. Assume that ω_m is the maximum angular speed of the USV. The next moment heading angle θ of the leader can be obtained through

$$\theta = \arctan\left(\frac{\mathrm{Im}(\mathbf{F}_{unit}(\mathbf{p}_{A0}))}{\mathrm{Re}(\mathbf{F}_{unit}(\mathbf{p}_{A0}))}\right), \tag{11}$$

where $\mathrm{Re}(\cdot)$ represents the real part, and $\mathrm{Im}(\cdot)$ represents the imaginary part.

The heading angle difference Δ_θ between two consecutive moments is

$$\Delta_\theta = \theta_{yaw} - \theta. \tag{12}$$

Update the heading angle θ_{yaw} using the maximum angular speed ω_m of the USV,

$$\theta_{yaw} = \begin{cases} \theta, & |\Delta_\theta| \leq \omega_m, \\ \theta_{yaw} + \omega_m, & |\Delta_\theta| > \omega_m. \end{cases} \tag{13}$$

$\mathbf{p}_{I0}$ not only represents the ideal position of the leader but also the actual position of the next moment of the leader. In other words, $\mathbf{p}_{I0} = \mathbf{p}_{A0}$. Assume that the linear speed of the USV formation is v_f, which is a scalar. $\mathbf{p}_{I0}$ can be obtained using the heading angles θ_{yaw} and v_f,

$$\mathbf{p}_{I0} = \mathbf{p}_{A0} + v_f(\cos\theta_{yaw}, \sin\theta_{yaw}). \tag{14}$$

4.2.3. The Ideal Position of the Follower

Now the ideal formation position of the follower can be updated based on the position and the heading angle of the leader using the virtual structure.

The ideal position $\mathbf{p}_{Ii}$ of the follower i can be obtained using

$$\mathbf{p}_{Ii} = \mathbf{p}_{I0} + l_i(\cos(\theta_{yaw} + \alpha_i), \sin(\theta_{yaw} + \alpha_i)). \tag{15}$$

The relationships among the formation representation $(\mathbf{p}_i, l_i)$ in the leader body coordinate system and the leader position $\mathbf{p}_{I0}$ and the follower position $\mathbf{p}_{Ii}$ are shown in Figure 7.

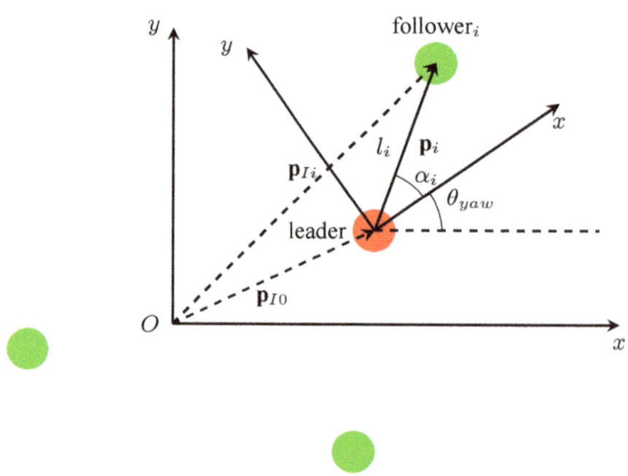

Figure 7. The relationship between the formation representations in the world coordinate system and in the leader body coordinate system.The red solid circle represents the leader, and the green solid circles represent the followers.

4.2.4. The Unit Vector of the Combined Force on the Follower

The repulsive force on the follower i is

$$\mathbf{F}_{rep}(\mathbf{p}_{Ai}) = \sum_{j=0}^{M-1} \mathbf{F}_{rep}(\mathbf{p}_{Ai}, \mathbf{p}_{Oj}) + \sum_{j=0, j\neq i}^{N-1} \mathbf{F}_{rep}(\mathbf{p}_{Ai}, \mathbf{p}_{Aj}). \tag{16}$$

We adjust the magnitude of the attractive force for the follower i through

$$\mathbf{F}_{att}(\mathbf{p}_{Ai}, \mathbf{p}_T) = \begin{cases} \frac{a\|\mathbf{F}_{rep}(\mathbf{p}_{Ai}, \mathbf{p}_T)\|}{\|\mathbf{F}_{att}(\mathbf{p}_{Ai})\|}\mathbf{F}_{att}(\mathbf{p}_{Ai}, \mathbf{p}_T), & \|\mathbf{F}_{att}(\mathbf{p}_{Ai}, \mathbf{p}_T)\| > \|\mathbf{F}_{rep}(\mathbf{p}_{Ai})\|, \\ \mathbf{F}_{att}(\mathbf{p}_{Ai}, \mathbf{p}_T) = \mathbf{F}_{att}(\mathbf{p}_{Ai}, \mathbf{p}_T), & \|\mathbf{F}_{att}(\mathbf{p}_{Ai}, \mathbf{p}_T)\| \leq \|\mathbf{F}_{rep}(\mathbf{p}_{Ai})\|. \end{cases} \tag{17}$$

where $\mathbf{p}_T$ is the position $\mathbf{p}_G$ of the target or the ideal position $\mathbf{p}_{Ii}$ of the follower i based on whether it is repelled by obstacles or not.

Now, we can calculate the unit vector of the combined force on the follower. From Equation (5), we can obtain the combined force on the follower i. If the follower is repelled by some obstacles, the combined force is

$$\mathbf{F}(\mathbf{p}_{Ai}) = \mathbf{F}_{att}(\mathbf{p}_{Ai}, \mathbf{p}_T) + \mathbf{F}_{rep}(\mathbf{p}_{Ai}). \tag{18}$$

Thus, the unit vector of $\mathbf{F}(\mathbf{p}_{Ai})$ is

$$\mathbf{F}_{unit}(\mathbf{p}_{Ai}) = \frac{\mathbf{F}(\mathbf{p}_{Ai})}{\|\mathbf{F}(\mathbf{p}_{Ai})\|}. \tag{19}$$

4.2.5. The Next Moment Position and the Heading Angle of the Follower

Assume that β_i represents the heading angle of the follower i at the moment. The next moment heading angle β of the follower i can be obtained through

$$\beta = \arctan\left(\frac{\mathrm{Im}(\mathbf{F}_{unit}(\mathbf{p}_{Ai}))}{\mathrm{Re}(\mathbf{F}_{unit}(\mathbf{p}_{Ai}))}\right). \tag{20}$$

The heading angle difference Δ_β between two consecutive moments is

$$\Delta_\beta = \beta - \beta_i. \tag{21}$$

Update the heading angle β_i using the maximum angular speed of the USV,

$$\beta_i = \begin{cases} \beta, & |\Delta_\beta| \leq \omega_m, \\ \beta_i + \omega_m, & |\Delta_\beta| > \omega_m. \end{cases} \tag{22}$$

When the heading angle of the USV formation changes significantly, this has a significant impact on the position changes of the follower at the edge of the formation. For the stability of the formation, we allow the follower's linear speed to be greater than the formation speed.

The distance d_i between the follower i and the traction object is

$$d_i = \|\mathbf{p}_{Ii} - \mathbf{p}_{Ai}\|, \tag{23}$$

or

$$d_i = \|\mathbf{p}_G - \mathbf{p}_{Ai}\|, \tag{24}$$

depending on whether the follower is repelled by obstacles or not.

Assume that the maximum linear speed of the USV follower is v_m, which is a scalar. We adjust the follower's speed v_i based on d_i,

$$v_i = \begin{cases} v_m, & d_i \geq v_m, \\ d_i, & d_i < v_m. \end{cases} \tag{25}$$

$\mathbf{p}_{Ii}$ can be obtained using the heading direction angle β_i,

$$\mathbf{p}_{Ii} = \mathbf{p}_{Ai} + v_i(\cos\beta_i, \sin\beta_i). \tag{26}$$

4.3. Algorithm

The formation control and obstacle avoidance algorithm is summarized in Algorithm 1.

Algorithm 1: Formation Control and Obstacle Avoidance Algorithm

---

Input: the goal position $\mathbf{p}_G$, the actual possition $\mathbf{p}_{Ai}$ of the USV i, the ideal position $\mathbf{p}_{Ii}$ of the USV i, the position $\mathbf{p}_{Oj}$ of the obstacle j, the formation parameters α_i and l_i, the heading direction θ_{yaw} of the formation, the formation linear speed v_f, the maximum angular speed ω_m of the USV, the maximum linear speed v_m of the follower, the attractive gain F_{att}, the repulsive gain F_{rep},the weighting coefficient a the attractive force

while *the leader hasn't reached the goal position or the followers hasn't reached their ideal position* **do**
    if *the leader hasn't reached the goal position* **then**
        Compute the combined force unit vector $\mathbf{F}_{unit}(A_0)$ of the leader;
        Update the heading angle θ_{yaw} of the formation according the maximum angular speed ω_m;
        Compute the ideal position $\mathbf{p}_{I0}$ of the leader with the formation linear speed v_f and θ_{yaw}, and let $\mathbf{p}_{A0} = \mathbf{p}_{I0}$;
        Compute the ideal position $\mathbf{p}_{Ii}$ of the follower i;
    end
    foreach *follower i* **do**
        if *the follower i hasn't reached their ideal position* **then**
            Update the heading direction β_i of the follower i;
            Compute the distance d_i between $\mathbf{p}_{Ai}$ and the traction object;
            Compute the linear speed v_i of the follower i according d_i and the maximum linear speed v_m of the follower;
            Update the actual position $\mathbf{p}_{Ai}$ of the follower i according v_i and β_i
        end
    end
end

---

5. Simulation Studies

For the sake of simplicity, we assume that USVs have the formation shown in Figure 8. The USV fleet consists of 6 USVs, with an equilateral triangle formation and three USVs evenly spaced on each edge.

The id of each USV is shown in Figure 8. The width of each USV is 0.8 m, and their length is 1.0 m. The formation parameters are shown in Table 1. Table 2 shows the performance parameters of the USVs.

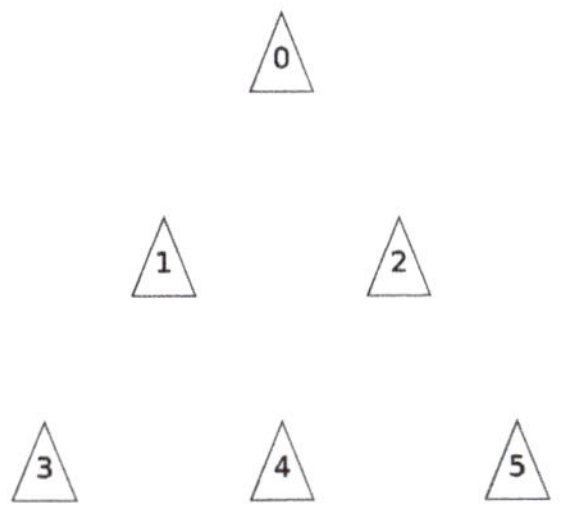

Figure 8. Formation of USVs.

Table 1. Formation parameters.

Parameters	Value	Parameters	Value
α_1	240°	l_1	3 m
α_2	300°	l_2	3 m
α_3	240°	l_3	6 m
α_4	270°	l_4	6 m
α_5	300°	l_5	$3\sqrt{3}$ m

Table 2. Performance parameters of the USV.

Parameters	Value	Parameters	Value
v_f	0.1 m/s	ω_m	15°
v_m	0.2 m/s		

5.1. Experiment without Obstacles

We first verified the formation maintenance effects of the algorithm when there is no obstacle.

In Figure 9, the black solid circle represents the starting position of the formation, the red solid circle represents the target position, the circle with numbers indicates the ideal position of the USV, and the solid triangle stands for the actual position of the USV. Figure 9b–d depicts the formations of USVs after 20, 40, and 80 iterations. From the figures, it can be observed that the USV formation reached stability after a brief adjustment.

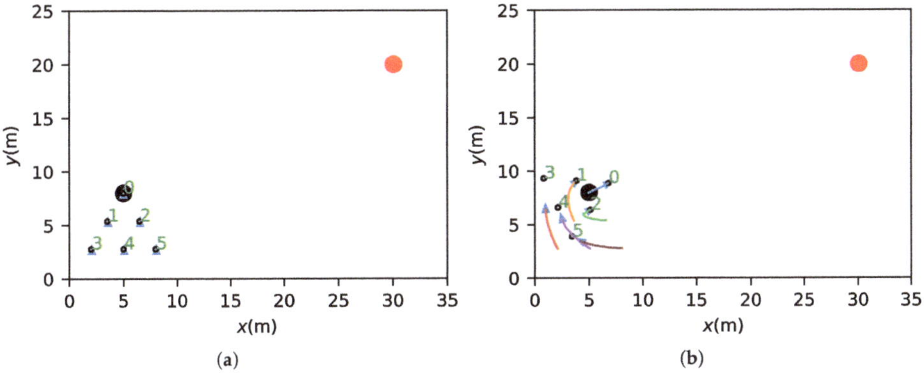

Figure 9. *Cont.*

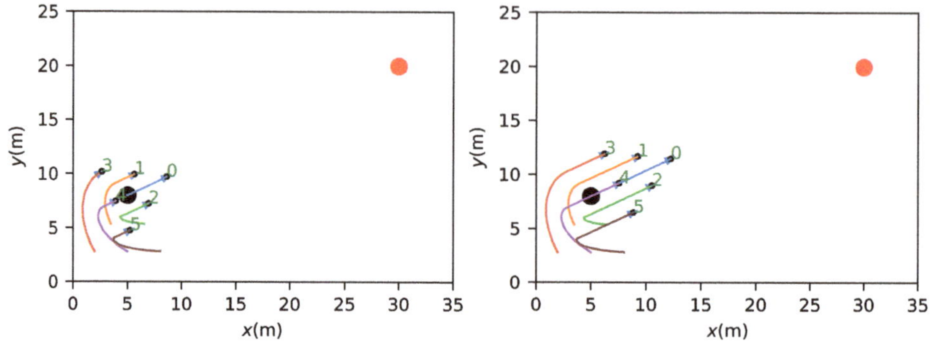

(c)

(d)

Figure 9. Formation of USVs during movement. (**a**) The initial formation of the USVs; (**b**) formation after 20 iterations; (**c**) formation after 40 iterations; (**d**) formation after 80 iterations.

5.2. Experiment with Obstacles

The parameters used in the APF approach in this simulation are shown in Table 3. The formation parameters are depicted in Table 1. The performance parameters of the USV are shown Table 2. The positions and radius of the obstacles are shown Table 4.

Table 3. APF parameters.

Parameters	Value	Parameters	Value
Goal attractive gain	5.0	Ideal position attractive gain	5.0
Obstacle repulsive gain	5.0	Limit distance	1.0 m
Attractive force weighting gain	1.0		

Table 4. Obstacle position and radius.

	Position	Radius
Obstacle 1	(15.0 m, 20.0 m)	3.0 m
Obstacle 2	(25.0 m, 10.0 m)	4.0 m
Obstacle 3	(45.0 m, 24.0 m)	1.5 m

In Figure 10, the black solid circle represents the starting position of the formation, the red solid circle indicates the target position, the circle with numbers stands for the ideal position of the USV, the solid triangle depicts the actual position of the USV, the blue solid circle represents the obstacle, and the dashed circle means the limit distance of the obstacle.

Figure 10b–f show the formations of the USVs after 150, 280, 350, 450, and 613 iterations. Figure 10f shows the final state of the USV formation after reaching its goal position. From the figures, it can be observed that the USV formation could effectively avoid obstacles and endeavored to maintain the stability of the formation during the movement.

The variation in the linear speed of the USVs over time is depicted in Figure 11. The leader moved at a constant linear speed, and the speed was 0 when it reached the target position. The linear speeds of follower 1, follower 2, and follower 5 exhibited rapid changes during specific time intervals when the USVs were maneuvering to avoid obstacles. The follower 4 encountered almost no obstacles throughout the entire process, so the time interval for rapid speed changes was very short. The time intervals during which the followers moved at their maximum speed corresponded to the adjustment periods of the formation. These can be observed from Figure 10.

In order to enhance the stability of the formation, this study allowed the maximum linear speed v_m of the followers to exceed the linear speed v_f of the leader (v_f is also the linear speed of the formation), and the linear speed v_i of the follower i was adjusted according to Equation (25). Figure 12 shows the variation in positional errors over time when the leader moved at a linear speed 0.1 m/s, the followers moved at the maximum linear speeds of 0.1 m/s (which is equal to v_f) and 0.2 m/s (which is greater than v_f), respectively.

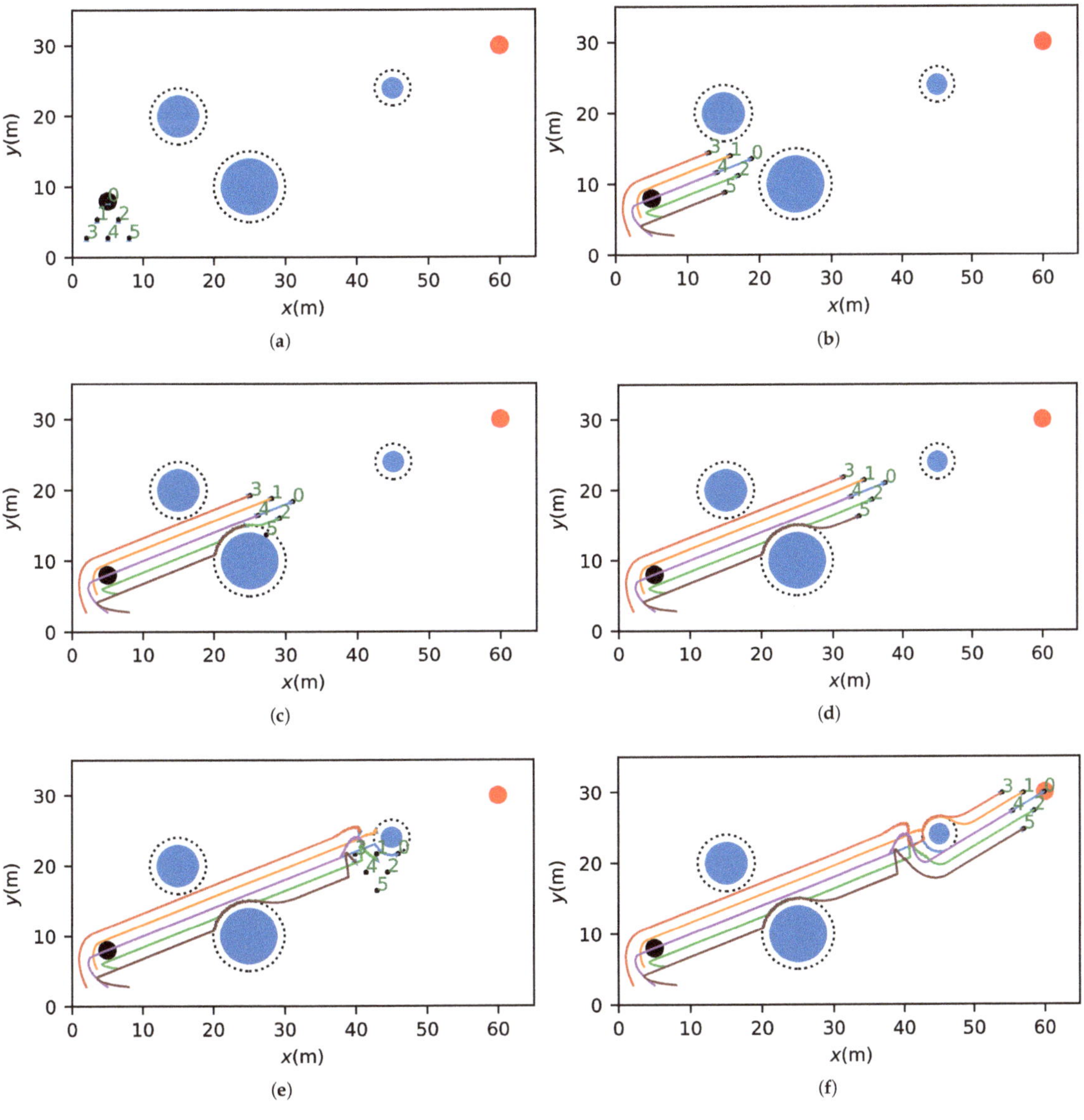

Figure 10. Formation of USVs during movement. (**a**) The initial formation of USVs; (**b**) formation after 150 iterations; (**c**) formation after 280 iterations; (**d**) formation after 350 iterations; (**e**) formation after 450 iterations; (**f**) formation after 613 iterations.

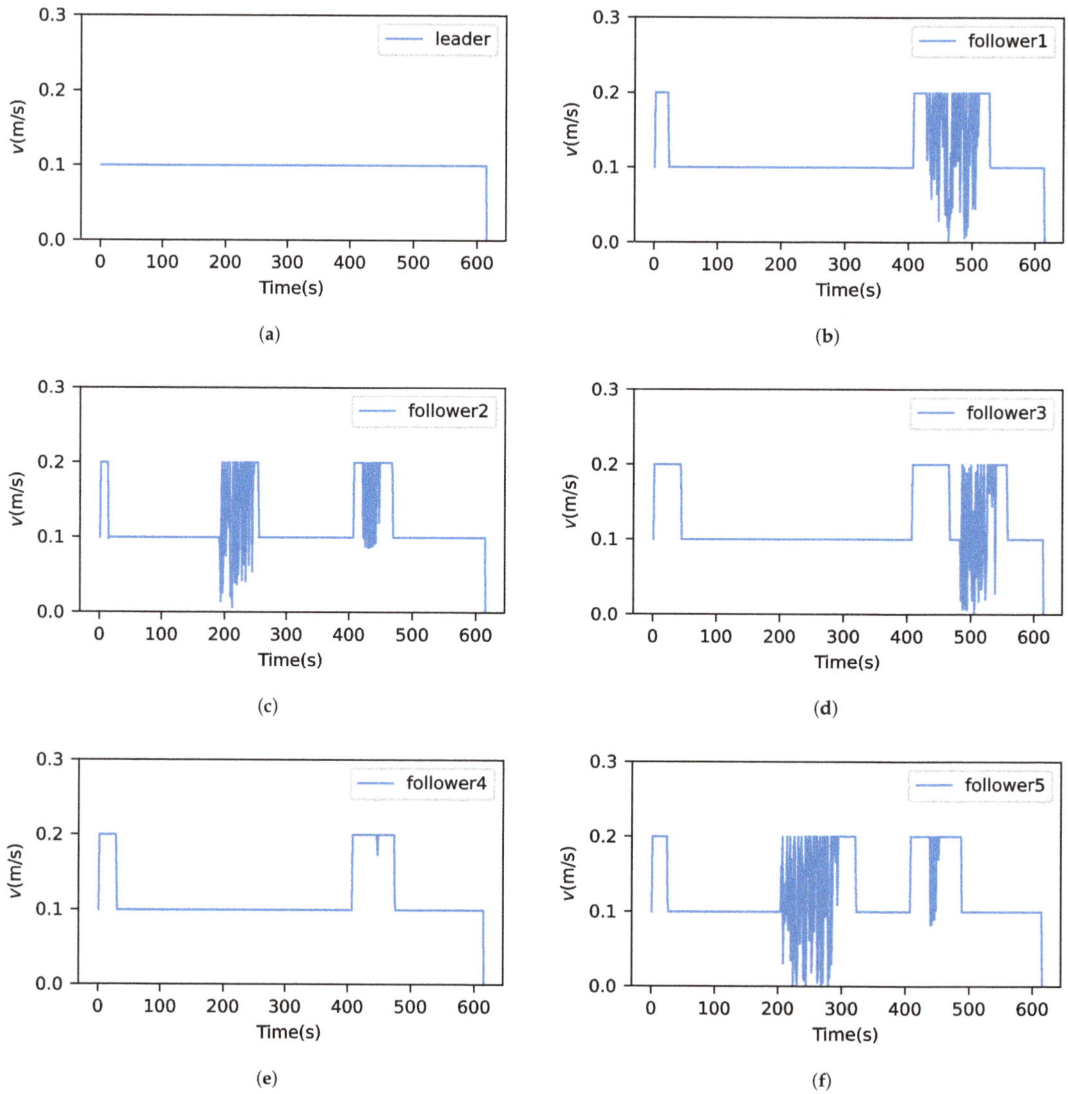

Figure 11. The varied linear speed of the USVs with respect to time. (**a**) Leader; (**b**) Follower 1; (**c**) Follower 2; (**d**) Follower 3; (**e**) Follower 4; (**f**) Follower 5.

As depicted in Figure 12, when the follower was restricted to moving at the maximum linear speed of the leader, the positional errors of the follower were large during almost the whole movement process. When the leader reached the target position, the followers required a long time to adjust to reach their respective ideal positions. When the linear speed of the follower surpassed that of the leader, the follower reached the ideal position more quickly. So when the formation changed slightly, the followers could be at the ideal positions. Under the constraints of two different maximum linear speeds, the times needed for the USV fleet to reach the target position and form a stable formation were not the same. Moreover, as evidenced by Figure 12; when the linear speed of the follower exceeded that of the leader, the time required was significantly reduced.

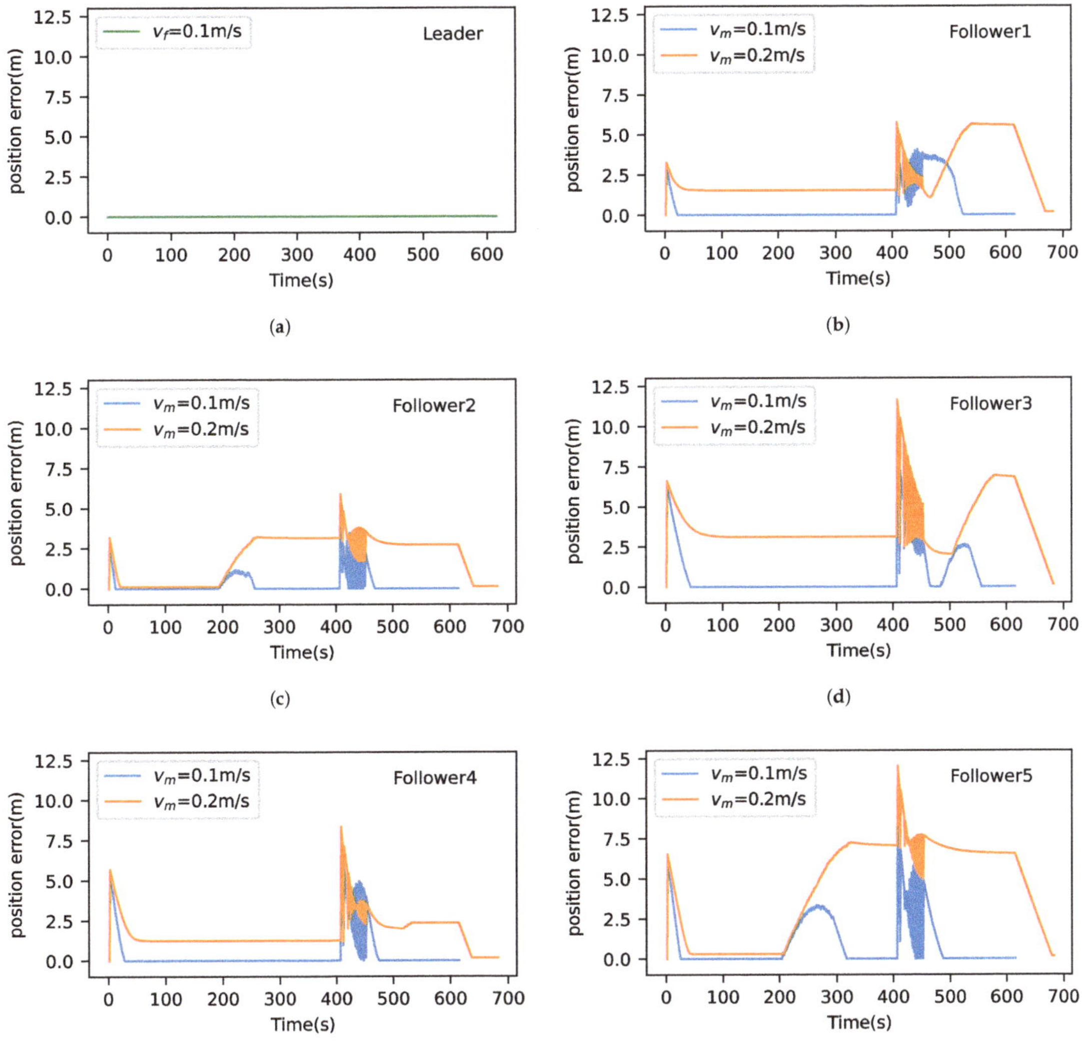

Figure 12. Comparison of positional errors of USVs at the different maximum linear speeds of 0.1 m/s and 0.2 m/s. (**a**) Position errors of the leader. The linear speed of the leader is v_f = 0.1 m/s. v_m = 0.1 m/s and v_m = 0.2 m/s are the maximum linear speeds of the followers, not the linear speed of the leader; (**b**) position errors of the follower 1; (**c**) position errors of the follower 2; (**d**) position errors of the follower 3; (**e**) position errors of the follower 4; (**f**) position errors of the follower 5.

6. Conclusions

In order to seek a simple and effective method to address the challenges of formation control and obstacle avoidance for multiple USVs, this study proposed a method that combines the artificial potential field method, virtual structure method, and leader–follower method. The artificial potential field method is used to tackle the obstacle avoidance problem for USVs, while the virtual structure method is utilized to generate the ideal position of the follower within the formation. The leader–follower method is employed to maintain the formation. The modifications were made to both the artificial potential field approach and the leader–follower scheme. To ensure the combined force deviates from obstacles and swiftly moves away from the obstacle area, the attractive force is constrained so as not to exceed the repulsive force when encountering obstacles. For effectively avoiding obstacles and maintaining formation, the followers select the target's position or their ideal position as the reference point based on whether they are repelled by obstacles or not. Additionally, followers can adjust their linear speed depending on their distance from the reference point, without exceeding their maximum speed, to better maintain the formation. Simulation experiments showed that this method could effectively avoid obstacles and maintain the formation. However, further research is needed to address several remaining issues, including incorporating the dynamic model of the USV, considering various types of obstacles, and refining the motion trajectory of the USV to achieve smoother movements.

Author Contributions: Conceptualization, G.L. and R.Z.; methodology, G.L. and R.Z.; software N.W. and F.L.; writing—original draft preparation, G.L.; writing—review and editing, G.L. and R.Z.; visualization, N.W. and F.L. All authors have read and agreed to the published version of the manuscript.

Funding: This research was funded by Basic Research Projects for Universities of Educational Department of Liaoning Province (grant number LJKMZ20220395).

Institutional Review Board Statement: Not applicable.

Informed Consent Statement: Not applicable.

Data Availability Statement: Data is contained within the article.

Conflicts of Interest: The authors declare no conflict of interest.

References

1. Niu, Y.; Zhang, J.; Wang,Y.; Yang, H.; Mu, Y. A review of path planning algorithms for USV. In Proceedings of the 2021 International Conference on Autonomous Unmanned Systems(ICAUS 2021), Changsha, China, 24–26 September 2021.
2. Tan, G.; Zhuang, J. ; Zou, J.; Wan, L. Multi-type task allocation for multiple heterogeneous unmanned surface vehicles (USVs) based on the self-organizing map. *Appl. Ocean Res.* **2022**, *126*, 103262. [CrossRef]
3. Wang, X.; Li, X.; Zheng, Z. Survey of developments on multi-agent formation control related problems. *Control Decis.* **2013**, *28*, 1601–1613.
4. Jin, K.; Wang, J.; Wang, H.; Liang, X.; Guo, Y.; Wang, M.; Yi, H. Soft formation control for unmanned surface vehicles under environmental disturbance using multi-task reinforcement learning. *Ocean Eng.* **2022**, *260*, 112035. [CrossRef]
5. MahmoudZadeh, S.; Abbasi, A.; Yazdani, A.; Wang, H.; Liu, Y. Uninterrupted path planning system for Multi-USV sampling mission in a cluttered ocean environment. *Ocean Eng.* **2022**, *254*, 111328. [CrossRef]
6. Miao, R.; Wang, L. Coordination of distributed unmanned surface vehicles via model-based reinforcement learning methods. *Appl. Ocean Res.* **2022**, *122*, 103106. [CrossRef]
7. Luo, D.; Zhang, H.; Xie, R.; Wu, S. Unmanned aerial vehicles swarm conflict based on multi-agent system. *Control Theory Appl.* **2015**, *30*, 1498–1504.
8. Tang, J.; Duan, H.; Lao, S. Swarm intelligence algorithms for multiple unmanned aerial vehicles collaboration: A comprehensive review. *Artif. Intell. Rev.* **2023**, *56*, 4295–4327. [CrossRef]
9. Chen, X.; Wei, X.; Xu, G. Multiple unmanned aerial vehicle decentralized cooperative air combat decision making with fuzzy situation. *J. ShangHai Jiao Tong Univ.* **2013**, *48*, 908–913, 921.
10. Jin, X.; Er, M.J. Dynamic collision avoidance scheme for unmanned surface vehicles under complex shallow sea Environments. *Ocean Eng.* **2020**, *218*, 108102. [CrossRef]
11. Pan, W.; Jiang, D.; Pang, Y.; Li, Y.; Zhang, Q. A multi-AUV formation algorithm combining artificial potential field and virtual structure. *Acta Armamentarii* **2017**, *38*, 326–334.
12. He, Y.; Mou, J.; Chen, L.; Zeng, Q.; Chen, P.; Zhang, S. Survey on hydrodynamic effects on cooperative control of maritime autonomous surface ships. *Ocean Eng.* **2021**, *235*, 109300. [CrossRef]
13. Feng, Z.; Hu, G.; Sun,Y.; Soon, J. An overview of collaborative robotic manipulation in multi-robot systems. *Annu. Rev. Control* **2020**, *49*, 113–127. [CrossRef]
14. Chen, H.; Wang, Y.; Xie, S.; Zhang, D.; Yan, H. Reinforcement learning-based close formation control for underactuated surface vehicle with prescribed performance and time-varying state constraints. *Ocean Eng.* **2022**, *256*, 11361. [CrossRef]
15. Woo, J.; Kim, N. Collision avoidance for an unmanned surface vehicle using deep reinforcement learning. *Ocean Eng.* **2020**, *199*, 107001. [CrossRef]

16. Kim, D.; Hirayama, K.; Park, G. Collision avoidance in multiple-ship situations by distributed local search. *J. Adv. Comput. Intell. Intell. Inform.* **2014**, *18*, 839–848. [CrossRef]
17. Ma, Y.; Zhao, Y.; Incecik, A.; Yan, X.; Li, Z. A collision avoidance approach via negotiation protocol for a swarm of USVs. *Ocean Eng.* **2021**, *224*, 108713. [CrossRef]
18. Khatib, O. Real-time obstacle avoidance for manipulators and mobile robots. *Int. J. Robtics Res.* **1986**, *5*, 90–98. [CrossRef]
19. Tan, G.; Zhuang, J.; Zou, J.; Wan, L. Coordination control for multiple unmanned surface vehicles using hybrid behavior-based method. *Ocean Eng.* **2021**, *232*, 109147. [CrossRef]
20. Nantogma, S.; Ran, W.; Yang, X.; Huang, X. Behavior-based genetic fuzzy control system for multiple USVs cooperative target protection. In Proceedings of the 2019 3rd International Symposium on Autonomous Systems (ISAS), Shanghai, China, 29–31 May 2019.
21. Li, Z.; Xian, B. Robust distributed formation control of multiple unmanned aerial vehicles based on virtual structure. *Control Theory Appl.* **2020**, *37*, 2423–2431.
22. Han, J.; Park, J.; Kim, T.; Kim, J. Precision navigation and mapping under bridges with an unmanned surface vehicle. *Auton. Robots* **2015**, *38*, 349–362. [CrossRef]
23. Lai, Y.; Li, R.; Shi, Y.; He, L. On the study of a multi-quadrotor formation control with triangular structure based on graph theory. *Control Theroy Appl.* **2018**, *32*, 1530–1537.
24. Liu, Y.; Yang, Y. Formation control of multi-robot research. *Control Eng. China* **2010**, *17*, 182–186.
25. Zhang, X.; Huang, Y.; Liu, G. Research on improved leader-following formation method. *Comput. Eng. Des.* **2010**, *32*, 2547–2549.
26. Hinostroza, M.A.; Xu, H.; Soares, C.G. Experimental results of the cooperative operation of autonomous surface vehicles navigating in complex marine environment. *Ocean Eng.* **2021**, *219*, 108256. [CrossRef]
27. Dong, Y.; Chen, J. Nonlinear observer-based approach for cooperative control of networked rigid spacecraft systems. *Automatica* **2021**, *128*, 109552. [CrossRef]
28. Li, R.; Zhang, L.; Han, L.; Wang, J. Multiple vehicle formation control based on robust adaptive control algorithm. *IEEE Intell. Transp. Sys. Mag.* **2017**, *9*, 41–51. [CrossRef]

Journal of
Marine Science and Engineering

Article

A Low-Cost and High-Precision Underwater Integrated Navigation System

Jiapeng Liu [1,2], Te Yu [1,2], Chao Wu [3,*], Chang Zhou [1,2], Dihua Lu [1,2] and Qingshan Zeng [1,2]

1 National Key Laboratory of Science and Technology on Underwater Acoustic Antagonizing, Shanghai 201108, China; liu_jiapeng@sjtu.edu.cn (J.L.); butalways@sjtu.edu.cn (T.Y.); alexioszhou@outlook.com (C.Z.); ludihua@sjtu.edu.cn (D.L.); qingshanzeng@sjtu.edu.cn (Q.Z.)

2 Shanghai Marine Electronic Equipment Research Institute, Shanghai 201108, China

3 School of Naval Architecture, Ocean & Civil Engineering, Shanghai Jiao Tong University, Shanghai 200240, China

* Correspondence: wuchaorr@sjtu.edu.cn

Abstract: The traditional underwater integrated navigation system is based on an optical fiber gyroscope and Doppler Velocity Log, which is high-precision but also expensive, heavy, bulky and difficult to adapt to the development requirements of AUV swarm, intelligence and miniaturization. This paper proposes a low-cost, light-weight, small-volume and low-computation underwater integrated navigation system based on MEMS IMU/DVL/USBL. First, according to the motion formula of AUV, a five-dimensional state equation of the system was established, whose dimension was far less than that of the traditional. Second, the main source of error was considered. As the velocity observation value of the system, the velocity measured by DVL eliminated the scale error and lever arm error. As the position observation value of the system, the position measured by USBL eliminated the lever arm error. Third, to solve the issue of inconsistent observation frequencies between DVL and USBL, a sequential filter was proposed to update the extended Kalman filter. Finally, through selecting the sensor equipment and conducting two lake experiments with total voyages of 5.02 km and 3.2 km, respectively, the correctness and practicality of the system were confirmed by the results. By comparing the output of the integrated navigation system and the data of RTK GPS, the average position error was 4.12 m, the maximum position error was 8.53 m, the average velocity error was 0.027 m/s and the average yaw error was 1.41°, whose precision is as high as that of an optical fiber gyroscope and Doppler Velocity Log integrated navigation system, but the price is less than half of that. The experimental results show that the proposed underwater integrated navigation system could realize the high-precision and long-term navigation of AUV in the designated area, which had great potential for both military and civilian applications.

Keywords: MEMS IMU; Kalman filter; underwater integrated navigation system

1. Introduction

In the field of marine-related research, the cooperation between an unmanned surface vehicle (USV) and an autonomous underwater vehicle (AUV) is widely used in both civilian and military strategic fields [1–4]. For the search mission for underwater targets, the collaborative operation mode of USV range coarse scanning and AUV approach reconnaissance has the advantages of a high search efficiency, flexible deployment and real-time information transmission. The underwater environment has characteristics such as strong closure and complexity, so the underwater navigation system is the key to AUVs executing underwater navigation tasks [5–7]. The AUV navigation method can be subdivided into five main branches: dead reckoning, vision-aided navigation, sonar-aided navigation, map matching navigation and acoustic navigation.

1.1. Dead Reckoning

A typical dead reckoning system is an SINS/DVL integrated navigation system. When the AUV starts from a determined initial position, a Strap-down Inertial Navigation System (SINS) is used to measure the vehicle's attitude, a Doppler Velocity Log (DVL) is used to measure the carrier's velocity and then the current position at that time is calculated. Liu et al. [8] proposed a Doppler shift-aided coupling method of an SINS/DVL integrated navigation-based dual adaptive factor and used a chi-square detection-aided dual factors' adaptive filter to suppress the outliers. Yuan et al. [9] proposed a new federated filtering algorithm based on the Sage–Husa adaptive Kalman filter to eliminate the influence of unknown or time-varying statistical characteristics. Shaukat et al. proposed a method for improving underwater navigation and vehicle positioning by using a multi-layer perceptron neural network trained using backpropagation. This method utilizes neural networks to enhance the noise processing ability of Kalman filtering for state estimation, which proposes a new improved underwater positioning algorithm. Guo et al. [10] proposed a square-root unscented information filter (SR-UIF) to solve the problem of low accuracy for the regular filter algorithm in SINS/DVL integrated navigation. For filtering algorithms, accuracy, operational speed, and adaptive adjustment ability are particularly important indicators. Jin et al. [11] proposed a novel data-driven approach enhancing a DVL/SINS integrated navigation system, by building a virtual beam predictor based on multi-output least-squares support vector regression (MLS-SVR), to improve the robustness and accuracy of UUV navigation with limited DVL beams. Zhang et al. [12] proposed a long short-term memory extended exponential weighted Kalman filter (LSTM-EEWKF) algorithm assisted by a long short-term memory (LSTM) neural network. Qin et al. [13] proposed a robust interactive multiple model (RIMM) algorithm to improve the performance of an INS/DVL integrated navigation system under a complex measurement environment. The existing filtering algorithms often have the problem of a high theoretical accuracy, but the computational complexity is large and difficult to apply in engineering practice.

1.2. Vision-Aided Navigation

Vision-aided navigation for AUV can generally be classified into two applications: visual SLAM and artificial mark recognition. Carrasco et al. [14] presents the application of an evolved stereo Graph-SLAM algorithm especially designed for underwater environments to improve the localization, navigation and control of the SPARUS II AUV. Stereo Graph-SLAM significantly improves positioning data due to additional pose constraints calculated from visual (stereo) loop closing. Drop et al. [15] proposed a navigation method for autonomous underwater robots based on visual odometry for small-scale motion. This method is based on the SLAM algorithm for stereo images and a loop detection algorithm. The operation of the position recognition algorithm is based on using a virtual network for coordinate binding, which is built during the vehicle movement. Xin et al. [16] proposed an end-to-end network for SLAM preprocessing in an underwater low-light environment to achieve low-light image enhancement. Xu et al. [17] integrated point features and object features to construct semantic landmarks. This proposed method can improve the performance of ORBSLAM2 in underwater scenarios. Most visual SLAM algorithms extract descriptors for matching after obtaining feature points of the video stream or use the sparse optical flow method for matching feature points before and after frames. After matching, the ICP and PnP methods are used to calculate the pose transformation matrix between frames, and some even use BA to optimize the pose transformation. Then, the pose changes are integrated, and position errors will gradually accumulate. Therefore, the above methods consider setting "landmarks" such as keyframes and eliminating accumulated errors through loop detection.

Jung et al. [18] designed artificial landmarks for camera positioning and proposed a new vision-based object detection technology, which is applied to the map-based Monte Carlo localization (MCL) algorithm. In the image processing step, a new correlation coefficient weighted sum, multi template-based target selection, and color-based image segmentation method were proposed to improve traditional methods. Duecker et al. [19] uses a camera to recognize the AprilTag marking system and obtain complete relative 3D pose information underwater. Yu et al. [20] proposed an ArUco code location determination method for AUV charging, based on deep learning and the law of refraction. Most of the above methods have been tested only in clean pools and simulated environments. While most water bodies have a high turbidity and short visual distance, it will result in a small viewing range for the camera.

1.3. Sonar-Aided Navigation

Sonar-aided navigation will not be affected by water turbidity. Leonard et al. [21] used the data obtained by a high-resolution array forward-looking sonar to extract feature points from the sonar image as features of the SLAM algorithm. They used the differential GPS in the experiment as a reference. The experimental results proved that the positioning results of the SLAM algorithm are better than those of INS/DVL. Palmer et al. [22] proposed an underwater SLAM algorithm based on a multi beam sonar. This algorithm combines the measurement of the seabed strip profile with the calculation of trajectory positioning results and implements the underwater SLAM algorithm based on point cloud registration considering uncertainty. The accuracy of the algorithm is verified based on the sea trial data of Girona 500. Daniel et al. [23] proposed an algorithm that uses acoustic shadow features to achieve side-scan sonar image matching. This algorithm achieves feature matching based on the geometric distribution characteristics of shadows in side-scan sonar images and their corresponding key position information. However, when the sonar detection direction is large, this matching method breaks down when the amplitude changes, causing the acoustic shadow to change significantly. Khater et al. [24] proposed a side-scan sonar image matching method based on SUSAN and Harris corner information. This method can achieve better matching results when the sonar image features are stable and uniform. However, usually, there are a large number of shadows and distortions on sonar images, and features are scarce and difficult to extract. Therefore, the generalization performance of this method is insufficient, and its application space in underwater scenes is limited. Zhou et al. [25] proposed using feature descriptors and style transfer methods based on deep learning to match side-scan sonar images. The experimental results show that even if the learned descriptors are trained based on optical image datasets, the matching effect is poor. It goes beyond traditional manual feature-based descriptors such as SIFT. When the style transfer method is introduced, the matching effect of side-scan sonar images is further improved. This research can provide a reference for underwater side-scan sonar image matching based on deep learning.

1.4. Matching Navigation

Matching navigation is a method that uses the inherent physical properties of the ocean such as gravity fields [26], magnetic fields [27], and seafloor terrain features [28] to perform positioning. The traditional matching algorithms are divided into two categories: single-point iteration represented by the Sandia Terrain-aided Navigation System (SITAN) and sequence matching represented by Terrain Contour Matching (TERCOM) and Iterative Closest Contour Point (ICCP). Zhao et al. [29] proposed a novel domain-center adaptive-transfer matching method to improve the matching efficiency and out-of-domain positioning reliability of underwater gravity matching navigation. Wang et al. [30] proposed a novel computationally efficient outlier-robust CKF-based matching algorithm to solve outlier interferences. Liu et al. [31] proposed a new gravity matching algorithm based on comprehensive features matching (CFM) to evaluate the similarity correlation between gravity measurement sequences and reference maps. Li et al. [32] proposed a new compensation method for an underwater magnetic vector measurement system that innovatively uses the geomagnetic total field as a benchmark. Liu et al. [33] proposed a fuzzy particle filter that dynamically estimates the variance of particle distribution under terrain gradients using fuzzy logic. Simulation experiments have shown that the algorithm still has good stability under various ocean current disturbances in the Arctic. However, this method must first conduct environmental mapping of each sea area, store it inside the AUV and compare it with the physical attributes collected by the sensor in real time during the navigation process.

1.5. Acoustic Navigaation

Acoustic navigation systems are generally categorized into three classes, i.e., Long Baseline (LBL) systems, Short Baseline (SBL) systems and Ultra Short Baseline (USBL) systems. LBL systems require the installation of acoustic beacons or acoustic signal responders with baseline distances ranging from hundreds of meters to several kilometers on the seabed, making deployment and retrieval difficult. SBL systems typically install three or more transducers on board the ship to form a transducer array, and the distance between the elements of the array needs to be strictly calibrated. USBL systems are installed in a transceiver to form an acoustic array. Typically, the distance between the acoustic array units is only a few centimeters, which is cheap and easy to install. In addition to the aforementioned systems, researchers have also proposed many different underwater acoustic positioning systems. Shi et al. [34] proposed the Acoustic Synthetic Baseline (ASBL) positioning technology, which can achieve target positioning through multi-frame acoustic distance estimates between the measured target and the underwater datum. He et al. [35] proposed an improved random drift particle swarm optimization (RDPSO)-aided approach for identifying the system-level installation parameters of SINS/USBL, leading to a good performance of the SINS/USBL integrated navigation system. Wang et al. [36] proposed an augmented underwater acoustic navigation with systematic error model based on a seafloor datum network to correct time delays and time-varying sound speed errors. Acoustic navigation systems are usually used as auxiliary methods for other navigation systems. The computational complexity and accuracy of the navigation system determine the efficiency and accuracy of underwater tasks. At the same time, AUVs are facing the requirements of swarm, intelligence, and miniaturization for military use, as well as the requirement of economy for civilian use. However, most underwater navigation systems of AUV use an optical fiber gyroscope (FOG) as the main equipment, DVL as the auxiliary equipment and a high-dimensional Kalman filter as the algorithm of the integrated navigation system [37]. It is obvious that mainstream underwater integrated navigation systems are expensive, bulky and heavy and cannot meet the growing requirements for military and civilian purposes.

The Micro-Electro-Mechanical Systems (MEMS) Inertial Measurement Unit (IMU) has the characteristics of a low price, small volume and light weight [38], while lightweight and inexpensive DVLs have also emerged in recent years. The integrated navigation system of MEMS IMU and DVL is also widely used, but due to the lack of auxiliary equipment to provide global positioning information, there is inevitably a problem of the navigation errors gradually accumulating. For the collaborative working mode of USV and AUV, a USBL known as underwater GPS is added, which can effectively eliminate accumulated errors in underwater positioning [39], useful in underwater navigation.

In this paper, we designed an integrated navigation system with a low cost, light weight, small volume and low computational complexity, which was mainly based on MEMS IMU, supplemented by DVL and USBL. Through correcting the sensor and fusing the data, a low-dimensional Extended Kalman filter was constructed to reasonably describe the system. According to the lake trial experiment, the performance of the system was the same as that of the expensive FOG/DVL integrated navigation system, whose average position error was 4.12 m, maximum position error was 8.53, average velocity error was 0.027 m/s and average yaw error was 1.41°. Its low price, light weight, small size, low computational complexity and high accuracy make it suitable for use without barriers in the vast majority of AUVs.

2. Sensor Devices of the Underwater Navigation System

IMU, DVL and USBL are the main equipment of the proposed underwater navigation system, and their data calculation and installation layout are key points of the system.

2.1. IMU

The underwater navigation system of an AUV usually uses a fiber optic gyroscope as the main navigation equipment. A fiber optic gyroscope utilizes the Sagnack effect to have a high accuracy, but its large volume and heavy weight make it difficult to apply to an AUV below 180 mm in diameter, and its expensive price also limits its use in underwater navigation. MEMS IMUs have significant advantages over a fiber optic gyroscope in terms of power consumption, weight and volume. Some MEMS IMUs carry magnetometers for auxiliary calibration, which have an accuracy level comparable to that of a fiber optic gyroscope. Some commercial MEMS IMUs can not only output the angular velocity and linear acceleration but also integrate magnetic information to output the directional angle, greatly improving its accuracy and convenience.

For some MEMS IMUs that cannot directly calculate the attitude angle, linear complementary filtering can be used for attitude calculation [40]. The stable attitude angle can be obtained by means of a three-axis accelerometer and three-axis magnetometer without an accumulated error. The calculation formula of attitude angles can be shown by:

$$\begin{cases} \theta_m = arcsin(\frac{-a_{x_bm}}{g}) \\ \phi_m = arctan(\frac{a_{y_bm}}{a_{z_bm}}) \\ \psi_m = arctan(\frac{m_{y_bm}cos\phi_m - m_{z_bm}sin\phi_m}{m_{x_bm}cos\theta_m + m_{y_bm}cos\phi_m sin\theta_m + m_{z_bm}cos\phi_m sin\theta_m}) \end{cases} \tag{1}$$

The attitude angle obtained from the above equation is the prior measurement angle. The classical complementary filtering formula of attitude angles can be shown by:

$$\begin{cases} \hat{\phi}(k) = \frac{\tau}{\tau+T_s}(\hat{\phi}(k-1) + T_s\omega_{x_bm}(k)) + \frac{T_s}{\tau+T_s}\phi_m(k) \\ \hat{\theta}(k) = \frac{\tau}{\tau+T_s}(\hat{\theta}(k-1) + T_s\omega_{y_bm}(k)) + \frac{T_s}{\tau+T_s}\theta_m(k) \\ \hat{\psi}(k) = \frac{\tau}{\tau+T_s}(\hat{\psi}(k-1) + T_s\omega_{z_bm}(k)) + \frac{T_s}{\tau+T_s}\psi_m(k) \end{cases} \tag{2}$$

where $\tau/(\tau+T_s)$ and $T_s/(\tau+T_s)$ denote the weight coefficients, and ω_{x_bm}, ω_{y_bm} and ω_{z_bm} denote the three-axis angular velocity. The weight coefficients are generally adjusted based on the accuracy performance of the gyroscope, accelerometers and magnetometers.

2.2. DVL

As Figure 1 illustrates, DVL usually adopts a four-beam Janus configuration for the velocity measurement, which involves configuring a pair of transducers in each of the four directions before and after the DVL, which can greatly eliminate the errors caused by the carrier's front and rear shaking. According to the Doppler frequency shift principle, the DVL velocity measurement formula can be calculated by [41]:

$$\begin{aligned} v_x &= \frac{c_0(f_{d1}-f_{d3})}{4f_0cos\alpha} \\ v_y &= \frac{c_0(f_{d2}-f_{d4})}{4f_0cos\alpha} \\ v_z &= -\frac{c0(f_{d1}+f_{d2}+f_{d3}+f_{d4})}{8f_0sin\alpha} \end{aligned} \tag{3}$$

where c_0 denotes the velocity of sound waves propagating in water, f_{d1}, f_{d2}, f_{d3} and f_{d4} denote the frequency shift of the four beams and f_{d1}, f_{d2}, f_{d3} and f_{d4} denote the velocity measured by DVL. The X and Y axes are shown in the figure, and the Z axis is perpendicular to the X and Y axes.

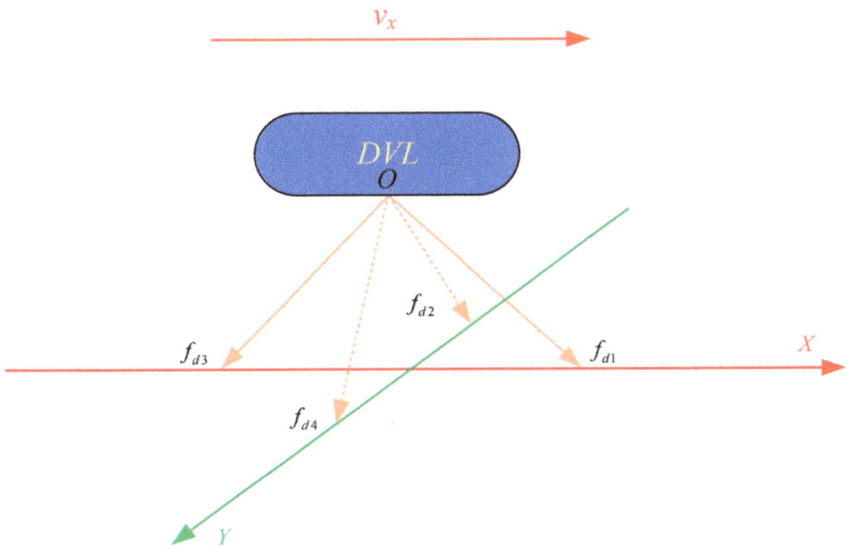

Figure 1. Schematic diagram of DVL velocity measurement.

2.3. USBL

The USBL system achieves the positioning function by calculating the relative position between the carrier and the transponder. The positioning system for the USBL uses a transmitter to transmit a fixed frequency acoustic signal to a transponder at a known location on the seabed. The transponder receives the acoustic signal and then sends an acoustic signal to the hydrophone fixed on the AUV. The system determines the tilt distance based on the time difference between the transmitted signal and the received signal and determines the azimuth angle based on the phase difference between the received signal and different hydrophones to achieve the positioning information for AUVs [42].

As shown in Figure 2, the three hydrophones are located on two perpendicular baselines (x-axis and y-axis); Among them, the hydrophone at the origin serves as the reference unit, the distance between each primitive is d and θ_{mx} and θ_{my} are the angles between the transmitted (received) sound wave and the X axis and Y axis, respectively. Assuming that the round-trip time of the sound wave is T, the slant distance R can be calculated by:

$$R = 0.5c_0T \tag{4}$$

The relationship between the phase difference R of the acoustic signal received by two hydrophones and the incident angle of the acoustic wave ϕ can be calculated by:

$$\phi = \frac{2\pi d}{\lambda} cos\theta_m \tag{5}$$

Substituting (4) for (5), the position of the transponder can be represented as:

$$X = R\cos\theta_{mx} = R\frac{\lambda\phi_{12}}{2\pi d}$$
$$Y = R\cos\theta_{my} = R\frac{\lambda\phi_{13}}{2\pi d} \quad (6)$$

where $\phi_{ij}(i,j = 1,2,3)$ denotes the phase difference between the received echoes of primitive i and primitive j. According to the experiment, the USBL system has a higher positioning accuracy in the X and Y directions but a lower accuracy in the Z direction.

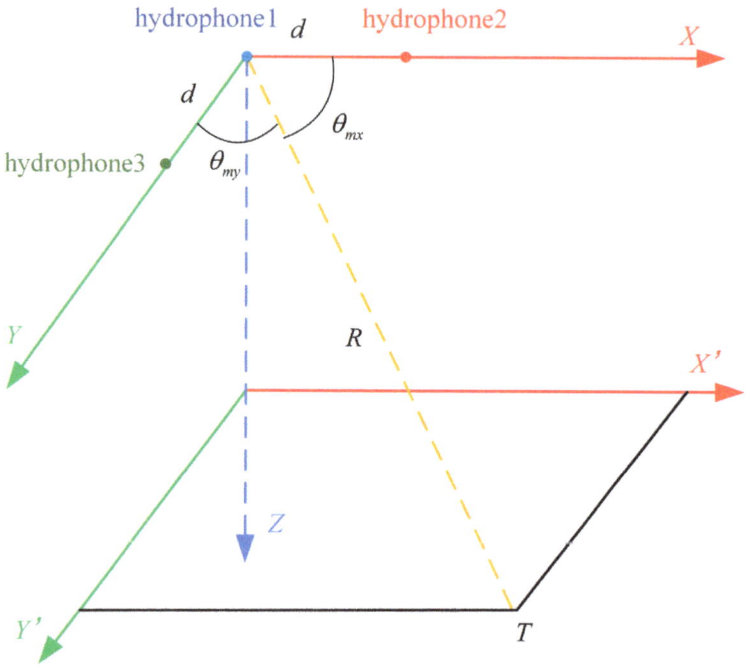

Figure 2. Schematic diagram of the USBL positioning measurement.

2.4. Equipment Layout and Frame Definition

The layout of relevant equipment and the definition of two reference frames are shown in Figure 3.

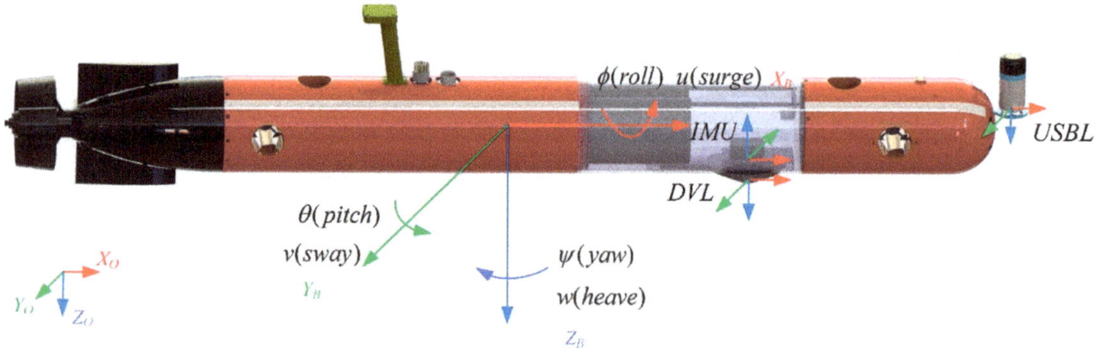

Figure 3. Equipment layout.

The navigation frame $\{O_N X_N Y_N Z_N\}$ has its origin on the surface and its axes pointing North, East and Down (NED reference frame); the body frame $\{O_b X_b Y_b Z_b\}$ is centered in the center of gravity of the AUV, with the x axis pointing in the direction of the forward motion of the vehicle, the z axis pointing down and the y axis completing a right-handed reference frame [43]. The main names and corresponding symbols in the body frame are shown in Table 1.

Table 1. Main names and symbols in the body frame.

Motion		X_b	Y_b	Z_b
linear	drift	X	Y	Z
	velocity	u	v	w
rotation	angle	θ	ϕ	ψ
	angular velocity	p	q	r

The IMU device uses the ENU reference frame, but the DVL device and the USBL device use the NED reference frame. For convenience, the integrated navigation system uses the NED coordinate system, and coordinate conversion is required for the data collected by the IMU. The data collected by the IMU, as described below, have undergone coordinate conversion. The lever arm between the IMU and DVL is measured as $l_{DVL}(0, 0, -0.05)$. In order to reduce the impact of the AUV metal shell on the USBL underwater acoustic signal, the USBL is selected to be installed at the head of the AUV, while the lever arm between the IMU and USBL is measured as $l_{USBL}(0.5, 0, 0.05)$. Due to the large lever arms of the USBL and IMU, the lever arm error cannot be ignored.

3. Algorithm of the Underwater Navigation System

3.1. Flow Chart of the Algorithm

The core of AUV's integrated navigation system is the filtering algorithm. In order to achieve high-precision navigation parameter estimation, the Kalman Filter (KF) has been widely valued due to its unique advantages. The KF method has the characteristics of being high-dimensional, non-stationary and time-varying, and it is a recursive algorithm that is very suitable for implementation on computers. Therefore, since its proposal, Kalman filters have received widespread attention in the field of engineering. However, KF is only suitable for linear system models. In engineering practice, integrated navigation systems always have certain nonlinear characteristics. If KF is used for filtering calculation, it will cause model approximation errors. Therefore, some improvement methods are constantly evolving. The application of the Extended Kalman Filter (EKF) is relatively widespread. The EKF algorithm performs first-order linearization truncation on the Taylor expansion of nonlinear state functions and measurement functions, transforming nonlinear filtering into a linear filtering problem. This algorithm is computationally simple and has a fast convergence speed.

Considering that there are many matrix multiplications in the EKF algorithm, and the time complexity of the matrix multiplication algorithm is $O(n^3)$, this means that as the matrix dimension increases, the computation time will increase by a power of three. Therefore, choosing a low-dimensional EKF will greatly reduce the computational complexity.

The flow chart of the algorithm is shown in Figure 4. IMU can pre-integrate its own gyroscope, accelerometer and magnetometer to directly calculate the attitude angle and velocity in the body frame. DVL provides velocity observation values after correction, and USBL provides position observation values after correction. The navigation information is ultimately output through the Extended Kalman filter.

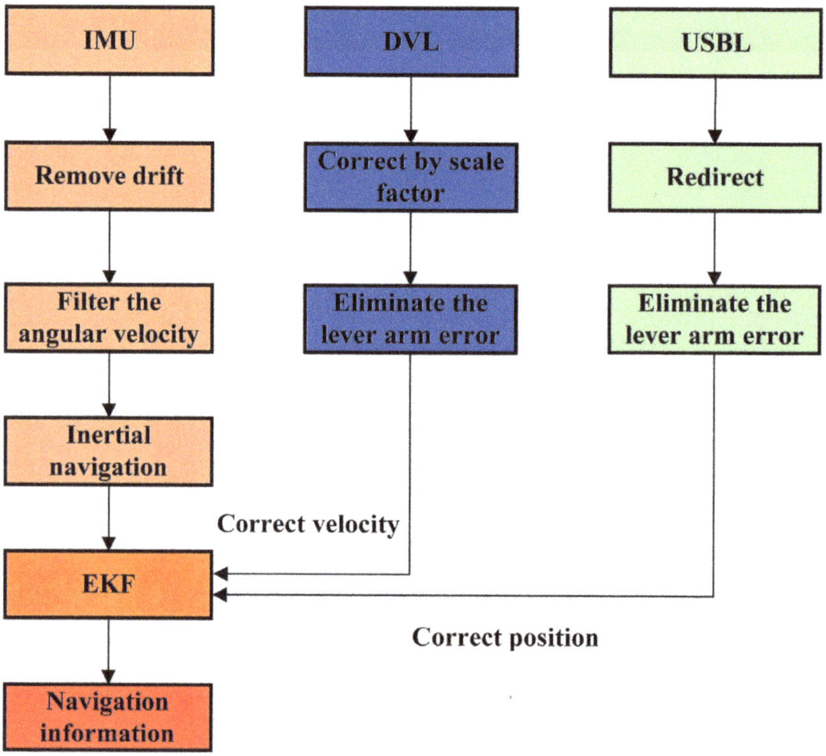

Figure 4. The flow chart of the algorithm.

3.2. Model Design

The Euler angles yaw, pitch and roll can be represented by ψ, θ and ϕ, respectively. From the navigation frame to the body frame, attitude changes are defined in order of the roll angle around the x axis, the pitch angle around the y axis and the yaw angle around the z-axis.

The attitude transformation matrices corresponding to each rotation are:

$$C_1^n = \begin{bmatrix} c\psi & -s\psi & 0 \\ s\psi & c\psi & 0 \\ 0 & 0 & 1 \end{bmatrix}, C_2^1 = \begin{bmatrix} c\theta & 0 & s\theta \\ 0 & 1 & 0 \\ -s\theta & 0 & c\theta \end{bmatrix}, C_b^2 = \begin{bmatrix} 1 & 0 & 0 \\ 0 & c\phi & -s\phi \\ 0 & s\phi & c\phi \end{bmatrix} \tag{7}$$

The attitude rotation matrix from the navigation frame to the body frame is:

$$C_b^n = C_1^n C_2^1 C_b^2 = \begin{bmatrix} c\psi c\theta & (c\psi s\theta s\phi - s\psi c\phi) & (c\psi s\theta c\phi + s\psi s\phi) \\ s\psi c\theta & (s\psi s\theta s\phi + c\psi c\phi) & (s\psi s\theta c\phi - c\psi s\phi) \\ -s\phi & c\phi s\theta & c\phi s\theta \end{bmatrix} \tag{8}$$

The navigation frame is defined as (ξ, η, ζ), and the body frame is defined as (x, y, z). They have the following transformation relationships:

$$\begin{pmatrix} \xi \\ \eta \\ \zeta \end{pmatrix} = \begin{bmatrix} c\psi c\theta & (c\psi s\theta s\phi - s\psi c\phi) & (c\psi s\theta c\phi + s\psi s\phi) \\ s\psi c\theta & (s\psi s\theta s\phi + c\psi c\phi) & (s\psi s\theta c\phi - c\psi s\phi) \\ -s\phi & c\phi s\theta & c\phi s\theta \end{bmatrix} \begin{pmatrix} x \\ y \\ z \end{pmatrix} = S \begin{pmatrix} x \\ y \\ z \end{pmatrix} \tag{9}$$

Considering that the depth of an AUV can be accurately measured using a depth gauge, the integrated navigation system does not need to fuse the depth data. The state variables of the IMU system can be represented as follows:

$$X = [x\ y\ u\ v\ w] \tag{10}$$

where, x, y is the position representation of the x axis and y axis in the navigation frame, and u, v, w is the linear velocity in the body frame. According to the motion equations of AUV, the equation of the state of the system can be expressed as:

$$\dot{X} = FX + W \tag{11}$$

where W denotes the process noise, and F is defined as follows:

$$F = \begin{bmatrix} 0\,0\; c\psi c\theta\; (c\psi s\theta\; s\phi\; - \; s\psi c\phi)\; (c\psi s\theta\; c\phi\; + \; s\psi s\phi) \\ 0\,0\; s\psi c\theta\; (s\psi s\theta\; s\phi\; + \; c\psi\; c\phi)\; (s\psi\; s\theta\; c\phi\; - \; c\psi s\phi) \\ 0\,0\quad 0 \quad\quad 0 \quad\quad 0 \\ 0\,0\quad 0 \quad\quad 0 \quad\quad 0 \\ 0\,0\quad 0 \quad\quad 0 \quad\quad 0 \end{bmatrix} \tag{12}$$

According to reference [44], the velocity measurement error of the DVL mainly comes from the installation error angle and scale factor error. If the velocity measured centered on DVL is expressed as $V_{DVL}^{raw} = [u_{DVL}^{raw}, v_{DVL}^{raw}, w_{DVL}^{raw}]$, we can multiply it by the scale factor δk to obtain the corrected DVL center measurement velocity as $\delta k V_{DVL}^{raw}$. The center of the IMU and the center of the DVL usually do not coincide. If the angular velocity output by the IMU is expressed as ω, then the lever arm error of the DVL can be expressed as $\omega \times l_{DVL}$. Eliminating the lever arm error from the velocity measured by the DLV, we can obtain the measured velocity centered on IMU, which can be calculated as:

$$\widetilde{V}_{DVL} = \delta k V_{DVL}^{raw} - \omega \times l_{DVL} \tag{13}$$

where $\widetilde{V}_{DVL}$ denotes the measured velocity centered on IMU, which can correct the velocity of the system. The measurement equation can be expressed by:

$$\begin{aligned} Z_{IMU/DVL} &= \begin{bmatrix} \widetilde{u}_{DVL} \\ \widetilde{w}_{DVL} \\ \widetilde{v}_{DVL} \end{bmatrix} = H_{IMU/DVL} X, \\ H_{IMU/DVL} &= \begin{bmatrix} 0\,0\,1\,0\,0 \\ 0\,0\,0\,1\,0 \\ 0\,0\,0\,0\,1 \end{bmatrix} \end{aligned} \tag{14}$$

For the measured value $P_{USBL}^{raw} = [x_{USBL}^{raw}, y_{USBL}^{raw}]$ of the USBL system, it is also necessary to eliminate the lever arm error. If the lever arm between IMU and USBL is expressed as l_{USBL} and the rotation matrix of the IMU output is C_b^n, the measurement position centered on IMU can be obtained by eliminating the lever arm error, which can be calculated as:

$$\widetilde{P}_{USBL} = P_{USBL}^{raw} - C_b^n \times l_{USBL} \tag{15}$$

where $\widetilde{X}_{USBL} = [\widetilde{x}_{USBL}, \widetilde{y}_{USBL}]$ denotes the measured position centered on IMU, which can correct the position of the system, and the measurement equation can be expressed by:

$$\begin{aligned} Z_{IMU/USBL} &= \begin{bmatrix} \widetilde{x}_{USBL} \\ \widetilde{y}_{USBL} \end{bmatrix} = H_{IMU/USBL} X, \\ H_{IMU/USBL} &= \begin{bmatrix} 1\,0\,0\,0\,0 \\ 0\,1\,0\,0\,0 \end{bmatrix} \end{aligned} \tag{16}$$

3.3. Extended Kalman Filter

We linearize the equation of the state of the system, which can be rewritten in the form of an extended Kalman filter:

$$\begin{cases} X_k = \Phi_{k-1} X_{k-1} + W_{k-1} \\ Z_k = H_k X_k + V_k \end{cases} \tag{17}$$

with:

$$\Phi_{k-1} = I + F_{k-1}\Delta t = \begin{bmatrix} 1\,0\; c\psi c\theta\Delta t\; (c\psi s\theta s\phi - s\psi c\phi)\Delta t\; (c\psi s\theta c\phi + s\psi s\phi)\Delta t \\ 0\,1\; s\psi c\theta\Delta t\; (s\psi s\theta s\phi + c\psi c\phi)\Delta t\; (s\psi s\theta c\phi - c\psi s\phi)\Delta t \\ 0\,0 \quad 1 \quad 0 \quad 0 \\ 0\,0 \quad 0 \quad 1 \quad 0 \\ 0\,0 \quad 0 \quad 0 \quad 1 \end{bmatrix} \tag{18}$$

where W denotes the process noise, subject to Normal distribution $W \sim (0, Q)$, and V denotes the measurement noise, subject to Normal distribution $V \sim (0, R)$. The filtering steps of the EKF algorithm are described as follows:

$$\begin{cases} X_{k|k-1} = \Phi_{k-1|k-1} X_{k-1|k-1} \\ P_{k|k-1} = \Phi_{k-1} P_{k-1|k-1} \Phi_{k-1}^T + Q \\ K_k = P_{k|k-1} H^T \left(H P_{k|k-1} H^T + R\right)^{-1} \\ X_{k|k} = X_{k|k-1} + K_k \left(Z_k - H X_{k|k-1}\right) \\ P_{k|k} = (I - K_k H) P_{k|k-1} \end{cases} \tag{19}$$

Considering that the data acquisition frequency of the DVL usually is 5–7 Hz, and that of the USBL is usually 1–2 Hz, filtering is carried out according to the principle of sequential filtering to ensure real-time filtering and reduce the calculation amount of the measurement update of the EKF algorithm [45]. The measurement equation can be written as:

$$\begin{bmatrix} Z_k^{(1)} \\ Z_k^{(2)} \end{bmatrix} = \begin{bmatrix} H_k^{(1)} \\ H_k^{(2)} \end{bmatrix} X + \begin{bmatrix} V_k^{(1)} \\ V_k^{(2)} \end{bmatrix} \tag{20}$$

where each parameter is defined as follows:

$$\begin{gathered} Z_k^{(1)} = \begin{bmatrix} x_k \\ y_k \end{bmatrix},\; Z_k^{(2)} = \begin{bmatrix} u_k \\ v_k \\ w_k \end{bmatrix}, \\ H_k^{(1)} = [I_{2\times2}\; 0_{2\times3}],\; H_k^{(2)} = [0_{3\times2}\; I_{3\times3}], \\ V_k^{(1)} \sim (0, R_k^{(1)}),\; V_k^{(2)} \sim (0, R_k^{(2)}), \end{gathered} \tag{21}$$

If the USBL position measurement value is obtained, the first sub measurement update will be performed as follows:

$$\begin{cases} K_k^{(1)} = P_k^{(0)} (H_k^{(1)})^T (H_k^{(1)} P_k^{(0)} (H_k^{(1)})^T + R_k^{(1)}) \\ \hat{X}_k^{(1)} = X_k^{(0)} + K_k^{(1)} (Z_k^{(1)} - H_k^{(1)} X_k^{(0)}) \\ P_k^{(1)} = (I - K_k^{(1)} H_k^{(1)}) P_k^{(0)} \end{cases} \tag{22}$$

If the DVL velocity measurement value is obtained, the second sub measurement update will be performed as follows:

$$\begin{cases} K_k^{(2)} = P_k^{(1)}(H_k^{(2)})^T(H_k^{(2)}P_k^{(1)}(H_k^{(2)})^T + R_k^{(2)}) \\ \hat{X}_k^{(2)} = X_k^{(1)} + K_k^{(2)}(Z_k^{(2)} - H_k^{(2)}X_k^{(1)}) \\ P_k^{(2)} = (I - K_k^{(2)}H_k^{(2)})P_k^{(1)} \end{cases} \tag{23}$$

Finally, the optimal estimation of the state variable is obtained as $X_k = \hat{X}_k^{(2)}, P_k = \hat{P}_k^{(2)}$.

3.4. Error Analysis

IMU, DVL and USBL, respectively, provide measurement values for angle, velocity and position, and error analysis is also conducted in terms of these three aspects. Although the EKF algorithm has been used for the optimal estimation of the motion state of AUVs, analyzing their motion equations can also elucidate the problem. The motion equation of AUV is as follows:

$$\begin{pmatrix} \xi \\ \eta \\ \zeta \end{pmatrix} = \begin{bmatrix} c\psi c\theta & (c\psi s\theta s\phi - s\psi c\phi) & (c\psi s\theta c\phi + s\psi s\phi) \\ s\psi c\theta & (s\psi s\theta s\phi + c\psi c\phi) & (s\psi s\theta c\phi - c\psi s\phi) \\ -s\phi & c\phi s\theta & c\phi s\theta \end{bmatrix} V dt \tag{24}$$

Using the yaw angle ψ as a typical example to analyze the impact of angles on the results, assuming there is a fixed deviation angle in the yaw angle: $\tilde{\psi} = \psi + \Delta\psi$, where $\tilde{\psi}$ denotes the measurement value, ψ denotes the true value and $\Delta\psi$ denotes the fixed deviation. The equation of motion can be rewritten as:

$$\begin{pmatrix} \tilde{\xi} \\ \tilde{\eta} \\ \tilde{\zeta} \end{pmatrix} = \begin{bmatrix} c(\psi+\Delta\psi)c\theta & (c(\psi+\Delta\psi)s\theta s\phi - s(\psi+\Delta\psi)c\phi) & (c(\psi+\Delta\psi)s\theta c\phi + s(\psi+\Delta\psi)s\phi) \\ s(\psi+\Delta\psi)c\theta & (s(\psi+\Delta\psi)s\theta s\phi + c(\psi+\Delta\psi)c\phi) & (s(\psi+\Delta\psi)s\theta c\phi - c(\psi+\Delta\psi)s\phi) \\ -s\phi & c\phi s\theta & c\phi s\theta \end{bmatrix} V dt \tag{25}$$

The above equation indicates that the added deviation will rotate the trajectory of the AUV with the deviation angle $\Delta\psi$. If the deviation angle $\Delta\psi$ is a small angle, the endpoint deviation (maximum error) dL will be the deviation $\Delta\psi$ times the total length L:

$$dL = \Delta\psi L \tag{26}$$

As stated in Section 3.2, the velocity measurement error of the DVL mainly comes from the installation error angle and scale factor error. The installation angle of DVL, as discussed above, will also cause the AUV's trajectory to rotate as a whole, so here we mainly discuss the scale factor error. Assuming there is a fixed scale factor error for velocity: $\widetilde{V} = (1 + \Delta k)V$, where $\widetilde{V}$ denotes the measurement value, V denotes the true value and Δk denotes the fixed scale factor error. The equation of motion can be rewritten as:

$$\begin{pmatrix} \tilde{\xi} \\ \tilde{\eta} \\ \tilde{\zeta} \end{pmatrix} = \begin{bmatrix} c\psi c\theta & (c\psi s\theta s\phi - s\psi c\phi) & (c\psi s\theta c\phi + s\psi s\phi) \\ s\psi c\theta & (s\psi s\theta s\phi + c\psi c\phi) & (s\psi s\theta c\phi - c\psi s\phi) \\ -s\phi & c\phi s\theta & c\phi s\theta \end{bmatrix} (1 + \Delta k) V dt \tag{27}$$

The above equation reveals that the scale factor error will cause the overall scaling of the AUV's trajectory. In special cases where the starting and ending points coincide, this error will not affect the position of the endpoint. In general, the maximum error will be calculated by the following equation:

$$dL = \Delta k L \tag{28}$$

The working principle of USBL ensures that the position error generated each time will not cause a long-term impact. The position error ΔP is also the global maximum error. The error analysis is shown in the Table 2:

Table 2. Error Analysis.

Equipment	Measurement Value	Error Source	Maximum Error
IMU	angle	deviation angle $\Delta\psi$	$\Delta\psi L$
DVL	velocity	scale factor error Δk	$\Delta k L$
USBL	position	position error ΔP	ΔP

The scale factor error from DVL can be calculated by fitting coefficients of experimental data. The position error from USBL can be reduced by setting an appropriate covariance matrix to minimize its impact. However, the angle deviation of IMU is difficult to reduce in practice.

4. Experiments and Results

In order to verify the effectiveness of the proposed underwater navigation system, we conducted experiments on the lake, as shown in Figure 5. A test ship was installed with a USBL (SeaTrac X150, Cumbria, UK) base station directly below the Real Time Kinematic GPS (RTK GPS, Shanghai, China) receiver (Sinognss M900, Shanghai, China). Another experimental ship was equipped with a mobile station of USBL (SeaTrac X110, Cumbria, UK), IMU (Xsens MTi-G-700, Enshed, The Netherlands) and DVL (WaterLinked DVL A125, Trondheim, Norway) to form an underwater integrated navigation system. The installation relationship of each piece of equipment on the ship is shown in Figure 6. The position data of the RTK GPS installed above the IMU is considered as the true data of the system. By comparing the position data received by the RTK GPS and the results of the integrated navigation system, we can quantitatively measure the performance of the system. The main equipment parameters are shown in Table 3, and it is obvious that the selected equipment was of a light weight and small volume, making it suitable for use on AUVs of any diameter. It should be emphasized that the horizontal accuracy of RTK GPS is $\pm(8 + 1 \times 10^{-6} \times D)$ mm. D means the distance between the base station and rover receiver. The accuracy of the RTK GPS is high enough to use its data as true values.

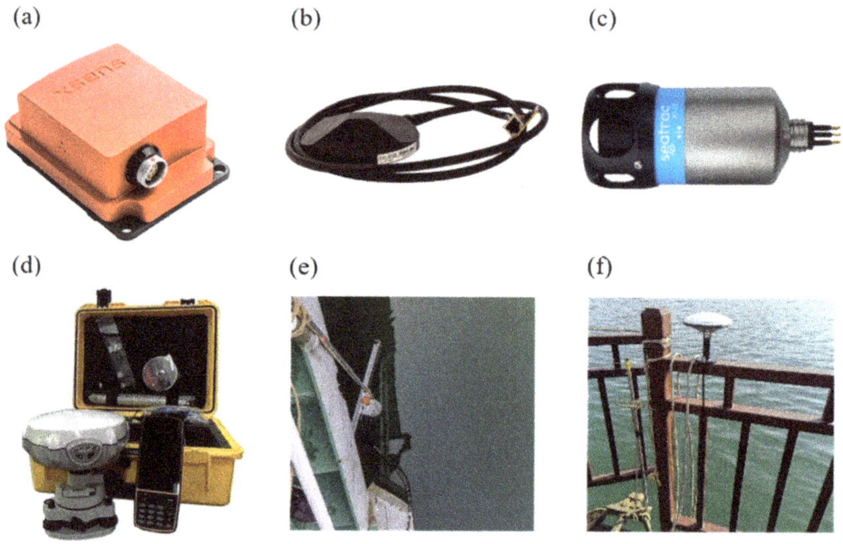

Figure 5. Equipment and experimental condition. (**a**) MEMS IMU equipment (weight: 55 g, dimensions: 57 mm × 41.90 mm × 23.60 mm, price: about USD 4000); (**b**) DVL equipment (weight: 0.75 kg, dimensions: Ø 125 mm × 30 mm, price: about USD 9900); (**c**) USBL equipment (weight: 690 g, dimensions: Ø 59 mm × 139 mm, price: about USD 27,000); (**d**) RTK GPS equipment; (**e**) Experiment platform; (**f**) RTK GPS receiver installed on the ship.

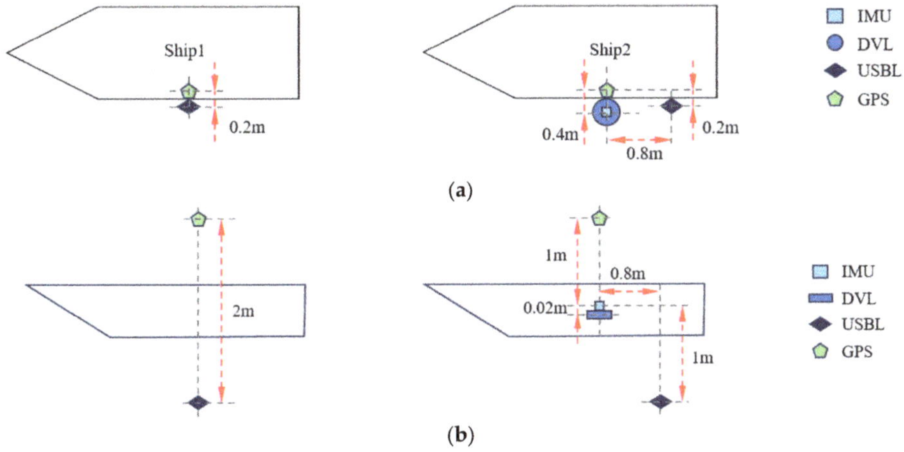

Figure 6. The installation relationship. (**a**) Vertical viewport; (**b**) Front viewport.

Table 3. Equipment parameters.

Equipment	Parameters	Values
IMU	In-run bias stability In-run bias stability	10 °/h 15 μg
DVL	Long-term accuracy	1.01% velocity
USBL	Range Resolution Angular Resolution	±0.1 m typ 2% of Acoustic Range ×(~±1°)
RTK GPS	Horizontal Accuracy	$\pm(8 + 1 \times 10^{-6} \times D)$ mm

The test ship conducted five experiments and sailed multiple trajectories. Considering that AUVs usually conduct routine inspections using a rectangular trajectory and fine detection using a lawnmower trajectory, these two trajectories are used as display results.

Figure 7 shows the experiment of a rectangular trajectory. In the rectangular trajectory, the duration is 2848 s, the total length is 3.2 km, the terminal deviation is 5.60 m and the ratio of the deviation to the total range is 0.18%. Figure 8 shows the experiment of the lawnmower trajectory. In the lawnmower trajectory, the duration is 4983 s, the total length is 5.3 km, the terminal deviation is 5.63 m and the ratio of the deviation to the total range is 0.11%. The result is very close to that of the most FOG/DVL integrated navigation.

As mentioned earlier, the position output by RTK GPS was taken as the true value. Figures 9 and 10 show the position error between the true position and the output of the navigation system. The average position error in the rectangular trajectory is 5.45 m, while the average position error in the lawnmower trajectory is 4.12 m. It can be seen that benefiting by the global positioning function of USBL, the position error of the integrated navigation system ultimately stabilizes within a range of approximately 6 m and does not diverge.

Considering that the position error output by RTK GPS was very small, the position difference divided by the time difference could be used as the true value of velocity. Figures 11 and 12 show the velocity error between the true velocity and the output of the navigation system. The jumping data in the figure are obvious abnormal data. For the sake of rigor, it is still necessary to display it. The average velocity error of the rectangular trajectory is 0.027 m/s, and the average velocity error of the lawnmower trajectory is 0.038 m/s. In most cases, the estimated velocity is very close to the true value.

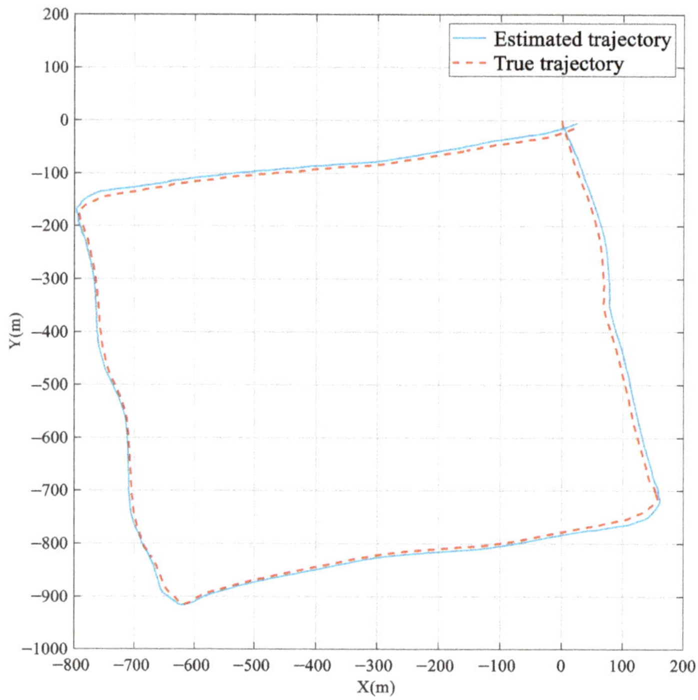

Figure 7. The experiment of a rectangular trajectory.

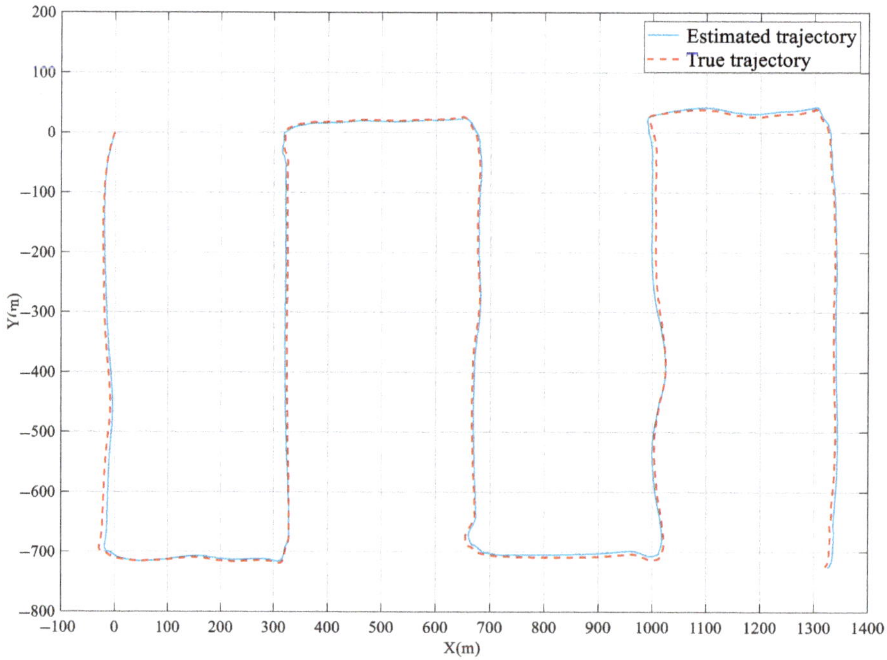

Figure 8. The experiment of a lawnmower trajectory.

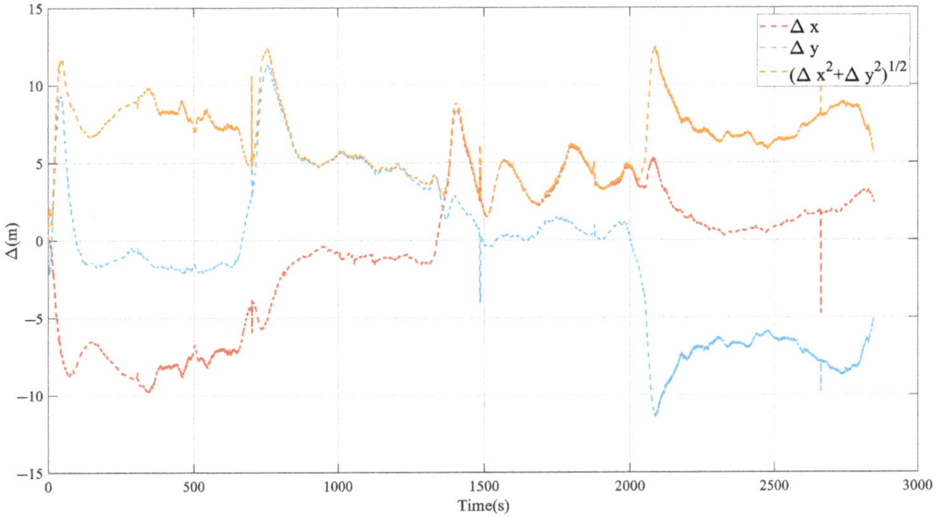

Figure 9. The position error between the true position and the output of the navigation system in the rectangular trajectory.

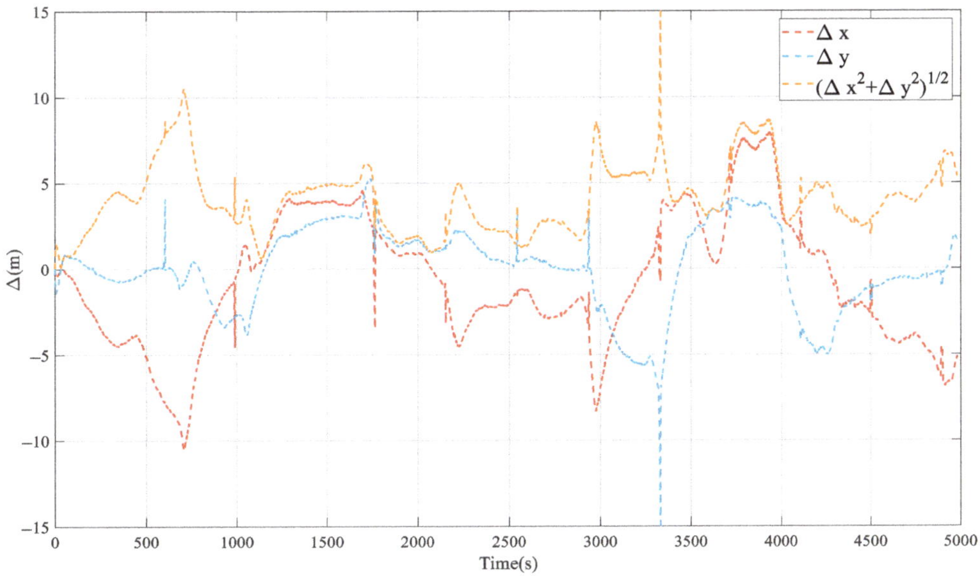

Figure 10. The position error between the true position and the output of the navigation system in the lawnmower trajectory.

Among the three directional angles of roll, pitch and yaw, the heading angle was usually the focus. The angle between the front and rear positions of RTK GPS was used as the true value of yaw. Figures 13 and 14 show yaw errors between the true yaw and the output of the navigation system, with an average heading error of 2° for rectangular trajectory tests and 1.41° for detection trajectory tests. Throughout the entire process, the filtered yaw was very close to the true value, without significant angle fluctuations.

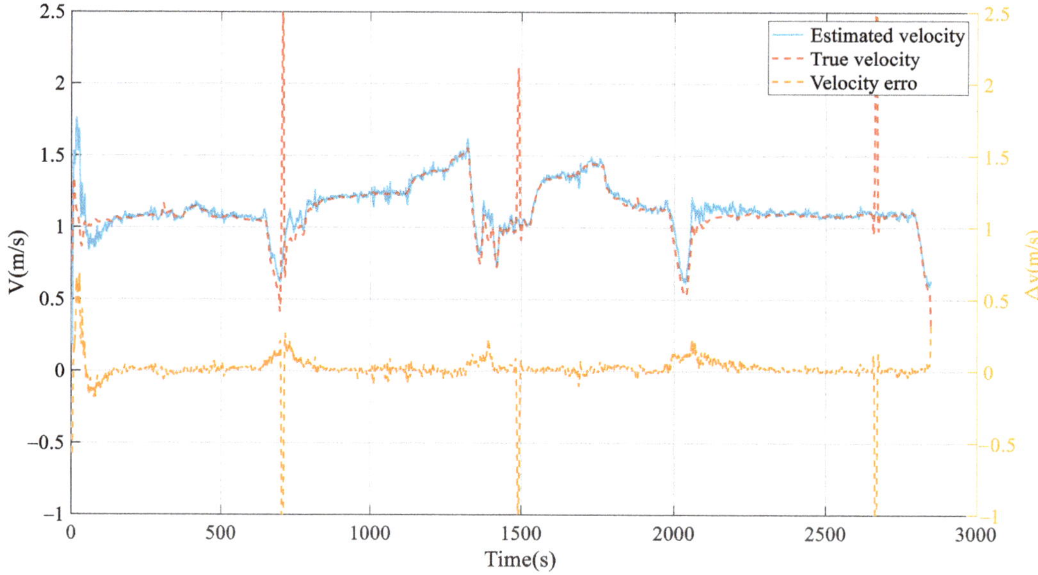

Figure 11. The velocity error between the true position and the output of the navigation system in the rectangular trajectory.

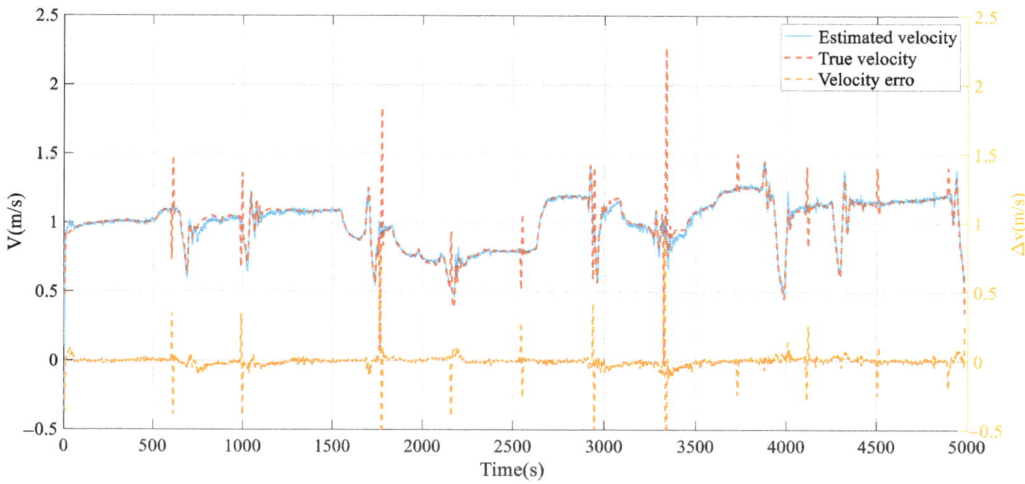

Figure 12. The velocity error between the true position and the output of the navigation system in the lawnmower trajectory.

The average results of multiple experiments are shown in theTable 4 and compared with a commercial SINS/DVL system. The models of SINS and DVL are BlueNaute Platinum and Nortek DVL500, respectively. Although the evaluation criteria for the accuracy of the two are different, they can still indicate the advantages and disadvantages of the two.

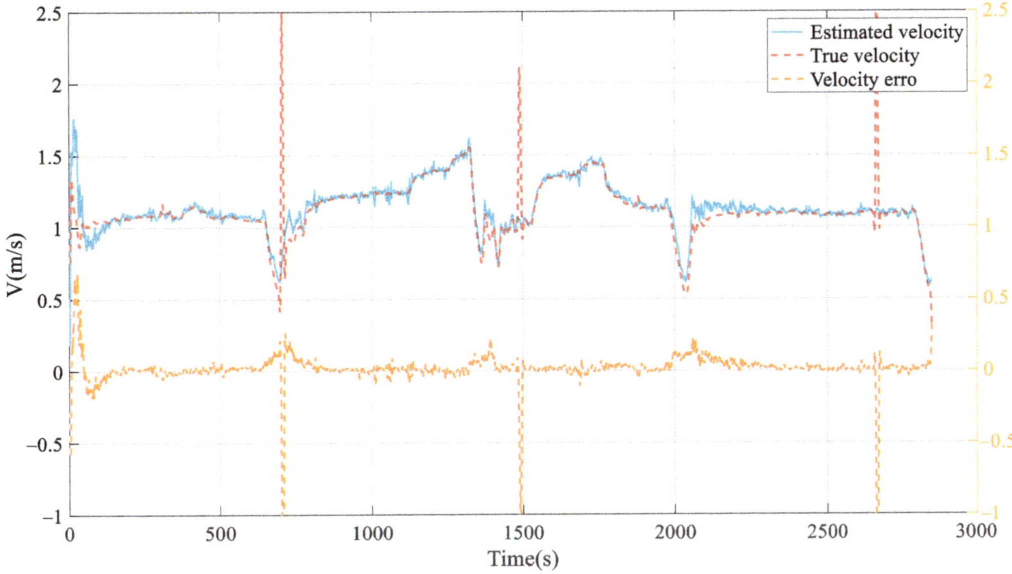

Figure 13. The yaw error between the true position and the output of the navigation system in the rectangular trajectory.

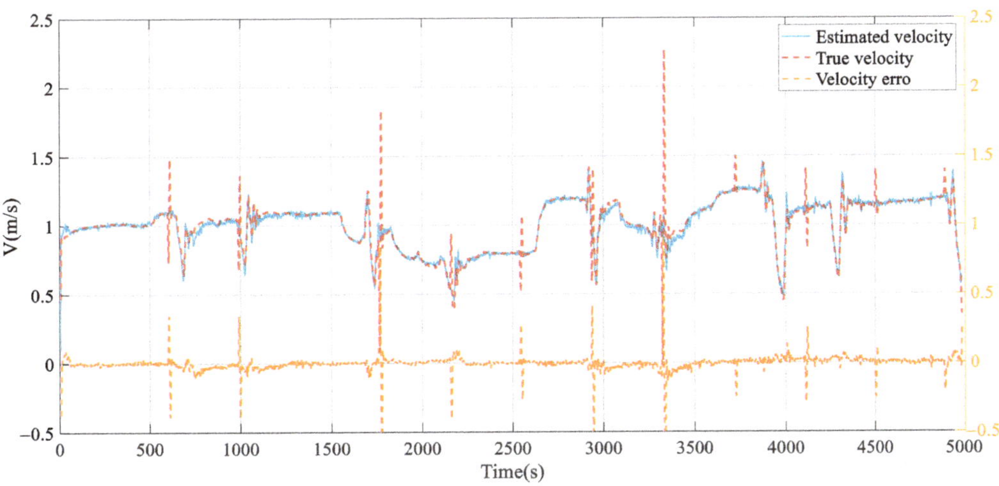

Figure 14. The yaw error between the true position and the output of the navigation system in the lawnmower trajectory.

It can be seen that thanks to USBL, the proposed system has an average position error of 4.12 m and a maximum position error of 8.53 m, which is smaller than the long-term error of SINS/DVL. The total weight of 1.495 kg is much smaller than the 8 kg of the commercial system, and the total price of USD 40,900 is also far smaller than the price of the commercial system. The dimensions also have advantages. Although there is a significant difference between the average velocity error of 0.024 m/s and the average heading angle error of 1.41° compared to those of the commercial system, their absolute values are still within the acceptable range.

Table 4. Comparison between the proposed system and a commercial system.

Parameters	Proposed System	Parameters	Commercial System
average position error	4.12 m	position error	0.3% of covered distance
maximum position error	8.53 m		
average velocity error	0.027 m/s	long-term accuracy	0.1%/±0.001 m/s
average yaw error	1.41°	heading dynamic accuracy	0.09 °/h
dimensions	IMU: 57 mm × 41.90 mm × 23.60 mm DVL: ∅ 125 mm × 30 mm USBL: ∅ 59 mm × 139 mm	dimensions	SINS: 208 mm × 275 mm × 136 mm DVL ∅ 186 mm × 203 mm
total weight	1.495 kg	total weight	8 kg
total price	about USD 40,900	total price	more than USD 100,000

It is obvious that although the proposed integrated navigation system has some disadvantages in terms of velocity error and yaw error, its size, weight and price make it achieve a more significant balance in terms of performance and cost. Although we have successfully validated the system in the same water area as other researchers, it is necessary to explore the impact of aquatic conditions such as the depth, salinity, temperature and pressure on the system. Considering the multiple influencing factors, using analysis methods such as analysis of variance will be one of the future works.

During the integration process of the entire system, there have been instances of DVL data anomalies, which have been effectively resolved through data validation and range limitations. The proposed system has significant advantages in terms of dimensions, weight and price compared to the traditional SINS systems. The most important error, position error, can remain non-divergent for a long time. Although the velocity and angle errors are not as good as those of traditional SINS systems, they are also within the allowable range. The system is very easy to integrate and deploy on different control boards, which makes it actually run on an AUV rather than simulations.

5. Conclusions

In this paper, a low-cost and high-precision underwater integrated navigation system was proposed to meet the requirements of AUV swarm, intelligence and miniaturization, which uses low-cost, lightweight and small-volume MEMS IMU, DVL and USBL. The integrated navigation algorithm fully considered the characteristics of hardware equipment and takes the position and velocity of the AUV as the state variables. Then, according to the motion equation of the vehicle, the state equation of the Kalman filter was constructed. Considering the source of the error, the integrated navigation algorithm eliminated the lever arm error from the raw positioning information to obtain the position measurement value and eliminated the scale error and lever arm error from the raw velocity measurement information to obtain the velocity measurement value. Then, a measurement equation of the Kalman filter was constructed. In order to reduce the computational complexity of the Kalman filter measurement update, and considering the low update frequency of the USBL, the sequential filter was used to update the measurement in order to obtain high real-time position and orientation information. Finally, a lake experiment was conducted, and the results confirmed that the proposed IMU/DVL/USBL integrated navigation system could precisely estimate the position, velocity and attitude of the vehicle. This proposed integrated navigation system has advantages such as a low cost, light weight, small volume, low computational cost and high precision, and it not only meets the requirements of collaborative operations between USV and AUV but also has high potential application value in military clusters and in the civilian detection of AUVs. Future work will focus on cluster navigation, on one hand, achieving better navigation accuracy by utilizing mutual positioning between AUVs. On the other hand, long-term reliability testing will be conducted to ensure productization.

Author Contributions: This study is the result of collaborative teamwork. Conceptualization, T.Y.; validation, J.L.; writing—original draft preparation, J.L.; writing—review and editing, C.Z.; visualization, D.L.; supervision, Q.Z. and C.W. All authors have read and agreed to the published version of the manuscript.

Funding: This work was supported by National Key Research and Development Program (2023YFC2809604) and the National Key Laboratory of Science and Technology on Underwater Acoustic Antagonizing (JCKY2023207CH12).

Institutional Review Board Statement: Not applicable.

Informed Consent Statement: Not applicable.

Data Availability Statement: Data are contained within the article.

Conflicts of Interest: The authors declare no conflicts of interest.

References

1. Ruscio, F.; Costanzi, R.; Gracias, N.; Quintana, J.; Garcia, R. Autonomous boundary inspection of *Posidonia oceanica* meadows using an underwater robot. *Ocean. Eng.* **2023**, *283*, 114988. [CrossRef]
2. Mancini, F.; Hollings, B.; Dieulangard, D.; Mangano, G.; Gjini, D.; Ridyard, D.; Hite, D.; Debens, H.; Manning, T. A New Generation of Autonomous Ocean-Bottom Nodes: Their Development and Applications. In Proceedings of the 84th EAGE Annual Conference & Exhibition, Vienna, Austria, 5–8 June 2023; European Association of Geoscientists & Engineers: Utrecht, The Netherlands, 2023; pp. 1–5.
3. Fan, S.; Zhang, X.; Zeng, G.; Cheng, X. Underwater ice adaptive mapping and reconstruction using autonomous underwater vehicles. *Front. Mar. Sci.* **2023**, *10*, 1124752. [CrossRef]
4. Wang, Q.; He, B.; Zhang, Y.; Yu, F.; Huang, X.; Yang, R. An autonomous cooperative system of multi-AUV for underwater targets detection and localization. *Eng. Appl. Artif. Intell.* **2023**, *121*, 105907. [CrossRef]
5. Guo, S.; Chen, M.; Pang, W. Planning for Autonomous Underwater Vehicles Based on an Improved Artificial Jellyfish Search Algorithm in Multi-Obstacle Ocean Current Environment. *IEEE Access* **2023**, *11*, 31010–31023. [CrossRef]
6. Li, P.; Liu, Y.; Yan, T.; Yang, S.; Li, R. A Robust INS/USBL/DVL Integrated Navigation Algorithm Using Graph Optimization. *Sensors* **2023**, *23*, 916. [CrossRef] [PubMed]
7. Saksvik, I.B. Topics in Guidance, Navigation & Control of Underwater Gliders. Master's Thesis, Oslo Metropolitan University, Oslo, Norway, 2022.
8. Liu, S.; Zhang, T.; Zhang, J.; Zhu, Y. A new coupled method of SINS/DVL integrated navigation based on improved dual adaptive factors. *IEEE Trans. Instrum. Meas.* **2021**, *70*, 8504211. [CrossRef]
9. Yuan, D.; Ma, X.; Liu, Y.; Yang, L.; Wu, Y.; Zhang, X. Reaearch on underwater integrated navigation system based on SINS/DVL/magnetometer/depth-sensor. In Proceedings of the OCEANS 2017-Aberdeen, Aberdeen, UK, 19–22 June 2017; IEEE: New York, NY, USA, 2017; pp. 1–6.
10. Guo, Y.; Wu, M.; Tang, K.; Zhang, L. Square-root unscented information filter and its application in SINS/DVL integrated navigation. *Sensors* **2018**, *18*, 2069. [CrossRef]
11. Jin, K.; Chai, H.; Su, C.; Xiang, M. A performance-enhanced DVL/SINS integrated navigation system based on data-driven approach. *Meas. Sci. Technol.* **2023**, *34*, 095120. [CrossRef]
12. Zhang, S.; Zhang, T.; Zhang, L. An underwater SINS/DVL integrated system outlier interference suppression method based on LSTM-EEWKF. *IEEE Sens. J.* **2023**, *23*, 27590–27600. [CrossRef]
13. Qin, X.; Zhang, R.; Wang, G.; Long, C.; Hu, M. Robust Interactive Multi-Model INS/DVL Intergrated Navigation System with Adaptive Model Set. *IEEE Sens. J.* **2023**, *23*, 8568–8580. [CrossRef]
14. Carrasco, P.L.N.; Bonin-Font, F.; Campos, M.M.; Codina, G.O. Stereo-vision graph-SLAM for robust navigation of the AUV SPARUS II. *IFAC-PapersOnLine* **2015**, *48*, 200–205. [CrossRef]
15. Drap, P.; Merad, D.; Hijazi, B.; Gaoua, L.; Nawaf, M.M.; Saccone, M.; Chemisky, B.; Seinturier, J.; Sourisseau, J.-C.; Gambin, T. Underwater photogrammetry and object modeling: A case study of Xlendi Wreck in Malta. *Sensors* **2015**, *15*, 30351–30384. [CrossRef]
16. Xin, Z.; Wang, Z.; Yu, Z.; Zheng, B. ULL-SLAM: Underwater low-light enhancement for the front-end of visual SLAM. *Front. Mar. Sci.* **2023**, *10*, 1133881. [CrossRef]
17. Xu, S.; Ma, T.; Li, Y.; Ding, S.; Gao, J.; Xia, J.; Qi, H.; Gu, H. An effective stereo SLAM with high-level primitives in underwater environment. *Meas. Sci. Technol.* **2023**, *34*, 105405. [CrossRef]
18. Jung, J.; Lee, Y.; Kim, D.; Lee, D.; Myung, H.; Choi, H.-T. AUV SLAM using forward/downward looking cameras and artificial landmarks. In Proceedings of the 2017 IEEE Underwater Technology (UT), Busan, Republic of Korea, 21–24 February 2017; IEEE: New York, NY, USA, 2017; pp. 1–3.

19. Duecker, D.A.; Eusemann, K.; Kreuzer, E. Towards an Open-Source Micro Robot Oceanarium: A Low-Cost, Modular, and Mobile Underwater Motion-Capture System. In Proceedings of the 2019 IEEE/RSJ International Conference on Intelligent Robots and Systems (IROS), Macau, China, 3–8 November 2019; IEEE: New York, NY, USA, 2019; pp. 8048–8053.
20. Yu, A.; Wang, Y.; Li, H.; Qiu, B. Automatic Alignment Method of Underwater Charging Platform Based on Monocular Vision Recognition. *J. Mar. Sci. Eng.* **2023**, *11*, 1140. [CrossRef]
21. Leonard, J.L.; Carpenter, R.N.; Feder, H.J.S. Stochastic mapping using forward look sonar. *Robotica* **2001**, *19*, 467–480. [CrossRef]
22. Palomer, A.; Ridao, P.; Ribas, D. Multibeam 3D underwater SLAM with probabilistic registration. *Sensors* **2016**, *16*, 560. [CrossRef]
23. Daniel, S.; Le Léannec, F.; Roux, C.; Soliman, B.; Maillard, E.P. Side-scan sonar image matching. *IEEE J. Ocean. Eng.* **1998**, *23*, 245–259. [CrossRef]
24. Khater, H.A.; Gad, A.S.; Omran, E.A.; Abdel-Fattah, A.A. Enhancement matching algorithms using fusion of multiple similarity metrics for sonar images. *World Appl. Sci. J.* **2009**, *6*, 759–763.
25. Zhou, X.; Yu, C.; Yuan, X.; Luo, C. Matching Underwater Sonar Images by the Learned Descriptor Based on Style Transfer Method. *J. Phys. Conf. Ser.* **2021**, *2029*, 012118.
26. Liu, F.; Qian, D.; Liu, F.; Li, Y. Integrated navigation system based on correlation between gravity gradient and terrain. In Proceedings of the 2009 International Joint Conference on Computational Sciences and Optimization, Sanya, China, 24–26 April 2009; IEEE: New York, NY, USA, 2009; pp. 289–293.
27. Rice, H.; Kelmenson, S.; Mendelsohn, L. Geophysical navigation technologies and applications. In Proceedings of the PLANS 2004. Position Location and Navigation Symposium (IEEE Cat. No. 04CH37556), Monterey, CA, USA, 26–29 April 2004; IEEE: New York, NY, USA, 2004; pp. 618–624.
28. Meduna, D.K.; Rock, S.M.; McEwen, R.S. Closed-loop terrain relative navigation for AUVs with non-inertial grade navigation sensors. In Proceedings of the 2010 IEEE/OES Autonomous Underwater Vehicles, Monterey, CA, USA, 1–3 September 2010; IEEE: New York, NY, USA, 2010; pp. 1–8.
29. Zhao, S.; Zheng, W.; Li, Z.; Zhu, H.; Xu, A. Improving Matching Efficiency and Out-of-Domain Positioning Reliability of Underwater Gravity Matching Navigation Based on a Novel Domain-Center Adaptive-Transfer Matching Method. *IEEE Trans. Instrum. Meas.* **2023**, *72*, 1001811. [CrossRef]
30. Wang, Z.; Huang, Y.; Wang, M.; Wu, J.; Zhang, Y. A computationally efficient outlier-robust cubature Kalman filter for underwater gravity matching navigation. *IEEE Trans. Instrum. Meas.* **2022**, *71*, 8500418. [CrossRef]
31. Liu, H.; Wu, L.; Bao, L.; Li, Q.; Zhang, P.; Wang, Y. Gravity matching navigation algorithm based on multiscale search and Hadamard transformed difference. *ISA Trans.* **2022**, *128*, 409–422. [CrossRef] [PubMed]
32. Li, S.; Cheng, D.; Wang, Y.; Zhao, J. Compensation Method for the Carrier Magnetic Interference of Underwater Magnetic Vector Measurement System. *IEEE Sens. J.* **2023**, *23*, 10694–10705. [CrossRef]
33. Liu, Y.; Zhang, G.; Huang, Z. Study on the Arctic Underwater Terrain-Aided Navigation Based on Fuzzy-Particle Filter. *Int. J. Fuzzy Syst.* **2021**, *23*, 1017–1026. [CrossRef]
34. Shi, Y.; Han, X.; Ma, L.; Guo, L. ASBL: Low-cost, small-sized AUV remote guidance method and experimental verification. *EURASIP J. Adv. Signal Process.* **2023**, *2023*, 36. [CrossRef]
35. He, H.; Tang, H.; Xu, J.; Liang, Y.; Li, F. A SINS/USBL System-Level Installation Parameter Calibration with Improved RDPSO. *IEEE Sens. J.* **2023**, *23*, 17214–17223. [CrossRef]
36. Wang, J.; Xu, T.; Liu, Y.; Li, M.; Li, L. Augmented Underwater Acoustic Navigation with Systematic Error Modeling Based on Seafloor Datum Network. *Mar. Geod.* **2023**, *46*, 129–148. [CrossRef]
37. Jalal, F.; Nasir, F. Underwater navigation, localization and path planning for autonomous vehicles: A review. In Proceedings of the 2021 International Bhurban Conference on Applied Sciences and Technologies (IBCAST), Islamabad, Pakistan, 12–16 January 2021; IEEE: New York, NY, USA, 2021; pp. 817–828.
38. Karal, Ö.; Kazdal, H. A New Fuzzy Logic-Based Adaptive Complementary Filter Algorithm for UAV Attitude Estimation. *Pamukkale Univ. J. Eng. Sci.* **2024**, *30*.
39. Pan, S.; Xu, X.; Zhang, L.; Yao, Y. A novel SINS/USBL tightly integrated navigation strategy based on improved ANFIS. *IEEE Sens. J.* **2022**, *22*, 9763–9777. [CrossRef]
40. Chang, C.-W.; Yan, J.-L.; Chang, C.-N.; Wen, K.-A. IMU-Based Real Time Four Type Gait Analysis and Classification and Circuit Implementation. In Proceedings of the 2022 IEEE Sensors, Dallas, TX, USA, 30 October–2 November 2022; IEEE: New York, NY, USA, 2022; pp. 1–4.
41. Karmozdi, A.; Hashemi, M.; Salarieh, H. Design and practical implementation of kinematic constraints in Inertial Navigation System-Doppler Velocity Log (INS-DVL)-based navigation. *Navig. J. Inst. Navig.* **2018**, *65*, 629–642. [CrossRef]
42. Wang, W.; Zhu, M.; Yang, B. Positioning Combination Method of USBL Using Four-Element Stereo Array. *Sensors* **2021**, *21*, 7722. [CrossRef] [PubMed]
43. Allotta, B.; Caiti, A.; Costanzi, R.; Fanelli, F.; Fenucci, D.; Meli, E.; Ridolfi, A. A new AUV navigation system exploiting unscented Kalman filter. *Ocean Eng.* **2016**, *113*, 121–132. [CrossRef]

44. Li, D.; Xu, J.; Zhu, B.; He, H. A calibration method of DVL in integrated navigation system based on particle swarm optimization. *Measurement* **2022**, *187*, 110325. [CrossRef]
45. Dong, Y.; Wang, D.; Zhang, L.; Li, Q.; Wu, J. Tightly coupled GNSS/INS integration with robust sequential kalman filter for accurate vehicular navigation. *Sensors* **2020**, *20*, 561. [CrossRef]

Journal of
Marine Science and Engineering

MDPI

Article

Quantification of Expert Knowledge in Describing COLREGs Linguistic Variables

Miho Kristić [1,*] and Srđan Žuškin [2,*]

[1] Maritime Department, University of Dubrovnik, Ćira Carića 4, 20000 Dubrovnik, Croatia
[2] Faculty of Maritime Studies, University of Rijeka, Studentska ulica 2, 51000 Rijeka, Croatia
* Correspondence: miho.kristic@unidu.hr (M.K.); srdan.zuskin@pfri.uniri.hr (S.Ž.); Tel.: +385-20445-700 (M.K.); +385-51338-411 (S.Ž.)

Abstract: The International Regulations for Preventing Collisions at Sea 1972 (COLREGs) have been the cornerstone of maritime navigation since their introduction. Knowledge and implementation of these rules are paramount in collision avoidance at sea. However, terms found in these rules are sometimes imprecise or fuzzy, as they are written by humans for humans, giving them some freedom in interpretation. The term *Very Large Ship* used in Rule 7 of the COLREGs is, by its nature, fuzzy. While human navigators understand this term's meaning, it could be challenging for machines or autonomous ships to understand such an unprecise expression. Fuzzy sets could easily describe unprecise terms used in maritime navigation. A fuzzy set consists of elements with degrees of membership in a set, making them perfect for interpreting some terms where boundaries are unclear. This research was conducted among 220 navigational experts to describe linguistic variables used in maritime regulations. This research consists of an internationally distributed questionnaire. Membership data were collected with the adapted horizontal method, and the results were statistically analyzed, followed by regression analyses to describe the range and shape of membership functions. A conceptual model of the implementation of linguistic variables is presented. The novelty of this study derives from the data collecting, modeling, and quantification of the important but neglected linguistic term *Very Large Ship* based on a large number of navigational experts. The same quantification method could be easily used for other COLREGs linguistic variables, which could easily lift barriers to advances in intelligent solutions based on fuzzy sets. The obtained quantified fuzzy sets can be used in decision support or control systems used by conventional or autonomous ships in the future.

Keywords: fuzzy set; collision avoidance; COLREGs; safety of navigation

1. Introduction

Although among the most important regulations for maritime navigators, the COLREGs contain numerous imprecise terms and definitions. Nevertheless, the implementation of these rules is a daily routine for navigational ranks. The obligation set by these rules to maintain proper lookout at all times means that the navigator has to follow the COLREGs from taking the watch until being relieved. During their education periods, professional navigators will extensively study collision regulations. However, misunderstanding rules is a serious problem [1,2]. Research by Mohović et al. showed gaps in understanding some parts of the COLREGs by navigators and students due to wrong interpretations of these rules [3]. Furthermore, the number of collisions is gradually increasing, and many of them result in catastrophic consequences [4]. The COLREGs Convention (1972) was created to update and replace the Collision Regulations of 1960 [5]. New regulations in 1972 implemented traffic separation schemes, which made traffic in congested areas safer.

These rules were written and designed for the human operation of ships, so they are based on human judgments of navigational situations and the common sense of navigators [6–9]. In the meantime, different maritime intelligent technologies have become more accepted and proposed [10–15]. While some terms used in the COLREGs are easily understandable to human navigators, explaining some of them mathematically can be difficult. Despite progress made in autonomous shipping, the International Maritime Organization (IMO) still plans to keep these rules as a reference point and keep them in their current form [16]. Rødseth et al. studied interactions between conventional and autonomous ships, and as a result, advocated changes in the COLREGs to improve safety at sea [17].

An analytical review of relevant literature by Chang et al. indicates that the present COLREGs represent a significant linguistic barrier to advances in autonomous navigation [18]. Additionally, the development of artificial intelligence solutions for collision avoidance is facing a lack of quantitative data regarding the COLREGs [19], so quantifying these rules is becoming imperative [20,21]. Research by Kim and Park recognized the limitations of terms used in the COLREGs and suggested quantifying COLREGs terms properly in time and with sufficient action using a Bayesian regularized artificial neural network [22].

Another study confirmed that there is a significant difference between a navigator's understanding of rules and collision-avoidance algorithms. Additionally, it was proven that navigators exhibit differing interpretations of collision regulation [23]. A study by Zhou et al. indicates that the COLREGs need to eliminate the uncertainty of interpretation as a barrier to the use of autonomous ships [24]. Research by Kim aimed to quantify COLREGs terms that were not previously quantified, such as "narrow channel" and "restricted visibility" [25], while Li suggested a model of autonomous collision avoidance under the constraint of the COLREGs [26]. Hagen et al. proposed a practical tool for the development of a collision avoidance system that is COLREGs-compliant [27].

Rule 7, covering the risk of collision, is among the most essential rules. The rule states that the risk of collision may exist even when the bearing of the approaching vessel changes if the approaching ship is very large. The term *Very Large Ship* used in this rule is challenging to define. When is the ship large enough to be considered very large? What is the limit or border between a *Not Very Large Ship* and a *Very Large Ship* crisp?

Hypothetically, let us consider that a *Very Large Ship* is anything above 200 m; in that case, a ship with a length of 199.99 m would not be very large, and a ship 200.01 m long would be very large (Figure 1a). Of course, this makes no sense by human reasoning, but it is the only possible solution based on the rules of classical binary logic. The solution to this problem is implementing fuzzy sets, in which elements can have a degree of membership (Figure 1b).

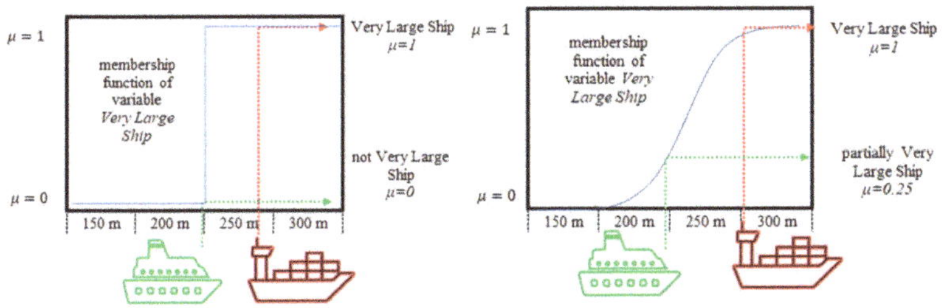

Figure 1. Membership example: (**a**) crisp set; (**b**) fuzzy set.

According to fuzzy set theory, the same element can partially belong to a given set and several fuzzy sets simultaneously. In our case, the ship can be large or very large, with a partial degree of membership in both sets. This is very close to human reasoning, so fuzzy theory and fuzzy logic are valuable tools for modeling variables humans use to describe surroundings. Fuzzy logic is a widely used method by authors researching maritime collision risk and avoidance [28–36], maritime fuzzy controllers [37], Fuzzy Fault Tree Analysis [38], and others. The essential part of classical fuzzy systems and the

Fuzzy Analytic Hierarchy Process (Fuzzy AHP) is the membership function, but there are difficulties related to defining parameters that define the shape of these functions [39]. In several previous studies, heuristic methods of assigning membership functions were used, but some approaches try to use probability distribution to describe membership functions [40]. Several studies have already described the linguistic variable *Ship's size* as part of creating fuzzy systems. However, the risk of collision for the purpose of intelligent solutions is mainly modeled by the use of only two variables, namely, *Distance to the Closest Point of Approach* (*DCPA*) and *Time to the Closest Point of Approach* (*TCPA*) [41], while not including linguistic variable *Very Large Ship*. That omission means that the rule, as stated in the COLREGs, is not adequately modeled. From the perspective of humans, avoiding an ultra-large tanker is not the same as avoiding a small tug boat. It is not only size that matters but maneuvering characteristics are also completely different. Sheng-Long Kao et al. (2007) defined the linguistic variable *Size of the Ship* in a range from 130 m to 250 m to create a fuzzy collision risk system [28], but the term *Very Large Ship* was not modeled. Linguistic variables are based on the Nippon Kaiji Kyokai classification society's (Class NK) ship data, and that paper does not show how the authors created membership functions. The authors created default triangular and partial trapezoidal membership functions to describe the terms *Small Ship*, *Medium Ship*, and *Large Ship*. As suggested by the authors, membership functions do not include ships below 130 m or above 250 m; thus, this linguistic variable does not include a full universe of discourse. Ren et al. [42] used the variable *Ship's Length* to modify the input variable *DCPA*. As a result, this linguistic variable was created using triangular and partially trapezoidal functions. Nevertheless, the variable *Very Large Ship*, as defined in the COLREGs, has not been quantified by previous research or used in modeling the risk of collision.

The quantified variable *Very Large Ship* could be further used in classical fuzzy systems or Fuzzy AHP to improve existing models or create new models that define Rule 7 adequately. Fuzzy AHP systems have been used in several recent studies as an alternative to classical fuzzy logic systems to describe the risk of collision [43–45]. Fuzzy AHP is used by Luo et al. to describe collision risk using basic triangular membership functions, and the risk of collision is described by the variables *DCPA* and *TCPA*. Gaussian membership functions have been used in research to overcome problems with basic triangular and trapezoidal membership functions [46,47]. However, research [48] has found that, on observed data, fuzzy logic system results are more conservative than those of Fuzzy AHP and the equal weights method. This could suggest that the classical fuzzy system is more appropriate for risk assessment. Another study's [49] verification results indicated that the classical fuzzy model performed better than AHP on a given data set.

Therefore, because the uncertainty of the interpretation of ship size needs to be eliminated to facilitate collision avoidance for autonomous ships, this paper deals with defining the range of the linguistic variable *Ship's Size* in the COLREGs. As defined in COLREGs Rule 7, the linguistic variable *Very Large Ship* might completely change the estimation of the risk of collision and, therefore, needs to be defined for an autonomous ship's collision avoidance algorithm to function adequately. According to that, the novelty of this study is the definition of the important COLREGs linguistic variable *Very Large Ship* through the quantification of expert knowledge, which was not adequately defined in previous studies, and in that way, this study opens up opportunities for further research and improvements of collision avoidance algorithms for autonomous ships. Once defined, linguistic variables could be further used in fuzzy systems, such as fuzzy decision support systems.

2. Method of Eliciting Expert Knowledge

Research data were collected using an internationally distributed online questionnaire. The questionnaire was prepared according to the observed problem, targeting randomly selected navigational experts from different shipping companies. It was prepared using the Lime Survey software (LimeSurvey Cloud Version 5.6.62, https://www.limesurvey.org/, accessed on 17 May 2024), an online open-source survey tool [50], and distributed among active sailing navigational experts. All questions were designed as closed-type questions due to a large sample size and to allow for easier data analysis.

However, eliciting expert knowledge is sometimes problematic, as the definition of an expert is quite broad. The literature uses the term expert very broadly and often describes anyone with any knowledge about a matter [51]. This makes sense with certain types of research and data collection, but in this paper, the questionnaire aims to consider participants with knowledge, skills, and navigation practice. This was achieved by disseminating a questionnaire among masters and navigational officers only, not including cadets, petty officers, or ratings. Additionally, it was intended to reach experts sailing on all ship types and sizes. The questionnaire was prepared by authors who are navigational experts, trying to resolve the meaning of some linguistic variables by examining a large number of navigational experts.

Research that aims to interpret COLREGs linguistic variables consists of two parts:

- Personal profile questions—PPQs.
- Fuzzy variable range questions—RFQs.

The first part of the questionnaire contains personal profile questions:

- PPQ1: Rank on board.
- PPQ2: Sea experience in years.
- PPQ3: Type of your last ship.
- PPQ4: Length overall in meters of your last ship.

The second part of the questionnaire consists of five questions:

- RFQ1: Please describe the linguistic variable *Very Small Ship* by recording minimum and maximum values in meters, in a range from 0 to 400 m.
- RFQ2: Please describe the linguistic variable *Small Ship* by recording minimum and maximum values in meters, in a range from 0 to 400 m.
- RFQ3: Please describe the linguistic variable *Medium Ship* by recording minimum and maximum values in meters, in a range from 0 to 400 m.
- RFQ4: Please describe the linguistic variable *Large Ship* by recording minimum and maximum values in meters, in a range from 0 to 400 m.
- RFQ5: Please describe the linguistic variable *Very Large Ship* by recording minimum and maximum values in meters, in a range from 0 to 400 m.

The standard membership estimation techniques are the horizontal and vertical methods, but both methods could have a lack of continuity as a drawback [52]. An adapted horizontal method was used by authors, where the drawback of continuity was handled by questioning experts about a full range of elements in a specific set. In the classical horizontal method, experts provide opinions if predefined elements belong to a specified set. Such a method was not practical for this research, as a number of predefined elements would be excessive for an effective questionnaire. The authors decided to question experts on the lower and upper limits of elements that belong to the set. Elements in the set are based on the obtained limits as follows:

$$X = \{x_{ll}, x_{ll+1}, x_{ll+2}, \ldots, x_{ul}\},$$

where x_{ll} is the lower bound of the set, and x_{ul} is the upper bound of the set. Given "n" number of experts, the ratio between the number of repeating elements and the total number of replies is the membership degree. When all experts agree, membership is equal to one. The Lime Survey software was chosen because it has enough flexibility to allow for some custom-made question types. This was especially important for questions that were supposed to define a specific fuzzy set with all its elements. To achieve this, a custom-made double slider available through the open-source website GitHub [53] was used (Figure 2). This tool assists in determining the lower and upper limits of some linguistic variables. Furthermore, this approach avoided leading experts toward a predefined range for any linguistic value. All linguistic values, from *Very Small Ship* to *Very Large Ship*, were examined on the same scale, from zero to four hundred meters.

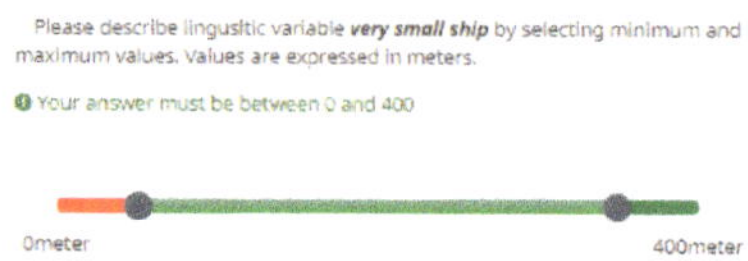

Figure 2. Custom-made double-sided slider used in the survey.

Respondents could move both sliders to a precision of 1 m, so the answers represented all values within the defined range with a step of 1 m. Descriptive statistics of the research data were created using MATLAB R2018a Update 6(9.4.0.949201) (https://www.mathworks.com/products/matlab.html, accessed on 17 May 2024), which is widely used to analyze data [54]. Outliers in the data were removed using the quartile method, removing elements that were more than 1.5 interquartile ranges above the upper quartile or below the lower quartile [55,56]. The obtained data were further used to form sets of linguistic values. The data were normalized for use from zero to one, as fuzzy membership functions are defined in the same range. Finally, appropriate curves were fitted to analyze data using regression analysis.

3. Results of Survey

The research results are presented as descriptive statistics for the first part of the questionnaire, which contains the personal profiles of the navigational experts. The second part consists of descriptive statistics on the linguistic variable and a curve-fitting method for analyzing data.

3.1. Personal Profile Questions

An online questionnaire was distributed internationally to active navigational ranks, and 220 active maritime experts participated in this research (Figure 3). This research contains answers from 113 Masters, 65 Chief Officers, 28 Second Officers, and 14 Third Officers. The largest group of respondents is Masters, with a share of 51%, followed by Chief Officers (30%). The total experience of the respondents closely corresponds to rank distribution; the most experienced participants with more than 15 years of experience occupy a share of 58.2%; those with less than 15 years but more than 10 years have a share of 16.3%; those with 5 to 10 years take a share of 15.5%; those with 3 to 5 years take up 4.5%; and finally, those with 1 to 3 years take up a share of 5.5%.

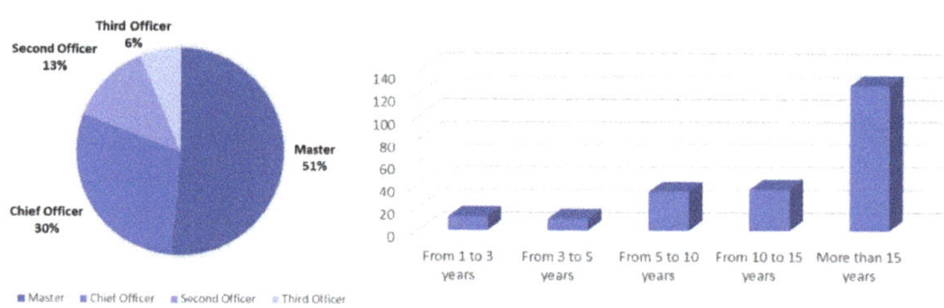

(a) (b)

Figure 3. Participants in the questionnaire (**a**) by navigational rank and (**b**) by their total experience at sea.

The remaining personal profile answers reveal the ship's type and the ship's length overall on which participants sail (Figure 4). The largest group of participants sailed on liquified natural gas (LNG) carriers (91), followed by cruisers (59), dry bulk carriers (27), container ships (22), tankers (13), and others (8). Participants, based on the length of the ship on which they serve, can be categorized into six groups: (i) 4.1% of respondents serve on ships from 100 to 150 m; (ii) 12.3% of respondents sail on ships from 150 m to 200 m; (iii) 17.3% of participants serve on ships from 200 m to 250 m; (iv) 46% of respondents represent the largest group, and they serve on ships from 250 m to 300 m; (v) 17.3% serve on ships from 300 m to 350 m; and (vi) last, the smallest group, 6%, are the navigators, who sail on the largest vessels, larger than 350 m.

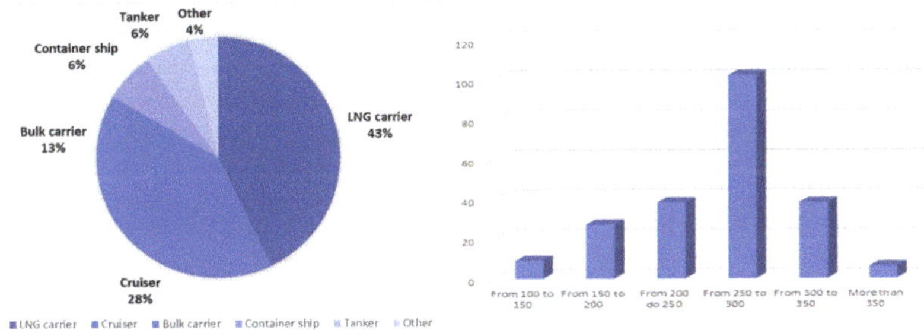

(**a**) (**b**)

Figure 4. Participants in the questionnaire (**a**) by type of ship and (**b**) by the length of their last ship.

3.2. Fuzzy Range Questions

The results of questions in the second part of the survey are summarized in Table 1, presenting respondents' answers on the lower and upper limits of the range of linguistic variables expressed in meters. Common expressions used in fuzzy theory for describing linguistic values were used: the linguistic value *Very Small Ship* is defined by the fuzzy value name *Very Low* (VL); *Small Ship* by the fuzzy value name *Low* (L); *Medium Ship* by the fuzzy value name *Medium* (M); *Large Ship by the fuzzy value name High* (H); and finally, *Very Large Ship* by the fuzzy value name *Very High* (VH). The abbreviations are as follows: standard deviation (SD), minimum value (Min), arithmetic mean (Mean), maximum value (Max), coefficient of Kurtosis (Ku), coefficient of skewness (Sk), interquartile range (IQR), and median (C).

Table 1. Descriptive statistics for linguistic values' boundaries.

Ship's Size	Range Limits	SD	Min	Mean	Max	Ku	Sk	IQR	C
VL	lower	10.3983	0	5.9659	41	4.8606	1.6843	10.5	0
	upper	20.2872	10	43.3450	101	3.2339	0.6615	24.25	45
L	lower	17.5196	0	39.0455	83	2.8815	−0.0821	25	41
	upper	28.0226	30	95.1724	180	3.1751	0.2994	24	100
M	lower	28.2761	30	95.1667	167	2.8211	0.1738	29	100
	upper	35.7669	100	185.0506	272	3.1328	−0.0768	45	187.5
H	lower	37.2397	80	187.2880	274	3.3676	−0.1965	41	193
	upper	36.8226	200	284.0000	350	2.8887	−0.2651	50	299
VH	lower	34.2945	197	279.2707	359	2.9865	−0.3982	50	297
	upper	0	400	400.0000	400	Na	Na	0	400

Since the answers determined the range limits, creating a set containing all elements of the set between limits was necessary, and we did so by discretizing continuous values to be discrete with an interval of 1 m. An interval of 1 m was also an interval used in the online questionnaire. After the discrete sets were created, the sets were analyzed with the results shown in Table 2. Elements of the descriptive statistics are as follows: standard deviation (SD), arithmetic mean (Mean), and median (C).

Table 2. Descriptive statistics for linguistic value data set.

Ship's Size	SD	Mean	C
VL	18.3734	26.729	24
L	27.1491	68.5861	67
M	39.8209	141.8283	140
H	45.3227	235.6199	236
VH	41.5732	336.1744	340

The analysis results show no significant difference between the median and arithmetic mean, indicating the data's lack of significant outliers and skewness. The distribution of data for the linguistic variable can be visually represented by histograms, as in Figure 5a–e.

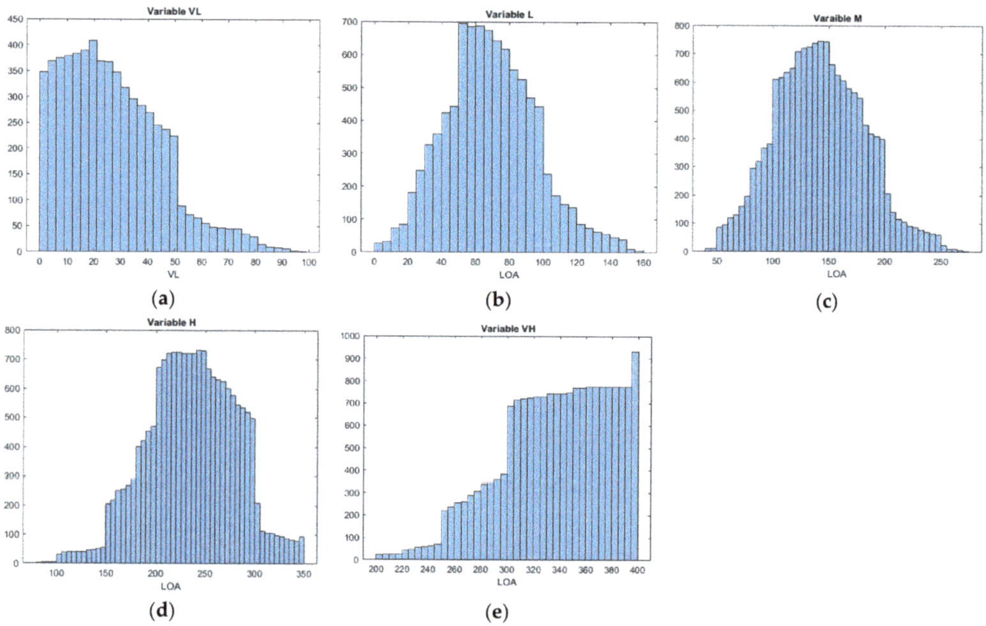

Figure 5. Distribution of data for linguistic values: (**a**) Very Low; (**b**) Low; (**c**) Medium; (**d**) High; (**e**) Very High.

It is visible on the obtained graphs that there is some fuzziness in the linguistic values, which was expected and needed to be further described by using the appropriate membership function. To use the results and define membership functions, the data obtained had to be normalized to a range between zero and one. The Min–Max method was used to convert values, as it is generally used as a normalization method for transferring values to a defined scale [57]. It can be expressed as

$$x' = \frac{x - \min(x)}{\max(x) - \min(x)} \tag{1}$$

where x' is the normalized value, x is the original value, $\min(x)$ is the minimum value in the set, and max x is the maximum value in a given set. As in the analyzed set, $\min(x)$ is equal to zero, and the formula can be reduced to

$$x' = \frac{x}{\max(x)} \tag{2}$$

By using a new normalized value, a graph showing rough fuzzy linguistic values could be represented as in Figure 6.

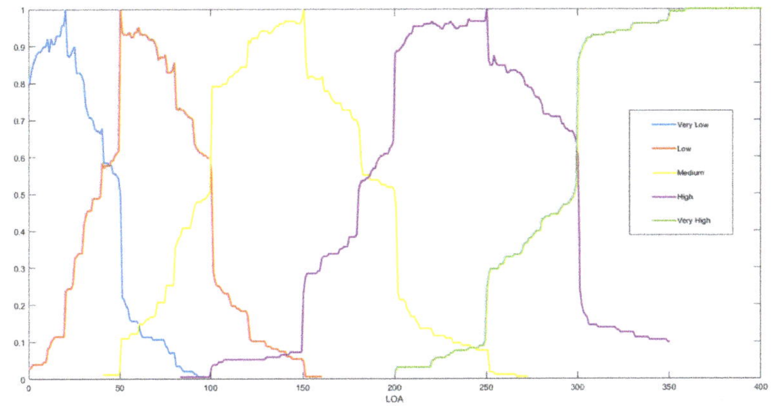

Figure 6. Graph of linguistic values in a full range of linguistic variables.

A so-defined linguistic value would not be practical for use in fuzzy systems. Therefore, normalized data will be further used to define membership functions by regression analysis and by assigning the best fit.

3.3. Regression Analysis to Obtain Membership Functions

The normalized data were further used to determine the range and shape of membership functions. Membership functions were modeled using the MATLAB curve-fitting tool, which allows for regression analysis using available linear and non-linear models [58]. Regression modeling uses statistical processing and analysis to mathematically describe dependencies between variables [59]. The results of the regression are presented and analyzed graphically, followed by an examination of goodness-of-fit figures. The MATLAB toolbox provides the sum of squares due to error, R-square, adjusted R-square, and root mean squared error as goodness-of-fit statistics.

Gaussian curves were used to obtain the best fit for the linguistic variables *Low*, *Medium*, and *High* (Figure 7a–c).

Gaussian curves for the linguistic variables *Low*, *Medium*, and *High* can be expressed by the expression

$$f(x) = a * e^{\left(-\frac{(x-b)}{c}\right)^2} \quad (3)$$

where a is amplitude, b is the centroid (location), and c is related to the peak width. Coefficients for the above variables are provided in Table 3.

Table 3. Coefficients for linguistic values Low, Medium, and High.

Coefficients	Low	Medium	High
a	1	1	1
b	66.65	139.7	237
c	36.2	57.5	66.89

After analyzing the data for the variables *Very Low* and *Very High*, a sigmoidal logistic curve achieved the best fit. A graphic representation of the fit reveals a good fit for these two values (Figure 8a,b).

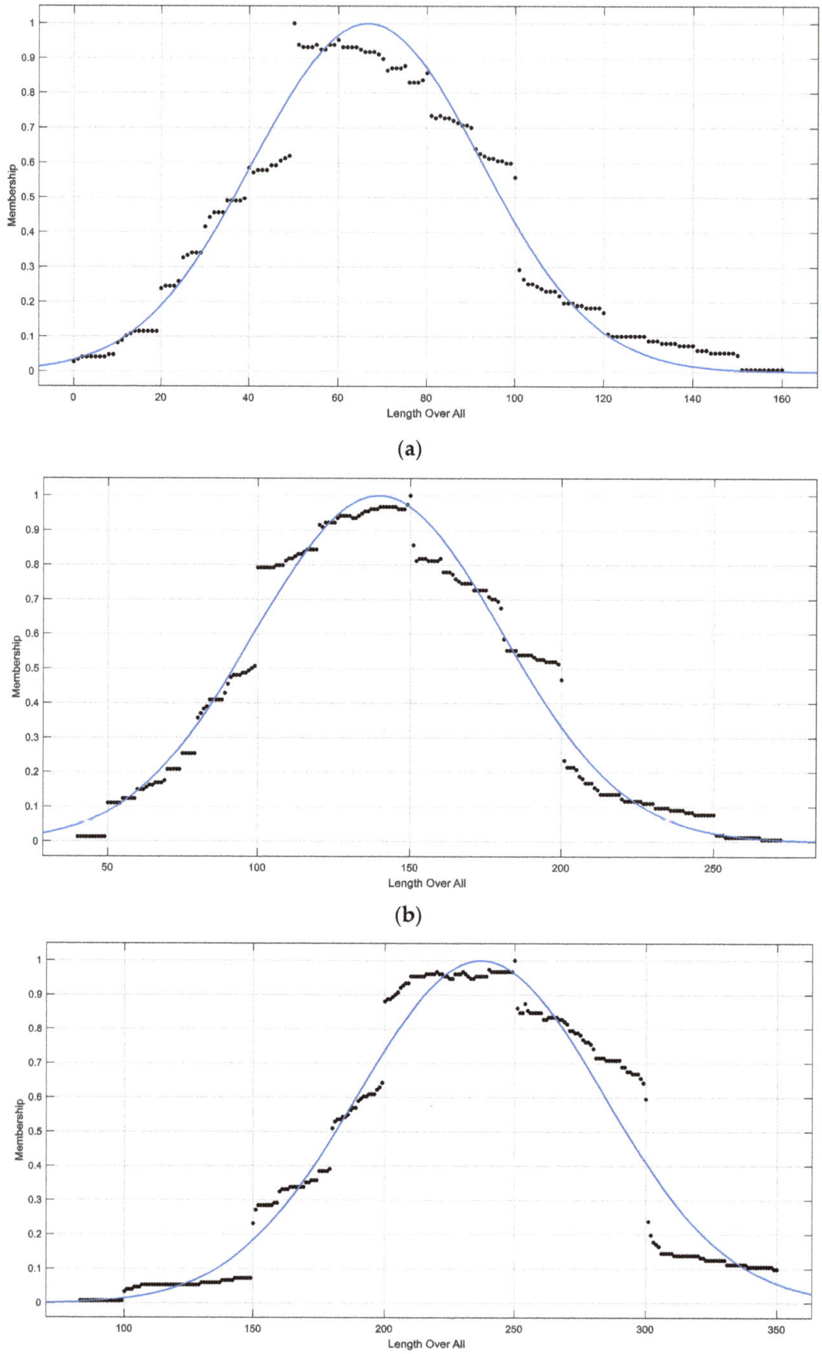

(a)

(b)

(c)

Figure 7. Fitting of Gaussian curve to linguistic values (**a**) Low, (**b**) Medium, and (**c**) High.

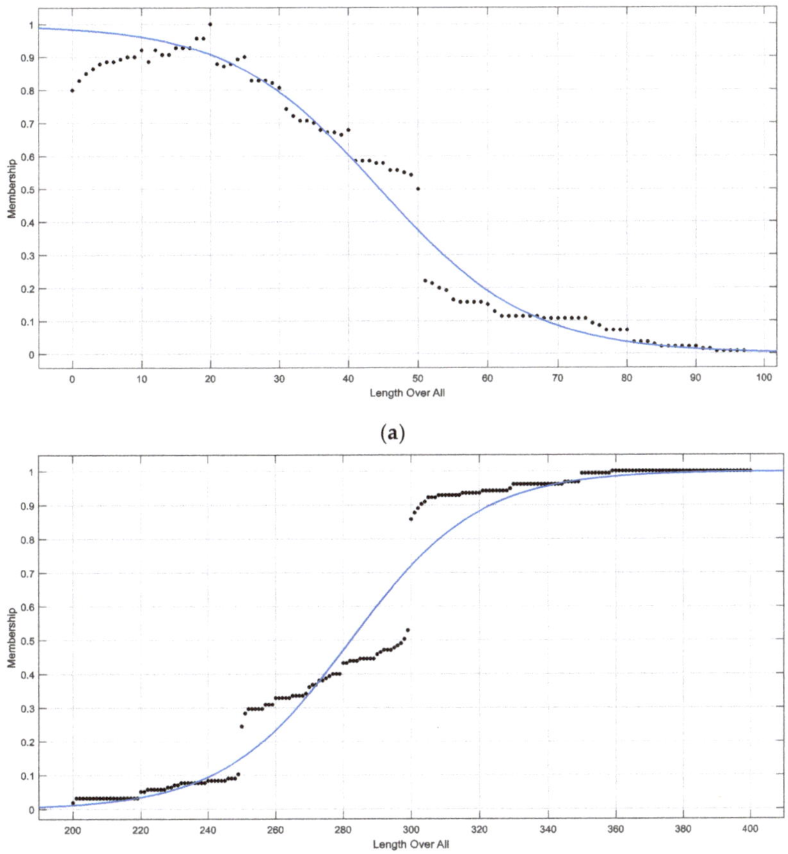

(**a**)

(**b**)

Figure 8. Fitting of sigmoidal logistic curve to linguistic values (**a**) Very Low and (**b**) Very High.

The sigmoidal logistic curve for linguistic variables *Very Low* and *Very High* can be expressed by the expression

$$f(x) = \frac{a}{1 + e^{-b(x-c)}} \tag{4}$$

where a is a parameter for the horizontal asymptote, b is a growth rate parameter, and c is the midpoint between the horizontal asymptotes and the inflection point. Coefficients for the above variables are provided in Table 4.

Table 4. Coefficients for the linguistic values Very Low and Very High.

Coefficients	Very Low	Very High
a	1	1
b	−0.0931	0.0535
c	44.587	282.1991

The goodness of fit for all the linguistic variables, after fitting the curves, is provided in Table 5, with SSE (sum of squares due to error), R-square (coefficient of determination), adjusted R-square, and RMSE (root-mean-square error).

Table 5. The goodness of fit for the linguistic values.

	Goodness of Fit			
Linguistic Variables	**SSE**	**R-Square**	**Adjusted R-Square**	**RMSE**
Very Low	0.3699	0.9716	0.9713	0.0621
Low	0.6204	0.9646	0.9644	0.0625
Medium	0.7338	0.9733	0.9732	0.0564
High	1.5820	0.9549	0.9547	0.0771
Very high	0.8787	0.9719	0.9717	0.0664

Among the linguistic values described by the Gaussian curve, the lowest SSE can be observed for the variable *Low* (*Small Ship*). Other statistics for the same variable indicate a good fit. For the other two variables described by the Gaussian curve, it can be seen that the appropriate SSE and other statistics also indicate a good fit. The same is visible when the graphical representation of fit is examined. For linguistic values described by the sigmoidal logistic curve, the linguistic value *Very Low* has a better fit with the lower SSE value than *Very High*, but not significantly. Altogether, the proposed curves fit the values properly, with the goodness-of-fit statistics showing excellent fitting.

3.4. Review of Obtained Results with Conceptual Model

After membership functions are defined, it is possible to review and evaluate membership in fuzzy sets for any element in a continuous range from 0 to 400 m. The degree of membership for some sizes of the ships in the fuzzy sets (*Very Low*, *Low*, *Medium*, *High*, and *Very High*) can be found in Table 6.

Table 6. Membership degree of some elements in the fuzzy sets.

	Length Overall (m)												
	20	**50**	**80**	**110**	**140**	**170**	**200**	**230**	**260**	**290**	**320**	**350**	**380**
Very Low	0.91	0.38	0.04	0	0	0	0	0	0	0	0	0	0
Low	0.19	0.81	0.87	0.24	0.02	0	0	0	0	0	0	0	0
Medium	0.01	0.09	0.34	0.77	1	0.76	0.33	0.08	0.01	0	0	0	0
High	0	0	0	0.03	0.12	0.37	0.74	0.99	0.89	0.53	0.21	0.06	0.01
Very High	0	0	0	0	0	0	0.01	0.06	0.23	0.6	0.88	0.97	1

As previously mentioned, fuzzy set theory allows some elements to have only partial membership in a given set and to be simultaneously present in two or more sets. Ships with a length overall (LOA) of 260 m are, for instance, members of three sets. Membership in *Medium Ship* is just 0.01, but membership in the set of *Large Ships* is 0.89. At the same time, this ship size is also a member of the *Very Large Ship* set with a membership of 0.23. Of course, due to the continuous nature of the linguistic variables, it is not possible to present each element in the table, but the membership of any ship by length overall can be calculated by the above-presented formulas and coefficients.

The implementation of the quantified linguistic value is defined by the COLREGs. The rules state that if during a determination of the risk of collision, it appears to the navigator that the risk does not exist, risk still has to be confirmed if the approaching ship is very large. This definition in the COLREGs indicates the importance of a ship's size in evaluating risk of collision.

The quantified linguistic value *Very Large Ship* could be easily used to improve existing collision avoidance systems or to create new collision avoidance fuzzy or Fuzzy AHP systems. Furthermore, this variable can be easily implemented as a new variable in a hierarchical fuzzy system. A hierarchical fuzzy system consists of two or more fuzzy systems that are arranged in a hierarchical tree structure (Figure 9).

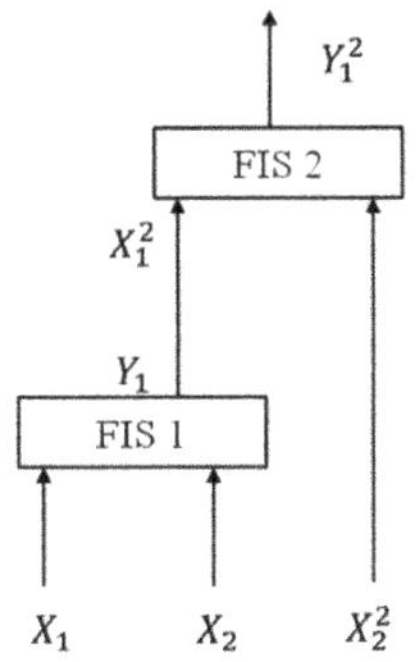

Figure 9. Fuzzy hierarchical concept.

In this tree structure, the outputs of lower-level fuzzy systems are used as inputs for higher-level fuzzy systems. In this way, hierarchical fuzzy systems manage to significantly reduce the number of fuzzy rules in multidimensional problems but still maintain the interpretability of the system [60,61].

4. Discussion

Advances in intelligent solutions in shipping and especially collision avoidance are obstructed by the fuzzy state of the rules for preventing collisions at sea. As stated in Rule 7, the risk of collision at sea is significantly affected by the other ship's size. The problem is that the expression *Very Large Ship* used in the mentioned rule is a linguistic variable, fuzzy by nature. Previous researchers have mostly used two important variables, CPA and TCPA, but neglected the influence of *Very Large Ship* size on the perception of risk. While this research's main objective was to model the linguistic term *Very Large Ship*, all other linguistic values of the linguistic variable *Ship's Size* were also modeled. This was achieved through the application of fuzzy theory based on the responses of 220 navigational experts. As the COLREGs are written by humans for humans, this research focused on modeling human expert knowledge to describe the linguistic variables.

The questionnaire targeted experienced seafarers only, considering that experience comes with time spent on board. The results show that most respondents were Masters, followed by Chief Officers, making up about 80% of the total respondents. Considering total experience at sea, respondents with more than ten years of experience at sea constituted more than 70%. This would indicate that the respondents' expertise is appropriate for creating a model based on expert knowledge. The respondents' profiles regarding the type and size of the ship reveal that groups and ship size do not reflect the actual distribution of the world fleet, with most respondents sailing on LNG carriers and on ships from 250 to 300 m. The mentioned characteristic of the sample could induce some skewness in the obtained results, so in future research, the authors will try to study different samples and seek differences.

The results of the second part of the questionnaire revealed significant facts. Firstly, the sigmoidal linguistic curve was the best fit for extreme values of the linguistic variables (*Very Low* and *Very High*). This was logically expected and confirmed as, by incrementing the ship's length, the membership value in the fuzzy set *Very High* can only increase or remain the same. It goes vice versa for the value *Very Low*. Furthermore, the Gaussian curve fitted the other three values as the best option. Also, as previous studies were mainly concentrated on creating a fuzzy system, the definition of a fuzzy membership function was not detailed enough. Expert knowledge cannot be modeled with the most straightforward membership function like a triangular function, as the result of this research confirmed. Fuzziness in the experts' knowledge was modeled with Gaussian and sigmoidal logistic functions, where the agreement of experts was expressed with a full membership value. At the same time, disagreement was modeled by partial membership in a fuzzy set. The membership value for various lengths of ships in the fuzzy sets was presented. For example, it was shown that a ship with a length of 260 m is a member of the *Very Large Ship* set, with a partial membership of just 0.23, while simultaneously, it would also be a member of the *Large Ship* fuzzy set at 0.89. This example presents the benefits of fuzzy sets compared with classical sets as a mathematically based representation of vagueness and uncertainty. Additionally, fuzzy sets and related fuzzy systems are easily understandable to humans, making them ideal solutions for resolving ambiguities in maritime terminology. Fuzzy systems are examples of *White Box* models, which are easily interpretable and, as such, suitable for applications that require transparency in prediction [62]. Since the COLREGs rules are written vaguely, relying on the navigator's knowledge, skills, and experience, modeling rules has to be achieved by using expert opinions. Fuzzy theory was developed to model such vague concepts, as it relies on human language, which has been successfully used for centuries.

Finally, a conceptual model of the implementation of a new linguistic variable was presented. Due to its specific nature, hierarchical fuzzy systems can easily adopt this new variable into their systems and allow for refined the modeling of collision risk in cases in which another vessel is very large. Other ship variables, including LOA, are easily available through the Automatic Identification System (AIS), a system used to exchange navigation information between ships [63].

A definition of linguistic variables in the full universe of discourse, from *Very Small Ship* to *Very Large Ship*, is required as an important step in developing a new fuzzy decision support system. Additionally, there is a very fine division in the five linguistic values: *Very Small Ship, Small Ship, Medium Ship, Large Ship*, and *Very Large Ship* help in achieving a more detailed and precise fuzzy system. Classical fuzzy systems and Fuzzy AHP systems depend on input variables, so the detailed definition of linguistic variables used in the COLREGs is essential in developing successful intelligent decision support systems for collision risk assessment. Decision support systems evolve gradually as additions to existing integrated navigational systems on ships and aim to improve the safety of navigation. Furthermore, the integration of decision support intelligent systems within integrated navigational systems is advancing toward autonomous navigation. The assessment of the risk of collision on humanless ships represents a specific problem, especially considering that the COLREGs will continue to have a primary role in collision avoidance. A fuzzy decision support system for collision risk assessment complying with the COLREGs requires the definition of other fuzzy variables influencing the risk of collision. As such, it could be a decision support system solution on ships with human crews and humanless ships. Future research will concentrate on the definition of other COLREGs linguistic variables by using fuzzy theory and expert opinion.

5. Conclusions

Fuzziness in the maritime world, particularly in rules, is often seen as a problematic issue, especially in intelligent solution advances. However, the theory of fuzzy sets is a great tool for modeling fuzzy or unprecise terms used in the human world. While the fuzzy theory approach has been used in significant fields of research and maritime study, modeling the membership functions of linguistic variables has not been the focus of previous studies. The shape and range of linguistic variables greatly influence the performance of fuzzy systems, and to express human reasoning, they have to be modeled on a navigational expert's opinion. In this research, the data used to define membership functions were collected using double-sided sliders. This approach to data collection is proposed when the sample size and the range of linguistic variables are large, followed by using an adapted horizontal method. This research aimed to elicit expert knowledge, which was achieved through the participation of 220 international navigational experts with significant experience in navigation. In future research, the authors plan to include ANOVA testing to study if there are statistically important differences between groups of respondents based on personal profiles or sailing areas. Fuzziness in responses between different groups is expected but normally handled by the inherent ability of fuzzy numbers. The responses were analyzed and statistically modeled to determine the appropriate shape of the linguistic variables. Gaussian and sigmoidal logistic curves proved to be the best fit for the linguistic values, confirming that simple basic curves like triangular curves are not the best option when defining expert opinions. Most importantly, the linguistic value *Very High* was mathematically defined, describing the unprecise term *Very Large Ship* used in the COLREGs, which has been previously neglected in creating collision avoidance systems. Additionally, all linguistic values within the linguistic variables were defined, resolving an interpretation of the linguistic variable *Ship's Size*. Using the same method, other linguistic variables abundantly present in the COLREGs could be modeled. Further research into quantifying methods for COLREGs terms is essential for developing intelligent solutions in maritime navigation. The obtained curves could be further used in various fuzzy systems, from control systems to decision support systems on conventional ships with human crews on board or autonomous ships. Finally, an implementation of quantified value within hierarchical fuzzy systems was presented as a simple and efficient method of including new variables in collision avoidance fuzzy systems

Author Contributions: Conceptualization, M.K. and S.Ž.; methodology, M.K. and S.Ž.; validation, M.K. and S.Ž.; formal analysis, M.K. and S.Ž.; writing—original draft preparation, M.K.; writing—review and editing, S.Ž.; visualization, M.K. and S.Ž.; supervision, S.Ž.; project administration, S.Ž. All authors have read and agreed to the published version of the manuscript.

Funding: This research was partially funded by European Union's Horizon Europe under the call HORIZON-CL5-2022-D6-01 (Safe, Resilient Transport and Smart Mobility services for passengers and goods), grant number 101077026, project name SafeNav. The views and opinions expressed are, however, those of the author(s) only and do not necessarily reflect those of the European Union or Executive Agency (CINEA). Neither the European Union nor the granting authority can be held responsible for them. Also, this study was partially funded by the University of Rijeka under the Faculty of Maritime Studies project ECDIS CoDe (UNIRI-ZIP-2103-10-22).

Institutional Review Board Statement: Not applicable.

Informed Consent Statement: Not applicable.

Data Availability Statement: Data are contained within the article.

Conflicts of Interest: The authors declare no conflicts of interest.

References

1. Martelli, M.; Žuškin, S.; Zaccone, R.; Rudan, I. A COLREGs-compliant decision support tool to prevent collisions at sea. *TransNav Int. J. Mar. Navig. Saf. Sea Transp.* **2023**, *17*, 347–353. [CrossRef]
2. Demirel, E.; Bayer, D. Further studies on the COLREGs (collision regulations). *TransNav Int. J. Mar. Navig. Saf. Sea Transp.* **2015**, *9*, 17–23. [CrossRef]
3. Mohović, D.; Mohović, R.; Barić, M. Deficiencies in learning COLREGs and new teaching methodology for nautical engineering students and seafarers in lifelong learning programs. *J. Navig.* **2016**, *69*, 765–776. [CrossRef]
4. Ahmed, Y.A.; Hannan, M.A.; Oraby, M.Y.; Maimun, A. COLREGs Compliant Fuzzy-Based Collision Avoidance System for Multiple Ship Encounters. *J. Mar. Sci. Eng.* **2021**, *9*, 790. [CrossRef]
5. International Maritime Organisation (IMO). Available online: https://www.imo.org/en/About/Conventions/Pages/COLREG.aspx (accessed on 6 March 2024).
6. Benjamin, M.R.; Curcio, J.A.; Leonard, J.J.; Newman, P.M. Navigation of unmanned marine vehicles in accordance with the rules of the road. In Proceedings of the 2006 IEEE International Conference on Robotics and Automation, Orlando, FL, USA, 15–19 May 2006.
7. Namgung, H.; Kim, J.S. Collision risk inference system for maritime autonomous surface ships using COLREGs rules compliant collision avoidance. *IEEE Access* **2021**, *9*, 7823–7835. [CrossRef]
8. Wróbel, K.; Gil, M.; Huang, Y.; Wawruch, R. The Vagueness of COLREG versus Collision Avoidance Techniques—A Discussion on the Current State and Future Challenges Concerning the Operation of Autonomous Ships. *Sustainability* **2022**, *14*, 16516. [CrossRef]
9. Zaccone, R.; Martelli, M. Interaction between COLREG-compliant collision avoidance systems in a multiple MASS scenario. *J. Phys. Conf. Ser.* **2023**, *2618/1*, 012006. [CrossRef]

10. Donnarumma, S.; Figari, M.; Martelli, M.; Zaccone, R. Simulation of the Guidance and Control Systems for Underactuated Vessels. In Proceedings of the 6th International Conference, MESAS 2019, Palermo, Italy, 29–31 October 2019. [CrossRef]
11. Martelli, M.; Faggioni, N.; Zaccone, R. Development of a navigation support system by means of a synthetic scenario. In Proceedings of the 18th International Congress of the Maritme Association of the Mediterranean (IMAM 2019), Varna, Bulgaria, 9–11 September 2019. [CrossRef]
12. Zhang, Y.; Zhang, D.; Jiang, H. A Review of Artificial Intelligence-Based Optimization Applications in Traditional Active Maritime Collision Avoidance. *Sustainability* **2023**, *15*, 13384. [CrossRef]
13. Fruzzetti, C.; Donnarumma, S.; Martelli, M. Dynamic target chasing: Parameters and performance indicators assessment. *J. Mar. Sci. Technol.* **2022**, *27*, 712–729. [CrossRef]
14. Wang, C.; Zhang, X.; Yang, Z.; Bashir, M.; Lee, K. Collision avoidance for autonomous ship using deep reinforcement learning and prior-knowledge-based approximate representation. *Front. Mar. Sci.* **2023**, *9*, 1084763. [CrossRef]
15. Zhang, X.; Wang, C.; Jiang, L.; An, L.; Yang, R. Collision-avoidance navigation systems for Maritime Autonomous Surface Ships: A state of the art survey. *Ocean Eng.* **2021**, *235*, 109380. [CrossRef]
16. International Maritime Organisation (IMO). *Outcome of the Regulatory Scoping Exercise for the Use of Maritime Autonomous Surface Ships (MASS)*; IMO: London, UK, 2021.
17. Rødseth, Ø.J.; Wennersberg, L.A.L.; Nordahl, H. Improving Safety of Interactions between Conventional and Autonomous Ships. *Ocean Eng.* **2023**, *284*, 115206. [CrossRef]
18. Chang, C.H.; Wijeratne, I.B.; Kontovas, C.; Yang, Z. COLREG and MASS: Analytical review to identify research trends and gaps in the Development of Autonomous Collision Avoidance. *Ocean Eng.* **2024**, *302*, 117652. [CrossRef]
19. Miyoshi, T.; Fujimoto, S.; Rooks, M.; Konishi, T.; Suzuki, R. Rules required for operating maritime autonomous surface ships from the viewpoint of seafarers. *J. Navig.* **2022**, *75*, 384–399. [CrossRef]
20. Xie, W.; Gang, L.; Zhang, M.; Liu, T.; Lan, Z. Optimizing Multi-Vessel Collision Avoidance Decision Making for Autonomous Surface Vessels: A COLREGs-Compliant Deep Reinforcement Learning Approach. *J. Mar. Sci. Eng.* **2024**, *12*, 372. [CrossRef]
21. He, Y.; Jin, Y.; Huang, L.; Xiong, Y.; Chen, P.; Mou, J. Quantitative analysis of COLREG rules and seamanship for autonomous collision avoidance at open sea. *Ocean Eng.* **2017**, *140*, 281–291. [CrossRef]
22. Kim, J.-K.; Park, D.-J. Determining the Proper Times and Sufficient Actions for the Collision Avoidance of Navigator-Centered Ships in the Open Sea Using Artificial Neural Networks. *J. Mar. Sci. Eng.* **2023**, *11*, 1384. [CrossRef]
23. Kim, J.K.; Park, D.J. Understanding of sailing rule based on COLREGs: Comparison of navigator survey and automated collision-avoidance algorithm. *Mar. Policy* **2024**, *159*, 105894. [CrossRef]
24. Zhou, X.-Y.; Huang, J.-J.; Wang, F.-W.; Wu, Z.-L.; Liu, Z.-J. A Study of the Application Barriers to the Use of Autonomous Ships Posed by the Good Seamanship Requirement of COLREGs. *J. Navig.* **2020**, *73*, 710–725. [CrossRef]
25. Kim, I. Systematization of Legal Procedures for Collision Avoidance between a Fully Autonomous Ship and a Traditional Manned Ship. *J. Mar. Sci. Eng.* **2023**, *11*, 1850. [CrossRef]
26. Li, Q. A Research on Autonomous Collision Avoidance under the Constraint of COLREGs. *Sustainability* **2023**, *15*, 2446. [CrossRef]
27. Hagen, I.B.; Vassbotn, O.; Skogvold, M.; Johansen, T.A.; Brekke, E. Safety and COLREGS Evaluation for Marine Collision Avoidance Algorithms. *Ocean Eng.* **2021**, *288*, 115991. [CrossRef]
28. Kao, S.L.; Lee, K.T.; Chang, K.Y.; Ko, M.D. A fuzzy logic method for collision avoidance in vessel traffic service. *J. Navig.* **2007**, *60*, 17–31. [CrossRef]
29. Chen, S.; Ahmad, R.; Lee, B.G.; Kim, D. Composition ship collision risk based on fuzzy theory. *J. Cent. S. Univ.* **2014**, *21*, 4296–4302. [CrossRef]
30. Olindersson, F.; Bruhn, W.; Scheidweiler, T.; Andersson, A. Developing a maritime safety index using fuzzy logics. *TransNav Int. J. Mar. Navig. Saf. Sea Transp.* **2017**, *11*, 469–475. [CrossRef]
31. Goerlandt, F.; Montewka, J.; Kuzmin, V.; Kujala, P. A risk-informed ship collision alert system: Framework and application. *Saf. Sci.* **2015**, *77*, 182–204. [CrossRef]
32. Shi, Z.; Zhen, R.; Liu, J. Fuzzy logic-based modeling method for regional multi-ship collision risk assessment considering impacts of ship crossing angle and navigational environment. *Ocean Eng.* **2022**, *259*, 111847. [CrossRef]
33. Cheng, Z.; Zhang, Y.; Wu, B.; Soares, C.G. Traffic-conflict and fuzzy-logic-based collision risk assessment for constrained crossing scenarios of a ship. *Ocean Eng.* **2023**, *274*, 114004. [CrossRef]
34. Brcko, T.; Androjna, A.; Srše, J.; Boć, R. Vessel Multi-Parametric Collision Avoidance Decision Model: Fuzzy Approach. *J. Mar. Sci. Eng.* **2021**, *9*, 49. [CrossRef]
35. Brcko, T.; Luin, B. A Decision Support System Using Fuzzy Logic for Collision Avoidance in Multi-Vessel Situations at Sea. *J. Mar. Sci. Eng.* **2023**, *11*, 1819. [CrossRef]
36. Lee, H.-J.; Park, D.-J. SASD Modeling Using an ANFIS to Prevent the Collision of MASS in Restricted Areas. *J. Mar. Sci. Eng.* **2022**, *10*, 961. [CrossRef]
37. Nguyen, V.S.; Im, N.K. Automatic ship berthing based on fuzzy logic. *Int. J. Fuzzy Log. Intell. Syst.* **2019**, *19*, 163–171. [CrossRef]
38. Zaib, A.; Yin, J.; Khan, R.U. Determining Role of Human Factors in Maritime Transportation Accidents by Fuzzy Fault Tree Analysis (FFTA). *J. Mar. Sci. Eng.* **2022**, *10*, 381. [CrossRef]
39. Chen, M.S.; Wang, S.W. Fuzzy clustering analysis for optimizing fuzzy membership functions. *Fuzzy Sets syst.* **1999**, *103*, 239–254. [CrossRef]

40. Hasuike, T.; Katagiri, H.; Tsubaki, H.; Tsuda, H. Constructing membership function based on fuzzy shannon entropy and human's interval estimation. In Proceedings of the 2012 IEEE International Conference on Fuzzy Systems, Brisbane, Australia, 10–15 June 2012.
41. Hu, L.; Hu, H.; Naeem, W.; Wang, Z. A review on COLREGs-compliant navigation of autonomous surface vehicles: From traditional to learning-based approaches. *J. Autom. Intell.* **2022**, *1*, 100003. [CrossRef]
42. Ren, Y.; Mou, J.; Yan, Q.; Zhang, F. Study on assessing dynamic risk of ship collision. In Proceedings of the First International Conference on Transportation Information and Safety, Wuhan, China, 30 June–2 July 2011.
43. Luo, J.; Wang, D.; Gao, Y. An improved AHP based fuzzy evaluation model for ship collision risk. *J. Intell. Fuzzy Syst.* **2021**, *41*, 4905–4913. [CrossRef]
44. Xiao, G.; Ren, B.; Tong, C.; Hong, X. A Quantitative Evaluation Method for Obstacle Avoidance Performance of Unmanned Ship. *J. Mar. Sci. Eng.* **2021**, *9*, 1127. [CrossRef]
45. Tonoğlu, F.; Atalar, F.; Başkan, İ.B.; Yildiz, S.; Uğurlu, Ö.; Wang, J. A new hybrid approach for determining sector-specific risk factors in Turkish Straits: Fuzzy AHP-PRAT technique. *Ocean Eng.* **2022**, *253*, 111280. [CrossRef]
46. Sahin, B.; Yip, T.L. Shipping technology selection for dynamic capability based on improved Gaussian fuzzy AHP model. *Ocean Eng.* **2017**, *136*, 233–242. [CrossRef]
47. Türk, A.; Özkök, M. A Comprehensive Risk Assessment Analysis of Accidental Falls in Shipyards Using the Gaussian Fuzzy AHP Model. *J. ETA Marit. Sci.* **2022**, *10*, 211–222. [CrossRef]
48. Baalousha, H.M.; Younes, A.; Yassin, M.A.; Fahs, M. Comparison of the Fuzzy Analytic Hierarchy Process (F-AHP) and Fuzzy Logic for Flood Exposure Risk Assessment in Arid Regions. *Hydrology* **2023**, *10*, 136. [CrossRef]
49. Pourghasemi, H.R.; Pradhan, B.; Gokceoglu, C. Application of fuzzy logic and analytical hierarchy process (AHP) to landslide susceptibility mapping at Haraz watershed, Iran. *Nat. Hazards* **2012**, *63*, 965–996. [CrossRef]
50. LimeSurvey. Available online: https://www.limesurvey.org/ (accessed on 15 December 2023).
51. Krueger, T.; Page, T.; Hubacek, K.; Smith, L.; Hiscock, K. The role of expert opinion in environmental modelling. *Environ. Model. Softw.* **2012**, *36*, 4–18. [CrossRef]
52. Ekel, P.Y.; Pedrycz, W.; Pereira, J.G., Jr. *Multicriteria Decision-Making under Conditions of Uncertainty: A Fuzzy Set Perspective*; John Wiley & Sons: Hoboken, NJ, USA, 2020.
53. GitHub. Available online: https://github.com/tpartner/LimeSurvey-Range-Slider-5x-6x (accessed on 26 March 2024).
54. MATLAB. Available online: https://matlab.mathworks.com/ (accessed on 1 March 2024).
55. Smiti, A. A critical overview of outlier detection methods. *Comput. Sci. Rev.* **2020**, *38*, 100306. [CrossRef]
56. Kwak, S.K.; Kim, J.H. Statistical data preparation: Management of missing values and outliers. *Korean J. Anesthesiol.* **2017**, *70*, 407. [CrossRef] [PubMed]
57. Mazziotta, M.; Pareto, A. Normalization methods for spatio-temporal analysis of environmental performance: Revisiting the Min–Max method. *Environmetrics* **2022**, *33*, e2730. [CrossRef]
58. The MathWorks, Inc. Available online: https://www.mathworks.com/help/pdf_doc/curvefit/index.html (accessed on 7 March 2024).
59. Kanovic, Z.; Vukelic, D.; Simunovic, K.; Prica, M.; Saric, T.; Tadic, B.; Simunovic, G. The Modelling of Surface Roughness after the Ball Burnishing Process with a High-Stiffness Tool by Using Regression Analysis, Artificial Neural Networks, and Support Vector Regression. *Metals* **2022**, *12*, 320. [CrossRef]
60. Kerr-Wilson, J.; Pedrycz, W. Generating a hierarchical fuzzy rule-based model. *Fuzzy Sets Syst.* **2020**, *381*, 124–139. [CrossRef]
61. Zhao, T.; Zhu, Y.; Xie, X. Topology structure optimization of evolutionary hierarchical fuzzy systems. *Expert Syst. Appl.* **2024**, *238*, 121857. [CrossRef]
62. Pintelas, E.; Livieris, I.E.; Pintelas, P. A Grey-Box Ensemble Model Exploiting Black-Box Accuracy and White-Box Intrinsic Interpretability. *Algorithms* **2020**, *13*, 17. [CrossRef]
63. Kristić, M.; Žuškin, S.; Brčić, D.; Car, M. Overreliance on ECDIS technology: A challenge for safe navigation. *TransNav Int. J. Mar. Navig. Saf. Sea Transp.* **2021**, *15*, 277–288. [CrossRef]

Journal of
Marine Science and Engineering

MDPI

Article

A Pruning and Distillation Based Compression Method for Sonar Image Detection Models

Chensheng Cheng, Xujia Hou, Can Wang, Xin Wen, Weidong Liu and Feihu Zhang *

School of Marine Science and Technology, Northwestern Polytechnical University, Xi'an 710072, China; chensheng.cheng@mail.nwpu.edu.cn (C.C.); hxj1363947894@mail.nwpu.edu.cn (X.H.); can.wang@mail.nwpu.edu.cn (C.W.); wenxin666@mail.nwpu.edu.cn (X.W.); liuwd@nwpu.edu.cn (W.L.)
* Correspondence: feihu.zhang@nwpu.edu.cn

Abstract: Accurate underwater target detection is crucial for the operation of autonomous underwater vehicles (AUVs), enhancing their environmental awareness and target search and rescue capabilities. Current deep learning-based detection models are typically large, requiring substantial storage and computational resources. However, the limited space on AUVs poses significant challenges for deploying these models on the embedded processors. Therefore, research on model compression is of great practical importance, aiming to reduce model parameters and computational load without significantly sacrificing accuracy. To address the challenge of deploying large detection models, this paper introduces an automated pruning method based on dependency graphs and successfully implements efficient pruning on the YOLOv7 model. To mitigate the accuracy degradation caused by extensive pruning, we design a hybrid distillation method that combines output-based and feature-based distillation techniques, thereby improving the detection accuracy of the pruned model. Finally, we deploy the compressed model on an embedded processor within an AUV to evaluate its performance. Multiple experiments confirm the effectiveness of our proposed method in practical applications.

Keywords: pruning; knowledge distillation; sidescan sonar; object detection; YOLOv7; embedded device

1. Introduction

In the field of object detection, deep learning has achieved tremendous success due to its powerful feature extraction capabilities, effectively driving the development of this domain. Currently, convolutional neural network based object detection models have been improving in performance as their network depth increases, as seen in RCNN [1,2], SSD [3,4], and YOLO [5,6]. However, as the network depth of these models continues to grow, the storage space required for the models and the computational resources needed for inference also increase, posing challenges for specific tasks that cannot utilize high-performance processors. Consequently, many researchers have begun to investigate how to ensure that models perform well on edge devices with limited computational resources [7,8].

Accurate target detection capabilities are very important for autonomous underwater vehicles (AUVs) performing underwater operations, but model deployment is constrained by computing resources [9,10]. The core devices responsible for data storage, data processing, system decision making, and power management are housed within a watertight electronics compartment. Due to the carrying capacity and overall space constraints of the AUV, the volume and weight of the watertight electronics compartment are limited, which prevents the use of large, high-performance processors. Therefore, researching network compression methods to strike a balance between detection accuracy and model size is crucial. This research aims to enable the successful deployment of powerful object detection networks on underwater embedded devices with limited computing and memory capacities, representing a highly significant practical problem worth investigating.

In the realm of model compression for deep neural networks, six primary research directions can be delineated: network pruning [11], lightweight network design [12], neural architecture search (NAS) [13], network quantization [14], knowledge distillation (KD) [15], and low-rank decomposition [16]. Network pruning involves deleting redundant parameters deemed unimportant through various evaluation criteria, leading to sparsity and consequent model compression. This method significantly reduces model size, lowers computational complexity, and enhances operational efficiency. In recent years, network pruning has gained popularity. Lightweight network design focuses on directly designing more compact convolutional kernels or methods to reduce the number of network structure parameters, thereby reducing model size and computational complexity. However, this may necessitate expertise to design effective lightweight network structures. NAS automates the search for optimal network architectures, improving model performance and efficiency. However, this method consumes significant computational resources, has a vast search space, and requires extensive training time and data, making it less accessible to general researchers. Network quantization converts high-precision operations into low-precision operations (floating-point to integer operations), reducing model storage space and computational load, and accelerating inference speed. However, quantization may sacrifice some model accuracy, requiring careful selection of quantization methods and parameters. KD transfers knowledge learned by a complex, parameter-heavy teacher network to a relatively simpler student network, achieving similar expressive capabilities while saving time and space costs. Due to its ability to significantly enhance small model performance, KD has become one of the mainstream methods in model compression. Low-rank decomposition aims to decompose convolutional kernels using matrix analysis methods like low-rank decomposition, approximating the original weight matrix by decomposing large convolutional kernel tensors into smaller ones, thus reducing the computational and parameter requirements of the model to some extent. However, implementing low-rank decomposition methods is not straightforward and involves higher computational costs.

In the field of underwater acoustic target detection, researchers have faced challenges due to scarce datasets and experimental difficulties, resulting in relatively limited studies on network compression. Tian et al. [17] optimized the YOLOv4 model using channel and layer pruning methods and successfully deployed it on underwater vehicles for target detection. It is worth noting that the pruned model retained only 7 % (from 245 M to 17 M) of the original parameters, ensuring real-time requirements for edge devices. Szymak et al. [18] adopted a more effective pre-training method to deploy the deep learning model on the Nvidia Jetson TX2 and realized target classification based on underwater videos. Ye et al. [19] introduced the Mobilenetv3 architecture into the backbone network of YOLOv4 to lightweight the network, reducing the model's parameter size to a quarter of the original, significantly alleviating storage pressure on edge devices. Yan et al. [20] improved the MobileNet-SSD benchmark network for underwater sonar target detection, greatly enhancing the network's inference speed. Hu et al. [21] integrated ShuffleNetv2 into YOLOv5s to make the model more compact, improving computational speed on embedded devices. Qin et al. [22] pruned the improved YOLOv7 using channel pruning methods, reducing model memory by 47.50% and increasing detection speed by nearly 2.5 times for forward-looking sonar target detection.

After considering usability, cost, and practical engineering demands, this study ultimately employed a combined approach of pruning and distillation for model compression. We focused on the YOLOv7 network model and successfully achieved substantial compression of the object detection network while maintaining detection accuracy. Furthermore, we validated the real-time performance and robustness of this lightweight network using both publicly available datasets and data we collected ourselves.

The main contributions of this paper are as follows:

(1) Addressing the challenge of deploying overly bulky detection models on edge devices, we introduce a dependency graph-based pruning method and efficiently implement pruning on the YOLOv7 model.

(2) To counter the significant drop in model accuracy after extensive pruning, we devise a fusion distillation method that combines output-based distillation with feature-based distillation, thereby enhancing the detection accuracy of pruned models.
(3) Deploying the compressed models on edge devices for testing, we validate the real-time detection performance of the methods utilized in this paper on embedded devices.

2. Related Work

2.1. Network Pruning

Currently, mainstream pruning methods can be divided into two categories: non-structural pruning and structural pruning. Non-structural pruning does not alter the network's structure but rather zeros out some of the weights. However, the distribution of these zeros is generally irregular, requiring specialized hardware and software acceleration for proper functioning. In contrast, structural pruning modifies the neural network's structure by physically removing grouped parameters based on predefined rules. This approach does not rely on specific AI accelerators or software, making it more widely applicable in engineering projects than non-structural pruning.

In the realm of non-structural pruning, the origins can be traced back to 1989 when LeCun et al. [23] balanced network complexity and training set errors using second-order derivative information to remove unimportant weights from the network. Their aim was to achieve a network with better generalization and improved inference speed. Subsequently, Han et al. [24] proposed a non-structured pruning method based on threshold weight values, analyzing weight parameters to prune the network structure. However, setting pruning thresholds requires a strong mathematical background and manual intervention, severely limiting its engineering application. In response to this challenge, Li et al. [25] introduced an optimization-based method for automatically adjusting thresholds, transforming the threshold adjustment problem into a constrained optimization problem and solving it using derivative-free optimization algorithms. Lee et al. [26] highlighted the importance of global pruning by studying hierarchical sparsity, introducing an adaptive magnitude pruning score that does not rely on manual hyperparameter tuning. Moreover, determining the importance of weights scientifically is a major research focus. Carreira et al. [27], for instance, framed pruning as an optimization problem, minimizing the loss of weights while satisfying basic pruning conditions and automatically learning the optimal number of weights to prune in each layer. Kwon et al. [28] proposed a novel representation scheme for sparsely quantized neural network weights, achieving high compression ratios through fine-grained and non-structural pruning while maintaining model accuracy.

In the realm of structural pruning, methods can be broadly categorized into neuron pruning, filter pruning, channel pruning, and layer pruning based on the granularity of pruning. Molchanov et al. [29] approximated neurons' contribution to the loss function using first- and second-order Taylor expansions and iteratively removed neurons with low scores. Wang et al. [30] proposed a layer-adaptive filter pruning method based on reducing structural redundancy, constructing a graph for each convolutional layer in CNNs to measure their redundancy. This method prunes unimportant filters in the most redundant layers rather than in all layers. To address the additional hyperparameters and long training cycles associated with filter pruning methods, Ruan et al. [31] introduced a novel dynamic and progressive filter pruning scheme (DPFPS), solving the optimization problem based on pruning ratios with an iterative soft-thresholding algorithm that exhibits dynamic sparsity. At the end of training, only redundant parameters need to be removed without additional stages. Ding et al. [32] proposed a lossless channel pruning method called ResRep, reducing CNN size by narrowing convolutional layer widths. However, premature pruning of some channels can severely impact model accuracy. To mitigate this issue, Hou et al. [33] suggested iteratively pruning and regrowing channels throughout the entire training process, reducing the risk of prematurely pruning important channels. Yang et al. [34] discovered that merely reducing CNN model size or computational complexity does not necessarily reduce energy consumption. To bridge the gap between CNN design and energy optimization, they proposed a CNN energy-aware pruning algorithm that uses CNN energy consumption directly to guide layer pruning processes.

2.2. Knowledge Distillation

Currently, knowledge distillation can be classified into three types based on the type of knowledge: output-based distillation, feature-based distillation, and relation-based distillation. Researchers need to choose the distillation method based on the specific task and may also consider combining different distillation methods for optimal results.

Buciluǎ et al. [35] were the first to propose a method for training fast and compact small models to mimic the performance of larger, slower, and more complex models. They achieved this by training the small model using data obtained from inference of the large model. Their method resulted in minimal performance loss for the small model, demonstrating that knowledge acquired from a large ensemble of models can be transferred to simpler, smaller models. Building upon this, Hinton et al. [36] introduced a more general solution termed "knowledge distillation". They defined soft targets (class probability distribution generated by the teacher network's Softmax layer) and hard targets (one-hot encoding of true labels) and introduced a "temperature" variable to control the smoothness of the class probability distribution. The student model was then trained to learn both soft and hard targets simultaneously, with the weighted average of the two target loss functions serving as the overall loss function, thereby improving the performance of the student model. Subsequently, Zhang et al. [37] proposed a deep mutual learning method that allows the student model to learn knowledge from multiple teacher models. To enhance the student model's understanding of the teacher models, Kim et al. [38] introduced a method of encoding and decoding the teacher network's outputs, leading to improved results for the student learning the decoded knowledge. In order to fully utilize the outputs of the teacher network, Zhao et al. [39] proposed a decoupled knowledge distillation (DKD) method, dividing the distillation loss into target class loss and non-target class loss, significantly enhancing the distillation performance.

The mentioned methods are all based on output distillation, yet some researchers argue that response-based distillation has certain limitations and propose feature-based distillation methods. Adriana et al. [40] argued that for deep neural networks, it is challenging to propagate supervisory signals all the way back to the front end, and constraining only at the label level is insufficient. Therefore, they propose the Fitnet method, introducing supervisory signals in the middle hidden layers of the network, using hints training to constrain the outputs of the intermediate layers of the two models to be as close as possible, allowing the student model to learn the intermediate layer features of the teacher model. Zagoruyko et al. [41] proposed attention transfer methods, enabling the student to learn the attention distribution of the teacher to enhance model performance. Huang et al. [42] introduced the neuron selectivity transfer (NST) method, enabling the student model to better mimic the neuron activation patterns of the teacher model, thereby improving knowledge distillation effectiveness. Ahn et al. [43] presented an information-theoretic framework for knowledge transfer, formulating knowledge transfer as maximizing mutual information between teacher and student networks, which has significant advantages in cross-heterogeneous network architecture knowledge transfer. To simplify the student model's learning process, Yang et al. [44] used an adaptive instance normalization method to perform feature statistics on each feature channel's data, making it easier for the student to capture the teacher model's higher-level feature representations.

In addition to output-based and feature-based distillation methods, some researchers believe that it is crucial for the student model to learn the relationships between network feature layers, known as relation-based distillation. Yim et al. [45] considered that a true teacher teaches the process of problem-solving, so they redefine knowledge distillation as learning the process of solving problems, specifically focusing on learning the relationships between network feature layers. They distill knowledge by fitting the relationships between teacher and student layers, allowing the student to learn the process of feature extraction from the teacher model. Subsequently, Peng et al. [46] proposed a distillation method based on correlation congruence, transferring knowledge by comparing the correlation of feature representations between the student and teacher models, aiming to align the correlation of the student model with that of the teacher model. Park et al. [47] introduced a relationship knowledge distillation (RKD) method, proposing distillation losses based on distance and angle constraints to transfer the mutual relationships between data instances. To further refine the knowledge of the teacher model, Liu et al. [48] introduced an instance relationship graph (IRG)-based method, modeling instance features, instance relationships, and feature space transformations, allowing the student's IRG to mimic the structure of the teacher's IRG. Zhou et al. [49] designed a student model based on graph neural networks (GNNs), leveraging GNNs to capture and transfer the global knowledge of the teacher model, effectively improving the performance of the student model.

In summary, unstructured pruning methods can achieve large-scale model pruning but require specialized hardware support for deployment. Current structured pruning methods are mostly designed for specific network architectures and necessitate manually crafted pruning rules, which are labor-intensive and less generalizable. In the realm of knowledge distillation, methods based on outputs and features are easier to implement, whereas those based on relationships are more complex and challenging to integrate with other distillation techniques. Additionally, due to the high cost and difficulty of underwater acoustic experiments, there remains a significant research gap in model compression for underwater acoustic target detection. Therefore, this paper employs an automated pruning scheme based on dependency graphs, combined with output- and feature-based distillation methods. This approach achieves substantial model compression while maintaining detection accuracy and has been validated for performance on embedded devices in underwater environments.

3. Methods

This study takes into account the characteristics of dense object detectors. Firstly, an automatic pruning method based on dependency graphs is employed to prune the YOLOv7 model. Subsequently, a combination of output-based binary cross-entropy classification distillation and feature-based mask generation distillation methods is applied to distill the pruned model, aiming to enhance its detection performance.

3.1. Automated Pruning Method Based on Dependency Graphs

Pruning is a highly challenging task for modern deep neural networks, given the complex interconnections between various modules in neural networks, as illustrated in Figure 1. When removing a channel, one must simultaneously consider its interdependencies with other layers; otherwise, we will end up with a damaged network. Additionally, manually analyzing dependency patterns for each module and designing pruning schemes individually is cumbersome and not conducive to widespread adoption. Therefore, this paper introduces a pruning method based on dependency graphs [50], which automatically constructs dependency relationships, groups all parameters according to these dependencies, uniformly evaluates the importance of grouped parameters, and ultimately achieves efficient pruning.

In addressing the complex dependencies among the modules of the detection network, we decompose the entire network model $\mathcal{M}(x;w)$ into more detailed components by layer. The network model can be represented as $\mathcal{M} = \{f_1, f_2, \ldots, f_L\}$, where f can be convolution layers, residual operations, activation functions, and batch normalization (BN) layers, among others. The inputs and outputs of the f_i layer are denoted as f_i^- and f_i^+, respectively, allowing for the use of different pruning strategies for the input and output parts of the same layer. Ultimately, the entire neural network can be expressed using Equation (1):

$$\mathcal{M} = \{f_1^-, f_1^+, f_2^-, f_2^+, \ldots, f_L^-, f_L^+\} \quad (1)$$

At this point, the dependencies within the entire neural network can be categorized into intra-layer dependencies and inter-layer dependencies, which can be represented using Equation (2):

$$(f_1^-, \underbrace{f_1^+) \leftrightarrow (f_2^-}_{Inter\text{-}layer\ Dep}, f_2^+) \cdots \leftrightarrow \underbrace{(f_L^-, f_L^+)}_{Intra\text{-}layer\ Dep} \tag{2}$$

Inter-layer dependencies always exist, while intra-layer dependencies exist only when the input and output share the same pruning scheme. For instance, the BN layer exhibits intra-layer dependencies, whereas the convolutional layer does not. According to these rules, we can build a dependency model, denoted as Equation (3):

$$D(f_i^-, f_j^+) = \underbrace{L\left[i - j = 1 \wedge f_i^- \leftrightarrow f_j^+\right]}_{Inter-layerDep} \vee \underbrace{L\left[i = j \wedge sch(f_i^-) = sch(f_j^+)\right]}_{Intra-layerDep} \tag{3}$$

In Equation (3), the symbols $\vee$ and $\wedge$ represent logical "or" and "and" operations, respectively. L is a decision function. When the inputs are adjacent layers, a dependency always exists, and L outputs "True". When the inputs are within the same layer and share the same pruning strategy, a dependency also exists, and L outputs "True". However, when the inputs are within the same layer but use different pruning strategies, no dependency exists, and L outputs "False". $sch(f_i^-)$ and $sch(f_j^+)$ respectively represent pruning schemes used for layers f_i^- and f_j^+. The first term checks for inter-layer dependencies caused by network connections, while the second term examines intra-layer dependencies introduced by shared pruning schemes between layer inputs and outputs. It is worth noting that $D(f_i^-, f_j^+)$ is a symmetric matrix, where $D(f_i^-, f_j^+) = D(f_j^+, f_i^-)$. Therefore, we can examine all input–output pairs to obtain the network's dependency graph.

After constructing the complete dependency graph, another challenge that arises is the group-level pruning problem. A common method for assessing the importance of group-level parameters is to compute an aggregate score. However, due to differences in data distribution and magnitude between different layers, the importance score of group parameters cannot be simply obtained by summing up the numerical values of each layer. Therefore, in the process of sparsification training, we introduce a simple regularization term for each parameter $w[k]$ with k prunable dimension, as shown in Equation (4):

$$\mathcal{R}(g, k) = \sum_{k=1}^{K} \gamma_k \cdot I_{g,k} = \sum_{k=1}^{K} \sum_{w \in g} \gamma_k \|w[k]\|_2^2 \tag{4}$$

Here, g represents a group of parameters with dependencies and can be expressed as $g = \{w_1, w_2, \ldots, w_g\}$, $I_{g,k} = \sum_{w \in g} \|w[k]\|_2^2$ represents the importance of the k prunable dimension, while γ_k represents the shrinkage strength applied to these parameters, which can be represented by Equation (5):

$$\gamma_k = 2^{\alpha(I_g^{\max} - I_{g,k})/(I_g^{\max} - I_g^{\min})} \tag{5}$$

The value of γ_k ranges between $\left[2^0, 2^\alpha\right]$. In the models and datasets used in this study, we found that the pruning performance was optimized when $\alpha = 2$. Therefore, in our experiments, the value of this parameter was set to 2. Based on the model construction and regularization methods mentioned above, we conducted sparsification training on the YOLOv7 model, removing unimportant parameters, and ultimately obtained a pruned lightweight network.

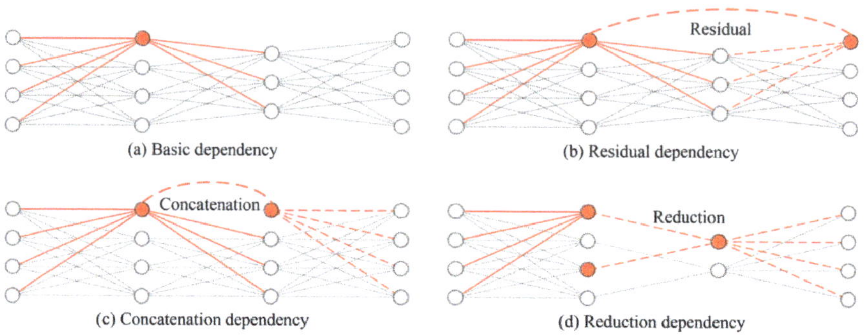

Figure 1. Complex dependencies in network models. In the figure, the solid red lines represent parameters directly connected to pruned neurons, while the dashed red lines indicate a special dependency on the pruned neurons, leading to the pruning of some indirect parameters.

3.2. Output Distillation Method Based on Binary Cross-Entropy and IoU

After pruning, the detection performance of the network tends to decrease, often requiring fine-tuning to restore its detection accuracy. However, the recoverable accuracy is usually limited. Therefore, this study employs a output distillation method [51] tailored for dense detectors like YOLOv7 to enhance the detection accuracy of the pruned model. The specific distillation method is illustrated in Figure 2.

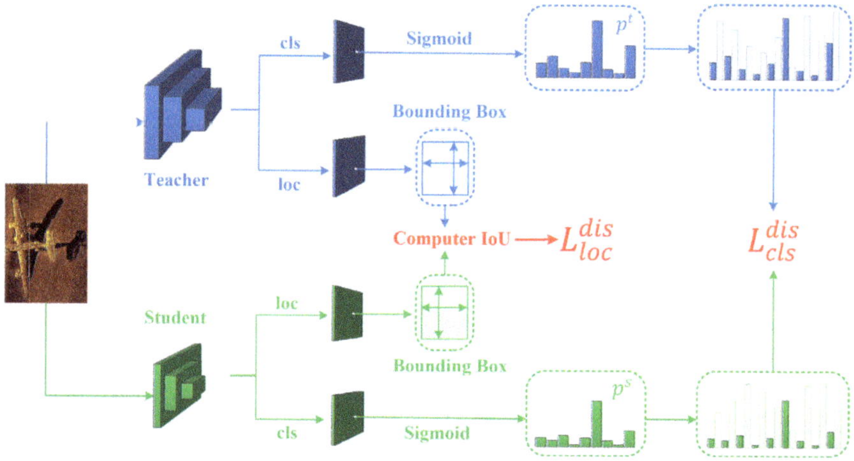

Figure 2. Schematic diagram of the method based on output distillation. For the task of object detection, separate loss functions are designed for classification and localization. In terms of classification distillation, the Sigmoid function is used instead of the Softmax function to process the classification logit map, and then the distillation loss is calculated based on binary cross-entropy. For localization distillation, the distillation loss function is constructed by calculating the IoU between the two models.

Dense detectors like YOLOv7 utilize the Sigmoid function for classification tasks in object detection, whereas the Softmax function is used in classification distillation tasks. This leads to a problem of inconsistent cross-task protocols in dense object detection. Specifically, when the classification distillation loss equals 0, there is an inconsistency between the scores of the student model and the teacher model in dense object detection, as illustrated in Figure 3.

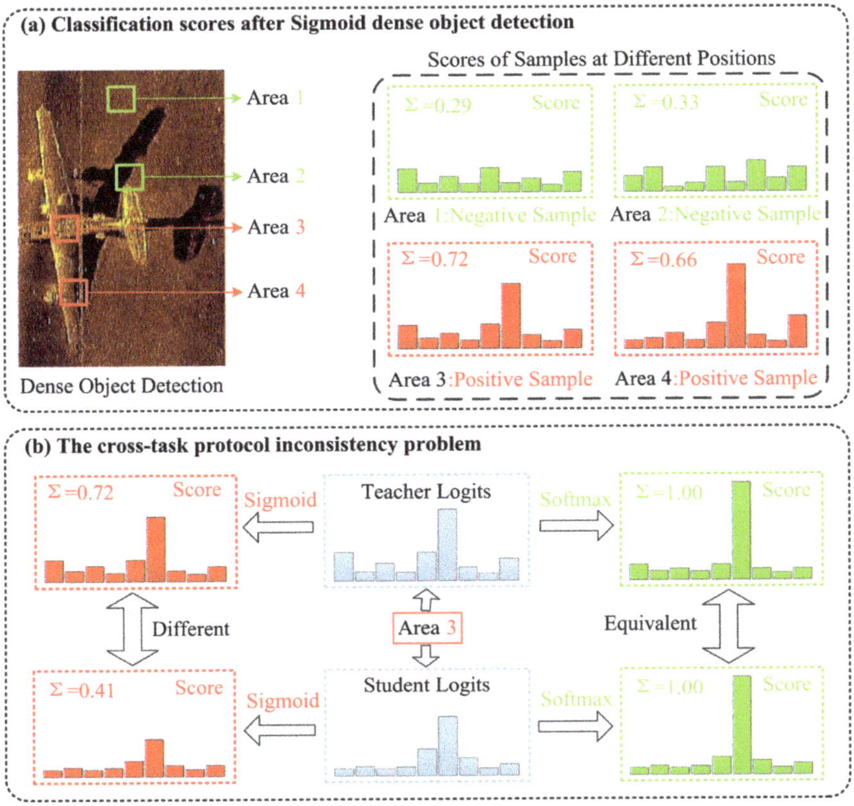

Figure 3. (**a**) In dense detectors like YOLOv7, the classification scores of different samples vary, which differs from image classification based on the Softmax function. (**b**) There is a cross-task protocol inconsistency issue. When the classification logit maps of a disparate teacher model and student model are processed using Softmax, they might produce identical scores, causing the student model to stop learning. However, using the Sigmoid function preserves the differences between them, allowing the student model to continue learning from the teacher model.

To address this issue, this study employs a binary cross-entropy distillation loss constructed from the classification scores of the teacher model and the student model. For dense detectors, an image is segmented into n regions. Assuming there are K classes, after passing through the Sigmoid function, we obtain $n * K$ classification scores. Let x be a sample, and p be the classification scores after Sigmoid activation. We can calculate the binary cross-entropy loss for the i-th feature map in the teacher and student networks for the j-th class as Equation (6):

$$\mathcal{L}_{BCE}(p^s_{i,j}, p^t_{i,j}) = -\left((1 - p^t_{i,j}) \cdot log(1 - p^s_{i,j}) + p^t_{i,j} \cdot log(p^s_{i,j})\right) \tag{6}$$

The $p^t_{i,j}$ represents the classification scores of the teacher network, while $p^s_{i,j}$ represents the classification scores of the student network. Based on the binary cross-entropy loss function mentioned above, the distillation loss for the class can be calculated using Equation (7):

$$\mathcal{L}^{dis}_{cls}(x) = \sum_{i=1}^{n}\sum_{j=1}^{K} \mathcal{L}_{BCE}(p^s_{i,j}, p^t_{i,j}) \tag{7}$$

Additionally, to enhance attention on important samples, this study introduces a weighted loss strategy for computing importance weight w, which can be represented as Equation (8):

$$w = \left|p^t - p^s\right| \tag{8}$$

The final class distillation loss can be represented as Equation (9):

$$\mathcal{L}_{cls}^{dis}(x) = \sum_{i=1}^{n}\sum_{j=1}^{K} w_{i,j} \cdot \mathcal{L}_{BCE}(p_{i,j}^{s}, p_{i,j}^{t}) \tag{9}$$

In terms of localization distillation, it is generally done by converting predicted boxes into probability distributions using the Softmax function, and then constructing a loss function between the teacher network and student network for distillation. This method requires the use of discrete position probability prediction heads, which are not commonly used in current dense detection models (such as the YOLO series of detectors). Therefore, in terms of localization, this paper introduces an unstructured localization distillation loss function based on intersection over union (IoU), which is driven by the IoU loss widely used in dense object detectors, to replace existing localization distillation losses.

Specifically, we obtain the information of detection boxes from the teacher model and student model. For a given input sample x, the predicted boxes on the i-th feature map of the teacher network and student network are denoted as b_i^t and b_i^s, respectively. Then, the IoU loss between them, denoted as u_i, can be calculated, and the localization loss can be expressed as Equation (10):

$$\mathcal{L}_{loc}^{dis}(x) = \sum_{i=1}^{n}(1 - u_i) \tag{10}$$

The final loss function based on the logic distillation part can be expressed as Equation (11):

$$\mathcal{L}_{logits}^{dis}(x) = \beta_1 \cdot \mathcal{L}_{cls}^{dis}(x) + \beta_2 \cdot \mathcal{L}_{loc}^{dis}(x) \tag{11}$$

where β_1 and β_2 represent the weights of classification distillation loss and localization distillation loss, respectively.

3.3. Feature Distillation Method Based on Mask Generation

The feature-based distillation method typically involves encouraging the student model to mimic the outputs of the teacher model as closely as possible, as the teacher model often has stronger feature representation capabilities. However, recent studies have shown that a distillation method forcing the student model to reconstruct the output features of the teacher model using partial pixels can enhance the feature representation capability of the student model more effectively than fully mimicking the outputs of the teacher model. Additionally, this approach brings the detection accuracy of the student model closer to that of the teacher model. Therefore, in this study, we employ a feature distillation method based on mask generation to distill the intermediate feature maps of the network model [52], as illustrated in Figure 4.

Firstly, we denote the feature maps of the l-th layer of the teacher model and the student model as $T^l \in R^{C\times H\times W}$ and $S^l \in R^{C\times H\times W}(l = 1, \ldots, L)$, respectively. Then, we set a random mask for the l-th layer feature map $M_{i,j}^l$:

$$M_{i,j}^{l} = \begin{cases} 0, & \text{if } R_{i,j}^{l} < \lambda \\ 1, & \text{Otherwise} \end{cases} \tag{12}$$

Here, $R_{i,j}^l$ represents the feature value at the i-th row and j-th column of the l-th layer feature map. Due to normalization, the values of $R_{i,j}^l$ range from 0 to 1. λ is a tunable hyperparameter that can represent the mask ratio.

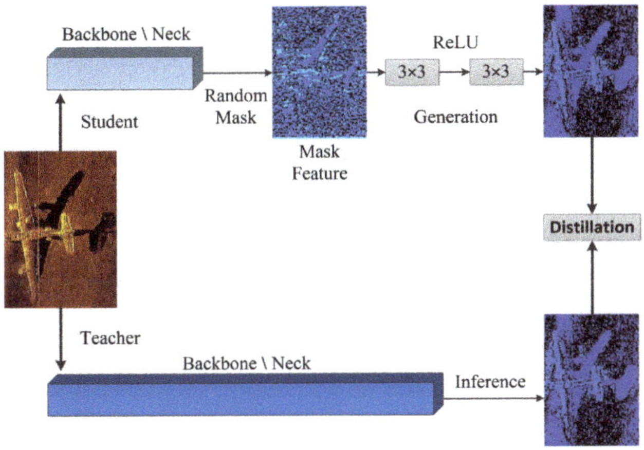

Figure 4. Diagram of the mask-generated distillation method. We use random masks to cover the feature maps of the student model and then utilize convolutional layers to forcefully generate the feature maps learned by the teacher model.

After obtaining the masked features, we use the mask of the l-th layer to cover the feature map of the student model's l-th layer. We then attempt to generate the teacher model's l-th layer feature map using the masked feature map. This process can be represented by Equations (13) and (14):

$$G(f_{cover}(S^l) \cdot M^l) \longrightarrow T^l \tag{13}$$

$$G(F) = W_{l2}(ReLU(W_{l1}(F))) \tag{14}$$

Here, F represents the feature map after masking, W_{l1} and W_{l2} are 3×3 convolutional layers, and ReLU is an activation layer between W_{l1} and W_{l2}. After computation, we obtain a feature map similar to the teacher model's l-th layer feature map. Based on this obtained feature map, we design the distillation loss function $L_{dis}(S,T)$ as Equation (15):

$$\mathcal{L}^{dis}_{features}(S,T) = \sum_{l=1}^{L}\sum_{k=1}^{C}\sum_{i=1}^{H}\sum_{j=1}^{W}\left(T^l_{k,i,j} - G\left(f_{cover}(S^l_{k,i,j}) \cdot M^l_{i,j}\right)\right)^2 \tag{15}$$

In this context, L represents the total number of layers to be distilled, C represents the number of channels in this layer, H and W denote the height and width of the feature maps, respectively. S represents the feature map generated by the student model, while T represents the feature map of the teacher model.

Combining the output distillation loss, the overall loss function of the distillation model is given by Equation (16):

$$L_{total} = \alpha \cdot \mathcal{L}^{dis}_{features}(S,T) + \beta \cdot \mathcal{L}^{dis}_{logits}(x) \tag{16}$$

4. Experiments and Analysis

To validate the effectiveness of the proposed lightweight method, this study conducted experiments on both the STCD dataset and a self-collected sidescan sonar dataset. Detailed analyses were performed on the detection accuracy and real-time performance of the model.

4.1. Experiment Platform

In this experiment, model training was conducted on a desktop computer running Ubuntu 20.04, with detailed specifications listed in Table 1. For testing, we used an embedded processor-Jetson AGX Orin, which can be installed inside the watertight compartment of an AUV. The installation position on the AUV is illustrated in Figure 5, and detailed specifications are provided in Table 2.

Table 1. Configuration of desktop computer.

Component	Specification
Operating System	Ubuntu 20.04 (64-bit)
Deep Learning Framework	Pytorch 1.11
Programming Language	Python 3.9
GPU Accelerated Environment	CUDA 11.3
Graphics Card (GPU)	Nvidia GeForce RTX 3090
Processor (CPU)	Platinum 8255C CPU @ 2.50 GHz

Table 2. Configuration of embedded processor.

Component	Specification
Operating System	Ubuntu 20.04 (64-bit)
Deep Learning Framework	Pytorch 1.12
Programming Language	Python 3.9
GPU Accelerated Environment	CUDA 11.3
Graphics Card (GPU)	NVIDIA Ampere architecture GPU
Processor (CPU)	8-core ARM Cortex-A78AE

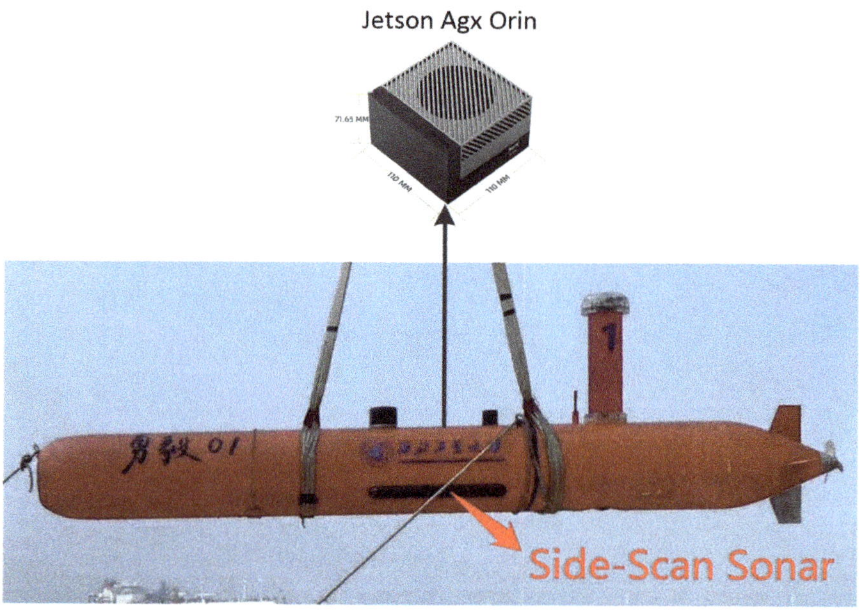

Figure 5. Schematic diagram of embedded processor installation location in AUV.

4.2. Dataset

The experiments in this paper utilized both the publicly available STCD dataset [53] and a self-collected dataset for testing. The STCD dataset is a public sidescan sonar dataset that includes three categories: ship, aircraft, and human, with a total of 497 images. In this study, the number of samples was increased to 692 through data augmentation and the addition of new data. The dataset was then split into training, validation, and test sets in a 7:1:2 ratio for experimentation. Our self-collected dataset was acquired during ocean experiments using a dual-frequency sidescan sonar. This dataset comprises three categories: cylinder, cone, and non-target. After data augmentation using a diffusion model [54], it contains a total of 975 images. This dataset was also divided into training, validation, and test sets in a 7:1:2 ratio. The detailed sample numbers of the two datasets are shown in Table 3.

Table 3. The number of samples in the STCD dataset and OUR dataset.

STCD Dataset [53]					OUR Dataset (256 × 256) [54]				
Category	**Train**	**Val**	**Test**	**Total**	**Category**	**Train**	**Val**	**Test**	**Total**
Ship	252	37	72	361	Cone	205	29	59	293
Aircraft	126	18	36	180	Cylinder	223	31	64	318
Human	105	16	30	151	Non-target	255	36	73	364

4.3. Model Evaluation Metrics

To assess the effectiveness of the proposed compression method for object detection models [55,56], this paper will analyze the experimental results using the following technical metrics.

Params and Size: "Params" represent the total number of parameters in the model, and "Size" denotes the actual size of the model. These two parameters are directly related to the disk space required to deploy the model. For embedded devices, the limited storage space makes it impractical to deploy models that require a large amount of disk space. Therefore, these two parameters are valuable metrics for evaluating model compression techniques.

GFLOPs: "GFLOPs" represent the number of floating-point operations that can be performed per second, reflecting the model's demand on hardware computational units. This metric is commonly used to assess the computational complexity or resource requirements of deep learning models. In model compression research, "GFLOPs" are often chosen as an evaluation metric. A higher "GFLOPs" value indicates a larger computational load and a greater need for computational resources.

Inter-time: "Inter-time" refers to the model's inference time, which is the time required for the model to process input data and generate output results, excluding the time for non-maximum suppression. This metric most directly reflects the model's inference speed. "Inter-time" is a crucial metric for evaluating the performance of compressed models, especially in real-time applications or scenarios requiring rapid responses.

mAP: mAP represents the overall performance of a model in detecting different categories of objects, comprehensively reflecting the model's accuracy and recall capability in object detection tasks. It is a widely adopted evaluation metric in object detection tasks. "mAP@0.5" denotes the mean average precision at an IoU threshold of 0.5, while "mAP@0.5:0.95" represents the average mAP value over a range of IoU thresholds from 0.5 to 0.95.

4.4. Model Training and Compression

First, the YOLOv7 model was trained using the aforementioned datasets. After multiple trials, the model with the best performance on the validation set was selected as the baseline model for subsequent pruning and distillation experiments. The PR curves of the baseline model on the two datasets are shown in Figure 6, and the detailed parameters and performance metrics are provided in Table 4.

Table 4. Performance metrics of the baseline model.

Dataset	Params (M)	Size (MB)	GFLOPs	Infer-Time (ms)	*mAP*@0.5 (%)	*mAP*@0.5:0.95 (%)
STCD (640 × 640)	36.5	71.4	103.2	9.7	92.84	71.01
OUR (256 × 256)	36.5	71.4	16.5	4.5	90.63	50.91

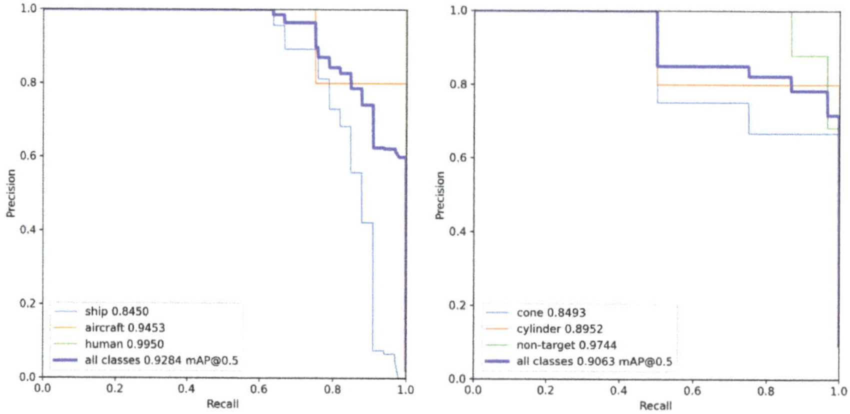

(a) (b)

Figure 6. (**a**) The PR curves of the STCD dataset. (**b**) The PR curves of the OUR dataset.

Then, the obtained baseline model is pruned. In pruning, this paper refers to GFLOP values and sets compression ratios of 2, 4, and 6 times. After sparse training, the model undergoes automatic compression to the specified ratio or reaches the maximum compression times before automatic fine-tuning.

Finally, to improve the accuracy of the pruned model, the method proposed in this paper, which fuses output and features, is used to distill the pruned model. The performance of the pruned and distilled models is compared with that of the baseline model, as shown in Table 5.

Table 5. Comparison of performance metrics between the pruned model, distilled model, and baseline model.

Dataset	Model	Params (M)	Size (MB)	GFLOPs	Infer-Time (ms)	*mAP*@0.5 (%)	*mAP*@0.5:0.95 (%)
STCD (640 × 640)	BaseLine	36.5	71.4	103.2	9.7	92.84	71.01
	Prune 2×	26.0	50.8	50.5	8.6	92.11	69.84
	KD 2×	26.0	50.8	50.5	8.6	92.90	69.97
	Prune 4×	13.3	25.8	26.0	7.8	85.86	62.13
	KD 4×	13.3	25.8	26.0	7.8	91.85	69.95
	Prune 6×	7.8	15.3	17.0	6.8	77.77	49.12
	KD 6×	7.8	15.3	17.0	6.8	86.21	59.07
OUR (256 × 256)	BaseLine	36.5	71.4	16.5	4.5	90.63	50.91
	Prune 2×	24.6	48.1	8.2	4.0	90.31	50.24
	KD 2×	24.6	48.1	8.2	4.0	90.61	50.88
	Prune 4×	13.0	25.1	4.1	3.4	86.14	47.20
	KD 4×	13.0	25.1	4.1	3.4	89.92	50.01
	Prune 6×	8.5	16.6	2.8	3.0	79.82	41.36
	KD 6×	8.5	16.6	2.8	3.0	84.57	45.26

Based on the results from the two datasets, it can be observed that under a compression ratio of 2 times, the detection accuracy of the pruned model is not significantly different from the baseline model. However, after applying the distillation method proposed in this paper, there is a slight improvement in the mAP metric. Under a compression ratio of 4 times, there is a noticeable decrease in detection accuracy for the pruned model. Using the distillation method proposed in this paper can improve the accuracy of the pruned model to a level close to that of the baseline model. Under a compression ratio of 6 times, there is a significant decrease in detection accuracy for the pruned model, and even with the improvement from the distillation method, its accuracy still differs significantly from the baseline model, indicating that it may not meet practical detection requirements.

Based on the above analysis, in practical engineering applications, it is advisable to choose the model compressed by a factor of 4 times. This choice ensures that the model's detection accuracy is maintained while maximizing compression to reduce model size and computational complexity, thus improving detection speed.

4.5. Model Performance Testing

To evaluate the real performance of the compressed models on embedded devices, we deployed both the baseline model and the distilled-enhanced compressed model on the Jetson AGX Orin for testing. The Jetson AGX Orin is a compact processor designed to be installed inside a 324-caliber AUV. In our testing experiments, we continued to use the two datasets mentioned earlier.

For the STCD dataset, the parameters and performance indicators of the compressed model compared to the baseline model are shown in Table 6. After distillation, the 4× compressed model reduced its parameter count by 23.2 M, model size by 45.6 MB, and computational load by 77.2 GFLOPs. The comparison of channel counts between the final compressed model and the baseline model is illustrated in Figure 7, demonstrating that our pruning method pruned most channels to varying degrees.

Table 6. Comparison of KD 4× model and baseline model on the STCD dataset.

Model	Params (M)	Size (MB)	GFLOPs	Infer-Time (ms)	*mAP*@0.5 (%)	*mAP*@0.5:0.95 (%)
Baseline	36.5	71.4	103.2	16.5	92.50	70.64
KD 4×	13.3	25.8	26.0	12.4	91.80	67.92

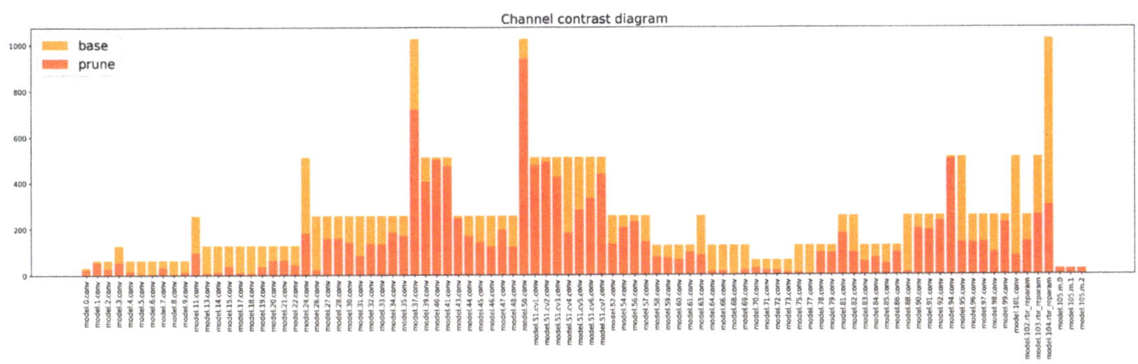

Figure 7. Comparison of channel numbers between the KD 4× model and the baseline model on the STCD dataset.

In terms of detection performance, the compressed model achieved an inference speedup of 4.1 ms compared to the baseline model. To ensure the rigor of the experiment, this inference time is the average result of 10 trials. The PR curve comparison between the compressed model and the baseline model on the test set is shown in Figure 8. The baseline model's $mAP@0.5$ score is 92.50%, while the compressed model, after distillation enhancement, achieved a $mAP@0.5$ score of 91.80%, with only a slight difference of 0.7%. The detection results comparison between the compressed model and the baseline model is illustrated in Figure 9.

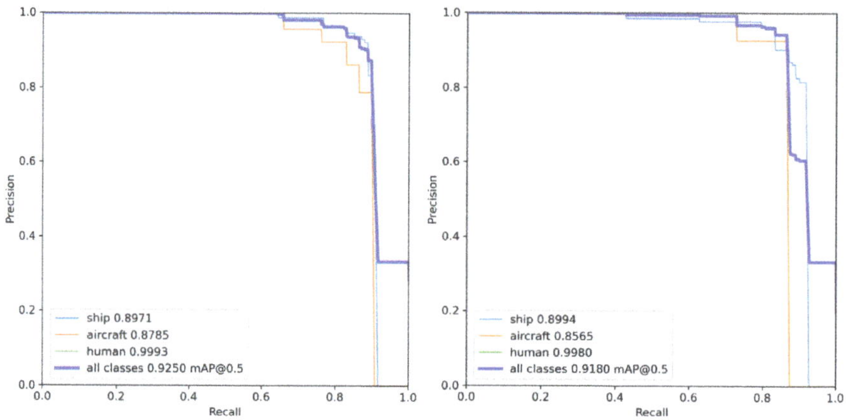

(a) (b)

Figure 8. PR Curves of the KD 4× model and baseline model on the STCD dataset. (**a**) The KD 4× model. (**b**) The baseline model.

(a) (b) (c)

Figure 9. Detection performance of the KD 4× model and baseline model on the STCD dataset. (**a**) Ground-truth labels. (**b**) The KD 4× model. (**c**) The baseline model.

For the OUR dataset, the parameters and performance metrics of the compressed model compared to the baseline model are shown in Table 7.

Table 7. Comparison of KD 4× model and baseline model on the OUR dataset.

Model	Params (M)	Size (MB)	GFLOPs	Infer-Time (ms)	$mAP@0.5$ (%)	$mAP@0.5:0.95$ (%)
Baseline	36.5	71.4	16.5	8.1	89.93	52.76
KD 4×	13.0	25.1	4.1	5.8	89.32	48.79

In terms of model parameters, the KD 4× compressed model reduced the parameter count by 23.5 M, the model size by 46.3 MB, and the computational load by 12.5 GFLOPs. The comparison of the number of channels per layer between the compressed model and the baseline model is shown in Figure 10. Similarly, most of the channels underwent various degrees of pruning.

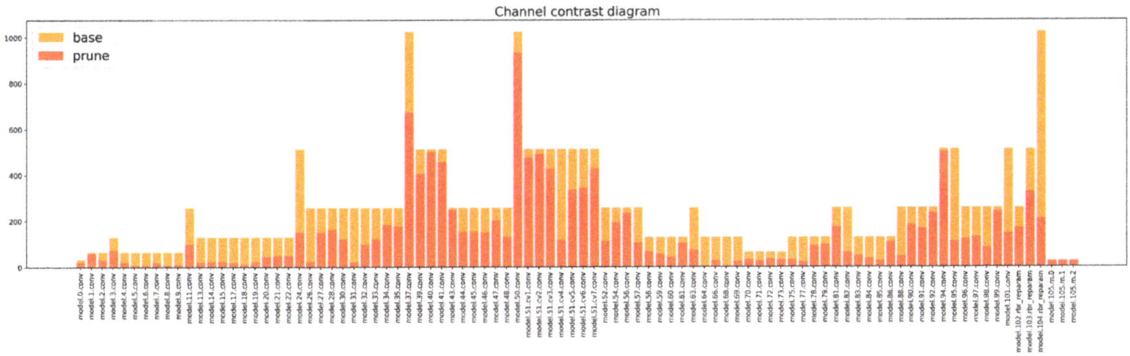

Figure 10. Comparison of channel numbers between the KD 4× model and the baseline model on the OUR dataset.

In terms of detection performance, the compressed model's average inference speed improved by 2.3 ms compared to that of the baseline model. The PR curves for the compressed model and the baseline model on the test set are shown in Figure 11. The $mAP@0.5$ for the baseline model is 89.93%, while the $mAP@0.5$ for the compressed model, after distillation enhancement, is 89.32%, with only a difference of 0.61% from the baseline model. The comparison of the detection results between the compressed model and the baseline model is shown in Figure 12.

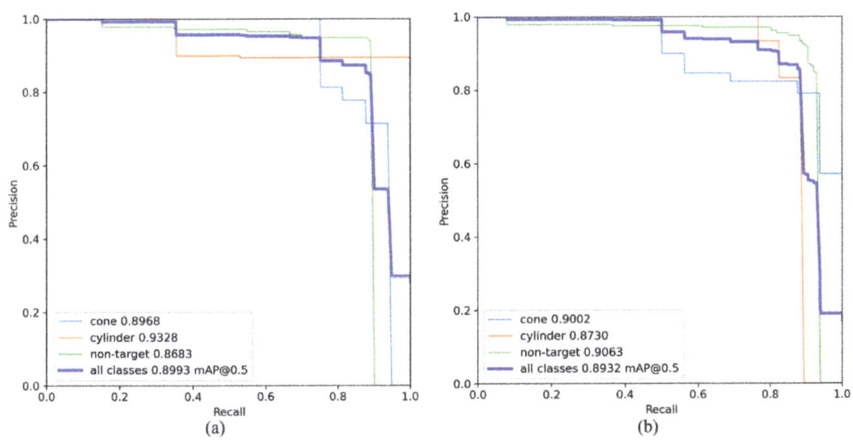

(a) (b)

Figure 11. PR Curves of the KD 4× model and baseline model on the OUR dataset. (**a**) The KD 4× model. (**b**) The baseline model.

Analyzing the experimental results reveals that the proposed method of combining automated pruning with distillation can effectively compress dense detectors like YOLOv7 while maintaining nearly unchanged detection accuracy. Additionally, it becomes apparent that the pruning ratios set based on GFLOPs do not result in proportional reductions in other model parameters, which is influenced by both network and hardware structures. In practical engineering applications, it is essential to balance detection accuracy and model parameters by selecting a suitable compression ratio based on actual detection needs and the hardware platform.

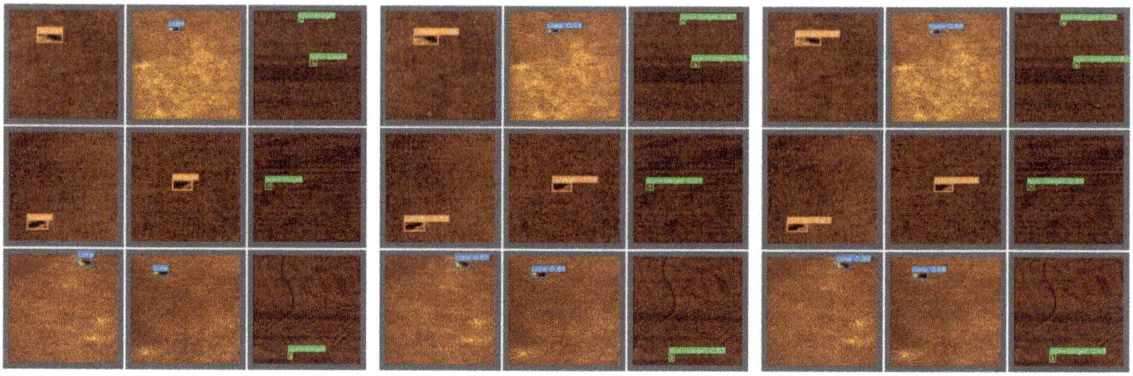

(a) (b) (c)

Figure 12. Detection performance of the KD 4× model and baseline model on the OUR dataset. (**a**) Ground-truth labels. (**b**) The KD 4× model. (**c**) The baseline model.

On the other hand, there are some considerations to note regarding the method presented in this paper. In terms of pruning, an automated pruning approach based on dependency graphs was used. This is a group-level pruning method that trims different dimensions across all parameter groups to achieve significant compression ratios. However, it may sometimes overly prune certain dimensions, such as reducing a dimension to 1, which can compromise final accuracy. In terms of distillation, the method employed combines output distillation and feature distillation, allowing the student network to learn more effectively from the teacher model. However, the distillation weight ratios need to be adjusted multiple times based on the dataset. For the dataset used in this study, the best results were obtained when the weights of output distillation and feature distillation together accounted for half of the total loss weight. In summary, excessive dimension pruning and improper distillation loss settings can lead to suboptimal results, but these issues can be resolved through multiple experiments.

5. Conclusions

This paper addresses the challenge of deploying large-scale detection models on edge devices by introducing a dependency graph-based pruning method, achieving efficient pruning on the YOLOv7 model. To counteract the significant accuracy drop caused by extensive pruning, we designed a hybrid distillation method combining output-based and feature-based distillation techniques to enhance the detection accuracy of the pruned model. Experimental validation demonstrates that our proposed method of combining pruning and distillation achieves a fourfold model compression with less than a 1% drop in accuracy. Finally, we deployed the compressed model on an embedded device within an AUV, reducing the model size and inference time while maintaining detection accuracy. These results confirm the effectiveness of our combined pruning and distillation approach.

Author Contributions: Conceptualization, C.C. and F.Z.; methodology, C.C., X.H. and C.W.; coding, C.C. and X.W.; designing the experiments, C.C. and X.H.; writing—original draft preparation, C.C.; writing—review and editing, C.C., F.Z. and W.L. All authors have read and agreed to the published version of the manuscript.

Funding: This study was supported by the National Key R&D Program of China (2023YFC2808400).

Institutional Review Board Statement: Not applicable.

Informed Consent Statement: Not applicable.

Data Availability Statement: The data presented in this study are available on request from the corresponding author.

Acknowledgments: We would like to acknowledge the facilities and technical assistance provided by the Key Laboratory of Unmanned Underwater Transport Technology.

Conflicts of Interest: The authors declare no conflict of interest.

References

1. Fan, Z.; Xia, W.; Liu, X.; Li, H. Detection and segmentation of underwater objects from forward-looking sonar based on a modified Mask RCNN. *Signal Image Video Process.* **2021**, *15*, 1135–1143. [CrossRef]
2. Zeng, L.; Sun, B.; Zhu, D. Underwater target detection based on Faster R-CNN and adversarial occlusion network. *Eng. Appl. Artif. Intell.* **2021**, *100*, 104190. [CrossRef]
3. Yin, Q.; Yang, W.; Ran, M.; Wang, S. FD-SSD: An improved SSD object detection algorithm based on feature fusion and dilated convolution. *Signal Process. Image Commun.* **2021**, *98*, 116402. [CrossRef]
4. Yundong, L.; Han, D.; Hongguang, L.; Zhang, X.; Zhang, B.; Zhifeng, X. Multi-block SSD based on small object detection for UAV railway scene surveillance. *Chin. J. Aeronaut.* **2020**, *33*, 1747–1755.
5. Cheng, C.; Wang, C.; Yang, D.; Wen, X.; Liu, W.; Zhang, F. Underwater small target detection based on dynamic convolution and attention mechanism. *Front. Mar. Sci.* **2024**, *11*, 1348883. [CrossRef]
6. Yu, Y.; Zhao, J.; Gong, Q.; Huang, C.; Zheng, G.; Ma, J. Real-time underwater maritime object detection in side-scan sonar images based on transformer-YOLOv5. *Remote Sens.* **2021**, *13*, 3555. [CrossRef]
7. Aguirre-Castro, O.A.; García-Guerrero, E.E.; López-Bonilla, O.R.; Tlelo-Cuautle, E.; López-Mancilla, D.; Cárdenas-Valdez, J.R.; Olguín-Tiznado, J.E.; Inzunza-González, E. Evaluation of underwater image enhancement algorithms based on Retinex and its implementation on embedded systems. *Neurocomputing* **2022**, *494*, 148–159. [CrossRef]
8. Jiang, M.; Li, R.; Liu, Q.; Shi, Y.; Tlelo-Cuautle, E. High speed long-term visual object tracking algorithm for real robot systems. *Neurocomputing* **2021**, *434*, 268–284. [CrossRef]
9. Tang, Y.; Wang, L.; Jin, S.; Zhao, J.; Huang, C.; Yu, Y. AUV-based side-scan sonar real-time method for underwater-target detection. *J. Mar. Sci. Eng.* **2023**, *11*, 690. [CrossRef]
10. Neves, G.; Ruiz, M.; Fontinele, J.; Oliveira, L. Rotated object detection with forward-looking sonar in underwater applications. *Expert Syst. Appl.* **2020**, *140*, 112870. [CrossRef]
11. He, Y.; Xiao, L. Structured pruning for deep convolutional neural networks: A survey. *IEEE Trans. Pattern Anal. Mach. Intell.* **2023**, *46*, 2900–2919. [CrossRef] [PubMed]
12. Zhou, Y.; Chen, S.; Wang, Y.; Huan, W. Review of research on lightweight convolutional neural networks. In Proceedings of the 2020 IEEE 5th Information Technology and Mechatronics Engineering Conference (ITOEC), Chongqing, China, 12–14 June 2020; pp. 1713–1720.
13. Ren, P.; Xiao, Y.; Chang, X.; Huang, P.Y.; Li, Z.; Chen, X.; Wang, X. A comprehensive survey of neural architecture search: Challenges and solutions. *ACM Comput. Surv.* **2021**, *54*, 1–34. [CrossRef]
14. Liang, T.; Glossner, J.; Wang, L.; Shi, S.; Zhang, X. Pruning and quantization for deep neural network acceleration: A survey. *Neurocomputing* **2021**, *461*, 370–403. [CrossRef]
15. Gou, J.; Yu, B.; Maybank, S.J.; Tao, D. Knowledge distillation: A survey. *Int. J. Comput. Vis.* **2021**, *129*, 1789–1819. [CrossRef]
16. Swaminathan, S.; Garg, D.; Kannan, R.; Andres, F. Sparse low rank factorization for deep neural network compression. *Neurocomputing* **2020**, *398*, 185–196. [CrossRef]
17. Tian, M.; Li, X.; Kong, S.; Yu, J. Pruning-based YOLOv4 algorithm for underwater gabage detection. In Proceedings of the 2021 40th Chinese Control Conference (CCC), Shanghai, China, 26–28 July 2021; pp. 4008–4013.
18. Szymak, P.; Piskur, P.; Naus, K. The effectiveness of using a pretrained deep learning neural networks for object classification in underwater video. *Remote Sens.* **2020**, *12*, 3020. [CrossRef]
19. Ye, X.; Zhang, W.; Li, Y.; Luo, W. Mobilenetv3-YOLOv4-sonar: Object detection model based on lightweight network for forward-looking sonar image. In Proceedings of the OCEANS 2021: San Diego–Porto, San Diego, CA, USA, 20–23 September 2021; pp. 1–6.
20. Yan, Z.; Shaochang, C. Lightweight improvement study of sonar image detection network. In Proceedings of the 2021 International Conference on Computer, Blockchain and Financial Development (CBFD), Nanjing, China, 23–25 April 2021; pp. 49–53.
21. Hu, S.; Liu, T. Underwater rescue target detection based on acoustic images. *Sensors* **2024**, *24*, 1780. [CrossRef]
22. Qin, K.S.; Liu, D.; Wang, F.; Zhou, J.; Yang, J.; Zhang, W. Improved YOLOv7 model for underwater sonar image object detection. *J. Vis. Commun. Image Represent.* **2024**, *100*, 104124. [CrossRef]
23. LeCun, Y.; Denker, J.; Solla, S. Optimal brain damage. *Adv. Neural Inf. Process. Syst.* **1989**, *2* 598–605.
24. Han, S.; Pool, J.; Tran, J.; Dally, W. Learning both weights and connections for efficient neural network. *Adv. Neural Inf. Process. Syst.* **2015**, *28*, 1135–1143.
25. Li, G.; Qian, C.; Jiang, C.; Lu, X.; Tang, K. Optimization based Layer-wise Magnitude-based Pruning for DNN Compression. In Proceedings of the Twenty-Seventh International Joint Conference on Artificial Intelligence (IJCAI-18), Stockholm, Sweden, 13–19 July 2018; pp. 2383–2389.
26. Lee, J.; Park, S.; Mo, S.; Ahn, S.; Shin, J. Layer-adaptive sparsity for the magnitude-based pruning. *arXiv* **2020**, arXiv:2010.07611.

27. Carreira-Perpinán, M.A.; Idelbayev, Y. "learning-compression" algorithms for neural net pruning. In Proceedings of the IEEE Conference on Computer Vision and Pattern Recognition, Salt Lake City, UT, USA, 18–23 June 2018; pp. 8532–8541.
28. Kwon, S.J.; Lee, D.; Kim, B.; Kapoor, P.; Park, B.; Wei, G.Y. Structured compression by weight encryption for unstructured pruning and quantization. In Proceedings of the IEEE/CVF Conference on Computer Vision and Pattern Recognition, Seattle, WA, USA, 13–19 June 2020; pp. 1909–1918.
29. Molchanov, P.; Mallya, A.; Tyree, S.; Frosio, I.; Kautz, J. Importance estimation for neural network pruning. In Proceedings of the IEEE/CVF Conference on Computer Vision and Pattern Recognition, Long Beach, CA, USA, 15–20 June 2019; pp. 11264–11272.
30. Wang, Z.; Li, C.; Wang, X. Convolutional neural network pruning with structural redundancy reduction. In Proceedings of the IEEE/CVF Conference on Computer Vision and Pattern Recognition, Nashville, TN, USA, 20–25 June 2021; pp. 14913–14922.
31. Ruan, X.; Liu, Y.; Li, B.; Yuan, C.; Hu, W. DPFPS: Dynamic and progressive filter pruning for compressing convolutional neural networks from scratch. *Proc. AAAI Conf. Artificial Intell.* **2021**, *35*, 2495–2503. [CrossRef]
32. Ding, X.; Hao, T.; Tan, J.; Liu, J.; Han, J.; Guo, Y.; Ding, G. Resrep: Lossless cnn pruning via decoupling remembering and forgetting. In Proceedings of the IEEE/CVF International Conference on Computer Vision, Montreal, BC, Canada, 11–17 October 2021; pp. 4510–4520.
33. Hou, Z.; Qin, M.; Sun, F.; Ma, X.; Yuan, K.; Xu, Y.; Chen, Y.K.; Jin, R.; Xie, Y.; Kung, S.Y. Chex: Channel exploration for cnn model compression. In Proceedings of the IEEE/CVF Conference on Computer Vision and Pattern Recognition, New Orleans, LA, USA, 18–24 June 2022; pp. 12287–12298.
34. Yang, T.J.; Chen, Y.H.; Sze, V. Designing energy-efficient convolutional neural networks using energy-aware pruning. In Proceedings of the IEEE Conference on Computer Vision and Pattern Recognition, Honolulu, HI, USA, 21–26 July 2017; pp. 5687–5695.
35. Buciluă, C.; Caruana, R.; Niculescu-Mizil, A. Model compression. In Proceedings of the 12th ACM SIGKDD International Conference on Knowledge Discovery and Data Mining, Philadelphia, PA, USA, 20–23 August 2006; pp. 535–541.
36. Hinton, G.; Vinyals, O.; Dean, J. Distilling the knowledge in a neural network. *arXiv* **2015**, arXiv:1503.02531.
37. Zhang, Y.; Xiang, T.; Hospedales, T.M.; Lu, H. Deep mutual learning. In Proceedings of the IEEE Conference on Computer Vision and Pattern Recognition, Salt Lake City, UT, USA, 18–23 June 2018; pp. 4320–4328.
38. Kim, J.; Park, S.; Kwak, N. Paraphrasing complex network: Network compression via factor transfer. *Adv. Neural Inf. Process. Syst.* **2018**, *31*, 2760–2769.
39. Zhao, B.; Cui, Q.; Song, R.; Qiu, Y.; Liang, J. Decoupled knowledge distillation. In Proceedings of the IEEE/CVF Conference on Computer Vision and Pattern Recognition, New Orleans, LA, USA, 18–24 June 2022; pp. 11953–11962.
40. Adriana, R.; Nicolas, B.; Ebrahimi, K.S.; Antoine, C.; Carlo, G.; Yoshua, B. Fitnets: Hints for thin deep nets. *Proc. ICLR* **2015**, *2*, 1.
41. Zagoruyko, S.; Komodakis, N. Paying more attention to attention: Improving the performance of convolutional neural networks via attention transfer. *arXiv* **2016**, arXiv:1612.03928.
42. Huang, Z.; Wang, N. Like what you like: Knowledge distill via neuron selectivity transfer. *arXiv* **2017**, arXiv:1707.01219.
43. Ahn, S.; Hu, S.X.; Damianou, A.; Lawrence, N.D.; Dai, Z. Variational information distillation for knowledge transfer. In Proceedings of the IEEE/CVF Conference on Computer Vision and Pattern Recognition, Long Beach, CA, USA, 15–20 June 2019; pp. 9163–9171.
44. Yang, J.; Martinez, B.; Bulat, A.; Tzimiropoulos, G. Knowledge distillation via adaptive instance normalization. *arXiv* **2020**, arXiv:2003.04289.
45. Yim, J.; Joo, D.; Bae, J.; Kim, J. A gift from knowledge distillation: Fast optimization, network minimization and transfer learning. In Proceedings of the IEEE Conference on Computer Vision and Pattern Recognition, Honolulu, HI, USA, 21–26 July 2017; pp. 4133–4141.
46. Peng, B.; Jin, X.; Liu, J.; Li, D.; Wu, Y.; Liu, Y.; Zhou, S.; Zhang, Z. Correlation congruence for knowledge distillation. In Proceedings of the IEEE/CVF International Conference on Computer Vision, Seoul, Republic of Korea, 27 October–November 2019; pp. 5007–5016.
47. Park, W.; Kim, D.; Lu, Y.; Cho, M. Relational knowledge distillation. In Proceedings of the IEEE/CVF Conference on Computer Vision and Pattern Recognition, Long Beach, CA, USA, 15–20 June 2019; pp. 3967–3976.
48. Liu, Y.; Cao, J.; Li, B.; Yuan, C.; Hu, W.; Li, Y.; Duan, Y. Knowledge distillation via instance relationship graph. In Proceedings of the IEEE/CVF Conference on Computer Vision and Pattern Recognition, Long Beach, CA, USA, 15–20 June 2019; pp. 7096–7104.
49. Zhou, S.; Wang, Y.; Chen, D.; Chen, J.; Wang, X.; Wang, C.; Bu, J. Distilling holistic knowledge with graph neural networks. In Proceedings of the IEEE/CVF International Conference on Computer Vision, Montreal, BC, Canada, 11–17 October 2021; pp. 10387–10396.
50. Fang, G.; Ma, X.; Song, M.; Mi, M.B.; Wang, X. Depgraph: Towards any structural pruning. In Proceedings of the IEEE/CVF Conference on Computer Vision and Pattern Recognition, Vancouver, BC, Canada, 17–24 June 2023; pp. 16091–16101.
51. Yang, L.; Zhou, X.; Li, X.; Qiao, L.; Li, Z.; Yang, Z.; Wang, G.; Li, X. Bridging Cross-task Protocol Inconsistency for Distillation in Dense Object Detection. In Proceedings of the IEEE/CVF International Conference on Computer Vision, Paris, France, 4–6 October 2023; pp. 17175–17184.
52. Yang, Z.; Li, Z.; Shao, M.; Shi, D.; Yuan, Z.; Yuan, C. Masked generative distillation. In *Computer Vision—ECCV 2022*; Springer: Cham, Switzerland, 2022; pp. 53–69.
53. Zhang, P.; Tang, J.; Zhong, H.; Ning, M.; Liu, D.; Wu, K. Self-trained target detection of radar and sonar images using automatic deep learning. *IEEE Trans. Geosci. Remote Sens.* **2021**, *60*, 1–14. [CrossRef]

54. Cheng, C.; Hou, X.; Wen, X.; Liu, W.; Zhang, F. Small-Sample Underwater Target Detection: A Joint Approach Utilizing Diffusion and YOLOv7 Model. *Remote Sens.* **2023**, *15*, 4772. [CrossRef]
55. Lyu, Z.; Yu, T.; Pan, F.; Zhang, Y.; Luo, J.; Zhang, D.; Chen, Y.; Zhang, B.; Li, G. A survey of model compression strategies for object detection. *Multimed. Tools Appl.* **2023**, *83*, 48165–48236. [CrossRef]
56. Zhou, Y.; Xia, L.; Zhao, J.; Yao, R.; Liu, B. Efficient convolutional neural networks and network compression methods for object detection: A survey. *Multimed. Tools Appl.* **2024**, *83*, 10167–10209. [CrossRef]

Journal of
Marine Science and Engineering

MDPI

Article

An Invariant Filtering Method Based on Frame Transformed for Underwater INS/DVL/PS Navigation

Can Wang, Chensheng Cheng, Chun Cao, Xinyu Guo, Guang Pan and Feihu Zhang *

School of Marine Science and Technology, Northwestern Polytechnical University, Xi'an 710072, China
* Correspondence: feihu.zhang@nwpu.edu.cn; Tel.: +86-029-88492611

Abstract: Underwater vehicles heavily depend on the integration of inertial navigation with Doppler Velocity Log (DVL) for fusion-based localization. Given the constraints imposed by sensor costs, ensuring the optimization ability and robustness of fusion algorithms is of paramount importance. While filtering-based techniques such as Extended Kalman Filter (EKF) offer mature solutions to nonlinear problems, their reliance on linearization approximation may compromise final accuracy. Recently, Invariant EKF (IEKF) methods based on the concept of smooth manifolds have emerged to address this limitation. However, the optimization by matrix Lie groups must satisfy the "group affine" property to ensure state independence, which constrains the applicability of IEKF to high-precision positioning of underwater multi-sensor fusion. In this study, an alternative state-independent underwater fusion invariant filtering approach based on a two-frame group utilizing DVL, Inertial Measurement Unit (IMU), and Earth-Centered Earth-Fixed (ECEF) configuration is proposed. This methodology circumvents the necessity for group affine in the presence of biases. We account for inertial biases and DVL pole-arm effects, achieving convergence in an imperfect IEKF by either fixed observation or body observation information. Through simulations and real datasets that are time-synchronized, we demonstrate the effectiveness and robustness of the proposed algorithm.

Keywords: AUV; navigation; invariant error; IEKF; DVL

1. Introduction

Autonomous Underwater Vehicles (AUVs) employ a diverse array of methods to determine their geographic location. This includes the use of inertial components, a Global Navigation Satellite System (GNSS) for satellite-based multipoint positioning, environmental perception through cameras and Doppler Velocity Logs (DVLs), acoustic ranging techniques for distance estimation, and pressure sensors for depth measurements. Given the limitations of electromagnetic wave propagation in the underwater environment, particularly about GNSS-based positioning and acousto-optic ranging, various fusion techniques utilizing an Inertial Navigation System (INS)/DVL have been utilized to improve the accuracy of positioning [1,2]. To overcome the nonlinear optimization challenges related to the fusion of multiple sensor sources, state estimation can be achieved using optimization methods such as the Extended Kalman Filter (EKF) or least squares techniques [3,4]. In a more complex manner, SLAM localization is achieved through data correlation and loopback detection, which can be accomplished by image sonar or underwater vision [5,6].

Nevertheless, conventional filtering techniques, which depend on the linearization of nonlinear problems, result in error accumulation, particularly noticeable in long-duration positioning tasks [7]. The property of Lie groups to maintain smooth structures on manifolds enables the effective description of robot positioning characteristics [8,9]. The Invariant Extended Kalman filter (IEKF) algorithm gives a novel solution [10,11], with applications extending to legged robots and beyond [12]. Unfortunately, for IEKF to attain trajectory-independent error with left–right invariance, the state must adhere to a group affine property. In the optimization problem of Earth frame-based navigation, researchers often introduce an auxiliary velocity vector, which leads to increased computational complexity [13–16].

Hence, drawing inspiration from [17], we propose a new method for representing observations in different coordinate systems using two frames, which enables the optimization of invariant filtering. This approach eliminates the need for auxiliary vectors and improves localization accuracy. Overall, the contributions of this paper can be delineated as follows:

- Utilizing a novel two-frame group filtering approach, along with a comprehensive formulation and validation, encompassing both fixed and body observations.
- Leveraging DVL and depth sensor data as primary observations, the methodology facilitates underwater navigation filtering, obviating the necessity for intricate transformations to satisfy group affine requirements.
- Algorithmic validation is conducted using simulated datasets and proprietary real-world underwater data, ensuring robustness and efficacy.

In the subsequent sections of this paper, we delve into specific aspects of our research. Section 2 provides an overview of recent advancements and limitations in underwater navigation optimization and the application of Lie group optimization. Section 3 presents the Lie group theory with a specific emphasis on the two-frame group theory, detailing its enhancements and application in underwater navigation. Section 4 outlines the validation process of both simulation and experimental data, followed by a comprehensive analysis and evaluation of the results in Section 5. Finally, Section 6 summarizes the whole paper and gives future research directions.

2. Related Works

2.1. Underwater Navigation Methods

With the advancement of marine science and technology, underwater navigation methods have diversified, ranging from traditional inertial navigation to acoustic localization, multi-sensor fusion navigation, and sense-calibrated navigation. For attitude estimation, Shi et al. achieved high accuracy for an INS on an underwater dynamic base using Mahalanobis distance-based outlier suppression and real-time measurement noise covariance estimation via variational Bayesian approximation [18]. Jorgensen et al. introduced a novel exogenous Kalman filter-based observer structure for attitude estimation utilizing Time Difference of Arrival (TDOA) measurements [19]. Subsequent research has focused on optimizing underwater multi-sensor fusion errors. Li et al. integrated robust navigation algorithms based on the maximum correlation entropy criterion and factor graph optimization [20]. Huang et al. used the SRAKF algorithm to compute the exact solution for the prediction error covariance matrix and measurement noise matrix, leveraging the traceless transform [21]. OuYang et al. posited an affine transformation correlation between the matched and real trajectories, estimating optimal positions using the Particle Swarm Optimization (PSO) algorithm [22]. Jin et al. formulated a mathematical model for precise alignment by utilizing the SQP algorithm to determine direction vectors and leading to rapid convergence to high-precision attitude estimation [23]. Russo et al. employed an expanded 1D convolutional layer to compress the input signal into a set of high-level features and utilized jump joins to enhance model capture [24]. This methodology facilitated attitude denoising that supports diverse filters and notably diminishes the Root Mean Square Error (RMSE).

Recently, underwater navigation has advanced to calibrate positioning errors using sensory information. Zhang et al. proposed an MBES subgraph construction method, solving error loopback issues with Frechet distance for effective SLAM data correlation [25]. Rahman et al. used a bag-of-words location recognition module for global relocation via geometric validation with PnP RANSAC [26]. Advanced deep learning methods have also been employed to adaptively learn time-varying parameters in underwater navigation. Ma et al. used deep learning to generate low-frequency position data and employed VB-based adaptive filters for navigation estimation [27]. He et al. designed a purely inertial deep learning navigation system using Convolutional Neural Network (CNN) and transformer networks for AUV motion states [28]. Li et al. achieved robust navigation during DVL failures using a NARX model [29]. Zhao et al. used a Long Short-Term Memory Network (LSTM) to process time series data, combining the accuracy of EKF with the real-time performance of federated learning (FL) to optimize underwater localization [30]. Qin et al. found that geometry-based methods typically outperform deep learning-based methods in positioning accuracy through experiments on the public underwater dataset AQUALOC [31]. Kim et al. improved INS fault diagnosis using the Nomoto model with enhanced preprocessing and input features [32].

In underwater navigation, filtering or optimization methods are employed for achieving optimal state estimation based on sensor measurements and a kinematics model. The current challenge in achieving high-precision underwater navigation is not merely selecting an optimization method and enhancing the filter's performance sensor characteristics. Costanzi et al. proposed a method for detecting and rejecting affected measurements based on single-axis fiber optic gyro (FOG) signals, utilizing complementary filters to address magnetic perturbations from locally indistinguishable sources [33]. Ngatini et al. developed an AUV position estimation approach employing an integrated Kalman filter and a fuzzy Kalman filter (FKF) [34]. Shariati et al. tackled nonlinear state and parameter estimation by proposing a method to identify the nonlinear dynamic model of an AUV, particularly in the presence of unmeasured states [35]. Pei et al. redefined the process model using the Lie group of the constant attitude matrix between two inertial systems as states and designed a state-dependent Lie group filter [36]. Xu et al. applied the conditional adaptive gain expansion Kalman filter (CAEKF) to various sensor noise models and introduced a confidence assessment method to smooth the positioning trajectory of the AUV [37]. Fossen employed a discrete-time, unit-quaternion error-state Kalman filter to determine the position of an AUV using range measurements from a hydroacoustic network, along with improved time-delay estimation and compensation methods [38]. Comparison and conclusion of different algorithms are shown in Table 1.

Table 1. Comparison of related filtering/optimization algorithms.

Author	Methodology	Conclusion
Shi et al. [18]	Variational Bayesian Robust Adaptive Kalman Filter (VBRAKF)	Error of 95.61% of KF
Huang et al. [21]	Statistical Regression Adaptive Kalman Filter (SRAKF)	93.83% improvement over UKF
Ngatini et al. [34]	Integrated Kalman Filter and Fuzzy Kalman Filter	Position error is 92% of FKF; angle error is 93% of FKF
Pei et al. [36]	State-Dependent Lie Group-based Filter	Accuracy is 70% improved over Quaternion KF
Xu et al. [37]	Conditional Adaptive gain Expansion Kalman Filter (CAEKF)	83% improvement over EKF
Proposed	Underwater Two-Frame Group IEKF	77% improvement over IEKF

2.2. Li Group Optimization and Application to Underwater Navigation

Utilizing non-Euclidean geometric structures to model rotation and attitude spaces, Lie groups offer a more precise characterization of kinematic constraints and attitude changes. The logarithmic mapping of Lie groups transforms elements into vectors within their tangent spaces. This mapping facilitates the conversion of group multiplication operations into vector addition operations, thereby simplifying the derivation process. Barrau, in his Ph.D. thesis, introduced the theory of detached invariant observers on Lie groups and proposed a framework that enables the optimization of nonlinear error variables controlled by linear differential equations. He further implemented local stability verification [10]. Subsequently, he proposed the invariant Kalman filtering framework [39] and demonstrated the log-linear characterization of the Inertial Measurement Unit (IMU) equation in discrete time. Additionally, he proposed a pre-integration formula based on the Lie exponential coordinates concerning noise and bias [14]. Chang et al. utilized the analytical and concrete form of manifold to derive log-linear differential equations in continuous time without any approximation and provide left-invariant and right-invariant errors within the matrix Lie group [40]. Li et al. used exponential formulas to correct the base and fuselage attitudes and linear formulas to correct the inertial navigation coefficient errors [41].

With advancements in Lie group theory in machine learning and optimization, it has become a powerful tool for optimizing inertial and fusion navigation systems. Barfoot applied Special Euclidean (*SE*) groups and Baker–Campbell–Hausdorff (BCH) formulations to link uncertainty with the transformation matrix, improving the description of noise-variable perturbations [42]. Brossard et al. related uncertainty to a matrix of five extended poses, providing detailed uncertainty propagation for IMU states and deriving accurate pre-integration formulas considering Earth's rotation [15]. Pardos used Sagittal Kinematics Dividing (SKD) analysis to model humanoid robots and employed Lie groups to solve the inverse kinematic problem [8]. Roux combined an incomplete RIEKF with a CNN to dynamically optimize the measurement noise covariance matrix, enhancing the estimation of the position, velocity, and direction of a projectile [43]. Hartley derived a continuous-time right-invariant EKF (RIEKF) for the IMU/contact process model, successfully applying it to the Cassie bipedal robots [12]. Chang developed right-vector and left-vector error state models that account for Earth's rotation and Coriolis effect, suited for arbitrarily unaligned attitude initialization, and applied them for pre-integration on the manifold [44].

Li group optimization theory has been widely applied to the initial error alignment and multi-sensor fusion problems of underwater sensor systems. Xu et al. introduced the $SO(3)$ group state-space model, utilizing the degree of variation of auxiliary parameters to detect outliers and proposed a DVL calibration method based on the $SO(3)$ [45]. Luo et al. devised a closed-loop scheme based on a linear state-space model to mitigate the effects of IMU bias, DVL lever arm, and mounting misalignment angle between the IMU and the DVL [46]. This scheme concurrently estimates and compensates for parameter errors and a body attitude matrix using Davenport's q method, thereby enhancing vector observation accuracy. Qian et al. developed an integrated navigation Lie group error model based on Euler angles utilizing the Proper Weighting and Conditioning Strategy (PWCS) matrix of the Lie group model [47]. Additionally, Potokar et al. extended the conventional invariant measurement model to accommodate measurements that capture only a subset of the 3D state. They incorporated additional information, such as underwater depth measurements, into the IEKF framework [48].

In this work, pre-integration-based synchronization is conducted on INS, DVL, and PS data, followed by the application of the TFG method for invariant filtering. We address the limitations associated with group affine in underwater navigation filtering by employing a novel approach that preserves positioning accuracy.

3. Methodology

3.1. Lie Group Theory

The concept of Lie group encompasses the notions of both group and smooth manifold and unifies them within a mathematical framework. In the process of robot state estimation, Lie groups, mainly the special orthogonal group $SO(3)$ and the special Euclidean group $SE(3)$, are employed to represent rotational and attitude transformations, respectively.

The orientation R of a rigid body in space is typically modeled as

$$\mathrm{SO}(3) := \{\mathbf{R} \in \mathbb{R}^{3\times 3} \mid \mathbf{R}^{\mathrm{T}}\mathbf{R} = \mathbf{I}_3, \det(\mathbf{R}) = 1\} \tag{1}$$

where $\mathbb{R}^{3\times 3}$ denotes the set of 3×3 vectors.

For the transformation matrix, which contains the rotation matrix R and the translation vector t, it is denoted by [49]

$$\mathrm{SE}(3) = \left\{ \boldsymbol{T} = \begin{bmatrix} \boldsymbol{R} & t \\ 0 & 1 \end{bmatrix} \in \mathbb{R}^{4\times4} \middle| \boldsymbol{R} \in SO(3), t \in \mathbb{R}^3 \right\} \tag{2}$$

For the extended pose containing position P, velocity V, and direction R, it is expressed by

$$SE_2(3) := \left\{ \boldsymbol{T} = \begin{pmatrix} \boldsymbol{R} & \boldsymbol{V} & \boldsymbol{P} \\ 0_{1,3} & 1 & 0 \\ 0_{1,3} & 0 & 1 \end{pmatrix} \middle| (\boldsymbol{R}, \boldsymbol{V}, \boldsymbol{P}) \in SO(3) \times \mathbb{R}^3 \right\} \tag{3}$$

where $SE_2(3)$ is called the "double direct isometric group" [50]. Matrix multiplication provides the group composition of two elements of $SE_2(3)$.

A Lie algebra resides within a vector space and corresponds to a Lie group structure. The Lie algebras $so(3)$ and $se(3)$ corresponding to $SO(3)$ and $SE(3)$ are, respectively,

$$so(3) = \{\phi \in \mathbb{R}^3, (\phi_\times) \in \mathbb{R}^{3\times3}\} \tag{4}$$

$$se(3) = \left\{ \xi = \begin{bmatrix} \phi \\ \rho \end{bmatrix} \in \mathbb{R}^6, \phi \in so(3), \rho \in \mathbb{R}^3, \xi^\wedge = \begin{bmatrix} (\phi_\times) & \rho \\ 0^T & 0 \end{bmatrix} \in \mathbb{R}^{4\times4} \right\} \tag{5}$$

where ϕ represents a three-dimensional rotation vector, $\times$ denotes the antisymmetric matrix operation applied to a three-dimensional rotation vector, and ρ is a three-dimensional translation vector. It is noteworthy that, in $se(3)$, the symbol $\wedge$ is extended to convert a six-dimensional vector to a four-dimensional matrix. Conversely, in $so(3)$, the antisymmetric matrix is commonly denoted by either $\wedge$ or $\times$.

As in classical Lie group theory, small perturbations of extended posets can be described in terms of elements of the Lie algebra $SE_2(3)$. The operator $\wedge$ similarly represents a linear mapping operation that transforms the elements $\xi := (\phi^T, \mathbf{v}^T, \rho^T)^T \in \mathbb{R}^9$ into elements of the Lie algebra, as follows:

$$\xi^\wedge := \begin{pmatrix} \phi \\ v \\ \rho \end{pmatrix}^\wedge = \begin{pmatrix} (\phi_\times) & v & \rho \\ \mathbf{0}_{1\times3} & 0 & 0 \\ \mathbf{0}_{1\times3} & 0 & 0 \end{pmatrix} \tag{6}$$

The exponential mapping conveniently maps small perturbations encoded in R^9 to $SE_2(3)$. For the matrix Lie group, it is defined as $\exp(\xi) := exp_m(\xi^\wedge)$, where $exp_m(\cdot)$ denotes the classical matrix exponential mapping.

The Lie algebra space of $SE_2(3)$ is equivalent to the nine-dimensional Euclidean space represented as an operation. Its Lie algebra and Lie group can be expressed by the exponential map $exp(\cdot)$ and the logarithmic map $\log(\cdot)$ as follows:

$$\chi = exp_m(\xi^\wedge) = \begin{bmatrix} \exp_m(\phi_\times) & J_l v & J_l \rho \\ 0_{1\times3} & 1 & 0 \\ 0_{1\times3} & 0 & 1 \end{bmatrix} \tag{7}$$

$$\xi = \log(\chi) = \begin{bmatrix} \theta a \\ J_l^{-1} v \\ J_l^{-1} \rho \end{bmatrix} \tag{8}$$

When θ is a small angle, $\exp_m(\phi_\times) \approx \mathrm{I}_3 + (\phi_\times)$. j_l and j_l^{-1} are given by the following equation:

$$J_l = \frac{\sin\theta}{\theta} I_3 + \left(1 - \frac{\sin\theta}{\theta}\right) aa^T + \left(\frac{1-\cos\theta}{\theta}\right)(a_\times) \tag{9}$$

$$J_l^{-1} = \frac{\theta}{2}\frac{\cot\theta}{2} I_3 + \left(1 - \frac{\theta}{2}\frac{\cot\theta}{2}\right) aa^T - \frac{\theta}{2(a_\times)} \tag{10}$$

The BCH formula shows that, for $\xi_1, \xi_2 \in \mathbb{R}^9$, there is $\exp(\xi_1)\exp(\xi_2) \approx \exp(\xi_1 + \xi_2)$. The so-called concomitant operator is defined by analogy with SE (3) as

$$Ad_T = \begin{pmatrix} R & 0_{3,3} & 0_{3,3} \\ (V)_\times R & R & 0_{3,3} \\ (P)_\times R & 0_{3,3} & R \end{pmatrix} \in \mathbb{R}^{9\times 9} \tag{11}$$

The system demonstrates a property known as group affine, as the deterministic dynamical model adheres to the following criterion:

$$f_u(\chi\tilde{\chi}) = f_u(\chi)\chi + \chi f_u(\tilde{\chi}) - \chi f_u(\tilde{I})\chi \tag{12}$$

Dynamic systems that adhere to the aforementioned relations are classified as exhibiting group affine dynamics. A notable characteristic of group affine dynamics is the independence of its error state model from the global state estimate. Specifically, both the left-invariant error and the right-invariant error remain unaffected by the trajectory.

3.2. Two Frame Groups

In addressing the multi-source data fusion navigation problem, particularly in fusion with other observation sensors, the formulation often encompasses various coordinate systems. These include the inertial coordinate system (i system), the navigation coordinate system (n system), and the Earth coordinate system (e system), among others [51]. For the characterization of underwater vehicle motion, a fixed navigation frame and a time-varying self-frame are typically defined [17]. Here, the state space is designated by

$$\chi = (R, x, X)^T \tag{13}$$

where the rotation matrix R_n signifies the transformation mapping vectors represented in the body frame to the world frame at time step n. Notably, x denotes variables within the fixed frame, while X represents variables within the body frame. In a product space $G \times V \times B$, where $R \in G$, $x \in V$, and $X \in B$,

Define the $*$ operator as of G on W, where the vector space $W = \mathbb{R}^{Nd}$; i.e., W is an N-vector tuple of $\mathbb{R}^d$, as follows:

$$R * (x^1, ..., x^N) := (Rx^1, ..., Rx^N) \tag{14}$$

Define the natural output in the fixed frame and body frame as follows:

$$\begin{cases} \text{Fixed-frame:} & h(R,x,X) = H_x^x + R * [H_X^X + B] \\ \text{Body-frame:} & H(R,x,X) = R^{-1} * [b - H^x x] - H_X^X \end{cases} \tag{15}$$

Define discrete-time natural vector dynamics as $\chi_n = f_n(\chi_n - 1)$, where

$$f_n\begin{pmatrix} R \\ x \\ X \end{pmatrix} = \begin{pmatrix} R \\ [F_n x + d_n] + R * [C_n X + U_n] \\ [\Phi_n X + \mathcal{D}_n] + R^{-1} * [\Gamma_n x + u_n] \end{pmatrix} \tag{16}$$

where the matrices $F_n : V \mapsto V$, $\phi_n : B \mapsto B$, $C_n : B \mapsto V$, and $\Gamma_n : V \mapsto B$ are all exchanged under the action of G.

Define the natural frame dynamics as $\chi_n = s_n(\chi_n - 1)$, where

$$s_n\begin{pmatrix} R \\ x \\ X \end{pmatrix} = \begin{pmatrix} s_n^R(\chi) \\ x \\ X \end{pmatrix} = \begin{pmatrix} O_n R \Omega_n \\ x \\ X \end{pmatrix} \tag{17}$$

where $\Omega_n \in G$ has known elements.

Define a natural two-frame observed system as follows:

$$\begin{aligned} \chi_n &= s_n \circ f_n(\chi_{n-1}) \\ y_n &= h_n(\chi_n) \quad \text{or} \quad \mathrm{Y}_n = \mathrm{H}_n(\chi_n) \end{aligned} \tag{18}$$

Define the group structure on the two-frame state space $G \times V \times B$ as follows:

$$\begin{pmatrix} R_1 \\ x_1 \\ X_1 \end{pmatrix} \bullet \begin{pmatrix} R_2 \\ x_2 \\ X_2 \end{pmatrix} = \begin{pmatrix} R_1 R_2 \\ x_1 + R_1 * x_2 \\ X_2 + R_2^{-1} * X_1 \end{pmatrix} \tag{19}$$

where those having the structure $G \times V \times B$ are denoted by $G^+_{V,B}$.

Define the group action on the output space in terms of the operator $\star$ as follows:

$$\begin{aligned} &G^+_{V,B} \times \mathrm{Y} \mapsto \mathrm{Y} \\ &(R, x, \mathbf{x}) \star \beta = H^x x + R * [H^x X + \beta] \end{aligned} \tag{20}$$

For $\chi_n = \phi_n(\chi_(n-1))$ and $\chi_1, \chi_2 \in G^+_{V,B}$, if $\varphi_n : G^+_{V,B} \to G^+_{V,B}$:

$$\phi_n(\chi_1 \bullet \chi_2) = \phi_n(\chi_1) \bullet \phi_n(Id)^{-1} \bullet \phi_n(\chi_2) \tag{21}$$

Then the dynamics are considered to satisfy the "group affine property".

3.3. Two Frame Groups IEKF

A linear observational system on a group is a dynamical system observed by measuring y_n or Y_n, denoted as follows:

$$\begin{cases} \text{Left action case}: & \chi_n = \phi_n(\chi_{n-1}) \quad y_n = \chi_n \star b \\ \text{Right action case}: & \chi_n = \phi_n(\chi_{n-1}) \quad \mathrm{Y}_n = \chi_n^{-1} \star b \end{cases} \tag{22}$$

where ϕ_n satisfies the group affine property.

Building alternative innovation terms,

$$Z_n = \hat{\chi}^{-1}_{n|n-1} \star y_n - b \quad \text{or} \quad z_n = \hat{\chi}_{n|n-1} \star \mathrm{Y}_n - b \tag{23}$$

The propagation and update steps of a linear observer are defined as follows:

$$\begin{cases} \text{Propagation step}: & \hat{\chi}_{n|n-1} = \phi_n(\hat{\chi}_{n-1|n-1}) \\ \text{Update step}: & \begin{cases} \hat{\chi}_{n|n} = \hat{\chi}_{n|n-1} \cdot L_n(Z_n) \\ \hat{\chi}_{n|n} = L_n(z_n) \cdot \hat{\chi}_{n|n-1} \end{cases} \end{cases} \tag{24}$$

The left-invariant error variable and the right-invariant error variable are denoted as follows:

$$\begin{cases} E_{n|n} = \hat{\chi}^{-1}_{n|n} \cdot \chi_n \\ e_{n|n} = \chi_n \cdot \hat{\chi}^{-1}_{n|n} \end{cases} \tag{25}$$

The updated terms for IEKF are as follows:

$$\begin{cases} L_n(Z_n) = \exp_{G^+_{V,B}}(K_n Z_n) \\ L_n(z_n) = \exp_{G^+_{V,B}}(K_n z_n) \end{cases} \tag{26}$$

where $exp_{G^{+}_{V,B}}$ is the exponential mapping of TFG and K_n represents Kalman gain.

The linearization error is defined as follows:

$$\begin{cases} \xi_{n|n-1} = A^s_n A^v_n \xi_{n-1|n-1} \\ \xi_{n|n} = (I - K_n H_n)\xi_{n|n-1} \end{cases} \tag{27}$$

where the Jacobi matrices as A^s_n, A^v_n, and H_n correspond to the frame dynamics, vector dynamics, and output mapping, respectively, tuned by the following Riccati equations.

The gain of TFG-IEKF is tuned using the following equation, and the final solution is shown in Figure 1.

$$\begin{cases} P_{n|n-1} = A^s_n A^v_n P_{n-1|n-1} (A^s_n A^v_n)^T + \hat{Q}_n \\ \mathbf{s}_n = H_n P_{n|n-1} H^T_n + \hat{N}_n \\ K_n = P_{n|n-1} H^T_n S^{-1}_n \\ P_{n|n} = (I - K_n H_n) P_{n|n-1} \end{cases} \tag{28}$$

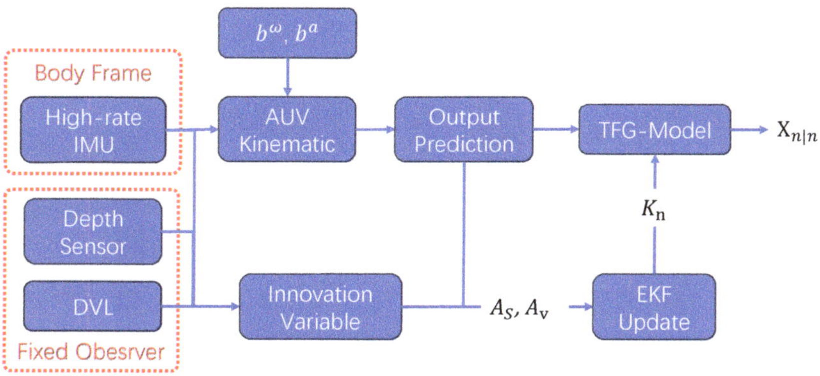

Figure 1. AUV is localized by multi-source data from underwater TFG-IEKF.

3.4. Algorithmic Implementation

The innovation variable is associated with the fixed frame output as follows:

$$\begin{cases} Z_n = \hat{R}^{-1}_{n|n-1} * (y_n - H^x \hat{x}_{n|n-1}) - H^X \hat{X}_{n|n-1} - B_n \\ z_n = \hat{R}_{n|n-1} * \left(Y_n + H^X \hat{X}_{n|n-1}\right) + H^x \hat{x}_{n|n-1} - b_n \end{cases} \tag{29}$$

After applying vector dynamics and frame dynamics and considering observations, respectively, the estimate of moment n is expressed as $\hat{\chi}_{n^-|n-1}$, $\hat{\chi}_{n|n-1}$ and $\hat{\chi}_{n|n}$. Then, the fixed frame observer of the left-invariant observer is implemented as follows:

$$\begin{cases} \hat{\chi}_{n-|n-1} = f_n(\hat{\chi}_{n-1|n-1}) \\ \hat{\chi}_{n|n-1} = s_n(\hat{\chi}_{n-|n-1}) \\ \hat{\chi}_{n|n} = \begin{pmatrix} \hat{R}_{n|n-1} L^R_n(Z_n) \\ \hat{x}_{n|n-1} + \hat{R}_{n|n-1} * L^x_n(Z_n) \\ L^X_n(Z_n) + L^R_n(Z_n)^{-1} * \hat{X}_{n|n-1} \end{pmatrix} \end{cases} \tag{30}$$

Similarly, under the observation of the body frame, the constant observation value is written as follows:

$$\begin{cases} \hat{\mathcal{X}}_{n-|n-1} = f_n\left(\hat{\mathcal{X}}_{n-1|n-1}\right) \\ \hat{\mathcal{X}}_{n|n-1} = s_n\left(\hat{\mathcal{X}}_{n-|n-1}\right) \\ \hat{\mathcal{X}}_{n|n} = \begin{pmatrix} L_n^R(z_n)\hat{R}_{n|n-1} \\ L_n^x(z_n) + L_n^R(z_n) * \hat{x}_{n|n-1} \\ \hat{X}_{n|n-1} + \hat{R}_{n|n-1}^{-1} * L_n^X(z_n) \end{pmatrix} \end{cases} \quad (31)$$

Rewrite the left invariant error $E_{n|n}$ using the TFG method as follows:

$$E_{n|n} = \begin{pmatrix} \hat{R}_{n|n}^{-1} R_n \\ \hat{R}_{n|n}^{-1} * (x_n - \hat{x}_{n|n}) \\ X_n - \left(R_n^{-1}\hat{R}_{n|n}\right) * \hat{X}_{n|n} \end{pmatrix} \quad (32)$$

Similarly, the right invariant error is represented in the body frame as follows:

$$e_{n|n} = \begin{pmatrix} R_n\hat{R}_{n|n}^{-1} \\ x_n - \left(R_n\hat{R}_{n|n}^{-1}\right) * \hat{x}_{n|n} \\ \hat{R}_{n|n} * \left(X_n - \hat{X}_{n|n}\right) \end{pmatrix} \quad (33)$$

3.5. Underwater Multi-Sensor TFG Model

This section provides details on the Multi-Sensor TFG Model for the AUV, bottoming velocity measurements via DVL and depth measurements via manometers, which perfectly matches the TFG assumptions. Based on the transformations of Equation (30) and considering the innovation variable (Equation (29)), the discrete-time dynamics are expressed as follows:

$$\begin{cases} R_{n^-|n-1} = R_{n-1|n-1} \\ v_{n^-|n-1} = v_{n-1|n-1} + \Delta t\left(g + R_{n-1|n-1}(a_n + b_{n-1}^a)\right) \\ p_{n^-|n-1} = p_{n-1|n-1} + \Delta t v_{n-1|n-1} \\ b_{n^\omega-|n-1}^\omega = b_{n-1|n-1}^\omega, b_n^a = b_{n-1|n-1}^a \end{cases} \quad (34)$$

$$\begin{cases} R_{n|n-1} = R_{n-|n-1}\exp(\Delta t[\omega_n + b_{n-1}^\omega]_x) \\ v_{n|n-1} = v_{n-|n-1} \\ p_{n|n-1} = p_{n-|n-1} \\ b_{n|n-1}^\omega = b_{n-|n-1}^\omega, \cdot b_n^a = b_{n-|n-1}^a \end{cases} \quad (35)$$

$$\begin{cases} \hat{R}_{n|n} = \hat{R}_{n|n-1}\exp_{SO(3)}(K_n^R z_n) \\ \hat{v}_{n|n} = \hat{v}_{n|n-1} + \hat{R}_{n|n-1} * \nu_3(K_n^R z_n)K_n^v z_n \\ \hat{p}_{n|n} = \hat{p}_{n|n-1} + \hat{R}_{n|n-1} * \nu_3(K_n^R z_n)K_n^p z_n \\ \hat{b}_{n|n}^\omega = \nu_3(-K_n^R z_n)K_n^{b^\omega} z_n + (\nu_3(-K_n^R z_n)K_n^R z_n)^{-1} * \hat{b}_{n|n-1}^\omega \\ \hat{b}_{n|n}^a = \nu_3(-K_n^R z_n)K_n^{b^a} z_n + (\nu_3(-K_n^R z_n)K_n^R z_n)^{-1} * \hat{b}_{n|n-1}^a \end{cases} \quad (36)$$

where $R_n \in G$ denotes the transformation of the body frame attached to the IMU mapped to the fixed frame of Earth at moment n, $p_n \in \mathbb{R}^3$ denotes the position of the body in space, $v_n \in \mathbb{R}^3$ denotes the velocity of the body, g is Earth's gravity vector, $a,\omega_n \in \mathbb{R}^3$ denote the accelerometer and gyroscope signals, b_n^a denotes the accelerometer bias, and b_n^ω denotes the gyroscope bias, $\nu_3(\xi) = I_3 + \frac{1-\cos(\|\xi\|)}{\|\xi\|^2}(\xi)_\times + \frac{\|\xi\|-\sin(\|\xi\|)}{\|\xi\|^3}(\xi)_\times^2$.

In this paper, we set $x_n = (p_n, v_n) \in \mathbb{R}^3 \times \mathbb{R}^3$, and $X_n = (b_n^\omega, b_n^a) \in \mathbb{R}^3 \times \mathbb{R}^3$. According to Equations (16) and (17), we obtain the following:

$$\begin{aligned}
\boldsymbol{F}_n &= \begin{pmatrix} \boldsymbol{I}_3 & \mathbf{0}_3 \\ \Delta t \boldsymbol{I}_3 & \boldsymbol{I}_3 \end{pmatrix} \\
\boldsymbol{C}_n &= \begin{pmatrix} 0_3 & \Delta t \boldsymbol{I}_3 \\ 0_3 & 0_3 \end{pmatrix} \\
d_n &= \begin{pmatrix} \Delta t g \\ 0 \end{pmatrix} \\
U_n &= \begin{pmatrix} \Delta t a_n \\ \mathbf{0}_{3,1} \end{pmatrix} \\
\Phi_n &= \begin{pmatrix} I_3 & 0_3 \\ 0_3 & I_3 \end{pmatrix} \\
\Gamma_n &= 0_6 \\
D_n &= u_n = 0_{6,1}
\end{aligned} \tag{37}$$

For observation, it is necessary to establish that the velocity measurements are of the fixed system and the IMU measurements belong to the body frame. Additionally, velocity and depth information are acquired within our body coordinate system via the DVL and PS, denoted as follows:

$$y_n = \begin{pmatrix} V_{3,1}^{DVL} \\ 0_{2,1} \\ Depth \end{pmatrix} \tag{38}$$

According to the formula, it is possible to obtain the following:

$$\begin{aligned}
H^x &= \begin{pmatrix} I_3 & 0_{3,1} & 0_{3,2} \\ 0_{1,3} & I_1 & 0_1 \\ 0_{2,3} & 0_{2,1} & 0_{2,2} \end{pmatrix} \\
H^X &= 0_{3,6} \\
B &= 0
\end{aligned} \tag{39}$$

The Jacobian matrix of $E_{n|n}$ for fixed frame observations is as follows:

$$\begin{aligned}
A_n^v &= \begin{pmatrix} \boldsymbol{I}_d & \mathbf{0}_{d,q} & \mathbf{0}_{d,r} \\ -(\mathbf{u}_n)_* & \boldsymbol{F}_n & \boldsymbol{C}_n \\ -(\mathbf{D}_n)_* & \boldsymbol{\Gamma}_n & \boldsymbol{\Phi}_n \end{pmatrix} \\
H_n &= \begin{pmatrix} -(\mathbf{B}_n)_* & \mathbf{H}^x & \mathbf{H}_\times \end{pmatrix} \\
A_n^s &= \begin{pmatrix} \boldsymbol{Ad}_{\boldsymbol{\Omega}_n}^{-1} & \mathbf{0}_{d,q} & \mathbf{0}_{d,r} \\ 0_{q,d} & (\boldsymbol{\Omega}_n^{-1} \boldsymbol{O}^{-1})* & \mathbf{0}_{q,r} \\ \mathbf{0}_{d,r} & \mathbf{0}_{q,r} & \boldsymbol{I}_r \end{pmatrix}
\end{aligned} \tag{40}$$

Proposition: the Jacobian of the above frame dynamics is written as follows:

$$A_n^v = \begin{pmatrix} \boldsymbol{I}_3 & \mathbf{0}_{3,3} & \mathbf{0}_{3,3} & \mathbf{0}_{3,3} & \mathbf{0}_{3,3} \\ -\Delta t(\boldsymbol{a}_n)_\times & \boldsymbol{I}_3 & \mathbf{0}_{3,3} & \mathbf{0}_{3,3} & \Delta t\boldsymbol{I}_3 \\ \mathbf{0}_{3,3} & \Delta t\boldsymbol{I}_3 & \boldsymbol{I}_3 & \mathbf{0}_{3,3} & \mathbf{0}_{3,3} \\ \mathbf{0}_{3,3} & \mathbf{0}_{3,3} & \mathbf{0}_{3,3} & \boldsymbol{I}_3 & \mathbf{0}_{3,3} \\ \mathbf{0}_{3,3} & \mathbf{0}_{3,3} & \mathbf{0}_{3,3} & \mathbf{0}_{3,3} & \boldsymbol{I}_3 \end{pmatrix}$$
$$H_n = \begin{pmatrix} \mathbf{0}_{3,3} & \begin{pmatrix} \boldsymbol{I}_4 & \\ & \mathbf{0}_{2,2} \end{pmatrix} & \mathbf{0}_{3,6} \end{pmatrix} \tag{41}$$
$$A_n^s = \begin{pmatrix} I_3 & \mathbf{0}_{3\times3} & \mathbf{0}_{3\times3} & M_1 & \mathbf{0}_{3\times3} \\ \mathbf{0}_{3\times3} & M_4 & \mathbf{0}_{3\times3} & \mathbf{0}_{3\times3} & \mathbf{0}_{3\times3} \\ \mathbf{0}_{3\times3} & \mathbf{0}_{3\times3} & M_4 & \mathbf{0}_{3\times3} & \mathbf{0}_{3\times3} \\ \mathbf{0}_{3\times3} & \mathbf{0}_{3\times3} & \mathbf{0}_{3\times3} & I_3 + (\hat{b}^w_{n-|n-1})_\times \cdot M_1 & \mathbf{0}_{3\times3} \\ \mathbf{0}_{3\times3} & \mathbf{0}_{3\times3} & \mathbf{0}_{3\times3} & (\hat{b}^a_{n-|n-1})_\times \cdot M_1 & I_3 \end{pmatrix}$$

where

$$\begin{cases} M_1 = \Delta t \hat{R}_{n|n-1} \bar{J} \hat{R}^T_{n-|n-1} \\ M_4 = (R_{n-|n-1} \cdot R_{n|n-1})^T \\ \bar{J} \;= I_3 - \dfrac{1-\cos(\|\mu\|)}{\|\mu\|^2}(\mu)_\times - \dfrac{\sin(\|\mu\|) - \|\mu\|}{\|\mu\|^3}(\mu)^2_\times \\ \mu \;:= \omega + \hat{b}^\omega \end{cases} \tag{42}$$

Proof. We study the effect of propagation through frame dynamics, $R_n = R_{n-1}\exp_m(\Delta t[\omega_n + b^\omega_{n-1}]_\times)$ and $\hat{R}_{n|n-1} = \hat{R}_{n-|n-1}\exp_m\left(\Delta t[\omega_n + \hat{b}^\omega_{n-|n-1}]_\times\right)$. After frame propagation, the error e^R is equal to

$$\begin{aligned} R_n \hat{R}^{-1}_{n|n-1} =& R_{n-1}\exp_m(\Delta t\left[\omega_n + b^\omega_{n-1}\right]_\times) \\ & \exp_m\left(-\Delta t\left[\omega_n + \hat{b}^\omega_{n-|n-1}\right]_\times\right)\hat{R}^T_{n-|n-1} \end{aligned} \tag{43}$$

For small error perturbations, this can be written as $\omega_n + b^\omega = \omega_n + \hat{b}^\omega + \delta b^\omega$, then:

$$\begin{aligned} \exp_m(\Delta t[\omega + b^\omega]_\times) &= \exp_m\left(\Delta t[\omega + \hat{b}^\omega + \delta b^\omega]_\times\right) \\ &\approx \exp_m(\Delta t[\omega + b^\omega]_\times)\exp_m(\Delta t[J\delta b^\omega]_\times) \end{aligned} \tag{44}$$

Thus, we can write $R = R\hat{R}^T\hat{R} = e^R\hat{R}$. These indices make $R_{n-1} = e^R_{n-|n-1}\hat{R}_{n-|n-1}$. Since we have that

$$\begin{aligned} e^R_{n|n-1} &= e^R_{n-|n-1}\hat{R}_{n|n-1}\exp_m(\Delta t[J\delta b^\omega]_\times)\hat{R}^T_{n|n-1} \\ &= e^R_{n-|n-1}\exp_m\left(\Delta t\left[\hat{R}_{n|n-1}J\delta b^\omega\right]_\times\right) \end{aligned} \tag{45}$$

Since the BCH values of both terms are small, they can be approximated by $\exp_m([\xi^R + \Delta t\hat{R}J\delta b^\omega]_\times)$, and we recall that $e^R = \exp_m([\xi^R]_\times)$. The right invariant error of the bias is defined as $\xi^\omega := \hat{R}(b^\omega - \hat{b}^\omega) = \hat{R}\delta b^\omega$.Then

$$\begin{aligned} \hat{R}_{n|n-1}J\delta b^\omega_{n-|n-1} &= \hat{R}_{n|n-1}J\hat{R}^T_{n-|n-1}\hat{R}_{n-|n-1}\delta b^\omega_{n-|n-1} \\ &= \left(\hat{R}_{n|n-1}J\hat{R}^T_{n-|n-1}\right)\xi^\omega_{n-|n-1} \end{aligned} \tag{46}$$

The left-invariant error is denoted as $e^x = \hat{R}^{-1} * (x - \hat{x})$. For example, for velocity v, we have $\hat{R}(v - \hat{v}) := \xi^v$. The frame dynamics are expressed as $\hat{R}^{-1}_{n|n-1}(v - \hat{v}) = \hat{R}^{-1}_{n|n-1}\hat{R}^T_{n-|n-1}\hat{R}_{n-|n-1}(v - \hat{v}) = M_4\xi^v$.

In the case of gyro bias error, it can be written as $e^{b^\omega} = b^\omega - (R_n^{-1}\hat{R}_{n|n})\hat{b}^\omega = b^\omega - e^{R^T}\hat{b}^\omega$. As shown in Equation (46) and based on the properties of skew-symmetric matrices, it can be obtained that $b^\omega - (e^R \exp_m[M_1\zeta^\omega]_\times)^T\hat{b}^\omega \approx b^\omega - e^{R^T}\hat{b}^\omega - [M_1\zeta^\omega]_\times\hat{b}^\omega = e^{b^\omega} + [\hat{b}^\omega]_\times M_1\zeta^{b^\omega}$. □

4. Experimental Setup

Given the intrinsic relationship between lateral stability and navigation accuracy, we have utilized a rotary-body underwater vehicle to conduct comprehensive navigation tests. The evaluation encompasses navigation data acquired from both simulated scenarios and real-world lake–sea navigation environments. Subsequently, all obtained results are compared with those generated by an EKF-based solver.

4.1. Comparison Term

This paper presents a comparative analysis of classical and advanced Lie group-based underwater localization techniques, all leveraging IMU and DVL data sources.

1. IEKF [39]: By exploiting the "log-linear" property of error evolution, this method exhibits enhanced filtering capabilities and accelerated convergence, particularly addressing the challenges posed by initial large errors while maintaining group affine satisfaction.
2. U/W-IEKF [48]: Tailored for IMU and DVL-equipped underwater vehicles, this approach integrates non-standard single-case measurements, such as depth readings from pressure transducers, into the Iterated Extended Kalman Filter framework via "pseudo" measurements. These pseudo measurements encapsulate current state estimates modeled with infinite covariance.
3. Underwater-TFG-IEKF: This method, proposed in our study, introduces novel techniques for underwater localization.

4.2. Simulation

The simulation data employed in this study was generated using HoloOcean 0.5.0 [52,53], a simulation environment built upon Unreal Engine 4. A torpedo-type AUV, equipped with an IMU, DVL, and PS, was employed. The sensor noise characteristics are detailed in Table 2. Specifically, the IMU data were sampled at a frequency of 200 Hz, DVL at 10 Hz, and pressure sensors at 100 Hz, with synchronization performed at the lowest frequency available. The simulated navigator and underwater environment are shown in Figure 2.

Table 2. Noise statistics for simulation.

Sensor	Noise Type	Noise Standard Bias
IMU	Angular velocity	0.00277 rad/s/Hz
	Linear acceleration	0.00123 m/s^2/Hz
	Angular velocity bias	0.00141 rad/s^2/ Hz
	Linear acceleration bias	0.00388 m/s^3/ Hz
DVL	Velocity per beam	0.02626 m/s
	Range per beam	0.1 m
PS	Depth sensor	0.255 m

The heading and depth adjustments of the navigator are achieved through control of the tail fin angle and thruster parameters. In simulation, four distinct trajectories are employed, and the navigation is conducted within an open water environment. As setup errors were superimposed on the trajectories during the simulation, the treatment of individual trajectories was insufficient to demonstrate the effectiveness of the filter. To address this issue, the algorithm was evaluated using multiple trajectories generated from identical inputs, with consistent propeller and rudder angles. Therefore, by varying the starting point, each trajectory set was subjected to 100 Monte Carlo simulations to eliminate the effects of random errors and biases on the filter.

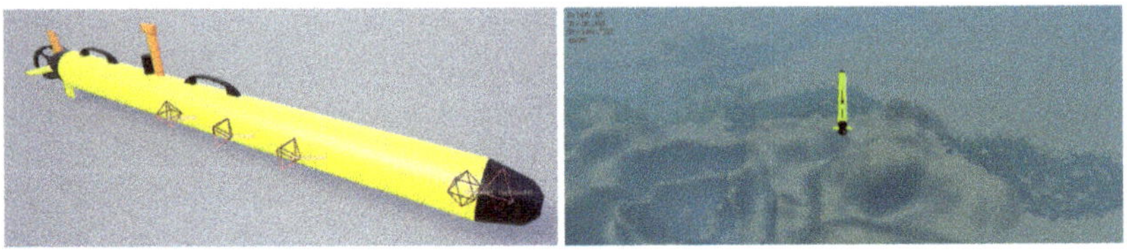

(a) (b)

Figure 2. HoloOcean simulation environment and vehicles, where (**a**) is the simulation environment of HoloOcean in open water, and (**b**) is the slewing body-type vehicle in HoloOcean [53].

The simulator delivers highly accurate positional data directly, effectively avoiding the inaccuracies and costs associated with acquiring actual positional data in real-world applications. To replicate real-world conditions, fixed-depth sailing is adopted for data generation. The location and depth of the starting point of each track are $(1.4, -100.5, -151.0)$, $(233.9, 41.7, -172.5)$, $(-659.3, -12.4, -126.0)$, and $(158.7209.9, -169.0)$, respectively. A comparative analysis of algorithmic outcomes across diverse trajectories is presented in Figure 3.

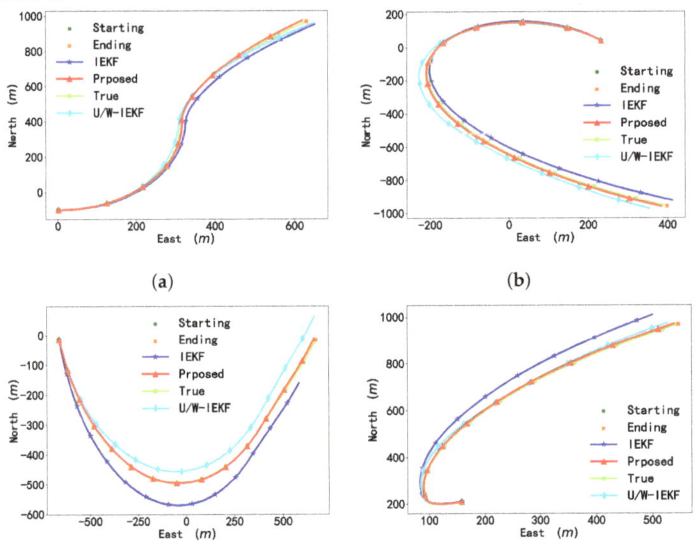

(a) (b)

(c) (d)

Figure 3. Comparison of localization results for four sets of trajectories in the simulation environment. The original data represent the online data after INS/DVL fusion by the AUV in a specific way (integrated within the Inertial Unit), with (**a**) trajectory 1, (**b**) trajectory 2, (**c**) trajectory 3, and (**d**) trajectory 4.

4.3. Natural Scenario Data

The dataset from a previous study [54] is employed in this research to conduct practical validation of the proposed algorithm using data gathered from natural environments. The validation process includes offline computation of global data, followed by comparison with the GNSS positions of the vehicle when it was not submerged in water. It is important to note that the GNSS positions are obtained in differential mode, and any errors associated with them are not considered in this study. Figure 4 illustrates both the vehicle and the acquisition environment employed for data collection. The INS and DVL are equipped with pole arm of dimensions $l = [0, 26, 9]$ mm. The mounting angle is negligible as it can be pre-measured, thus eliminating its effect. The GPS trajectory differs from the true trajectory only in terms of time delay, and the overall consideration of the trajectory allows the pole arm between the GPS and the body to be ignored.

(a) (b)

Figure 4. Vehicle used for data acquisition and data acquisition environment, where (**a**) shows the AUV for data acquisition in a lake, and (**b**) shows the data acquisition in a marine environment.

For the navigator data collected at different latitudes and salinities, the proposed algorithm is used to perform a multi-sensor fusion of waypoint projection. The algorithm is set up with uniform parameters where $Q = diag([I_3, I_3, I_3, 0.1 \cdot I_3, 0.001 \cdot I_3])$, $Cov_{DVL} = 0.6 \cdot I_3$, $N = diag(Cov_{DVL}, I_2, 0.01)$. The trajectory results are shown in Figure 5. The true position is converted from latitude and longitude to a local position using the European Petroleum Survey Group (EPSG) database [55], starting at $(0, 0)$ in meters.

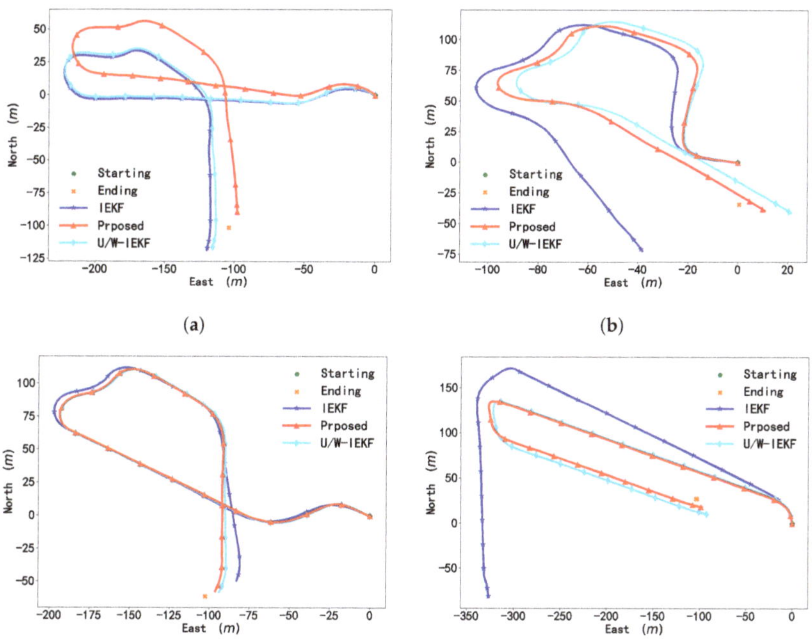

(a) (b)

(c) (d)

Figure 5. Comparison of REs between the proposed algorithm and the state-of-the-art Underwater IEKF method for each set of trajectories at different time scales. with (**a**) scenario 1, (**b**) scenario 2, (**c**) scenario 3, and (**d**) scenario 4.

5. Results and Discussion

The comparison of simulation results (Figure 3) reveals that the proposed TFG-IEKF algorithm exhibits superior localization accuracy across various trajectories compared with the underwater IEKF algorithm, which similarly employs Lie group optimization techniques. This superiority can be attributed to the sequential solving of IEKF, where the solving procedure data are conducted only upon input of a particular sensor class, whereas the proposed algorithm optimizes all data types within a uniform framework following frequency synchronization.

The algorithm's effectiveness is demonstrated by its ability to manage random data errors. We examine the role of the algorithm on trajectory accuracy using RMSE. The results are averaged over multiple times for the same trajectory, and the results are shown in Table 3. The quantitative analysis reveals that the proposed algorithm improves positional accuracy by 77.00% compared with the classical IEKF algorithm and by 57.21% compared with the U/W-IEKF method. It should be noted that the U/W-IEKF method is not stable enough, especially for trajectory 2, resulting in a large error. Excluding trajectory 2, the proposed method shows a 76.01% improvement over the U/W-IEKF method. Regarding attitude accuracy, the proposed method improves by 74.52% over the IEKF method and by 7.32% over the U/W-IEKF method. In conclusion, while the U/W-IEKF method is relatively effective for attitude optimization, the proposed algorithm provides significant improvements in both positional and attitude accuracy.

Table 3. The RMSE of the AUV position and attitude.

Trajectory No.		Traj.1	Traj.2	Traj.3	Traj.4
Position (m)	IEKF	56.5914	69.6770	73.3606	48.9478
	U/W-IEKF	50.8412	93.0147	72.7105	45.4038
	Proposed	**11.6488**	**16.8054**	**20.5661**	**9.4366**
Attitude (rad)	IEKF	−0.0249	0.0238	−0.0208	0.0082
	U/W-IEKF	0.0136	0.0205	0.0023	−0.0142
	Proposed	**−0.0057**	**0.0074**	**−0.0066**	**−0.0013**

To assess the robustness of the algorithm throughout the trajectory optimization process, we employ metrics such as Relative Error (RE) [56]. The set $\tilde{S} = \{d_k\}_k^K$, $d_k = \{\hat{x}_s, \hat{x}_s\}$, of K pairs of states is chosen from $\hat{X}$ based on some criteria (e.g., distances along the trajectory). Then RE is denoted by the following:

$$RE = \{\|p_e - \Delta R_k \hat{p}'_e\|_2\}_{k=1}^{K-1} \tag{47}$$

where K is the distance or time interval. In this paper, the time step size criterion $K = \{3, 7, 13, 19, 29, 37\}s$ is selected. The comparison results between the proposed algorithm and the U/W-IEKF method are shown in Figure 6. It can be observed that, although the proposed algorithm exhibits significant uncertainty in the short term, this uncertainty decreases over time and consistently outperforms the U/W-IEKF method.

For real underwater navigation data, the position of AUV upon surfacing and acquiring the Real-Time Kinematic (RTK) signal is regarded as the definitive endpoint. The total navigation error for the entire voyage is computed using the relative metric of horizontal position error (Hori-Error) percentage (PEP) [57] as follows:

$$PEP = \frac{Hori\text{-}Error}{D} \times 100\% \tag{48}$$

where D is the Total Distance Traveled for each trajectory.

We computed the total distance traveled for each trajectory, along with the PEP error at the endpoint. The calculated results are presented in Table 4. The results demonstrate that the proposed algorithm achieves superior accuracy across most trajectories, with the U/W-IEKF method showing the second-best performance. In certain trajectories, both the proposed algorithm and the U/W-IEKF method yield the same results. Overall, the proposed algorithm enhances accuracy by 77.73% compared with the IEKF method and by 52.69% compared with the U/W-IEKF method.

Based on the results of simulation and dataset optimization, the proposed algorithm exhibits significant benefits in improving the precision of underwater navigation. State updating is achieved by relying on Equations (34)–(36) without the need for auxiliary vectors. Evaluation with the latest U/W-IEKF method demonstrates comparable performance. However, as depicted in Figure 6, the proposed algorithm shows dispersion in positioning accuracy over short time intervals, yet it demonstrates enhanced optimization capability over longer durations. This illustrates the algorithm's stability throughout the optimization process. Subsequent research efforts will focus on conducting rigorous stability analyses and validation to substantiate the algorithm's efficacy.

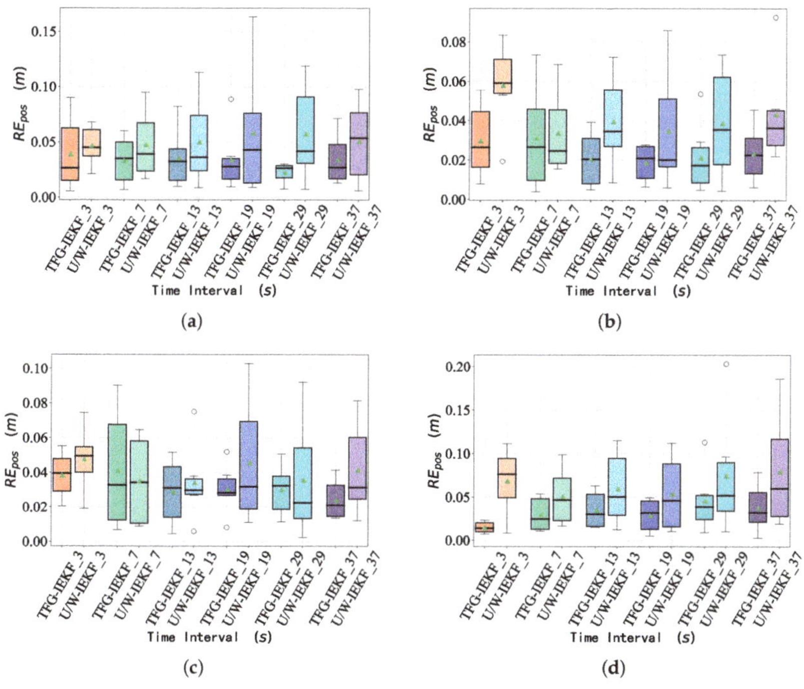

Figure 6. Various types of AUV tracks in different regions, with (**a**) RE of Traj. 1, (**b**) RE of Traj. 2, (**c**) RE of Traj. 3, and (**d**) RE of Traj. 4. In the figure, circles indicate outliers, triangles indicate means, and the horizontal line in the box indicates the median.

The algorithm presented exhibits improved performance compared with the current underwater Lie groups' filtering algorithm when tested on actual datasets. However, the algorithm's performance is constrained in a few trajectories, likely due to interference from water currents or significant inertial misalignments during specific intervals. Notably, the proposed algorithm achieves high accuracy without necessitating intricate transformations. Furthermore, the algorithm demonstrates better outcomes with ideal data (errors following a Gaussian distribution) in comparison with actual navigation data. Nevertheless, the algorithm's enhanced accuracy across all trajectories highlights its efficacy.

Table 4. PEP for multiple trajectories in the dataset.

Traj. Seq.	Total Mileage (m)	PEP		
		IEKF	U/W-IEKF	Proposed
20210428_1_1	791.4829	0.0247	**0.0052**	**0.0052**
20210605_0_0	753.1149	0.0653	0.0208	**0.0192**
20220718_1_1	409.2852	0.1083	0.1010	**0.0148**
20220718_2_1	451.1789	0.0307	0.0141	**0.0122**
20220719_1_2	235.1641	0.0414	0.0249	**0.0197**
20220816_0_4	1138.8411	0.1127	**0.0143**	**0.0142**

6. Conclusions

This paper addresses the challenge of multi-sensor fusion in underwater navigation by employing Lie group theory for invariant error-based fusion. The proposed algorithm circumvents the necessity of auxiliary vectors, a common limitation in conventional methods to meet group affine constraints, thereby facilitating predictions and updates within both fixed and body frameworks. The algorithm achieves a 77% improvement in accuracy over the standard IEKF and a 52% improvement over U/W-IEKF by incorporating fixed inputs from DVL and depth data through simulations and actual data analyses. Future work will concentrate on enhancing the algorithm's adaptability to navigators with diverse dynamics and exploring the impact of parameter settings. Similar to AI-IMU research, investigating dynamic parameters through deep learning will be a compelling avenue.

Author Contributions: Conceptualization, C.W. and F.Z.; methodology, C.W. and C.C. (Chensheng Cheng); software, C.W. and C.C. (Chun Cao); validation, C.C. (Chensheng Cheng) and X.G.; formal analysis, C.C. (Chensheng Cheng) and C.C. (Chun Cao); investigation, C.C. (Chensheng Cheng); resources, F.Z. and G.P.; data curation, C.W. and C.C. (Chun Cao); writing—original draft preparation, C.W.; writing—review and editing, C.C. (Chun Cao); visualization, C.W. and C.C. (Chun Cao); supervision, F.Z.; project administration, F.Z.; funding acquisition, F.Z. All authors have read and agreed to the published version of the manuscript.

Funding: This work was supported by the National Key R&D Program of China (2023QYXX).

Institutional Review Board Statement: Not applicable.

Informed Consent Statement: Not applicable.

Data Availability Statement: The data presented in this study are available on request from the corresponding author.

Acknowledgments: The authors gratefully acknowledge the support provided by the Key Laboratory of Unmanned Underwater Transport Technology during the data collection process and the assistance of the research team members.

Conflicts of Interest: The authors declare no conflicts of interest.

Abbreviations

The following abbreviations are used in this manuscript:

IEKF	Invariant Extended Kalman filter
AUV	Autonomous Underwater Vehicle
DVL	Doppler Velocity Log
IMU	Inertial Measurement Unit
SO	special Orthogonal group
SE	special Euclidean groups
BCH	Baker–Campbell–Hausdorff

References

1. Yan, J.; Ban, H.; Luo, X.; Zhao, H.; Guan, X. Joint Localization and Tracking Design for AUV with Asynchronous Clocks and State Disturbances. *IEEE Trans. Veh. Technol.* **2019**, *68*, 4707–4720. [CrossRef]
2. Naus, K.; Piskur, P. Applying the Geodetic Adjustment Method for Positioning in Relation to the Swarm Leader of Underwater Vehicles Based on Course, Speed, and Distance Measurements. *Energies* **2022**, *15*, 8472. [CrossRef]
3. Huang, H.; Chen, X.; Zhang, B.; Wang, J. High accuracy navigation information estimation for inertial system using the multi-model EKF fusing adams explicit formula applied to underwater gliders. *ISA Trans.* **2017**, *66*, 414–424. [CrossRef] [PubMed]
4. Kim, T.; Kim, J.; Byun, S.W. A Comparison of Nonlinear Filter Algorithms for Terrain-referenced Underwater Navigation. *Int. J. Control. Autom. Syst.* **2018**, *16*, 2977–2989. [CrossRef]
5. Cheng, C.; Wang, C.; Yang, D.; Liu, W.; Zhang, F. Underwater SLAM Based on Forward-Looking Sonar. In Proceedings of the Cognitive Systems and Signal Processing, Zhuhai, China, 25–27 December 2020; Sun, F., Liu, H., Fang, B., Eds.; Springer: Singapore, 2021; pp. 584–593.
6. Bonin-Font, F.; Burguera Burguera, A. NetHALOC: A learned global image descriptor for loop closing in underwater visual SLAM. *Expert Syst.* **2021**, *38*, e12635. [CrossRef]
7. Liu, R.; Liu, F.; Liu, C.; Zhang, P. Modified Sage-Husa Adaptive Kalman Filter-Based SINS/DVL Integrated Navigation System for AUV. *J. Sens.* **2021**, *2021*, 1–8. [CrossRef]
8. Pardos, J.; Balaguer, C. RHO humanoid robot bipedal locomotion and navigation using Lie groups and geometric algorithms. In Proceedings of the 2005 IEEE/RSJ International Conference on Intelligent Robots and Systems, Edmonton, AB, Canada, 2–6 August 2005; pp. 3081–3086. [CrossRef]
9. Barfoot, T. *State Estimation for Robotics*; Cambridge University Press: Cambridge, UK, 2017; pp. 1–368. [CrossRef]
10. Barrau, A. Non-Linear State Error Based Extended Kalman Filters with Applications to Navigation. (Filtres de Kalman étendus Reposant sur une Variable d'erreur non linéaire Avec Applications à la Navigation). Ph.D. Thesis, Mines Paristech, Paris, France, 2015.
11. Barrau, A.; Bonnabel, S. The Invariant Extended Kalman Filter as a Stable Observer. *IEEE Trans. Autom. Control.* **2017**, *62*, 1797–1812. [CrossRef]
12. Hartley, R.; Ghaffari, M.; Eustice, R.M.; Grizzle, J.W. Contact-aided invariant extended Kalman filtering for robot state estimation. *Int. J. Robot. Res.* **2020**, *39*, 402–430. [CrossRef]
13. Chang, L.; Qin, F.; Xu, J. Strapdown Inertial Navigation System Initial Alignment Based on Group of Double Direct Spatial Isometries. *IEEE Sens. J.* **2022**, *22*, 803–818. [CrossRef]
14. Barrau, A.; Bonnabel, S. A Mathematical Framework for IMU Error Propagation with Applications to Preintegration. In Proceedings of the 2020 IEEE International Conference on Robotics and Automation (ICRA), Paris, France, 31 May–31 August 2020; pp. 5732–5738. [CrossRef]
15. Brossard, M.; Barrau, A.; Chauchat, P.; Bonnabel, S. Associating Uncertainty to Extended Poses for on Lie Group IMU Preintegration with Rotating Earth. *IEEE Trans. Robot.* **2022**, *38*, 998–1015. [CrossRef]
16. Tang, H.; Xu, J.; Chang, L.; Shi, W.; He, H. Invariant Error-Based Integrated Solution for SINS/DVL in Earth Frame: Extension and Comparison. *IEEE Trans. Instrum. Meas.* **2023**, *72*, 9500617 . [CrossRef]
17. Barrau, A.; Bonnabel, S. The Geometry of Navigation Problems. *IEEE Trans. Autom. Control.* **2023**, *68*, 689–704. [CrossRef]
18. Shi, W.; Xu, J.; Li, D.; He, H. Attitude Estimation of SINS on Underwater Dynamic Base with Variational Bayesian Robust Adaptive Kalman Filter. *IEEE Sens. J.* **2022**, *22*, 10954–10964. [CrossRef]
19. Jørgensen, E.K.; Fossen, T.I.; Bryne, T.H.; Schjølberg, I. Underwater Position and Attitude Estimation Using Acoustic, Inertial, and Depth Measurements. *IEEE J. Ocean. Eng.* **2020**, *45*, 1450–1465. [CrossRef]
20. Li, P.; Liu, Y.; Yan, T.; Yang, S.; Li, R. A Robust INS/USBL/DVL Integrated Navigation Algorithm Using Graph Optimization. *Sensors* **2023**, *23*, 916. [CrossRef] [PubMed]
21. Huang, H.; Tang, J.; Zhang, B. Positioning Parameter Determination Based on Statistical Regression Applied to Autonomous Underwater Vehicle. *Appl. Sci.* **2021**, *11*, 7777. [CrossRef]
22. OuYang, M.; Fan, H.; Du, P. Particle Swarm Optimization for Underwater Gravity-Matching: Applications in Navigation and Simulation. *Int. Arch. Photogramm. Remote Sens. Spat. Inf. Sci.* **2022**, *XLVI-3/W1-2022*, 149–154. [CrossRef]
23. Jin, K.; Chai, H.; Su, C.; Xiang, M.; Hui, J. An optimization-based in-motion fine alignment and positioning algorithm for underwater vehicles. *Measurement* **2022**, *202*, 111746. [CrossRef]
24. Russo, P.; Di Ciaccio, F.; Troisi, S. DANAE++: A Smart Approach for Denoising Underwater Attitude Estimation. *Sensors* **2021**, *21*, 1526. [CrossRef]
25. Zhang, D.; Chang, S.; Zou, G.; Wan, C.; Li, H. A Robust Graph-Based Bathymetric Simultaneous Localization and Mapping Approach for AUVs. *IEEE J. Ocean. Eng.* **2024**, *49*, 1 –21. [CrossRef]
26. Rahman, S.; Li, A.Q.; Rekleitis, I. SVIn2: An Underwater SLAM System using Sonar, Visual, Inertial, and Depth Sensor. In Proceedings of the 2019 IEEE/RSJ International Conference on Intelligent Robots and Systems (IROS), Macau, China, 3–8 November 2019; pp. 1861–1868. [CrossRef]

27. Ma, H.; Mu, X.; He, B. Adaptive Navigation Algorithm with Deep Learning for Autonomous Underwater Vehicle. *Sensors* **2021**, *21*, 6406. [CrossRef] [PubMed]
28. He, Q.; Yu, H.; Fang, Y. NINT: Neural Inertial Navigation Based on Time Interval Information in Underwater Environments. *IEEE Sens. J.* **2024**, *24*, 21719. [CrossRef]
29. Li, D.; Xu, J.; He, H.; Wu, M. An Underwater Integrated Navigation Algorithm to Deal With DVL Malfunctions Based on Deep Learning. *IEEE Access* **2021**, *9*, 82010–82020. [CrossRef]
30. Zhao, W.; Zhao, S.; Zhang, G.; Liu, G.; Meng, W. FL-EKF Based Cooperative Localization Method for Multi-AUVs. *IEEE Internet Things J.* **2024**, *11*, 1–12 . [CrossRef]
31. Qin, J.; Li, M.; Li, D.; Zhong, J.; Yang, K. A Survey on Visual Navigation and Positioning for Autonomous UUVs. *Remote Sens.* **2022**, *14*, 3794. [CrossRef]
32. Kim, B.; Joe, H.; Yu, S.C. High-precision Underwater 3D Mapping Using Imaging Sonar for Navigation of Autonomous Underwater Vehicle. *Int. J. Control. Autom. Syst.* **2021**, *19*, 3794. [CrossRef]
33. Costanzi, R.; Fanelli, F.; Monni, N.; Ridolfi, A.; Allotta, B. An Attitude Estimation Algorithm for Mobile Robots under Unknown Magnetic Disturbances. *IEEE/ASME Trans. Mechatron.* **2016**, *21*, 1900–1911. [CrossRef]
34. Ngatini.; Apriliani, E.; Nurhadi, H. Ensemble and Fuzzy Kalman Filter for position estimation of an autonomous underwater vehicle based on dynamical system of AUV motion. *Expert Syst. Appl.* **2017**, *68*, 29–35. [CrossRef]
35. Shariati, H.; Moosavi, H.; Danesh, M. Application of particle filter combined with extended Kalman filter in model identification of an autonomous underwater vehicle based on experimental data. *Appl. Ocean. Res.* **2019**, *82*, 32–40. [CrossRef]
36. Pei, F.; Xu, H.; Jiang, N.; Zhu, D. A novel in-motion alignment method for underwater SINS using a state-dependent Lie group filter. *Navigation* **2020**, *67*, 451–470. [CrossRef]
37. Xu, C.; Xu, C.; Wu, C.; Qu, D.; Liu, J.; Wang, Y.; Shao, G. A novel self-adapting filter based navigation algorithm for autonomous underwater vehicles. *Ocean. Eng.* **2019**, *187*, 106146. [CrossRef]
38. Fossen, T.I. Feedback error-state Kalman filter with time-delay compensation for hydroacoustic-aided inertial navigation of underwater vehicles. *Control. Eng. Pract.* **2023**, *138*, 105603. [CrossRef]
39. Barrau, A.; Bonnabel, S. Invariant Kalman Filtering. *Annu. Rev. Control. Robot. Auton. Syst.* **2018**, *1*, 237–257. [CrossRef]
40. Chang, L.; Luo, Y. Log-Linear Error State Model Derivation without Approximation for INS. *IEEE Trans. Aerosp. Electron. Syst.* **2023**, *59*, 2029–2035. [CrossRef]
41. Li, J.; Zhang, S.; Yang, H.; Jiang, Z.; Bai, X. A Fast Continuous Self-Calibration Method for FOG Rotational Inertial Navigation System Based on Invariant Extended Kalman Filter. *IEEE Sens. J.* **2023**, *23*, 2456–2469. [CrossRef]
42. Barfoot, T.D.; Furgale, P.T. Associating Uncertainty with Three-Dimensional Poses for Use in Estimation Problems. *IEEE Trans. Robot.* **2014**, *30*, 679–693. [CrossRef]
43. Roux, A.; Changey, S.; Weber, J.; Lauffenburger, J.P. CNN-based Invariant Extended Kalman Filter for projectile trajectory estimation using IMU only. In Proceedings of the 2021 International Conference on Control, Automation and Diagnosis (ICCAD), Grenoble, France, 3–5 November 2021; pp. 1–6. [CrossRef]
44. Chang, L.; Di, J.; Qin, F. Inertial-Based Integration With Transformed INS Mechanization in Earth Frame. *IEEE/ASME Trans. Mechatron.* **2022**, *27*, 1738–1749. [CrossRef]
45. Xu, B.; Guo, Y. A Novel DVL Calibration Method Based on Robust Invariant Extended Kalman Filter. *IEEE Trans. Veh. Technol.* **2022**, *71*, 9422–9434. [CrossRef]
46. Luo, L.; Huang, Y.; Zhang, Z.; Zhang, Y. A New Kalman Filter-Based In-Motion Initial Alignment Method for DVL-Aided Low-Cost SINS. *IEEE Trans. Veh. Technol.* **2021**, *70*, 331–343. [CrossRef]
47. Qian, L.; Qin, F.; Li, K.; Zhu, T. Research on the Necessity of Lie Group Strapdown Inertial Integrated Navigation Error Model Based on Euler Angle. *Sensors* **2022**, *22*, 7742. [CrossRef]
48. Potokar, E.R.; Norman, K.; Mangelson, J.G. Invariant Extended Kalman Filtering for Underwater Navigation. *IEEE Robot. Autom. Lett.* **2021**, *6*, 5792–5799. [CrossRef]
49. Luo, Y.; Lu, F.; Guo, C.; Liu, J. Matrix Lie Group-Based Extended Kalman Filtering for Inertial-Integrated Navigation in the Navigation Frame. *IEEE Trans. Instrum. Meas.* **2024**, *73*, 9503916. [CrossRef]
50. Luo, Y.; Wang, M.; Guo, C. The Geometry and Kinematics of the Matrix Lie Group $SE\_K(3)$. *arXiv* **2020**, arXiv:2012.00950.
51. Ashkenazi, V. Coordinate Systems: How to Get Your Position Very Precise and Completely Wrong. *J. Navig.* **1986**, *39*, 269–278. [CrossRef]
52. Potokar, E.; Lay, K.; Norman, K.; Benham, D.; Neilsen, T.B.; Kaess, M.; Mangelson, J.G. HoloOcean: Realistic Sonar Simulation. In Proceedings of the 2022 IEEE/RSJ International Conference on Intelligent Robots and Systems (IROS), Kyoto, Japan, 23–27 October 2022; pp. 8450–8456. [CrossRef]
53. Potokar, E.; Ashford, S.; Kaess, M.; Mangelson, J.G. HoloOcean: An Underwater Robotics Simulator. In Proceedings of the 2022 International Conference on Robotics and Automation (ICRA), Yokohama, Japan, 13–17 May 2022; pp. 3040–3046. [CrossRef]
54. Wang, C.; Cheng, C.; Yang, D.; Pan, G.; Zhang, F. Underwater AUV Navigation Dataset in Natural Scenarios. *Electronics* **2023**, *12*, 3788. [CrossRef]
55. Nicolai, R.; Simensen, G. The New EPSG Geodetic Parameter Registry. In Proceedings of the 70th EAGE Conference and Exhibition incorporating SPE EUROPEC 2008, Rome, Italy, 9–12 June 2008. [CrossRef]

56. Zhang, Z.; Scaramuzza, D. A Tutorial on Quantitative Trajectory Evaluation for Visual(-Inertial) Odometry. In Proceedings of the 2018 IEEE/RSJ International Conference on Intelligent Robots and Systems (IROS), Madrid, Spain, 1–5 October 2018; pp. 7244–7251. [CrossRef]
57. Tang, J.; Bian, H.; Ma, H.; Wang, R. Initialization of SINS/GNSS Error Covariance Matrix Based on Error States Correlation. *IEEE Access* **2023**, *11*, 94911–94917. [CrossRef]

Journal of
*Marine Science
and Engineering*

MDPI

Article

An Improved NSGA-II Algorithm for MASS Autonomous Collision Avoidance under COLREGs

Zuopeng Liang, Fusheng Li and Shibo Zhou *

Navigation College, Jimei University, Xiamen 361021, China; 202311823009@jmu.edu.cn (Z.L.);
15859226806@139.com (F.L.)
* Correspondence: zhoushibo@jmu.edu.cn

Abstract: Autonomous collision avoidance decision making for maritime autonomous surface ships (MASS), as one of the key technologies for MASS autonomous navigation, is a research hotspot focused on by relevant scholars in the field of navigation. In order to guarantee the rationality, efficacy, and credibility of the MASS autonomous collision avoidance scheme, it is essential to design the MASS autonomous collision avoidance algorithm under the stipulations of the Convention on the International Regulations for Preventing Collisions at Sea (COLREGs). In order to enhance the autonomous collision avoidance decision-making capability of MASS in accordance with the relevant provisions of COLREGs, an improved NSGA-II autonomous collision avoidance decision-making algorithm based on the good point set method (GPS-NSGA-II) is proposed, which incorporates the collision hazard and the path cost of collision avoidance actions. The experimental results in the four simulation scenarios of head-on situation, overtaking situation, crossing situation, and multi-ship encounter situation demonstrate that the MASS autonomous collision avoidance decision making based on the GPS-NSGA-II algorithm under the constraints of COLREGs is capable of providing an effective collision avoidance scheme that meets the requirements of COLREGs in common encounter situations and multi-ship avoidance scenarios promptly, with a promising future application.

Keywords: MASS; COLREG; good point set; NSGA-II algorithm; autonomous collision avoidance

1. Introduction

The integration of cutting-edge technologies, including artificial intelligence, big data, communications, and the Internet of Things, with traditional ships has led to the emergence of MASS, a technology that revolutionises the conventional ship navigation model. MASS controls the ship's navigation process through advanced sensors and machine learning algorithms, which enable the navigation of ships in a more efficient and safer manner [1–3]. Currently, MASS is still in the experimental phase, and its full implementation is constrained by several challenges. One such challenge is the development of an autonomous system capable of avoiding collisions with other vessels in the complex maritime environment [4]. The issue of autonomous collision avoidance among ships is not only concerned with intelligent collision avoidance path planning, but also needs to comply with the relevant provisions of COLREGs [5,6]. COLREGs are universally recognised and commonly adhered to maritime navigation rules, and also form a supporting document for the MASS autonomous navigation decision-making system. Only by incorporating the relevant provisions of the COLREGs into the design of the MASS autonomous collision avoidance system can a safe and effective autonomous collision avoidance decision-making system be constructed [7,8].

The traditional autonomous collision avoidance system for ships employs a geometric collision avoidance method, which provides advice on the avoidance programme for vessels in accordance with COLREGs based on the relative positional relationships between vessels [9–11]. MASS is mainly land-based or unmanned, and the traditional collision avoidance decision-making methods can no longer adapt to the requirements of autonomous collision avoidance for MASS. In light of the autonomous collision avoidance problem of MASS, various solutions have been proposed by researchers, including the artificial potential field method [12,13], heuristic algorithms [14,15], path planning algorithm [16,17], reinforcement learning approaches [18], velocity obstacles (VO) [19], and so on. Heuristic algorithms and VO are the two most commonly employed methods.

Heuristic algorithms employ the cognitive processes of human intelligence to address complex problems, which employ a combination of reasoning and searching to rapidly identify optimal solutions to search problems, and therefore have a wide range of applications in MASS autonomous collision avoidance decision-making research [20]. Liu et al. [21] proposed an algorithm based on a combination of an enhanced bacterial foraging algorithm and a particle swarm algorithm with global search capability to optimise the collision avoidance path. The algorithm was subsequently integrated with a hybrid optimisation algorithm and a collision risk model to assess the collision risk of the ship and facilitate autonomous collision avoidance in the MATLAB simulation environment. Liu et al. [22] proposed a beetle antennae search algorithm based on multi-ship encounters, which combined COLREGs and ship maneuvering performances. The algorithm was validated through case studies using voyage loss as the objective function, and the results demonstrated that this algorithm is a viable solution for the problem of multi-ship encounters. Zhou et al. [23] employed real-time ship navigation data to construct a state set for deep reinforcement learning. Subsequently, a deep reinforcement learning collision avoidance algorithm was proposed, based on COLREGs' constraints, and its efficacy was verified through simulation tests. Finally, the efficacy of the algorithm was demonstrated in two-vessel and multi-vessel rendezvous scenarios. Zheng et al. [24] proposed an algorithm for ship collision avoidance based on an improved particle swarm algorithm. This algorithm determines the optimal steering angle for collision avoidance between ships and has been verified as a feasible solution. Xin et al. [25] proposed an intelligent collision avoidance algorithm based on COLREGs. This algorithm automatically extracts ship state features using deep neural networks, ensuring that the ship navigates to the target and dynamically avoids obstacles through a reward and punishment mechanism. Xie et al. [26] proposed a real-time collision avoidance prediction optimisation strategy based on a simplified three-degree-of-freedom hydrodynamic model for the real-time prediction of ship states and collision risks. The objective was to minimise safety and economic costs. Huang et al. [27] proposed a Softmax deep double deterministic strategy gradient multi-ship coordinated collision avoidance model, with an improved double-delay deep deterministic strategy gradient algorithm to address the issue of autonomous collision avoidance among multiple ships. Approaches based on heuristic algorithms consider the planning of collision avoidance behaviour as an optimisation problem and transform other dynamical equations and collision avoidance requirements, among other factors, into constraint relations or optimisation terms. Such approaches exhibit superior real-time performance, yet their optimisation problems are non-convex. The optimisation results are contingent upon the initial conditions, and the outcomes are inherently stochastic. And, there is a risk of violating the COLREGs regulations.

The VO algorithm was first proposed by Fiorini et al. [28] in 1998 for the collision avoidance planning of mobile robots, with the advantages of a simple principle and fast computation speed. In addition to evaluating the risk of collision, VO also provides collision avoidance methods, making it widely used in path planning and collision avoidance studies in MASS. He et al. [29] integrated VO and COLREGs with a comprehensive analysis of the ship and other obstacles to transform the actual collision avoidance problem into a convex optimisation problem. This approach provides real-time collision avoidance reference information for MASS. Zhao Guixiang et al. [30] proposed an enhanced VO method suitable for unmanned ships, addressing the shortcomings of lengthy collision avoidance routes, extended avoidance times, and the overlooking of avoidance responsibility in emergency situations when the conventional VO method is employed for local path planning. They developed a collision avoidance model for unmanned ships by incorporating the Quaternion's ship domain into the calculation of the collision avoidance timing of VO. Xu et al. [31] devised an alternating local path planning method based on VO and an enhanced APF algorithm. By modifying the algorithm's parameters, it is possible to achieve precise collision avoidance for both static and dynamic obstacles. Hong et al. [32] constructed a model of an obstacle-enclosed rectangle and an unmanned ship relative motion model. They then improved the VO to achieve unmanned ship path planning in complex environments. Zhu et al. [33] employed ellipses to represent safe distances from obstacles, overcoming the limitations of excessive constraints on VO and the inapplicability of representing obstacles as circles. Kuwata et al. [34] integrated COLREGs with VO for unmanned ship path planning. Zhang et al. [35] proposed a hybrid algorithm of APF and VO, which considered COLREGs with respect to the give-way ship clause constraints. Emil et al. [36] determined which side of an obstacle an unmanned ship passes from based on COLREGs and developed a VO collision avoidance algorithm based on the extended domain, whereby an extended domain was assigned to a dynamic obstacle. The velocity obstacle method is not well suited to deal with nonlinear motion and high dynamics, the presence of crossover phenomena in motion trajectories, or in dealing with velocity-independent obstacles. In order to address these limitations, the algorithms can be enhanced by combining, for instance, the dynamic window method and the potential-field-based method. CHO et al. [37] proposed an automatic collision avoidance algorithm based on the use of VO technology for ships that were in compliance with the COLREGs. This algorithm was applied to the case of a multi-ship rendezvous. In a study by LIU et al. [38], a speed-derived heading obstacle algorithm was proposed to satisfy COLREGs constraints based on VO. The determination of the obstacle extension domain in the VO method is more subjective and often does not take into account the size, speed, manoeuvrability, etc., of the vessel, which can lead to excessively long collision avoidance paths and times. In addition, VO methods based on COLREGs usually ignore the constraints of the duty to avoid close-quarters situations.

From the above studies on MASS autonomous collision avoidance, it can be seen that the current MASS autonomous collision avoidance algorithms do not fulfill the collision avoidance responsibilities required by the COLREGs well, and thus the collision avoidance schemes provided do not comply with the requirements of the COLREGs or the ordinary practice of seamen. Some researchers considered the constraints of the COLREGs and designed an autonomous collision avoidance problem based on the aforementioned constraints. Wang et al. proposed an enhanced APF algorithm integrating DCPA and TCPA in the repulsion field, which enhances the collision avoidance capability in the case of dynamic obstacles. However, the algorithm does not take into account the impact of wind and currents on the movement of the ship, which limits its applicability and generality [39]. Xu et al. and Kim et al., respectively, improved the DWA algorithm, which significantly improved the ability of the predicted trajectories to closely follow the extraction points of the global paths, resulting in a better match between the locally planned paths and the globally optimal routes, but the simulation environments were not constructed with the complexity of multi-ship encounters in mind [40,41]. Zhang et al. proposed a finite-time adaptive event-triggered control method with consideration of COLREGs. As their potential field function did not fully meet expectations, the steering method used only the direction of the resultant quantity and ignored the consideration of its magnitude [42]. Bayrak et al. devised an RRT-based local path planning algorithm compliant with COLREGs for ship encounters [43]. Although the above researchers took into account the constraints of COLREGs, the experimental simulations were poor, and the models and methods for solving multi-ship encounter situations remain challenging dilemmas. The NSGA-II algorithm, as a classical heuristic algorithm, has received considerable attention for its efficiency in solving complex optimisation problems. It is suitable for dealing with multivariate and multi-constrained collision avoidance path planning problems, and thus has a wide range of applications in MASS collision avoidance decision making [44–46]. However, the NSGA-II algorithm is susceptible to premature convergence and the formation of local optimal solutions, which constrains the applicability of the NSGA-II algorithm in practical MASS autonomous collision avoidance. To address this issue, this paper employs the concept of good point set theory to optimise the initial population selection of an NSGA-II algorithm. On this basis, an autonomous collision avoidance decision-making method for MASS based on the improved NSGA-II algorithm under COLREGs is proposed. The method fully considers the requirements of COLREGs with regard to ship manoeuvring and realises the autonomous collision avoidance of MASS under the constraints of COLREGs.

2. Collision Risk Assessment Modeling

The term 'risk of collision' refers to the intermediate state between safe navigation and a collision accident. It is one of the key standards used to assess the degree of collision risk between two ships. Article 7(1) of COLREGs stipulates that 'Every vessel shall use all available means appropriate to the prevailing circumstances and conditions to determine if risk of collision exists. If there is any doubt such risk shall be deemed to exist'. It is crucial to evaluate the risks of a ship collision during the design phase of an MASS autonomous collision avoidance algorithm. Seafarers utilise a combination of Automatic Identification System (AIS), radar, and other equipment, as well as their own experience, to assess the risk of collision when two ships are encountered. This approach has been proven effective in the seafarer-led ship navigation mode. However, in the context of MASS autonomous collision avoidance, the absence of a seafarer's observation and judgement means that the traditional method is no longer suitable for the needs of ship collision avoidance. A great deal of research has been conducted by scholars on the risk of collision. One common practice in this field is to measure the collision risk of a ship using the Collision Risk Index (CRI), which is shown in Equation (1).

$$\mathrm{CRI} = \{\mathrm{S}, \mathrm{DCPA}, \mathrm{TCPA}, \mathrm{K}, \mathrm{Q}\} \tag{1}$$

where distance to closest point of approach (DCPA) represents the distance between the two ships during the encounter, time of closest point of approach (TCPA) denotes the time at which the two ships reached the closest encounter point, S denotes the distance between the two ships, K represents the speed ratio between the two ships, and Q denotes the true bearing of the other ship relative to the present ship. The CRI methodology incorporates a wide range of variables for evaluating collision risk, in particular the parameters DCPA and TCPA, which are required by the COLREGs and adopted by seafarers as a matter of common practice. This integrated assessment method provides a more accurate reflection of the actual collision risk between ships and also offers a standarised quantitative index of collision risk. Such a system would facilitate the consistent application of assessment criteria to different vessels and navigational conditions, thereby reducing the potential for human error in judgement.

Due to the inherent ambiguity and uncertainty associated with CRI, the affiliation function is often employed when assessing the risk of collision between ships using DCPA and TCPA. This approach allows for a more comprehensive consideration of the safety of collision avoidance in different scenarios [47,48].

2.1. Distance between Two Vessels

The distance between two vessels is calculated according to the Equation (2).

$$\mathrm{S} = \sqrt{(\mathrm{X_A} - \mathrm{X_B})^2 + (\mathrm{Y_A} - \mathrm{Y_B})^2} \tag{2}$$

where (X_A, Y_A), (X_B, Y_B) represent the position coordinates of the ship.

If there is a risk of collision, ships will take appropriate collision avoidance actions. The timing of collision avoidance actions is generally determined by the distance between the ships. In the event that a ship takes collision avoidance action, there is a designated action node, which is defined as the point at which the distance between the two ships is less than S_1. This is considered to be the point at which the ship is bound to collide.

Similarly, there exists a safe distance, S_2, for ships in the process of rendezvous, i.e., there is no danger of collision between two ships when the distance between them is greater than S_2. The calculation of S_1 and S_2 is shown in Equations (3) and (4).

$$S_1 = \begin{cases} 1.1 - \frac{Q}{180^\circ} \times 0.2 & 0^\circ \leq Q \leq 112.5^\circ \\ 1.0 - \frac{270^\circ - Q}{180^\circ} \times 0.8 & 112.5^\circ \leq Q \leq 247.5^\circ \\ 1.1 - \frac{360^\circ - Q}{180^\circ} \times 0.4 & 247.5^\circ \leq Q \leq 360^\circ \end{cases} \quad (3)$$

$$S_2 = 2 \times S_1 \quad (4)$$

The ship distance affiliation function is employed to quantify the probability of a collision between two ships in the process of approaching each other. Through the ship distance affiliation function, the actual distance between the two ships can be converted into an affiliation value between 0 and 1, which indicates the degree of collision risk at a specific distance. This is demonstrated in Equation (5). The likelihood of collision is inversely proportional to the affiliation value, with a closer value indicating a higher likelihood and a further value indicating a lower likelihood.

$$\mu_S = \begin{cases} 1 & S \leq S_1 \\ \left(\frac{S_2 - S}{S - S_1}\right)^2 & S_1 < S \leq S_2 \\ 0 & S > S_2 \end{cases} \quad (5)$$

2.2. DCPA and TCPA

The distance between the other ship and our ship during the encounter is referred to as DCPA, while TCPA represents the time required for the target ship to move from its current position to the nearest point of encounter. Both DCPA and TCPA have been widely used in nautical practice and the research of autonomous collision avoidance for ships [49–51]. In the context of nautical practice, the ship pilot acquires information regarding the movement of the ship in question and other ships in the vicinity through the use of navigational aids, such as AIS or radar. This information is then used to determine the potential for a collision between the ships in question, with the aim of steering clear of any such danger. During the navigation process of a ship, the DCPA and TCPA are employed to quantify the collision risk in both the spatial and temporal dimensions. If both DCPA and TCPA are below a specified threshold, it is deemed that there is a collision risk. In such instances, the ship must alter course or speed in order to avoid a potential collision, in accordance with COLREGs. V_t represents the speed of the approaching ship, V_0 indicates the speed of the ship, and V_r denotes the relative speed of the two vessels. Figure 1 illustrates the schematic diagram of DCPA.

Modern navigation systems frequently possess the capacity to automatically calculate and monitor TCPA and DCPA, effectively integrating existing technical systems and applying them to real-time navigation [52,53]. The information is then analysed to determine the potential for collision between the ships, and appropriate avoidance actions are implemented in response. DCPA and TCPA are employed to quantify the collision hazard in both the spatial and temporal dimensions, respectively, while the ship is underway. In the event that both the DCPA and TCPA fall below a specified threshold, it is deemed that a collision is imminent. In such a scenario, the ship should steer or change speed to avoid a collision in accordance with the principles of 'early, large, wide and clear' as outlined in the COLREGs.

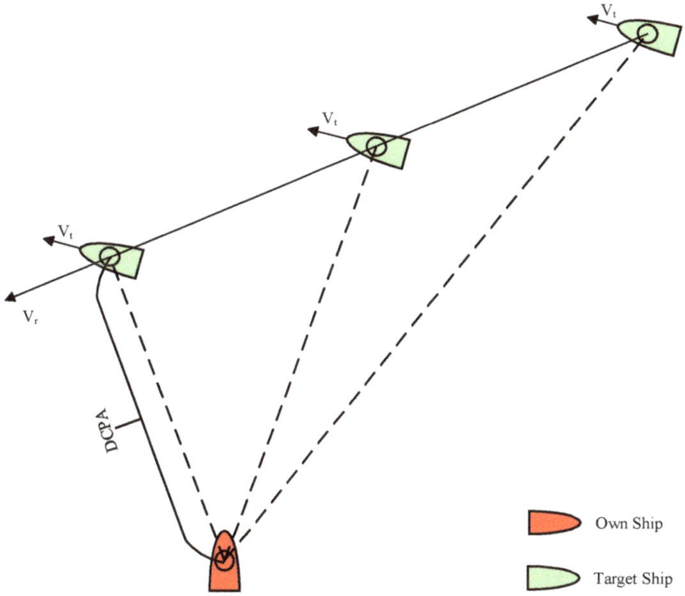

Figure 1. Schematic diagram of DCPA.

The expressions for DCPA and TCPA affiliation functions are provided in Equations (6) and (7). The DCPA's affiliation function is employed to convert DCPA values into collision risk, with a value between 0 and 1. When the DCPA value is minimal, the affiliation is proximate to 1, signifying a markedly elevated spatial collision risk. Conversely, when the DCPA value surpasses a predefined threshold, the affiliation diminishes to 0, indicating that the spatial dimension of the collision risk is negligible. The TCPA affiliation function is analogous to the DCPA affiliation function in that it is employed to assess collision risk in the time dimension.

$$\mu_{DCPA} = \begin{cases} 1 & DCPA \leq S_1 \\ \frac{1}{2} - \frac{1}{2}\sin[\frac{\pi}{S_2 - S_1}(DCPA - \frac{S_1+S_2}{2})] & S_1 < DCPA \leq S_2 \\ 0 & DCPA > S_2 \end{cases} \tag{6}$$

$$\mu_{TCPA} = \begin{cases} 1 & TCPA \leq t_1 \\ \frac{t_2 - TCPA}{t_2 - t_1} & t_1 < TCPA \leq t_2 \\ 0 & TCPA > t_2 \end{cases} \tag{7}$$

where t_1 represents the latest opportunity for a ship to take collision avoidance action. This is defined as the time required for a ship to travel from a distance of S_1 to the nearest encounter point. The latest time for a ship to take collision avoidance action is used to characterise the collision risk of a ship in the time dimension, forming an imminent situation. t_2 is the time required for the ship to manoeuvre to the nearest encounter point, starting from the time when the two ships are at a distance of S_r ($S_r \geq S_2$). The ship manoeuvre time is employed to quantify the probability that the two ships will have sufficient temporal separation to avoid a collision. If TCPA is greater than t_2, it is assumed that the two ships have sufficient time to perform adequate collision avoidance behaviours. The calculation of t_1 and t_2 is presented in Equations (8) and (9).

$$t_1 = \begin{cases} \frac{\sqrt{S_1^2 - DCPA^2}}{v_r} & DCPA \leq S_1 \\ \frac{DCPA - S_1}{v_r} & DCPA > S_1 \end{cases} \tag{8}$$

$$t_2 = \frac{\sqrt{S_r^2 - DCPA^2}}{v_r} \quad S_r \geq S_2 \tag{9}$$

2.3. Velocity Ratio of Two Ships K

The velocity ratio is defined as the ratio of the speeds of two ships. This parameter reflects the speed of the relative motion of the two ships, which is of great significance to the collision avoidance decision. The affiliation function of the speed ratio of the two vessels is used to evaluate the impact of collision risk based on relative speed. The calculation of the affiliation function of the ship speed ratio is shown in Equation (10).

$$\mu_{K_i} = 1/[1 + (^2 / K_i \sqrt{K_i^2 + 2K_i \sin C + 1})] \tag{10}$$

where $\sin C = |\sin(|C_R - C_0|)|$, C_R and C_0 are the heading of the vessel and the target vessel, respectively.

2.4. True Bearing Q

The true bearing describes the position of a ship relative to another ship and takes values from 0° to 360°. The affiliation function of the true bearing converts this true bearing, Q, into an affiliation value reflecting the degree of collision risk, which is between 0 (no risk) and 1 (maximum risk). This value is used to characterise the effect of the relative position between two ships on the collision risk. The literature [54] specifies the affiliation function of the true bearing Q. The calculation is shown in Equation (11), which is based on the results of this research.

$$\mu_{Q_i} = [\cos(Q_i - 19°) + \sqrt{440/289 + \cos^2(Q_i - 19°)}]/2 - \frac{5}{17} \tag{11}$$

2.5. Crash Hazard Evaluation Set

The varying influence of each factor on CRI necessitates the application of a weighting system to determine the value of CRI. The literature [47] has conducted a comprehensive analysis and calculation of these weights, and the resulting Equation (12) illustrates the calculation of CRI.

$$CRI_i = 0.40\mu_{DCPA} + 0.367\mu_{TCPA} + 0.167\mu_S + 0.033\mu_{K_i} + 0.033\mu_{Q_i} \tag{12}$$

The value of CRI is within the range of [0, 1], with the lower limit of the interval representing the absence of danger of collision, and the upper limit representing the inevitability of collision. Different CRI values represent varying degrees of collision danger, and the actions required by the ship are also different. See Table 1 for details. When CRI ≤ 0.4, both ships are in the free action phase. When $0.4 \leq CRI < 0.6$, the vessel on the right should take action as soon as possible. When the collision danger reaches 0.6, the vessel on the right should take action immediately, as failure to do so may result in the vessel on the left taking action alone. When the collision danger reaches 0.8, the two vessels should take the action that is most helpful for collision avoidance, as shown in Table 1. The cross-encounter posture of two ships is used as an illustrative example in Figure 2, which depicts the ship's actions under different collision hazards.

Table 1. Collision risk evaluation set.

Collision Risk Evaluation Set	CRI
Very safe	$0 \leq CRI < 0.2$
Safe.	$0.2 \leq CRI < 0.4$
Normal.	$0.4 \leq CRI < 0.6$
Dangerous.	$0.6 \leq CRI < 0.8$
Very dangerous.	$0.8 \leq CRI \leq 1$

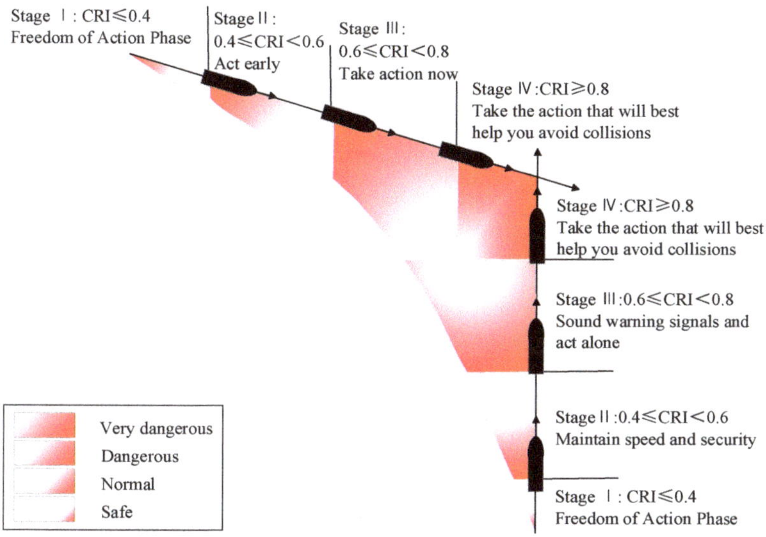

Figure 2. Collision hazards at different stages of navigation.

3. Constraints from COLREGs

3.1. Encounter Situation Classification

In accordance with COLREGs, the encounter situation between two ships can be classified into three categories: head-on, overtaking, and crossing situation. These categories are illustrated in Figure 3. The relevant provisions for ships in these encounter situations are outlined in Table 2.

Table 2. Compliance rules and articles in various encounter situations according to the COLREGS.

Encounter Situation	The Relevant Provisions That Apply the Collision Avoidance Rules
Overtaking (Give-way vessel)	Chapter 2, Articles 6, 8, 13, 16, and 18; Chapter 3—Bugle Lights and Types; and Chapter 4—Acoustic and Light Signals.
Overtaking (Stand-on vessel)	Chapter 2, Articles 6, 8, 13, 17, and 18; Chapter 3—Bugle Lights and Types; and Chapter 4—Acoustic and Light Signals.
Head-on	Chapter 2, Articles 6, 8, 14, 16, and 18; Chapter 3—Bugle Lights and Types; and Chapter 4—Acoustic and Light Signals.
Crossing (Give-way vessel)	Chapter 2, Articles 6, 8, 15, 16, and 18; Chapter 3—Bugle Lights and Types; and Chapter 4—Acoustic and Light Signals.
Crossing (Stand-on vessel)	Chapter 2, Articles 6, 8, 15, 17, and 18; Chapter 3—Bugle Lights and Types; and Chapter 4—Acoustic and Light Signals.

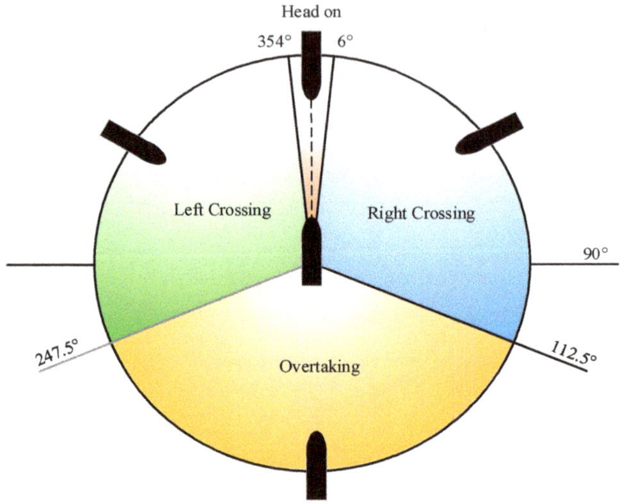

Figure 3. Ship encounter dynamics.

In different encounter situations, ships should take appropriate collision avoidance actions.

1. Head-on: Two ships in the opposite or nearly opposite direction of the encounter, which constitutes a risk of collision, constitute the situation of the encounter. The opposite heading means that the difference between the bow direction of the two ships is 180°. A near-opposite heading usually means that the crossing angle of the two ships is approximately 6°. Once an encounter has been established, each vessel must turn to the right in order to pass on the port side of the other vessel, as required by the COLREGs. The combined effect of the actions taken by the two vessels will not necessarily result in the passage of the two vessels at a safe distance. Rather, each vessel must take an early and substantial right-hand turn, and the actions of each vessel will result in the passage of the two vessels at a safe distance.

2. Overtaking: In accordance with the COLREGs, a vessel is considered to be in the overtaking situation when it is catching up with another vessel from a direction greater than 22.5° directly behind the transverse of the other vessel. This occurs when the vessel in the overtaking situation is in such a position with respect to the vessel it is overtaking that, at night, it is only possible to see the tail light of the vessel in overrun, but not either of its port lights. In the event that two vessels have constituted an overtaking situation, the overtaking vessel is obliged by the COLREGs to yield to the vessel being overtaken. Furthermore, any subsequent change of bearing between the two vessels does not relieve the overtaking vessel of the duty to yield to the vessel being overtaken until it has finally sailed through to yield.

3. Crossing: This is defined as a bow-to-bow crossing of two vessels, i.e., a bow-to-bow crossing greater than 6° of broadside, but less than 112.5° of broadside. With regard to the distance at which a cross-encounter situation is deemed to commence, it is widely acknowledged that a cross-encounter situation commences when one vessel is able to visually discern the masthead light of the other vessel. In accordance with the requirements of COLREGs, when two vessels constitute a cross-encounter situation, the vessel on the port side shall give way to the vessel on the starboard side, which is the giving wayvessel. Conversely, when there is another vessel on the starboard side of the vessel, the vessel on the port side is the straight vessel, and the other vessel shall give way to the vessel.

3.2. Responsibilities for Collision Avoidance between Vessels

According to the various provisions of COLREGs, the responsibilities for collision avoidance between vessels can be categorized as follows:

1. Stand-on vessel responsibilities: A stand-on vessel should not impede the passage or safe transit of another vessel. According to Rule 17 of COLREGs, the stand-on vessel has different responsibilities and obligations at various stages, which include maintaining course and speed, taking independent manoeuvering actions when necessary, and performing actions that most effectively aid in collision avoidance.

2. Give-way vessel responsibilities: A give-way vessel should take early and substantial action to keep well clear of the other vessel. In addition to taking early, substantial, and clear actions, the give-way vessel must adhere to other COLREGs provisions. For instance, in a crossing situation, the give-way vessel should avoid crossing ahead of the stand-on vessel.

The constraints on the collision avoidance actions of MASS for the three cases of head-on, overtaking, and crossing are illustrated in Figure 4.

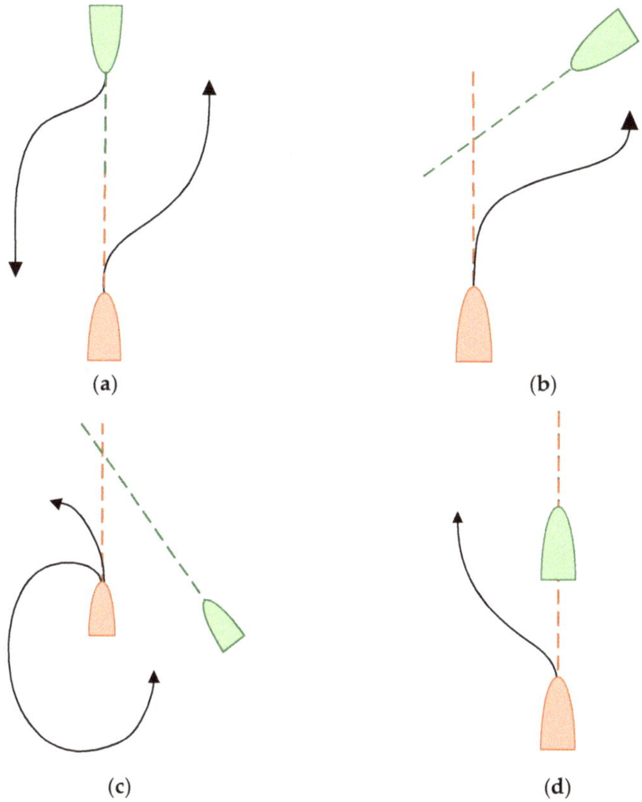

Figure 4. Collision avoidance obligations of MASS in head-on, overtaking, and crossing situations. (**a**) Head-on situation, (**b**) small-angle crossing situation, (**c**) large-angle crossing situations, and (**d**) overtaking situation.

When two vessels are in a head-on situation, the relative speed is high, and the time available for judgment, consideration, and evasive action is limited. Consequently, COLREGs stipulate that each ship must turn to the right to pass port to port. Furthermore, each ship must make a substantial right turn at the earliest possible opportunity, with the objective of ensuring that the two ships pass at a safe distance from each other.

When two vessels are in a crossing situation, one ship shall give way to the other, and in order for the giving way ship to be able to assume the obligation to stand-on ship, she must be able to ascertain the position and dynamics of the stand-on ship, as well as whether or not it is stabilising at a certain speed in a certain direction.

When the two vessels are in an overtaking situation, the relative speed of the two ships is relatively low, the transfer is small, the time spent in parallel is considerable, and the probability of a collision is high. Consequently, when there is uncertainty as to whether or not overtaking is occurring, it is prudent to assume that overtaking is taking place. COLREGs stipulate that in the case of an overtaking situation, the obligation to give way continues until the ship has passed and it is clear. It is worth noting that an overtaking situation can easily be confused with a wide-angle crossing situation, and when there is doubt as to whether it constitutes an overtaking, it should be assumed that the ship is overtaking and appropriate action should be taken accordingly. While Regulation 13 of the COLREGs provides comprehensive guidance on the pursuit of ships, it does not explicitly delineate the side from which the pursuing ship is to be pursued. It is important to note that overtaking from the starboard side of the vessel being pursued, in particular from the vicinity of an azimuth of approximately 22°, is a risky manoeuvre. It is considered risky practice among seafarers to overtake a vessel from a position 5° aft of the right main transverse. Such a manoeuvre can easily be confused with a wide-angle cross-encounter, as defined in COLREGs Regulation 14. This can result in uncoordinated actions between the pursuing vessel and the vessel being pursued, increasing the risk of collision. In light of the aforementioned requirements pertaining to the common practice of seafarers, this paper imposes restrictive constraints on the actions of the pursuing vessel in an overrun situation. Specifically, it requires the pursuing vessel to overrun from the port side of the vessel being overrun, with the aim of enhancing the safety, reasonability, and reliability of the collision avoidance decision.

4. MASS Collision Avoidance Decision-Making Model Based on GPS-NSGA-II Algorithm

4.1. NSGA-II Algorithm

The non-dominated Sorting Genetic Algorithm II (NSGA-II) is a multi-objective optimisation algorithm developed by Kalyanmoy Deb et al. It is based on the traditional genetic algorithm framework and is specifically designed for solving problems with multiple conflicting objectives. Collision avoidance between ships necessitates the avoidance actions of two or more ships, which must consider both the potential for collisions between the ships and the constraints of the COLREGs when taking the avoidance actions. Consequently, this is a typical multi-objective optimisation problem. NSGA-II is more suitable for this kind of problem than traditional genetic algorithms, due to its high efficiency in dealing with multi-objective problems.

One of the advantages of NSGA-II is its rapid non-dominated sorting algorithm. The algorithm employs a Pareto dominance-based ranking mechanism to effectively manage the population structure of the set of collision avoidance schemes by ranking and stratifying all unknown collision avoidance schemes according to their non-dominated relations in the avoidance space. The NSGA-II algorithm classifies unknown collision avoidance schemes into dominated tiers according to the number of other collision avoidance schemes that they are dominated by. The collision avoidance schemes within each tier are not dominated by those in other tiers, thus ensuring the diversity and quality of the collision avoidance schemes. Pareto, the sorting mechanism works as follows:

In the context of a certain collision avoidance scheme X_i within the MASS collision avoidance process, whose objective function is (f_{i1}, f_{i2}), if another collision avoidance scheme $X_j = (f_{j1}, f_{j2})$ that satisfies the following conditions: $\{f_{j1} \leq f_{i1}, f_{j2} \leq f_{i2}\}$ and $\{f_j < f_{i1}$ or $f_{j2} < f_{i2}\}$, then X_i is said to dominate X_j (denoted as $X_j \prec X_i$). For the sake of argument, consider five avoidance schemes, A, B, C, D, and E. According to the definition of domination, there are two sets of domination relations, namely {(C A, B)} and {(D A, E)}, as illustrated in Figure 5. According to the Pareto ordering mechanism, points C and D are assigned to the ParetoLv.1 stratification because they have no dominated elements. The curve where the set of solutions of Lv.1 is located is called the Pareto frontier curve. There is no superiority or inferiority of the points on the Pareto frontier; rather, they are simply evaluated differently. The function value of f_1 for the C solution is the smallest, while the function value of f_2 for the C solution is the largest among all solutions under the same stratification. The curve representing the Pareto frontier is defined as the Pareto frontier curve. Consequently, the evaluation of avoidance solutions may vary according to the specific requirements.

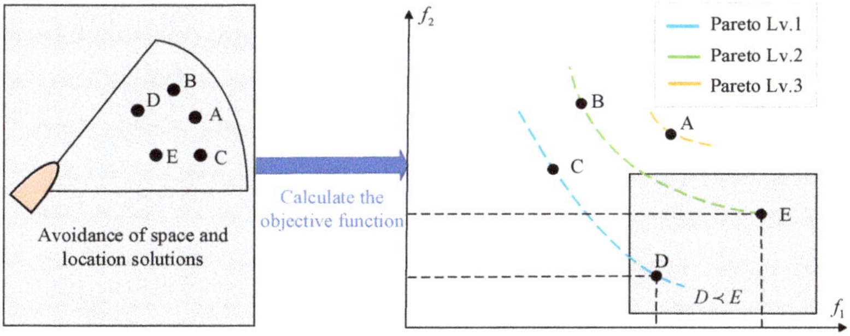

Figure 5. Presentation of a schematic diagram of non-dominated sorting.

Furthermore, NSGA-II employs an elite strategy to perpetuate the inheritance of superior individuals. This entails selecting the most effective avoidance schemes from the current set and the set of offspring schemes to form a new population. This process ensures the continued evolution of the algorithm, while maintaining the stability and excellence of the avoidance schemes.

In Figure 5, points C and D are not dominated by any solution, thus the number of dominated is 0, which is located in ParetoLv.1. Points B and E are dominated by points C and D, respectively, and the number of domination points is 1, which is located in ParetoLv.2. Point A is dominated by points C and D, and the number of domination points is 2, which is located in ParetoLv.3. In a similar encounter scenario, there are numerous collision avoidance strategies. When multiple targets must be considered simultaneously and the diversity of collision avoidance strategies is maintained, the NSGA-II algorithm is more adept at evaluating whether a collision avoidance strategy aligns with the ordinary practice of seamen from multiple perspectives rather than traditional genetic algorithms.

4.2. An Improved NSGA-II Algorithm Based on a Good Point Set

The outcome of NSGA-II is influenced by the initial population. If the initial population is randomly generated to solve the collision avoidance scheme, there may be a phenomenon of population aggregation, which may result in the selection of the collision avoidance scheme falling into the local optimum. This may prevent the identification of the optimal collision avoidance scheme by traversing the whole situation. Nevertheless, if the initial collision avoidance scheme is highly similar to the optimal collision avoidance scheme, not only can the convergence speed be enhanced, but the optimisation accuracy can also be improved, and the optimal collision avoidance scheme can be obtained in a shorter time under the condition of a certain number of iterations. As the position of the optimal collision avoidance scheme in the initial population is unknown when the NSGA-II is employed to identify the optimal collision avoidance scheme, a uniform distribution is employed to distribute the initial collision avoidance scheme of the population within the permissible range. This is completed to ensure that the first generation of collision avoidance scheme is as close to the optimal value as possible.

The good point set (GPS), proposed by Hua Luogeng et al. [55], is a method for selecting points in a multidimensional space that can be used to design a set of sampling point sets. The uniformity of the GPS allows for the initialisation of the population of

NSGA-II, which can improve the diversity of the population and make the initial population more ergodic. This, in turn, can better achieve the purpose of a global optimality search.

The principle of the good point set is as follows: let G_s be the unit cube in the s-dimensional Euclidean space. If $r \in G_s$, then the good point set can be expressed as Equation (13).

$$P_n(k) = \{(\{r_1^{(n)} \cdot k\}, \{r_2^{(n)} \cdot k\}, \ldots, \{r_s^{(n)} \cdot k\}), 1 \le k \le n\} \tag{13}$$

The deviation $\varphi(n)$ in Equation (13) is such that $\varphi(n) = C(r, \varepsilon)n^{-1+\varepsilon}$ is satisfied, where it is a constant related only to r and ε (ε is any positive number). In this case, $P_n(k)$ is said to be the set of good points, and r is the good point. $\{r_s^{(n)} \cdot k\}$ represents the fractional part, and n represents the number of points, taking $r = \{2\cos(2\pi k/p), 1 \le k \le s\}$ (p is the smallest prime number satisfying $(p-3)/2 \ge s$). The mapping of this to the search space is illustrated by Equation (14).

$$x_i(j) = (j_{ub} - j_{1b})\{r_j^{(i)} \cdot k\} + j_{1b} \tag{14}$$

The location of the jth dimension is defined by the upper and lower bounds, denoted by j_{ub} and j_{1b}, respectively.

Figure 6 illustrates the initial population distribution of 250 populations generated using the good point set and random method within the range of [0, 1].

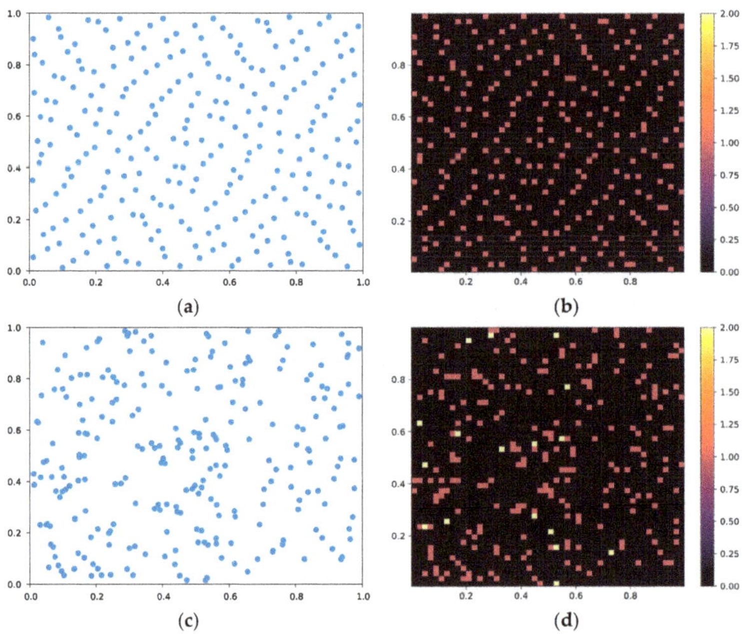

Figure 6. Initial population generated by the good point set and randomly. (**a**) Good points set, (**b**) good points set density, (**c**) randomly, and (**d**) randomly density.

It can be observed from the figure that the distribution of populations generated using good point sets is relatively uniform in space, with a low probability of population aggregation. This advantage is not limited to two-dimensional space, it also applies to high-dimensional problems. The construction of the good point set has no relation to the number of dimensions. Mapping it to the target solution space can still result in a very uniform distribution. The corresponding algorithm performs more stably and always maintains a good diversity of the initial population, thus avoiding the phenomenon of precocity and ultimately converging to the global optimum.

The following steps are to be taken in order to optimise the NSGA-II population using the set of good points:

1. The parameter space must be determined and the range of values for each parameter specified;
2. An initial population must be created using the good point set generator. The dimensions of the problem and the required population size must be taken into account in order to generate the corresponding number of points;
3. The generated sequence of points must be mapped to the actual parameter space;
4. Utilise the mapped points as the initial population for the NSGA-II algorithm.

Following optimisation of the NSGA-II algorithm population using the good point set, the algorithm exhibits a reduction in the number of iterations and a more uniform distribution of solutions. This is beneficial for the algorithm to explore the solution space more extensively at an early stage, compared with a randomly generated population. The process of GPS-NSGA-II algorithm is illustrated in Figure 7.

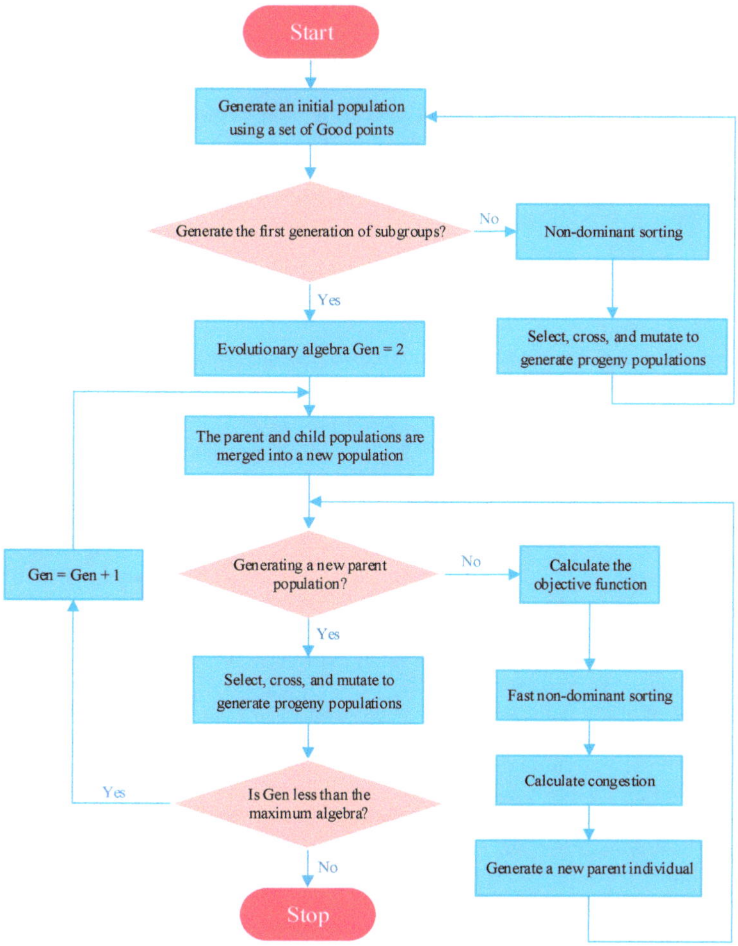

Figure 7. GPS-NSGA-II algorithm flow.

4.3. Parameter Coding and Fitness Evaluation Function

4.3.1. Parameter Encoding and Iterative Approach

The selection of an appropriate coding method is a crucial aspect of the design of a mass collision avoidance scheme, as it directly impacts the efficacy of the system's operation and the effectiveness. The most common coding methods include binary coding, real-valued coding, permutation coding, and tree coding. Real-valued coding is employed to encode the pertinent parameters within the MASS collision avoidance algorithm, as it facilitates the handling of constraints on complex decision variables and offers a greater advantage in terms of search efficiency.

1. The parameters of own ship and the target ship are as follows: The OS is recorded as the own ship, while the TS is recorded as the target ship. The attributes of the own ship are the ship's position coordinates (X_{os}, Y_{os}), the own ship's course $Course_{os}$, and the own ship's speed Vos. The attributes of the target ship are (X_{ts}, Y_{ts}), Coursets, and V_{ts}.

2. With regard to the coding of the encounter situation, it should be noted that the ship designated as the 'give-way' ship will generate an avoidance space in advance. This space encompasses all the points that the ship is about to move to in order to avoid a collision. It is important to ensure that the setup of the avoidance space is not overly expansive, as this will have a detrimental impact on the iteration efficiency and increase the computation time. The polar coordinate system is employed to represent the entire space, with ρ_{os} designated as the angle of rotation of the polar axis with the north direction serving as the reference. r_{os} represents the polar radius, and the avoidance space can be expressed as Space $= \{0 \le \rho_{os} \le 50,\ 0.5 \le r_{os} \le 1\}$. Each position point in the avoidance space represents a specific avoidance scheme.

Assuming that the coordinates of the ship are (X_{next}, Y_{next}), the speed of the ship in the collision avoidance process can be calculated by Equation (15).

$$V_{os} = \frac{\sqrt{(X_{next} - X_{os})^2 + (Y_{next} - Y_{os})^2}}{0.25/60} \tag{15}$$

The course of the ship during navigation is shown in Equation (16).

$$Course_{os} = \begin{cases} 90^\circ - \arctan(\frac{Y_{next} - Y_{os}}{X_{next} - X_{os}}) & X_{next} > X_{os} \\ 270^\circ - \arctan(\frac{Y_{next} - Y_{os}}{X_{next} - X_{os}}) & X_{next} < X_{os} \end{cases} \tag{16}$$

Once the original population has been generated using the set of good points, it is necessary to map the population using a linear transformation. This is achieved by setting $(0 \le \rho_{os} \le 50) \rightarrow (0 \le \rho'_{os} \le 1)$ and $(0.5 \le r_{os} \le 1) \rightarrow (0 \le {r_{os}}' \le 1)$. The individuals of the population are coded as (X_a, X_b, Y_a, Y_b), where X_a represents the integer part of the X coordinate of the collision avoidance scheme, X_b represents the fractional part of the X coordinate of the collision avoidance scheme, Y_a represents the integer part of the Y coordinate of the collision avoidance scheme, and Y_b represents the fractional part of the Y coordinate of the collision avoidance scheme.

The initial stage of the iterative process is to select the sorted population, utilising a roulette wheel to identify the parents of the subsequent generation that is prepared for cross mutation. The advantage of employing a roulette for selection is that even individuals situated at the lower end of the sorting list still have the opportunity to be selected, thereby ensuring diversity of the population. The crossover method is selected as a single-point random crossover, which is the classical crossover method employed in genetic algorithms. This method randomly selects a point on the coding of the parent individuals as the crossover point, and then exchanges part of the genes of the two parent individuals at this point in order to generate two new offspring individuals. The single-point random crossover operation of real-valued coding can increase the genetic diversity of the population and effectively improve the search ability of the algorithm.

4.3.2. Adaptation Evaluation Function

The adaptation degree is the primary indicator of the efficacy of individual collision avoidance schemes. The superiority or inferiority of these schemes is determined by the magnitude of their adaptation degree. The fitness function serves as the criterion for distinguishing between optimal and suboptimal individual collision avoidance schemes among all collision avoidance schemes. Therefore, the selection of an appropriate fitness function is crucial for achieving optimal global optimisation performance of the autonomous collision avoidance algorithm. Inadequate selection of the fitness function may result in the identification of a local optimal collision avoidance scheme rather than a global optimal solution. In this paper, we determine the fitness function from two perspectives: the ship collision risk change function ΔCRI and the path optimisation cost function (POCF), which considers the steering angle of the ship and the sailing distance. The CRI formula is presented in Equation (17), and the POCF calculation is shown in Equation (18).

$$\Delta \mathrm{CRI} = \mathrm{CRI_j} - \mathrm{CRI_i} \tag{17}$$

$$\mathrm{POCF} = \rho_{os} + \sqrt{(X_{next} - X_{os})^2 + (Y_{next} - Y_{os})^2} \tag{18}$$

where $\mathrm{CRI_i}$ represents the collision risk between the two ships in the absence of collision avoidance measures. The relevant calculation is based on Equation (12), while $\mathrm{CRI_j}$ denotes the collision risk between the two ships following the implementation of collision avoidance measures. The difference in collision risk before and after the collision avoidance measures are employed serves to indicate the impact of the aforementioned measures. If the value is negative, it indicates that the collision risk between the ships after taking collision avoidance action has decreased, thereby demonstrating the effectiveness of the collision avoidance action taken.

In Equation (18), (X_{os}, Y_{os}) and (X_{next}, Y_{next}) represent the position of the ship prior to the collision avoidance action and the position of the ship following the completion of the collision avoidance action, respectively. The parameter ρ_{os} denotes the change in heading direction relative to the ship's heading direction, with the latter serving as the reference. The schematic of the parameter is illustrated in Figure 8.

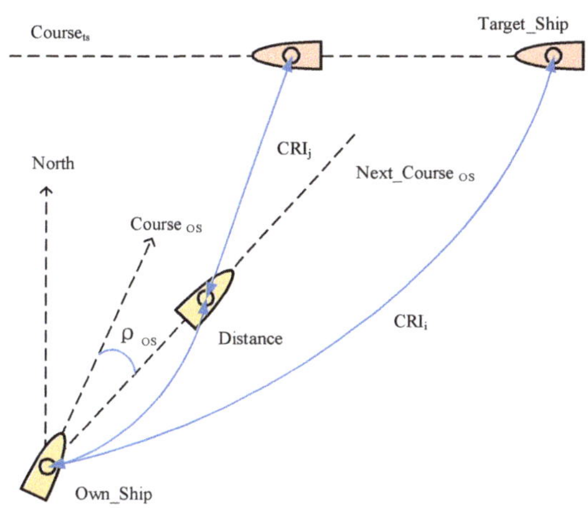

Figure 8. A schematic representation of the parameters of the fitness function.

The rationale behind the adoption of ΔCRI and POCF as the adaptation evaluation functions in the GPS-NSGA-II algorithm is that these two objective functions consider both the safety and the manoeuvring cost of the collision avoidance scheme. Under the constraint of the two objective functions, the position point with the smallest value in the position set of the Pareto front is selected as the moving point of the ship, which represents the optimal solution in all the collision avoidance position sets.

4.4. Autonomous Collision Avoidance Decision Making for MASS under COLREGs

The autonomous collision avoidance decision-making process of MASS based on the GPS-NSGA-II algorithm under COLREGs is shown in Figure 9. This process is divided into three parts: data processing, constraint from COLREGs, and the algorithm controller. The operation of this collision avoidance decision is contingent upon the satisfaction of three conditions:

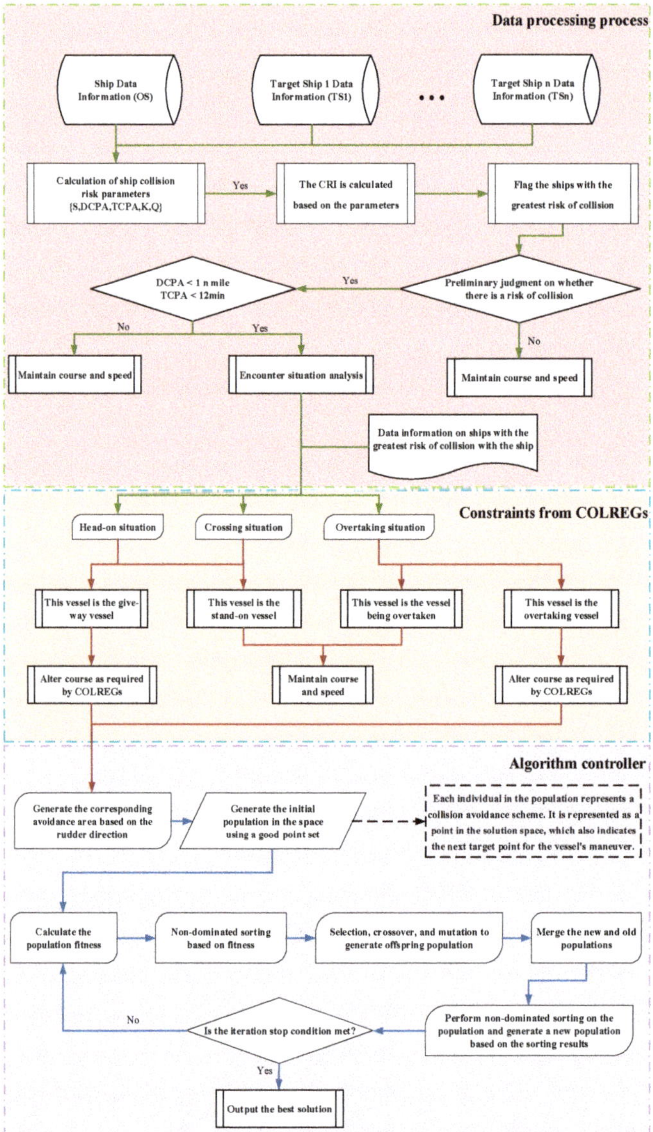

Figure 9. Flowchart of MASS autonomous collision avoidance process.

1. The sea conditions were favourable with good visibility;

2. The navigable waters are sufficiently wide for MASS to take collision avoidance action when a collision hazard exists;

3. The MASS responds in a timely manner, and the stand-on vessels take action in accordance with Article 17 of the COLREGs.

In accordance with Section 3.2, the principal responsibility and obligation of the stand-on vessel is to maintain course and speed. Consequently, when the stand-on vessel is required to take action, it can be inferred that the give-way vessel has not complied with the COLREGs. One of the preconditions of the COLREGs-based MASS collision avoidance model proposed in this paper is that "The MASS responds in a timely manner, and the stand-on vessels take action in accordance with the Article 17 of COLREGs". Therefore, the situation requiring action by the stand-on vessel does not align with the experimental configuration. Consequently, in the experiment, the stand-on vessel only bears the responsibility and obligation to maintain direction and speed.

The following steps are to be taken in order to implement MASS autonomous collision avoidance decision making under COLREGs constraints:

Step 1: Obtain the ship's navigational data through radar, AIS, and other navigational aids, including the ship's initial position, relative distance, speed, relative bearing, heading, and other relevant data.

Step 2: The data obtained are used to calculate the parameter, including TCPA, DCPA, and the collision risk. The initial evaluation of the collision risk is conducted, and a decision is made as to whether the ship will take collision avoidance action according to the collision risk. If there is no risk of collision between the two ships, then no collision avoidance action is required, and the own ship will maintain its course and speed.

Step 3: Determine whether the ship has reached the predefined threshold for initiating collision avoidance measures. In accordance with the ordinary practices of seamen, in open water, the threshold for initiating collision avoidance action is typically deemed to have been reached if the conditions set out in {DCPA < 1 n mile, TCPA < 12 min} are met.

Step 4: If the threshold for taking action is reached, an analysis of the encounter situation between the own ship and the ship with the highest risk of collision must be conducted. The own ship's obligations, with regard to avoidance in different encounter situations, are delineated in accordance with the COLREGs.

Step 5: In the event that the own ship is required to steer in order to avoid a collision, the steering avoidance space must be set in accordance with the avoidance responsibilities specified in the COLREGs. If, however, the own ship is not required to steer in order to avoid a collision, it is the responsibility of the own ship to keep her course and speed.

Step 6: Code the collision avoidance space.

Step 7: Generate the collision avoidance scheme set using the good point set theory within the collision avoidance scheme space.

Step 8: Initiate the algorithm controller to call upon the GPS-NSGA-II collision avoidance decision-making algorithm to select the optimal collision avoidance scheme.

Step 9: Collision avoidance scheme legitimacy check to ascertain whether the given scheme complies with the COLREGs.

Step 10: Repeat steps 1–9 until the ship has successfully passed and cleared all the target ships.

Step 11: Upon completion of the aforementioned collision avoidance scheme, the own ship is permitted to resume its scheduled route.

In the third step, we qualify the values of the DPCA and the TCPA. The DCPA and TCPA are the two most significant and frequently utilised indicators for the assessment of potential collision risks between two ships. The dimensions of the vessel, its location in the waterway, and the circumstances of its proximity to other maritime crafts all contribute to the values of DCPA and TCPA. Therefore, relevant international conventions, such as IMO and COLREGs, do not give a quantitative index for the values of DCPA and TCPA. In accordance with the customary procedures observed by maritime professionals in open waters, when the DCPA between two ships exceeds one nautical mile and the TCPA exceeds 12 min, collision avoidance measures may be temporarily discontinued. Consequently, the evaluation of whether the DCPA is less than 1 nautical mile and the TCPA is less than 12 min serves as an effective criterion for measuring the collision risk of ships in both spatial and temporal dimensions, which is also consistent with the standards set by the customary practices of seafarers.

5. Simulation Experiment and Result Analysis

5.1. Simulation Test Scenarios

In order to assess the efficacy of the MASS autonomous collision avoidance decision-making model proposed in this paper, three representative test scenarios in two-ship encounters, cross-situation, crossing situation, and overtaking situation have been devised. Additionally, in order to evaluate the efficacy of the autonomous collision avoidance decision-making model in multiple-ship encounters, a scenario of autonomous collision avoidance in three-ship encounters has been constructed. The initial state parameters of the ships and target ships are presented in Table 3.

Table 3. Parameter table of the initial state of the ship.

Scenario	Ship	Initial Position	Course (°)	Speed (knots)	Encounter Situation
1	OS TS	(3.5, 1) (3.5, 6)	0 180	10 10	Head-on situation
2	OS TS	(3.5, 1) (7.5, 5)	0 270	10 10	Crossing situation
3	OS TS	(3, 1.5) (3, 3)	0 0	20 10	Overtaking situation
4	OS TS1 TS2	(4, 2) (8, 8) (12, 6)	0 243 270	10 11 12	Multi-ship situation

The test scenario is set in open water with good visibility. All ships comply with the COLREGs. The stand-on ships keep her course and speed. The give-way ships take avoidance actions as early as possible. The ship operation process does not take into account the change of speed. The ships resume their original route after avoidance is completed and then carry out the next avoidance operation according to the risk of collision.

5.2. Collision Avoidance Actions in a Head-On Situation

Figure 10 presents a schematic diagram of the posture of the two ships at the critical node in the head-on situation. The blue line indicates the trajectory of the own ship, while the orange line indicates the trajectory of the target ship. Figure 10a depicts the initial position of the two ships, demonstrating that there is no imminent danger of collision. However, should the two ships continue to adhere to their original sailing plan, the distance between them will gradually diminish, eventually leading to a pair encounter situation until a collision accident occurs. Upon reaching the (3.5, 1.5) coordinate position, the own ship encounters the target ship at the (3.5, 5.5) coordinate position, thereby forming a head-on situation. At this juncture, the DCPA and TCPA of the two ships reach the threshold of taking avoidance action, as depicted in Figure 10b.

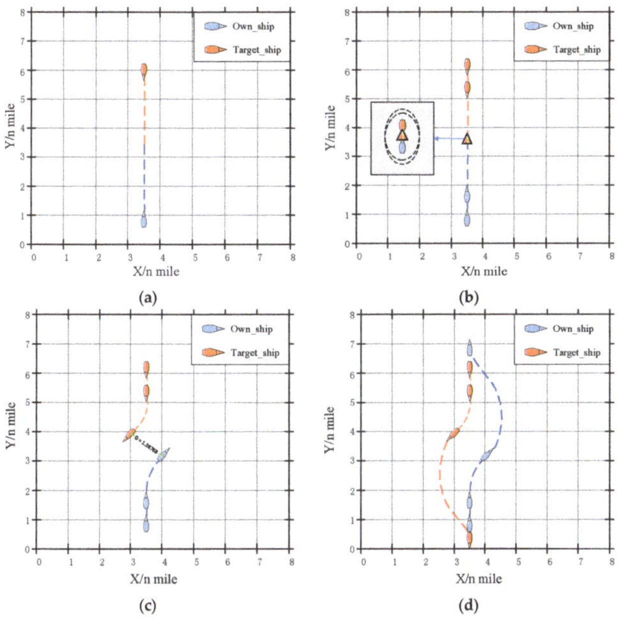

Figure 10. Key scenario diagrams of the collision avoidance process for a head-on situation. (**a**) The initial situation, (**b**) thresholds for action, (**c**) avoiding collision successfully, and (**d**) resume to original route.

In accordance with COLREGs Article 14, each of the two ships turns to the right. Once the two ships have initiated collision avoidance manoeuvers, the own ship successfully avoids the target ship at t = 675 s. At this point, the distance between the two ships is 1.39 n mile, as shown in Figure 10c. Subsequently, the efficacy of the avoidance action is validated, and the original route is resumed in accordance with the requirements of the COLREGs, as shown in Figure 10d.

The alterations in the distance between the vessels and the DCPA throughout the collision avoidance procedure between the two vessels in a head-on configuration are depicted in Figure 11.

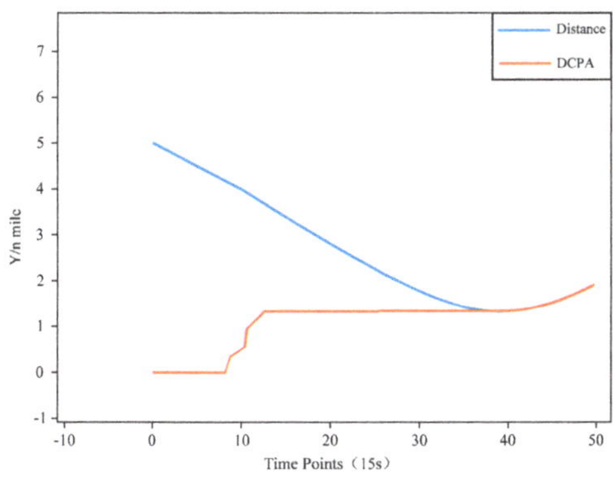

Figure 11. Variation curves during the collision avoidance process for head−on situation.

5.3. Collision Avoidance Actions in Crossing Situation

Figure 12 presents a schematic diagram of the posture of the two ships at the critical node in the crossing situation. The blue line indicates the trajectory of the ship, while the orange line indicates the trajectory of the target ship. Figure 12a depicts the initial position of the two ships, demonstrating that there is no imminent danger of collision. However, if the two ships continue to navigate according to their original plan, as the distance between them decreases, a crossing situation will arise until a collision accident occurs. Upon reaching (3.5, 3) and (5.5, 5), respectively, the ships encounter each other in a crossing situation. At this juncture, the DCPA and TCPA of the two ships reach the threshold for taking avoidance action, as depicted in Figure 12b. In accordance with COLREGs Article 15, the own ship will alter course to the right. Once the ship has initiated collision avoidance action, at t = 1080 s, the own ship has successfully avoided the target ship, with a distance of 0.81 n mile between the two ships, as shown in Figure 12c. Thereafter, the avoidance action is verified and the original route is resumed in accordance with the COLREGs, as shown in Figure 12d. The change in distance between the two ships and the change in DCPA during the collision avoidance process between the two ships in a crossing situation is illustrated in Figure 13.

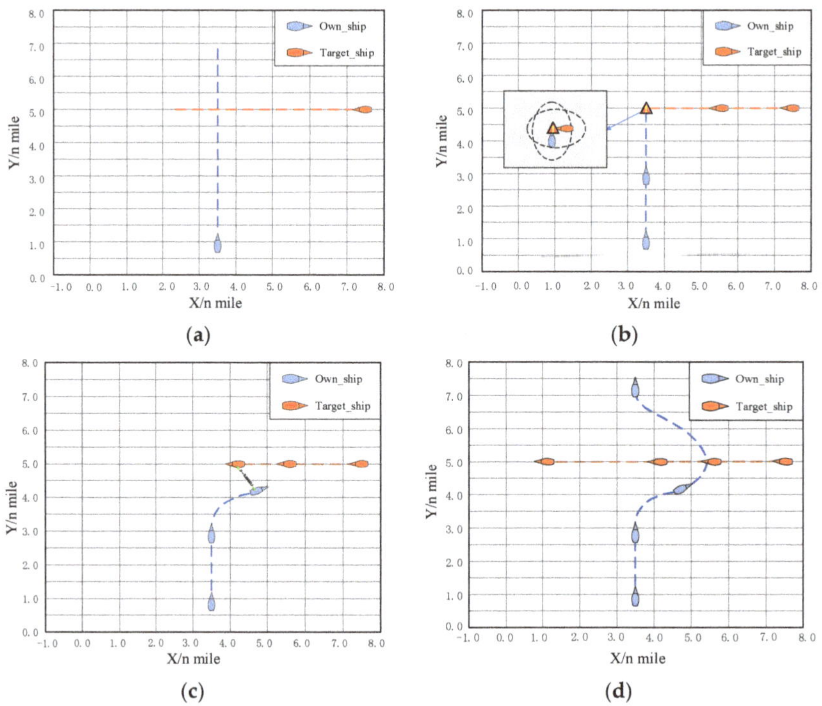

Figure 12. Key scenario diagrams of the collision avoidance process for crossing situation. (**a**) The initial situation, (**b**) thresholds for action, (**c**) avoiding collision successfully, and (**d**) resume to original route.

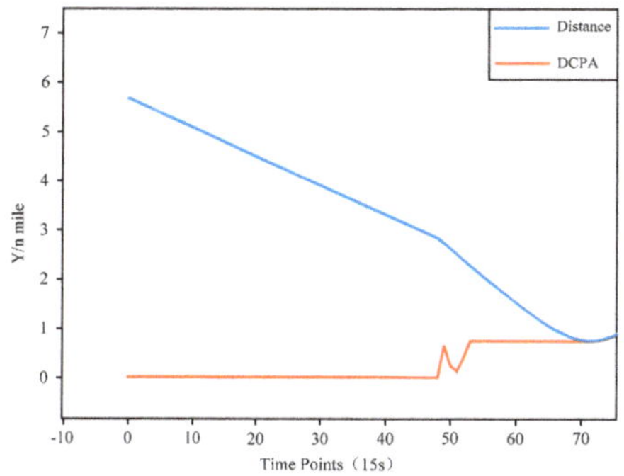

Figure 13. Variation curves during the collision avoidance process for the crossing situation.

5.4. Collision Avoidance Actions in Overtaking Situation

Figure 14 presents a schematic diagram of the posture of the two ships at the critical node in overtaking situation. The blue line indicates the trajectory of own ship, while the orange line indicates the trajectory of the target ship. Figure 14a depicts the initial position of the two ships, wherein there is no danger of collision. However, should the two ships continue to navigate according to their original plan, as the distance between them decreases, an overtaking situation will be constituted, which may culminate in a collision. Upon the own ship reaching (3, 2.5) and the target ship reaching (3, 3.5), an overtaking situation is formed. At this juncture, the DCPA and TCPA of the two ships reach the threshold of taking avoidance action, as depicted in Figure 14b. In accordance with COLREGs Article 13, the ship initiates a turn to avoid collision. Following the implementation of collision avoidance actions, the own ship successfully avoids the target ship at t = 255 s, with a distance of 0.89 n mile between the two ships, as illustrated in Figure 14c. Thereafter, the efficacy of the avoidance action is validated and the original route is resumed in accordance with the COLREGs, as depicted in Figure 14d. The alterations in the distance between the ships and the DCPA during the collision avoidance process between the two ships in the overtaking situation are illustrated in Figure 15.

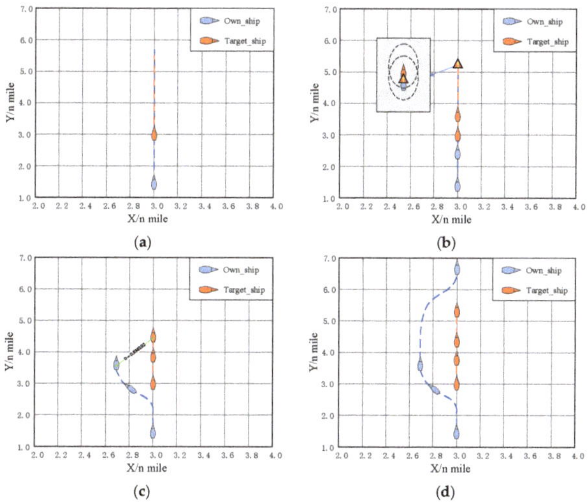

Figure 14. Key scenario of collision avoidance process for overtaking situation. (**a**) The initial situation, (**b**) thresholds for action, (**c**) avoiding collision successfully, and (**d**) resume to original route.

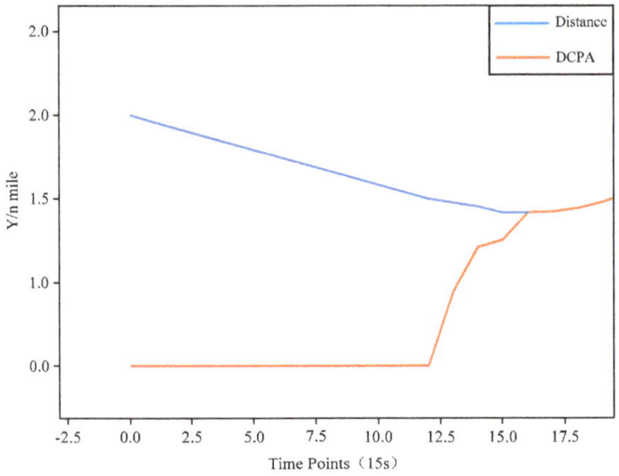

Figure 15. Variation curves during the collision avoidance process for the overtaking situation.

5.5. Collision Avoidance Actions in a Multi-Ship Situation

In the experiment of a multi-ship encounter, the objective of our algorithmic design is to address the issue of collision avoidance between the own ship and other target ships. The potential for collision avoidance between target ships is not within the scope of this investigation. As the number of target ships increases, it is inevitable that the issue of collision avoidance between target ships will arise, which falls beyond the remit of this paper. Furthermore, when the number of target ships is excessive, some ships will deviate from the COLREGs when taking action. The collision avoidance actions taken by these ships are unpredictable, which renders the constraints of the COLREGs ineffective. Accordingly, the objective of the three-ship encounter scenario presented in this paper is to ensure that all vessels adhere to the COLREGs guidelines for collision avoidance, thereby preventing any deviation from these regulations. In subsequent research, we will direct our attention to the issue of vessels failing to comply with the COLREGs in their collision avoidance manoeuvers. Finally, in the conclusion of this paper, we will present this issue as a potential avenue for future research.

In the multi-ship encounter experiment, the blue line represents the trajectory of the own ship, the orange line indicates the trajectory of target_ship1, and the green line indicates the trajectory of target_ship2, as shown in Figure 16a. The own ship first forms a collision risk with target_ship1, and the two ships will collide if they continue to follow the original route. Upon reaching (4, 4), target ship No. 1 proceeded to (6, 7), at which point the DCPA and TCPA of the two ships reached the threshold for avoidance action. This resulted in the formation of a crossing situation, in accordance with the requirements of the COLREGs. The own ship was required to alter course to starboard, while target_ship1 was instructed to keep her course and speed. As illustrated in Figure 16b, at t = 870 s, the own ship successfully avoided target_ship1, and at this time, the distance between the two ships was 2.63 n mile. The own ship continues sailing and has a risk of collision with target_ship2. At this time, the two ships form a crossing situation. In accordance with the requirements of the COLREGs, the own ship will alter its course to starboard, as shown in Figure 16c. At t = 1365s, own ship successfully avoids target_ship2 at a distance of 1.28 nm. Once the collision avoidance procedure has been completed, own ship will gradually turn and resume its original route, as illustrated in Figure 16d. The statistics of DCPA parameters between the own ship and the target ships during the collision avoidance in the multi-ship situation are presented in Figure 17.

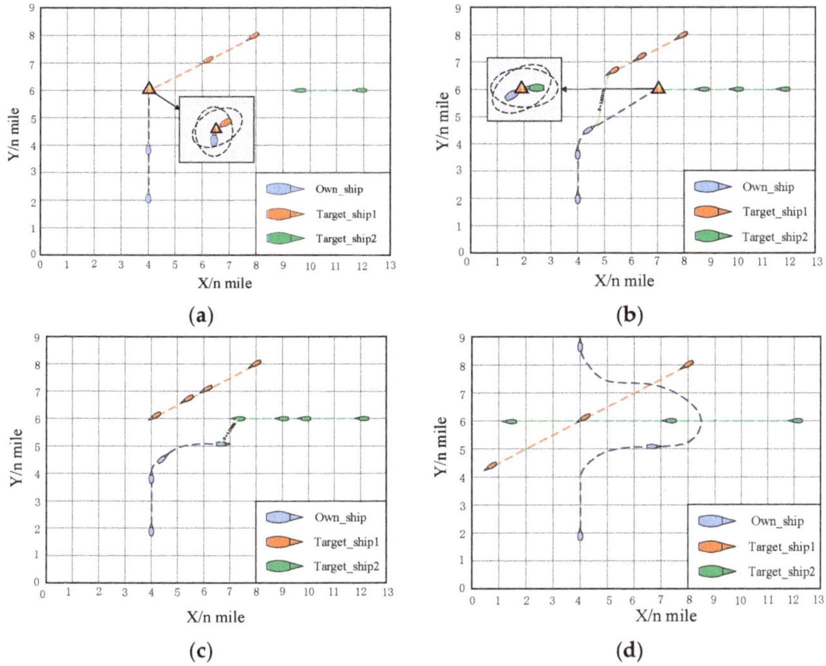

Figure 16. Key scenario diagrams of the collision avoidance process for the muti-ship situation. (**a**) Initial situation, (**b**) successful avoidance of target_ship1, (**b**) successful avoidance of target_ship2, and (**d**) resume to original route.

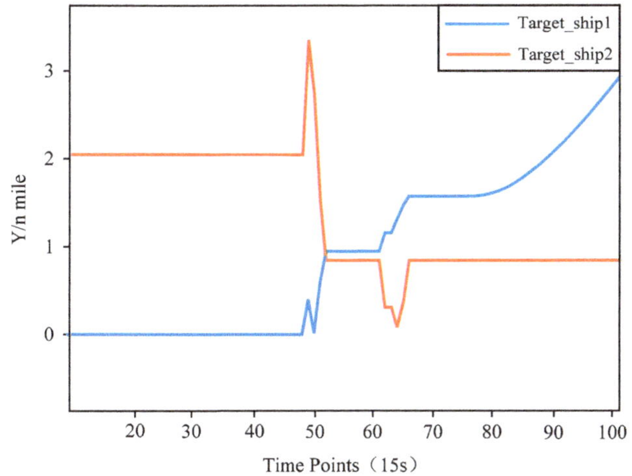

Figure 17. Variation curves during the collision avoidance process for the muti-ship situation.

6. Conclusions

The completion of the collision avoidance decision by the ship's autonomy system places significant demands on the effectiveness and reliability of the decision-making process. This paper takes the autonomous collision avoidance of MASS as its research object. The responsibility for avoiding collisions between ships is determined in accordance with the COLREGs, and an evaluation function for the adaptability of MASS autonomous collision avoidance is constructed based on the collision risk of ships (CRI) and the economic cost function (POCF) of collision avoidance actions. On this basis, a MASS autonomous collision avoidance decision algorithm (GPS-NSGA-II) is proposed for MASS autonomous collision avoidance, which incorporates a good point set improvement NSGA-II under COLREGs constraints. The MASS automatic collision avoidance decision-making model, as proposed in this paper, offers the following advantages: Firstly, the MASS fully considers the constraints of the COLREGs in the autonomous collision avoidance process, taking actions in strict accordance with the requirements of the COLREGs. Consequently, the autonomous collision avoidance scheme of the MASS satisfies the requirements of the COLREGs. Secondly, the MASS collision avoidance scheme is designed to consider both the collision risk between ships and the path cost of the avoidance scheme. This ensures that the autonomous collision avoidance scheme of MASS aligns with the actual collision avoidance manoeuvres of the ships. The simulation experiments conducted in a variety of scenarios demonstrate that the MASS autonomous collision avoidance decision-making model proposed in this paper is an effective solution to the autonomous collision avoidance problem of MASS in open water, with significant practical value. However, the MASS autonomous collision avoidance decision-making model proposed in this paper does not consider the possibility of MASS deviating from the COLREGs, nor is it applicable to the autonomous collision avoidance of MASS in complex waters with spatial limitations. Further research is required to better meet the needs of MASS autonomous collision avoidance in different waters.

Author Contributions: Conceptualization, S.Z. and F.L.; methodology, S.Z.; software, Z.L.; validation, S.Z.; formal analysis, F.L.; investigation, F.L.; resources, Z.L.; writing—original draft preparation, Z.L.; writing—review and editing, S.Z.; visualization, Z.L.; supervision, S.Z. project administration, F.L. All authors have read and agreed to the published version of the manuscript.

Funding: This research received no external funding.

Institutional Review Board Statement: Not applicable.

Informed Consent Statement: Not applicable.

Data Availability Statement: Data and results supporting the overview can be obtained from the corresponding author upon reasonable request.

Conflicts of Interest: The authors declare no conflicts of interest.

References

1. Wei, J.B.; Zhang, J.Q.; Liu, Z.; Qu, J.; Sui, B.; Zhang, Y. Path-Following and Obstacle-Avoidance Control of USV Based on Finite-Distance Convergence. *J. Mar. Sci. Eng.* **2024**, *12*, 34. [CrossRef]
2. Lin, C.Y.; Zhen, R.; Tong, Y.T.; Yang, S.H.; Chen, S.K. Regional ship collision risk prediction: An approach based on encoder-decoder LSTM neural network model. *Ocean Eng.* **2024**, *296*, 117019. [CrossRef]
3. Chen, X.Q.; Liu, S.H.; Zhao, J.; Wu, H.; Xian, J.; Montewka, J. Autonomous port management based AGV path planning and optimization via an ensemble reinforcement learning framework. *Ocean Coastal Manag.* **2024**, *251*, 107087. [CrossRef]
4. Chang, C.-H.; Wijeratne, I.B.; Kontovas, C.; Yang, Z. COLREG and MASS: Analytical review to identify research trends and gaps in the Development of Autonomous Collision Avoidance. *Ocean Eng.* **2024**, *302*, 117652. [CrossRef]
5. Wen, N.F.; Long, Y.D.; Zhang, R.B.; Liu, G.; Wan, W.; Jiao, D. COLREGs-Based Path Planning for USVs Using the Deep Reinforcement Learning Strategy. *J. Mar. Sci. Eng.* **2023**, *11*, 2334. [CrossRef]
6. Johansen, T.A.; Perez, T.; Cristofaro, A. Ship Collision Avoidance and COLREGS Compliance Using Simulation-Based Control Behavior Selection With Predictive Hazard Assessment. *IEEE Trans. Intell. Transp. Syst.* **2016**, *17*, 3407–3422. [CrossRef]
7. Zhang, Y.-J.; Zhai, P.-Y. Research progress and trend of autonomous collision avoidance technology for marine ships. *J. Dalian Marit. Univ.* **2022**, *48*, 1–11.
8. Chi, X.M.; Liu, X.M. Design of Ship Intelligent Collision Prevention System Based on Computer Vision. *J. Coast. Res.* **2019**, *97*, 242–247. [CrossRef]
9. Chen, X.Q.; Lv, S.Y.; Shang, W.L.; Wu, H.F.; Xian, J.F.; Song, C.C. Ship energy consumption analysis and carbon emission exploitation via spatial-temporal maritime data. *Appl. Energy* **2024**, *360*, 122886. [CrossRef]
10. Guo, J.; Li, L.N.; Gao, J.J.; Li, G. Early Warning of Collision and Avoidance Assitance for River Ferry. *Navig. China* **2020**, *43*, 8–14.
11. Wu, Z.L. Quantification of action to avoid collision. *J. Navig.* **1984**, *37*, 420–430.

12. Lyu, H.G.; Yin, Y. COLREGS-Constrained Real-time Path Planning for Autonomous Ships Using Modified Artificial Potential Fields. *J. Navig.* **2019**, *72*, 588–608. [CrossRef]
13. Lee, S.M.; Kwon, K.Y.; Joh, J. A fuzzy logic for autonomous navigation of marine vehicles satisfying COLREG guidelines. *Int. J. Control Autom. Syst.* **2004**, *2*, 171–181.
14. Meng, W.L.; Gong, Y.; Xu, F.; Tao, P.; Bo, P.; Xin, S. Efficient path planning for AUVs in unmapped marine environments using a hybrid local-global strategy. *Ocean Eng.* **2023**, *288*, 116227. [CrossRef]
15. Kathen, M.J.T.; Samaniego, F.P.; Flores, I.J.; Reina, D.G. An Informative Path Planner for a Fleet of Autonomous Surface Vehicles With Heterogeneous Sensing Capabilities Based on Multi-Objective PSO. *IEEE Access* **2023**, *11*, 110943–110966. [CrossRef]
16. Zhen, R.; Gu, Q.Y.; Shi, Z.Q.; Suo, Y.F. An Improved A-Star Ship Path-Planning Algorithm Considering Current, Water Depth, and Traffic Separation Rules. *J. Mar. Sci. Eng.* **2023**, *11*, 1439. [CrossRef]
17. Gu, Q.Y.; Zhen, R.; Liu, J.L.; Li, C. An improved RRT algorithm based on prior AIS information and DP compression for ship path planning. *Ocean Eng.* **2023**, *279*, 114595. [CrossRef]
18. Hadi, B.; Khosravi, A.; Sarhadi, P. Cooperative motion planning and control of a group of autonomous underwater vehicles using twin-delayed deep deterministic policy gradient. *Appl. Ocean Res.* **2024**, *147*, 103977. [CrossRef]
19. Du, Z.; Li, W.F.; Shi, G.Y. Multi-USV Collaborative Obstacle Avoidance Based on Improved Velocity Obstacle Method. *J. Dalian Marit. Univ.* **2024**, *43*, 1–7. [CrossRef]
20. Longo, G.; Martelli, M.; Russo, E.; Merlo, A.; Zaccone, R. Adversarial waypoint injection attacks on Maritime Autonomous Surface Ships (MASS) collision avoidance systems. *J. Mar. Eng. Technol.* **2023**, *23*, 184–195. [CrossRef]
21. Liu, H.D.; Deng, R.; Zhang, L.Y. The Application research for Ship Collision Avoidance with Hybrid Optimization Algorithm. In Proceedings of the IEEE International Conference on Information and Automation (ICIA), Ningbo, China, 1–3 August 2016; pp. 760–767.
22. Liu, Z.; Huang, L.W.; Zhang, K.; Hao, G. Decision-making Approach for Multi-ship Collision AvoidanceBased on Beetle Antennae Search Algorithm. *J. Wuhan Univ. Technol. (Transp. Sci. Eng.)* **2021**, *45*, 1000–1004.
23. Zhou, S.L.; Yang, X.; Liu, K.Z.; Xiong, Y.; Wu, X.L.; Liu, J.J.; Wang, W.Q. COLREGs-Compliant Method for Ship Collision Avoidance Based on Deep Reinforcement Learning. *Navig. China* **2020**, *43*, 27–32.
24. Zheng, Y.S.; Zhang, X.G.; Shang, Z.J.; Guo, S.; Du, Y. A Decision-Making Method for Ship Collision Avoidance Based on Improved Cultural Particle Swarm. *J. Adv. Transp.* **2021**, *31*, 898507. [CrossRef]
25. Xu, X.L.; Lu, Y.; Liu, X.C.; Zhang, W. Intelligent collision avoidance algorithms for USVs via deep reinforcement learning under COLREGs. *Ocean Eng.* **2020**, *217*, 107704. [CrossRef]
26. Xie, S.; Chu, X.M.; Zheng, M.; Liu, C. Ship predictive collision avoidance method based on an improved beetle antennae search algorithm. *Ocean Eng.* **2019**, *192*, 106542. [CrossRef]
27. Huang, R.X.; Luo, L.; Yang, M.; Liu, W.Q. Multiship Coordinated Collision Avoidance Decision Based on lmproved Twin Delayed Deep Deterministic Policy Gradient. *Comput. Sci.* **2023**, *50*, 269–281.
28. Fiorini, P.; Shiller, Z. Motion Planning in Dynamic Environments Using Velocity Obstacles. *Int. J. Robot. Res.* **1998**, *17*, 760–772. [CrossRef]
29. He, X.; Shi, Z.Y.; Zhong, Y.S. Multi-USV Cooperative Collision Avoidance Based on Velocity Obstacle. *Acta Aeronaut. Astronaut. Sin.* **2023**, *44*, 387–397.
30. Zhao, G.X.; Wang, C.X.; Wang, H.P.; Li, Y.M. Local path planning for unmanned surface vehicle usingimproved velocity obstacle method. *Syst. Eng. Electron.* **2023**, *45*, 3975–3983.
31. Xu, X.-Q.; Yang, J.-D.; Mao, Y.; Chen, X. Study on Path Planning of USV Based on VelocityObstacle and Improved Artificial Potential Field Algorithm. *J. Wuhan Univ. Technol.* **2022**, *44*, 96–102.
32. Hong, X.-B.; Xu, Z.-P.; Wei, X.-Y.; Zhu, K.C.; Chen, Y.M. Dynamic obstacle avoidance of unmanned surface vehicle based on improved speed obstacle method. *Opt. Precis. Eng.* **2021**, *29*, 2126–2139. [CrossRef]
33. Zhu, X.M.; Yi, J.J.; Ding, H.K.; He, L. Velocity Obstacle Based on Vertical Ellipse for Multi-Robot Collision Avoidance. *J. Intell. Robot. Syst.* **2020**, *99*, 183–208. [CrossRef]
34. Kuwata, Y.; Wolf, M.T.; Zarzhitsky, D.; Huntsberger, T.L. Safe Maritime Navigation with COLREGS Using Velocity Obstacles. In Proceedings of the IEEE/RSJ International Conference on Intelligent Robots and Systems, San Francisco, CA, USA, 25–30 September 2011.
35. Zhang, L.; Mou, J.M.; Chen, P.F.; Li, M. Path Planning for Autonomous Ships: A Hybrid Approach Based on Improved APF and Modified VO Methods. *J. Mar. Sci. Eng.* **2021**, *9*, 761. [CrossRef]
36. Thyri, E.H.; Breivik, M. Partly COLREGs-compliant collisionavoidance for ASVs usingencounter-specific velocity obstacles. *IFAC-PapersOnLine* **2022**, *55*, 37–43. [CrossRef]
37. Cho, Y.; Han, J.; Kim, J. Efficient COLREG-Compliant Collision Avoidance in Multi-Ship Encounter Situations. *IEEE Trans. Intell. Transp. Syst.* **2022**, *23*, 1899–1911. [CrossRef]
38. Liu, J.P.; Zhao, B.G.; Li, L.B. Collision Avoidance for Underactuated Ocean-Going Vessels Considering COLREGs Constraints. *IEEE Access* **2021**, *9*, 145943–145954. [CrossRef]
39. Wang, Z.K.; Namkyun, I. Enhanced artificial potential field for MASS's path planning navigation in restricted waterways. *Appl. Ocean Res.* **2024**, *149*, 104052. [CrossRef]

40. Kim, H.G.; Yun, S.J.; Choi, Y.H.; Ryu, J.K.; Suh, J.H. Collision Avoidance Algorithm Based on COLREGs for Unmanned Surface Vehicle. *J. Mar. Sci. Eng.* **2021**, *9*, 863. [CrossRef]
41. Xu, D.H.; Yang, J.; Zhou, X.Q.; Xu, H. Hybrid path planning method for USV using bidirectional A* and improved DWA considering the manoeuvrability and COLREGs. *Ocean Eng.* **2024**, *298*, 117210. [CrossRef]
42. Zhang, G.Q.; Han, J.; Zhang, W.D.; Yin, Y.; Zhang, L. Finite-time adaptive event-triggered control for USV with COLREGS-compliant collision avoidance mechanism. *Ocean Eng.* **2023**, *285*, 115357. [CrossRef]
43. Bayrak, M.; Bayram, H. COLREG-Compliant Simulation Environment for Verifying USV Motion Planning Algorithms. In Proceedings of the OCEANS Conference, Limerick, Ireland, 5–8 June 2023.
44. Zeng, Y.; Zhang, J.F.; Zhang, M.Y.; Zhang, D. Collision Avoidance Decision-Making Based on Particle Swarm Optimization and Genetic Algorithm. *Navig. China* **2020**, *43*, 1–6.
45. Fu, Z.J.; Wang, H.J.; Gu, Y.M.; Li, C.; Tong, H.; Wang, H. Method for Collision Avoidance by USV Based on Improved Genetic Algorithm. In Proceedings of the Global OCEANS Singapore-U.S. Gulf Coast Conference, Electr Network, 5-30 October 2020.
46. Ni, S.K.; Liu, Z.J.; Cai, Y.; Wang, X. Modelling of ship's trajectory planning in collision situations by hybrid genetic algorithm. *Pol. Marit. Res.* **2018**, *25*, 14–25. [CrossRef]
47. Liu, S.; Zhang, L.Y. Study and Simulation on Automation Multi-ships Collision Avoidance Strategy. *J. Comput. Theor. Nanosci. (CTN)* **2016**, *13*, 194–210. [CrossRef]
48. Xu, Y.M.; Lyu, J.H.; Liu, J.L.; Li, L.H.; Guan, H.X. Multi-vessel intelligent collision avoidance decision-making based on CSSOA. *Chin. J. Ship Res.* **2023**, *18*, 88–96.
49. Lu, C.-W.; Hsueh, C.K.; Chuang, Y.L.; Lai, C.M.; Yang, F.S. Marine Collision Avoidance Route Planning Model for MASS Based on Domain-Based Predicted Area of Danger. *J. Mar. Sci. Eng.* **2023**, *11*, 1724. [CrossRef]
50. Zhen, R.; Shi, Z.Q.; Gu, Q.Y.; Yang, S.H. A novel deterministic search-based algorithm for multi-ship collaborative collision avoidance decision-making. *Ocean Eng.* **2024**, *292*, 116524. [CrossRef]
51. Li, L.N.; Chen, G.Q.; Yang, L.B.; Xu, C.L.; Wang, X.H.; Wen, T. Test and Application of Personifying Intelligent Decision-Making Algorithm for Vessel Collision Avoidance. *Navig. China* **2022**, *45*, 1–7.
52. Zhou, Y.; Du, W.J.; Liu, J.; Li, H.Q.; Grifoll, M.; Song, W.J.; Zheng, P.J. Determination of Ship Collision Avoidance Timing Using Machine Learning Method. *Sustainability* **2024**, *16*, 16114626. [CrossRef]
53. Li, W.F.; Zhong, L.F.; Xu, Y.; Shi, G.Y. Collision Risk Index Calculation Based on an Improved Ship Domain Model. *J. Mar. Sci. Eng.* **2022**, *10*, 10122016. [CrossRef]
54. Chen, D.J.; Wan, X.C.; Dai, C.; Mou, J.M. A Research on AIS-based Embedded System for Ship Collision Avoidance. In Proceedings of the 3rd Int Conference Transportation Information Safety, Wuhan, China, 25–28 June 2015; pp. 512–517.
55. Yan, S.Q.; Yang, P.; Zhu, D.L.; Wu, F.X.; Yan, Z. Improved sparrow search algorithm based on good point set. *J. Beijing Univ. Aeronaut. Astronaut.* **2023**, *9*, 790–2798.

Journal of
Marine Science and Engineering

MDPI

Article

Side-Scan Sonar Image Generator Based on Diffusion Models for Autonomous Underwater Vehicles

Feihu Zhang *, Xujia Hou, Zewen Wang, Chensheng Cheng and Tingfeng Tan

School of Marine Science and Technology, Northwestern Polytechnical University, Xi'an 710072, China; hxj1363947894@mail.nwpu.edu.cn (X.H.); wzw828@mail.nwpu.edu.cn (Z.W.); chensheng.cheng@mail.nwpu.edu.cn (C.C.); ttf@mail.nwpu.edu.cn (T.T.)

* Correspondence: feihu.zhang@nwpu.edu.cn

Abstract: In the field of underwater perception and detection, side-scan sonar (SSS) plays an indispensable role. However, the imaging mechanism of SSS results in slow information acquisition and high complexity, significantly hindering the advancement of downstream data-driven applications. To address this challenge, we designed an SSS image generator based on diffusion models. We developed a data collection system based on Autonomous Underwater Vehicles (AUVs) to achieve stable and rich data collection. For the process of converting acoustic signals into image signals, we established an image compensation method based on nonlinear gain enhancement to ensure the reliability of remote signals. On this basis, we developed the first controllable category SSS image generation algorithm, which can generate specified data for five categories, demonstrating outstanding performance in terms of the Fréchet Inception Distance (FID) and the Inception Score (IS). We further evaluated our image generator in the task of SSS object detection, and our cross-validation experiments showed that the generated images contributed to an average accuracy improvement of approximately 10% in object detection. The experimental results validate the effectiveness of the proposed SSS image generator in generating highly similar sonar images and enhancing detection accuracy, effectively addressing the issue of data scarcity.

Keywords: side-scan sonar; diffusion model; image generation; deep learning; gain enhancement; object detection

1. Introduction

The oceanic realm accounts for approximately two-thirds of the Earth's surface area, yet human exploration has only mapped about five percent of it [1]. Consequently, the majority of oceanic zones remain obscure, abysmal, and unknown to humankind. To gain further insights into the oceans, more in-depth research on underwater environmental perception technologies is imperative [2]. At present, underwater environment detection technology mainly includes acoustic-based methods and optical-based methods. Due to the significant attenuation of electromagnetic waves under water, optical signals only have the capability to propagate over several tens of meters subaqueously. Should water clarity be compromised, the effective range diminishes further, rendering optical detection principally suitable for proximate inspection and verification [3]. Acoustic detection, however, leverages the propagation laws of sound signals underwater for exploration and monitoring purposes. It boasts a notable advantage over optical detection in terms of range; thus, the bulk of underwater detection is predominantly conducted via acoustic means. A Sound Navigation and Ranging (sonar) device is a type of sensor engineered based on the transduction of sound waves and information processing exploiting the propagation and reflection properties of acoustics in aquatic environments. Within this spectrum of tools, SSS operates by projecting fan-shaped beams of acoustic pulses laterally downwards, and due to its high resolution and cost-effective nature, it constitutes a vital element of underwater acoustic detection. SSS is extensively applied in marine cartography [4,5], subaquatic geological surveying [6], underwater target detection and identification [7], and dam foundation inspection [8]. Moreover, with the continued evolution of scientific technology, deep learning-based SSS detection methods are increasingly superseding traditional expert system-based approaches, thereby consistently increasing the precision of detection [9]. Nevertheless, data-driven detection models require an abundance of sample data and labels to assure performance efficacy. During SSS data acquisition, the imaging principle dictates that a subsequent acoustic wave is emitted only upon receipt of the returning echo signal at the current moment, resulting in a data interval per ping typically exceeding 0.2 s. This entails that during the data collection process, approximately five pings of sonar data are generated per second. Acquiring a sonar image with a height of 400 pixels requires a minimum of 80 s, which is considerably longer compared with the time required for forward-looking sonars or optical cameras. Additionally, the carrier platform for a sonar sensor may also host other hydroacoustic devices, such as Doppler velocimeters or Ultra-Short Baseline (USBL) positioning systems. If the operational frequencies of these devices are integer multiples of the working frequency of the SSS sensor, interference may occur, resulting in striping noise artifacts on the sonar imagery.

In this study, we initially built an AUV-based SSS data acquisition device to achieve efficient data acquisition. The AUV possessed a streamlined shape to significantly diminish drag during subaquatic navigation and was powered by an electric propulsion system to further reduce noise emissions from the platform. An SSS sensor was mounted on the AUV, a vehicle used to perform constant-depth cruising missions in diverse marine areas to collect data. Subsequently, for the raw SSS data, we developed a nonlinear gain enhancement algorithm to compensate for the image features at the far end that are generally weakened due to the propagation loss of sound waves underwater, thereby obtaining sonar images with high contrast and clarity. Building on this, a sonar image dataset containing five categories of information, which can provide authentic raw data for subsequent research based on SSS imagery, was constructed. Furthermore, we developed a diffusion model-based controllable category sonar image generator that obviates the need for training additional classifiers. By balancing the labeled conditional diffusion model with the unlabeled non-conditional diffusion model, we achieved the generation of both high-quality and diverse images.Our study effectively addresses the present insufficiency of SSS image datasets, supplying numerous exemplary samples for data-driven detection models. The results of experiments conducted on publicly available target detection models demonstrate the efficacy of the proposed SSS image generator in enhancing object detection accuracy. In summary, the main contributions of this study are as follows:

- We established an SSS data collection platform based on an AUV and collected a large number of raw sonar data from different marine areas. We developed a nonlinear gain enhancement algorithm suitable for compensating for spherical wave propagation loss, achieving balanced sonar image processing and improving the quality of SSS imaging.
- We created a five-category SSS image dataset that includes common seabed backgrounds and targets. Based on this dataset, we established a controllable category SSS image generator that can generate images of specified categories without relying on additional classifiers, effectively expanding the SSS image dataset.
- We conducted both quantitative and qualitative evaluations of the SSS image generator, using the FID, the IS, and Haralick texture features to assess the model. The generative model was also applied to the task of target detection, and the cross-validation results demonstrate its positive impact on improving detection accuracy, providing data support for subsequent SSS image-based research.

2. Related Work

2.1. Basic Image Generation Model

The objective of image generation tasks is to obtain the probabilistic statistical distribution of the original data. Early image generation efforts predominantly relied on feature representation methods to generate images which were only capable of handling simple and regular image generation tasks [10]. The advent of deep neural networks, with their exceptional feature-learning capabilities and nonlinear expression abilities, has significantly advanced the development of generative models [11]. To date, the primary deep learning-based generative models include Variational Autoencoders (VAEs) [12], Generative Adversarial Networks (GANs) [13], autoregressive models [14], and denoising diffusion models.

VAEs, deep generative models based on the autoencoder structure [15], introduce constraints on the latent space during training, such as the Kullback–Leibler (KL) divergence, to ensure that the latent variables follow a Gaussian distribution. However, constraining the latent space to conform to a normal distribution is often overly challenging. Therefore, Oord et al. [16] developed the Vector-Quantized Variational Autoencoder (VQVAE), which constrains the VAE's latent space to satisfy a discrete distribution and employs an autoregressive model to model the discrete codes, thus partially addressing the issue of fitting the discrete space distribution. Subsequent work [17] developed hierarchical quantized VAEs to encode images, which can more comprehensively address the issue of fitting the discrete space distribution and diminish the decoder's burden in reconstructing the image, thereby enabling the generation of more realistic images.

GANs, previously the most widely used generative models, consist of a generator and a discriminator. The former aims to fit the data distribution, while the latter distinguishes between real and generated data. However, due to the effective fitting capabilities of both the generator and the discriminator and the inability to guarantee optimal convergence during iterations, these networks do not reach the Nash equilibrium state during actual training [18]. Additionally, GANs often face issues such as training instability, mode collapse, and gradient vanishing problems. To address these issues, Mao et al. [19] developed the Least Squares GAN (LSGAN) to enhance the stability of adversarial training and improve the quality of the synthesized images. The authors in [20] established WGAN, further analyzed the reasons for training instability, and introduced the Lipschitz continuity constraint on the discriminator, effectively enhancing training stability. DCGAN [21] is the first network in which deep convolutional layers and batch normalization were introduced to improve network stability. The authors in [22] developed SAGAN by incorporating the self-attention mechanism into a GAN, which enhanced the network's representational capacity, improving the quality and diversity of the generated images. Furthermore, PGGAN [23] represents a progressive growth strategy for image generation, increasing the resolution, thereby making it possible to successfully generate 1024 × 1024-resolution images for the first time. Inspired by image style transfer, StyleGAN [24] is based on adjustable instance normalization to inject modulation signals, which greatly enhances the network's capabilities. StyleGAN2 [25], an improvement on the latter, enables end-to-end training and achieves high-quality, high-resolution image generation.

Autoregressive models are commonly used generative models based on the idea that subsequent variables in a sequence can be regressed from preceding variables. Different autoregressive models employ various network structures for generation. PixelRNN [26] uses the temporal concept of Recurrent Neural Networks (RNNs) for the prediction process, where the RNN model is composed of multiple Long Short-Term Memory (LSTM) layers. PixelCNN [27] approximates the RNN structure with masked convolutional layers and eliminates the pooling layers. Subsequent work on PixelCNN further enhanced its computational efficiency and generation results [28]. In recent years, the Transformer [29] structure has allowed for significant advances in natural language generation by introducing the attention mechanism, which significantly strengthens the fitting capabilities of autoregressive models. In the field of image generation, DALL-E [29] leverages the VQVAE to encode images into discrete latent codes and then uses an autoregressive model based on the Transformer structure to fit the distribution of these discrete codes. However, this approach faces the following issues: (1) The efficiency of serially generating conditional distribution probabilities is lower than directly generating joint distributions and sampling, resulting in significant time consumption for image generation. (2) The assumption that each position in the autoregressive model depends on the previous positions is unreasonable for images, as image patches are often related to various positions. Once a position is generated, it cannot be modified based on subsequent positions, causing any errors in earlier positions to propagate and affect subsequent results, leading to cumulative errors.

Denoising diffusion models are the latest advancement in image generation, comprising a forward noise diffusion Markov process and a reverse denoising Markov process [30]. By incrementally adding noise in multiple steps, this approach decomposes the complex distribution fitting problem into several simpler sub-problems. During the training phase, the denoising network updates its parameters to fit the data distribution of the noise-added process at each step. In the image generation phase, noise signals are first sampled from a known distribution and then progressively denoised through multiple steps by the denoising network, thereby sampling from the complex image distribution. This method has garnered widespread attention since its introduction, with researchers exploring its application in tasks such as unconditional image generation [31,32], class-conditional generation [33], and text-to-image translation [34], achieving state-of-the-art results across these domains.

2.2. SSS Image Generation Model

Due to the difficulty of data acquisition in side-scan sonar and the sparsity of the target distribution, an important research direction in the field of SSS is image generation. To mitigate the scarcity of SSS data, Jiang [35] proposed a GAN-based image generation model that rapidly generates side-scan sonar images. However, this method can only generate single-channel grayscale images, and since the generator and discriminator are not equal in fitting ability, it is difficult to reach the Nash equilibrium, resulting in model collapse. Bore [36] created a generation model with the environment as the input and the side-scan sonar image as the output based on the conditional generation of adversarial networks. However, this method has significant restrictions on implementation. Song [37] suggested using Extreme Learning Machines for sonar image segmentation and synthesis, employing a multi-path convolutional neural network to learn different image features, thereby progressively synthesizing detailed side-scan sonar images. Wang [38] employed transfer learning to introduce a style transfer approach between optical and SSS images, effectively enhancing classification accuracy. Ge [39] also proposed the use of synthetic data and transfer learning to convert optical images into side-scan sonar images for classification. In the case of zero samples, Xu [40] proposed a multi-feature fusion self-attention network (MFSANet) to generate SSS images of new categories, which transformed the problem into a traditional supervised learning problem. By taking the optical image as the input to the network, the Simplified Self-Attention Module (SSAM) is used to model the acoustic image, so as to effectively generate virtual SSS samples. However, this transfer learning-based method requires an additional photo, such as one taken on land, to be used as a specification. In many cases, this extra requirement limits the application of the method.

With recent breakthroughs in diffusion models [30–33], more and more researchers are turning to this approach for image generation. Yang [41] was the first to utilize a diffusion model for SSS image modeling: they transformed the sonar image into random noise with a Gaussian distribution and iteratively refined this noise in a reverse process to reconstruct new samples that matched the prior data distribution. Similarly, Zhang [42] and Cheng [43] applied diffusion models to expand the collected sonar data and created a hybrid dataset composed of real and generated samples. Upon experimenting with various mainstream target detection algorithms, they concluded that the generated samples could effectively improve the accuracy of the detection models considered.

However, most of the aforementioned methods have been applied to the SCTD dataset [44]. A significant portion of images in the SCTD dataset are not raw data (having undergone text editing, watermarking, etc.), which disrupts the statistical distribution of authentic data. In addition, previous work has not been able to achieve category-specific image generation, which means that the generated images often require further manual selection before they can be employed for subsequent detection tasks. Therefore, the primary limitations of the current side-scan sonar image generation task include the following: (1) the existing public datasets do not truly reflect the data distribution of sonar data, thus hindering the development of generation models; (2) generators based on diffusion models can only produce uncategorized images and necessitate extensive human effort to further filter the generated images.

3. Materials and Methods

3.1. Data Acquisition and Processing

The quality of SSS data is influenced by various factors, including the motion of the carrier, its speed, and its altitude above the seabed. Using an AUV equipped with an SSS sensor for data collection offers advantages such as constant speed, continuously adjustable altitude above the seabed, and minimal noise interference. Therefore, to obtain reliable and sufficient underwater sonar images, we selected an AUV with a diameter of 324 mm, equipped with the SS4590 module for SSS data collection, as shown in Figure 1. The AUV had a modular design, allowing each section to be detached and replaced. The front segment was equipped with a USBL for underwater communication. The forward and aft auxiliary propulsion sections enabled the AUV to move horizontally and vertically, respectively. The tail was fitted with rudders and a primary propulsor, providing the AUV with a maximum cruising speed of 12 knots. The battery section contained lithium-ion batteries, which could power the AUV for 10 h at a cruising speed of 3 knots. Additionally, the AUV was equipped with a Global Positioning System (GPS) module and inertial navigation components for self-localization. The sonar module section carried the SS4590 SSS sensor, which could emit both 450 kHz and 900 kHz frequency CW or Chirp signals, with a maximum single-side detection range of 150 m.

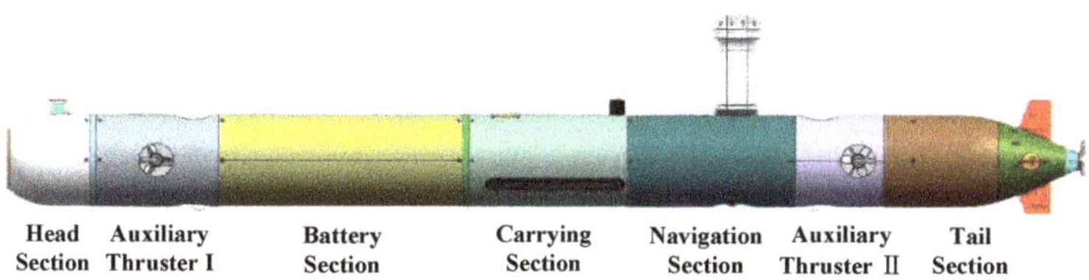

(a) AUV model diagram.

(b) AUV being deployed in the test area.
Figure 1. AUV overall layout structure and layout drawing.

To ensure the collection of representative and abundant data, we selected two marine areas for our data acquisition missions. Additionally, we placed cylindrical objects with a base diameter of 100 cm and a height of 150 cm on the seabed to serve as targets. Due to the side-scan sonar sensor's installation angle of 20 degrees horizontally downwards on the AUV, there were blind spots in its field of view; therefore, it was necessary to plan the AUV's route during data acquisition to ensure that the sonar beams could cover the blind spots to avoid missing any targets. The AUV performed back-and-forth patrolling detection tasks in the two preset marine areas (as shown in Figure 2); to collect clear images, its cruising speed was set to 3 knots and the depth to 10 m. The distance between the return paths was set to 100 m to guarantee that the sonar beams could cover the blind spots. The experimental areas selected in the marine environment had depths ranging from 30 to 50 m, ensuring the acquisition of seabed lines and background, along with seabed terrain that exhibited significant depth variations. During the collection process, Nvidia Jetson AGX Orin edge computing devices were utilized to control the SSS sensor and store data. The AUV navigated for 1 h and 40 min across the two marine areas, respectively, generating 2.29 GB of XTF files.

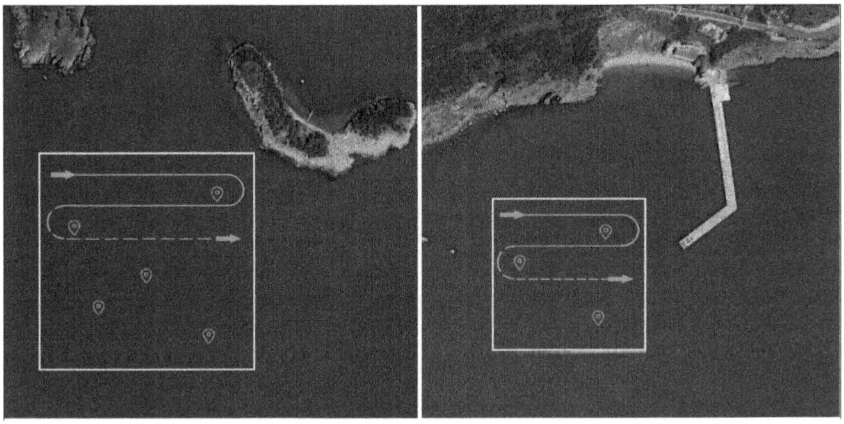

Figure 2. Schematic diagram of AUV path planning in experimental sea areas.

After parsing the XTF files, it was necessary to convert the acoustic signals into image signals. An SSS device emits sound waves from both sides of a moving platform, covering a broad area and receiving echo signals, which are then processed and transformed into intensity values. SSS images are created based on the variation in the grayscale intensity of the pixels within each scan line, forming a grayscale contrast [45]. Areas with stronger grayscale intensity depict the target image's geometric shapes, and the intensity of the image's grayscale directly corresponds to the amplitude variations of the echo signal, which are primarily associated with seabed topography, geomorphological features, and sediment types. The original data sampling precision was 16-bit, and sonar data imaging typically requires quantizing the received signal strength into a grayscale range of 0~255. At this point, a model needed to be established to convert the acoustic intensity information into the grayscale information describing the image. The quantization formula for SSS data with 16-bit sampling precision is

$$G = \frac{GB - GB_{min}}{GB_{max} - GB_{min}}(G_{max} - G_{min}) \tag{1}$$

$$G = C \times e^{\frac{now-close}{far-close}} \times G \tag{2}$$

In Equation (1), G represents the quantized grayscale data; GB represents the pre-quantization echo data; G_{max} and G_{min} are the maximum and minimum values of the grayscale image, respectively, and GB_{max} and GB_{min} are the maximum and minimum values of the echo data, respectively. In Equation (2), C is a constant, and $e^{\frac{now-close}{far-close}}$ stands for the distance dimension, where the gain is normalized in the exponential domain.

Equations (1) and (2) enhance the echo intensity while converting it into grayscale levels, making it suitable for sonar data with weaker sampling echo intensity. The advantage of this technique lies in its ability to effectively compensate for the echo intensity in the far field area of the image, thus making the targets in the distant area of the image distinctly visible. Once the data were converted into image signals, the images were cropped into 256 × 256 pixel segments. These segments were then categorized into five classes based on their specific content: targets, seabed lines, seabed backgrounds, seabed terrain, and images with interference. The interference was caused by the AUV's DVL and Ultra-Short USBL, which can generate acoustic interference when their operating frequencies are harmonically related to the frequency of the SSS device. This interference appeared in the images as regular patterns of lines.

Following the aforementioned data collection and processing procedures, the final authentic dataset comprised images for each category, as indicated in Table 1. Figure 3 provides a partial visual display of the dataset.

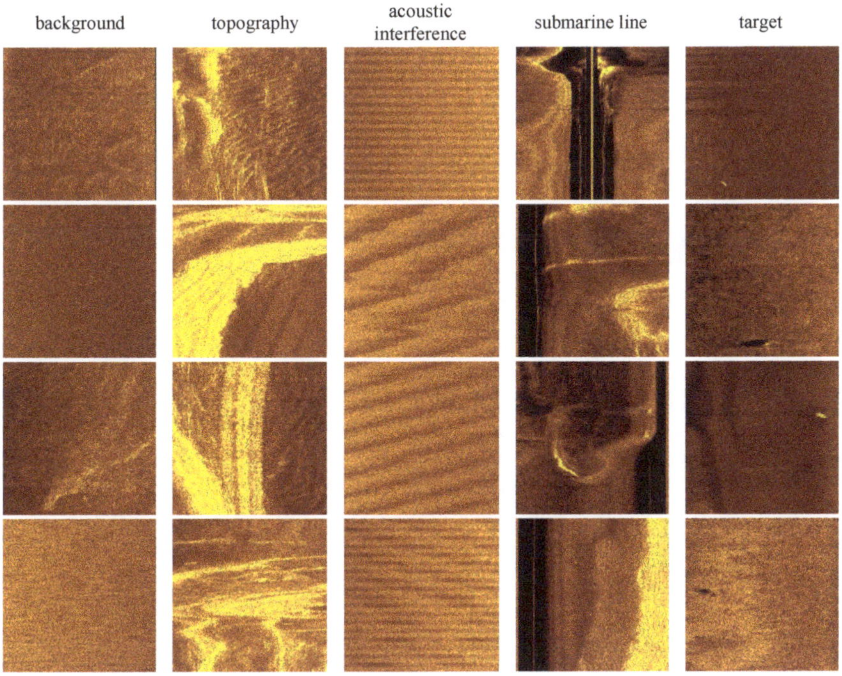

Figure 3. Five-category SSS dataset based on raw images.

Table 1. Categories and quantities of SSS dataset images.

Background	Topography	Acoustic Interference	Submarine Line	Target
1317	342	402	1712	399

3.2. SSS Diffusion Model

Diffusion models have recently become popular machine learning algorithms, demonstrating powerful representational learning capabilities. They have been widely applied in bioinformatics [46], object detection [47], and image reconstruction [48]. A diffusion model consists of a forward diffusion process and a reverse diffusion process, as shown in Figure 4. In the forward process, noise is progressively added to the initial real data, and in the reverse process, the original image is gradually recovered from the noise. Our diffusion model can be divided into two components: the objective modeling of the diffusion process and the parameter estimation network.

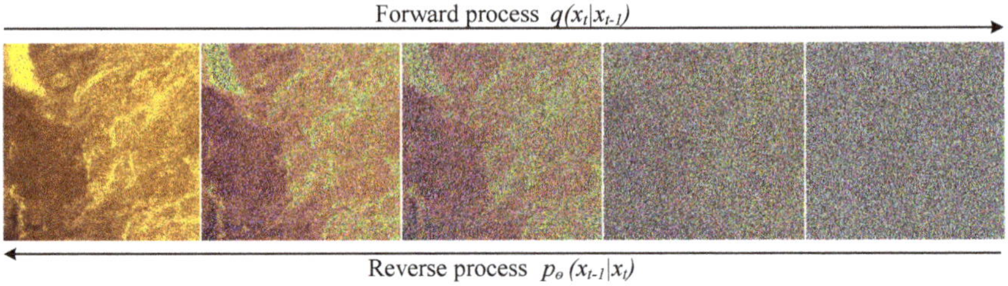

Figure 4. Schematic of the principle of diffusion model.

3.2.1. Target Modeling

Given the initial SSS raw image data ($x \sim q(x)$), the forward diffusion process consists of T steps, with each step adding Gaussian noise to the data from the previous step (x_{t-1}), and this process constitutes a Markov chain. According to the modeling of denoising diffusion probabilistic models (DDPMs) [30], the key equation for the forward process is as follows:

$$q(x_t|x_{t-1}) = N\left(x_t; \sqrt{1-\beta_t}x_{t-1}, \beta_t I\right) \tag{3}$$

$$q(x_{1:T}|x_0) = \prod_{t=1}^{T} q(x_t|x_{t-1}) \tag{4}$$

$$q(x_t|x_0) = N\left(x_t; \sqrt{\overline{\alpha}_t}x_0, (1-\overline{\alpha}_t)I\right) \tag{5}$$

In the equation, the standard deviation of the noise added at each step is determined by a fixed value, β_t, and the mean is determined by β_t and the current data, x_t, at time t. Moreover, we specify that $\alpha_t = 1 - \beta_t$ and $\overline{\alpha_t} = \prod \alpha_i$.

In the reverse process, the goal is to recover the original data from the Gaussian noise. It is assumed that this is also a Gaussian distribution, but since it is not feasible to sequentially fit the distribution, a parameterized distribution is constructed for estimation. The reverse process is still a Markov chain, and its core formulas are as follows:

$$p_\theta(x_{0:T}) = p(x_T)\prod_{i=1}^{T} p_\theta(x_{t-1}|x_t) \tag{6}$$

$$p_\theta(x_{t-1}|x_t) = N\left(x_{t-1}; \mu_\theta(x_t, t), \sum_\theta(x_t, t)\right) \tag{7}$$

$$q(x_{t-1}|x_t, x_0) = N(x_{t-1}; \tilde{\mu}(x_t, x_0), \tilde{\beta}_t I) \tag{8}$$

$$\tilde{\beta}_t = \frac{1-\overline{\alpha}_{t-1}}{1-\overline{\alpha}_t} \cdot \beta_t \tag{9}$$

$$\tilde{\mu}_t(x_t, x_0) = \frac{1}{\sqrt{\alpha_t}}\left(x_t - \frac{\beta_t}{\sqrt{1-\overline{\alpha}_t}}\epsilon\right) \tag{10}$$

It can be observed that the variance in the posterior diffusion conditional probability $q(x_{t-1}|x_t, x_0)$ does not contain any unknown parameters, while the mean includes a term with a stochastic parameter, ϵ. Therefore, by continuously fitting a neural network to make distribution $p_\theta(x_{t-1}|x_t)$ as close to $q(x_{t-1}|x_t, x_0)$ as possible, $p_\theta(x_{t-1}|x_t)$ can be used for generation. It is important to note that the target of neural network estimation can be noise $\epsilon_\theta(x_t)$ or the initial x_0, or it can be predicted score $\nabla_{x_t} \log p_\theta(x_t)$ [49].

To achieve controllable category image generation, in addition to the given original data, x, there is also the corresponding category information, y. Assuming that the forward noise addition process is still a Markov chain, its data distribution is $x \sim q(x)$. According to the derivation by Dhariwal [33], it is evident that the forward process of the conditional diffusion model is completely identical to that of a DDPM, which is

$$\hat{q}(x_t|x_{t-1}, y) = \hat{q}(x_t|x_{t-1}) = q(x_t|x_{t-1}) \quad (11)$$

During the reverse denoising process, according to Bayes' theorem, the following relationship holds:

$$p(y|x_t) = \frac{p(x_t|y)p(y)}{p(x_t)} \quad (12)$$

We note that $p(y)$ serves as the prior distribution; by taking the logarithm of the above equation, deriving with respect to x_t, and then incorporating the score function, we obtain

$$\nabla_{x_t} \log p(y|x_t) = -\frac{1}{\sqrt{1-\overline{\alpha}_t}}(\epsilon_\theta(x_t|y) - \epsilon_\theta(x_t)) \quad (13)$$

By substituting the aforementioned equation into Dhariwal's classifier gradient, the estimated parameters can be derived:

$$\hat{\epsilon}_\theta(x_t|y) = \epsilon_\theta(x_t) + s[\epsilon_\theta(x_t|y) - \epsilon_\theta(x_t)] \quad (14)$$

Thus, two models are trained: one is an unconditional generation model, like DDPMs, and the other is a conditional generation model. The final results can be obtained by using linear extrapolation from both the conditional and the unconditional generation models. The quality of the generated output can be adjusted with the guidance coefficient, s, which controls the balance between the realism and the diversity of the generated samples.

3.2.2. Parameter Estimation Network

We previously discussed controllable category generation based on diffusion models, where the core is to train a noise prediction model. Since noise and the original data are in the same dimensionality, we can opt for an autoencoder architecture [50] for the noise prediction model. Specifically, we utilized a U-Net model [51] that incorporates attention mechanisms and residual blocks, as shown in Figure 5.

The U-Net is composed of an encoder, a decoder, and their cross-layer connections. The encoder consists of four stages, each of which includes two residual blocks, a linear attention layer, and a downsampling module to reduce the size of the feature space. The linear attention mechanism was utilized because its time and memory consumption are linearly related to the sequence length, which is much more efficient than traditional attention mechanisms with quadratic complexity. After passing through the linear attention layer, the data are normalized with RMSNorm [52] to further improve computational efficiency. With each stage in the encoder, the dimensions of the height and width of the feature space are halved, while the channels are doubled. The structure of the decoder mirrors the encoder, except for the final downsampling module, which is replaced with an upsampling module to reverse the operation. In the decoder module, U-Net also introduces skip connections, which concatenate the features of the same dimensionality obtained from the encoder. This facilitates network optimization by allowing for the flow of information from previous layers to the following layers, which helps to reconstruct the finer details in the output image.

To differentiate among different time steps, inspired by the concept of positional encoding in Transformers [29], we employed sinusoidal position embeddings to encode the time (t). This enables the model to recognize which time step's noise it is predicting in the batch. Furthermore, to achieve controllable category generation, it is necessary to input the category information (y). Hence, an embedding is utilized to embed y, and during the training process, it is optimized to maximize the encoding differences among different categories. As a result, we can train a single shared U-Net model to generate images of different categories. Specifically, in the various residual blocks of U-Net, both the time (t) and category information (y) are introduced.

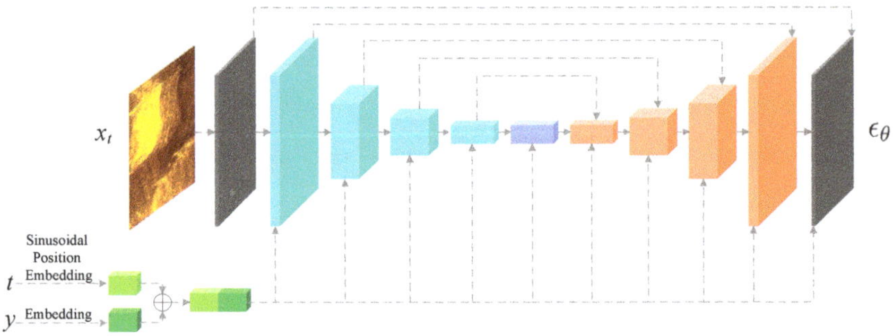

Figure 5. Schematic diagram of parameter estimation network structure of controllable category SSS image generator.

4. Experiments and Analysis

In this section, we present the experimental validation of the proposed method. First, we compared the imaging results of the SSS device, analyzing the impact of different gain factors on the nonlinear gain enhancement algorithm to determine suitable parameters. Next, we tested the proposed SSS image generator, examining the effects of different scale factors on the FID and IS. Additionally, we used Haralick texture features to further statistically assess the similarity between the real and generated images. Finally, we applied the generator to an object detection task to verify the extent to which the generator improved detection performance. For the experiments described in this chapter, the hardware used included four RTX 3090 GPUs (NVIDIA Corporation, Santa Clara, CA, USA), 256 GB of RAM, and an Intel Xeon Silver 4210 processor (Intel Corporation, Santa Clara, CA, USA). The software environment consisted of the Ubuntu 20.04 LTS operating system, CUDA 11.6, Python 3.9, PyTorch 1.12.1, and Matlab 2018. The data in the experiment were derived from the sea trial data collected as described in Section 3.1.

4.1. SSS Imaging Results

The sonar images were obtained through the quantization of the original acoustic signals, and the quality of imaging significantly impacted the subsequent experiments. Therefore, an initial analysis of the imaging results was conducted, presenting the original image and amplitude images with different C values, as shown in Figure 6.

By parsing the original data according to the Extended Triton Format (XTF) [53] protocol and converting them into image signals by using Equation (1), we obtained the initial image, as shown in Figure 7a. It was observed that at longer propagation distances, the echo intensity decreased, resulting in darker images in distant areas, which hindered the identification of detailed information within the images. To address this issue, we applied nonlinear gain enhancement to the images based on Equation (2), experimenting with five different values of C. As shown in the remaining subfigures of Figure 7, the brightness of the images increased with larger values of C. However, the images exhibited an overexposure-like effect when $C = 4e$, leading to reduced contrast. At a C value of $2e$, the intensity in distant areas was effectively compensated, making targets in those areas more discernible. When the C value was less than e, it was difficult to identify any effective information in the images. Based on this analysis, we selected $C = 2e$ as the default value for image conversion in the subsequent experiments.

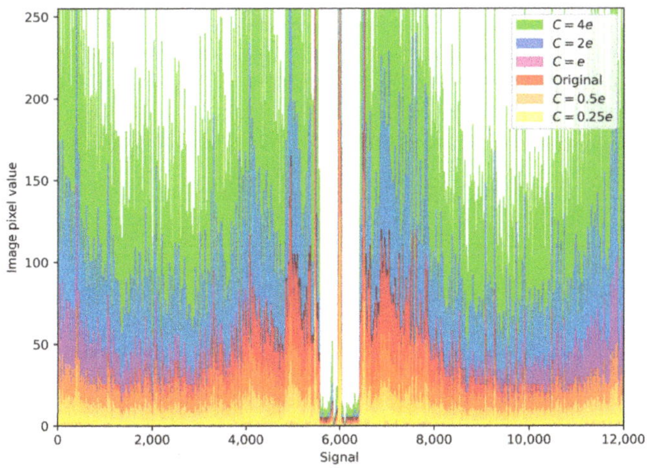

Figure 6. Amplitude comparison plots for different values of C and raw data.

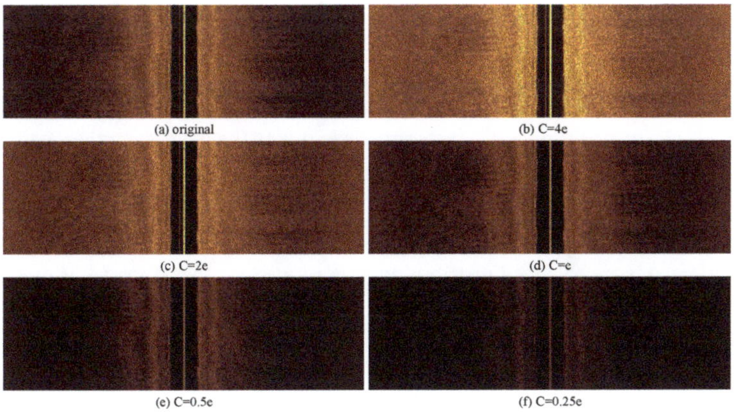

Figure 7. Comparison of imaging results with different gain values.

4.2. SSS Image Generation

Based on the analysis of the original image in the previous section, training was conducted according to the method described in Section 3.2. Using the data in Table 1 as the dataset, the images were randomly selected into training and testing sets with a ratio of 0.8:0.2. After 40,000 iterations, we obtained the weight files for the model. By utilizing this trained model, we generated 20,000 images each for the experiments by using different scale factors (s); Figure 8 shows some samples of the generated images. To quantitatively assess the performance of our generative model, we first selected the evaluation metrics. In the domain of generative models, the FID and the IS [54] are commonly used as evaluation criteria. The IS evaluates the model by measuring the diversity and quality of the generated images, using a pre-trained Inception network [55] for classifying the generated images and calculating the entropy of their distribution. The FID measures the statistical difference between the generated and real images in the feature space, extracting image feature vectors with a pre-trained Inception network and calculating the Fréchet distance between these features for real and generated images. The FID not only considers pixel-level differences but also considers the overall statistical distribution in the feature space, making it more sensitive to capturing subtle differences in the generated images. However, considering that our study focused on the generation of side-scan sonar images, using the original Inception pre-trained network directly would have rendered the results meaningless. Therefore, we fine-tuned the existing pre-trained models to obtain classification networks based on InceptionV3 and ResNet [56], which were used as feature extraction networks to calculate the FID and the IS; the results are shown in Table 2.

Table 2. Generation evaluation metric scores for the model. A lower FID score and a higher IS are better.

Scale	FID ↓		IS ↑	
	InceptionV3	**ResNet**	**InceptionV3**	**ResNet**
$s = 0.0$	17.5257	27.7368	2.6612	2.5764
$s = 0.2$	17.5456	28.2699	2.8712	2.8424
$s = 0.4$	17.6471	29.3880	3.1743	3.1884
$s = 0.5$	17.5502	31.2412	3.4934	3.6395
$s = 0.6$	17.5741	32.4902	3.6551	3.9083
$s = 0.8$	17.7355	32.6387	3.7988	4.0885
$s = 1.0$	17.7328	33.2389	3.8120	4.1173
$s = 2.0$	18.2058	32.7953	3.8921	4.1597
$s = 4.0$	19.6987	33.3931	3.8971	4.2231
$s = 6.0$	20.0235	33.4207	3.9059	4.2294
$s = 8.0$	20.1990	32.9504	3.9128	4.2487

According to Equation (14), the essence of controlled category generation is achieved through the linear extrapolation of conditional and unconditional generative models. It was observed that as the value of s increased, both the FID and IS metrics exhibited a gradual increase. The increase in the FID was primarily due to the increase in the s value making the model more inclined towards conditional generation. As the s value increased, the model's controllability was enhanced and its uncertainty was reduced, leading to a decrease in the diversity of the generated images. The increase in the IS was due to the increase in the s value allowing the model to generate images of specified categories more reliably, resulting in a more balanced distribution of generated image categories, thereby increasing entropy. Additionally, it was noted that when $S = 0$, the model degenerated into an unconditional generative model; when S was between 0 and 1, the model essentially combined conditional and unconditional generative models; when S was greater than 1, it manifested as the conditional generative model minus the unconditional generative model.

After considering the FID and IS results comprehensively, we chose $S = 0.6$ as the scale factor for the generative model. Based on this, further statistical analysis was conducted on the generated images. We calculated eight Haralick texture features [57] between the real and generated images to quantify the texture differences in the two-dimensional space between image pairs. Specifically, we first established the Gray-Level Co-occurrence Matrices (GLCMs) [58] for the two sets and then calculated angular second moment, contrast, correlation, inverse difference moment, sum entropy, entropy, difference variance, and difference entropy. Subsequently, we employed the Multidimensional Scaling (MDS) method [59] to measure the texture dissimilarities in two dimensions. To minimize the impact of outliers in both sets on the overall results, we fitted the results with ellipses in a 95% confidence interval. The final results are shown in Figure 9. It can be observed that almost all ellipses representing the set of generated images encompass the ellipses representing the set of real images and their centroids are very close. This indicates that the generated images and the real images had high similarity in Haralick texture features.

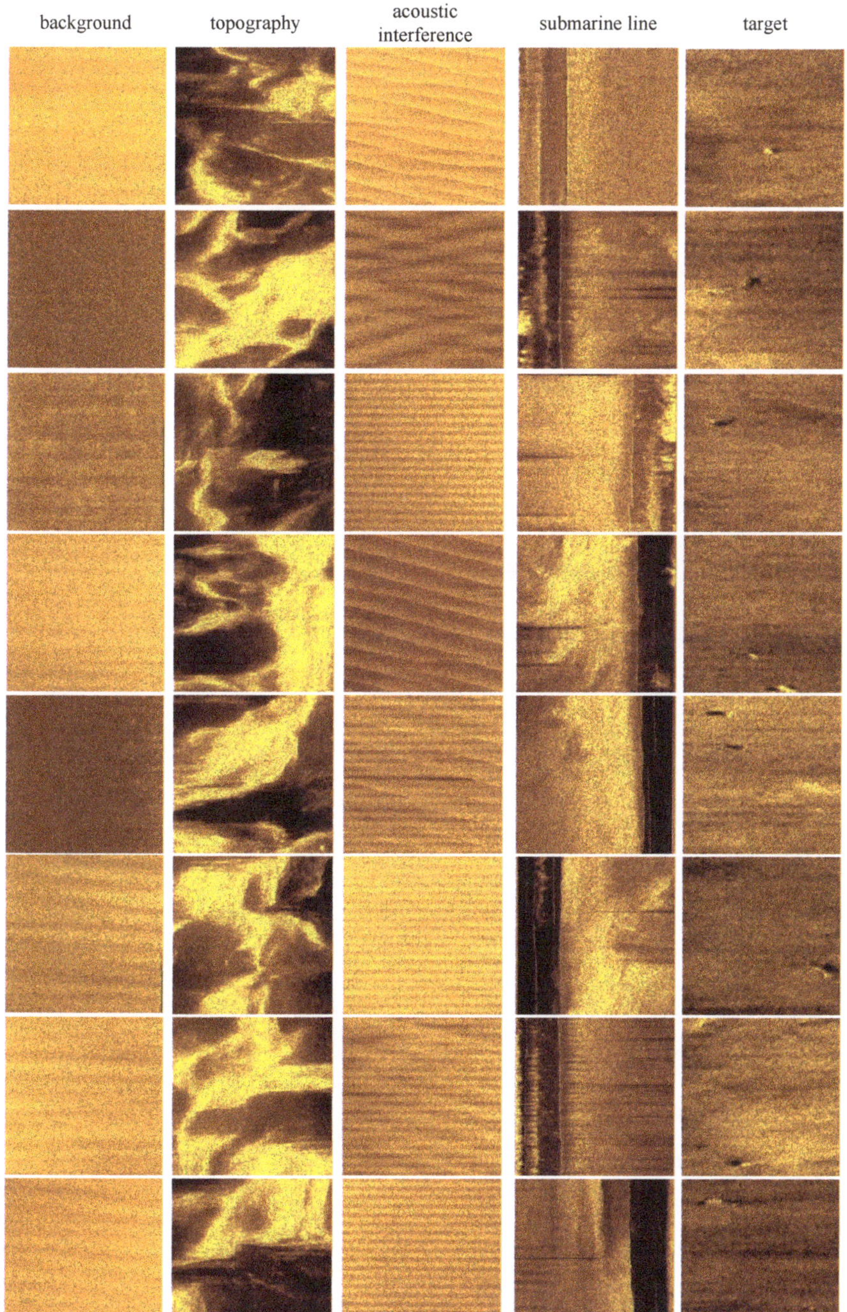

Figure 8. Generated SSS images of five categories ($s = 0.6$).

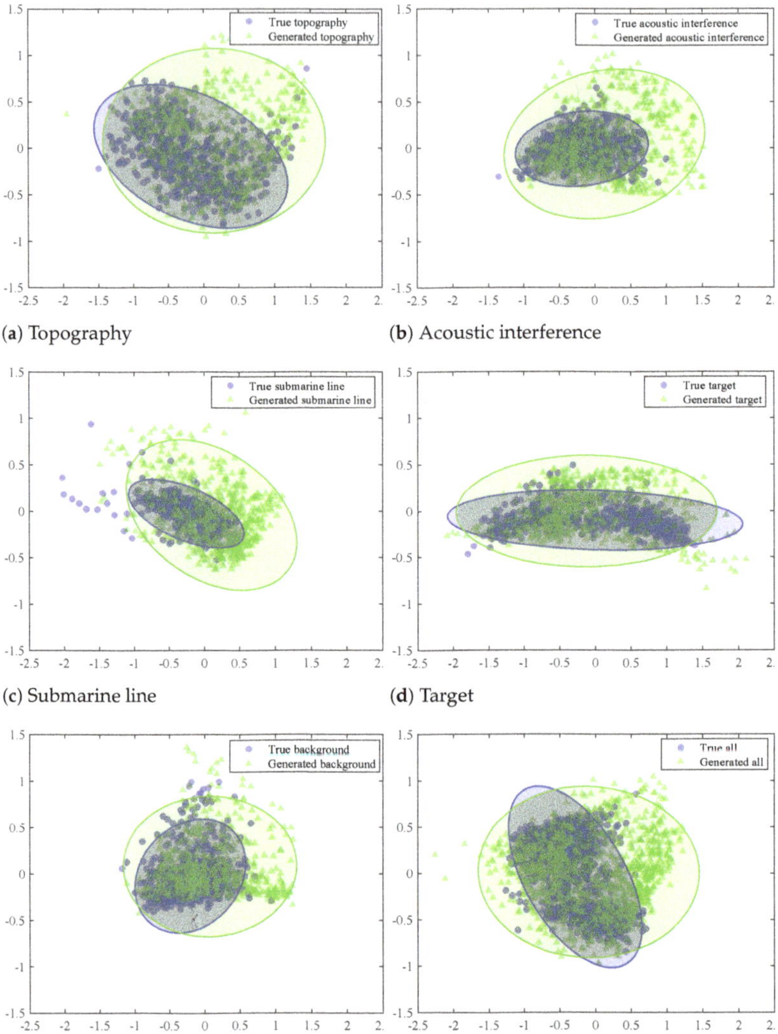

(**a**) Topography (**b**) Acoustic interference (**c**) Submarine line (**d**) Target (**e**) Background (**f**) All

Figure 9. Relative texture dissimilarity between generated and original data.

4.3. SSS Image Detection

An important application of generative models is supplementing the original dataset to address the issue of insufficient data volume. Therefore, in this section, we describe the use of the generated images to augment the dataset for object detection purposes. Our generative model is capable of specifically generating sonar images of five categories, among which the "target" category is the most commonly used for detection. Accordingly, we first generated a substantial number of "target" category images by using the generative model. To explore the impact of varying quantities of generated images on object detection accuracy in real images, we prepared six different datasets, as shown in Table 3. All the original data in the table were derived from the target category in Table 1. The test and validation sets for all datasets were sourced from real images in order to genuinely reflect the effect of different training sets on enhancing detection accuracy.

Table 3. The six kinds of datasets used for the object detection experiments. The data of dataset B were all real data. The training set of dataset A was entirely derived from the generated data. The training sets of the remaining datasets were mixed datasets.

Dataset	Training Set		Test Set		Validation Set	
	Original	Generated	Original	Generated	Original	Generated
A	0	1500	103	0	153	0
B	143	0	103	0	153	0
C	143	300	103	0	153	0
D	143	500	103	0	153	0
E	143	800	103	0	153	0
F	143	1000	103	0	153	0

In our object detection experiments, we utilized the state-of-the-art (SOTA) YOLOv10 detection model [60]. Due to its exceptional performance and low computational power consumption, the YOLO series [61] has been a significant paradigm in the field of object detection, now updated to its tenth version. YOLOv10 introduces a novel real-time object detection method by improving the model architecture and eliminating Non-Maximum Suppression (NMS). Based on varying computational power requirements, YOLOv10 is further divided into six distinct model variants. We conducted cross-experiments on these six model variants and six datasets under the same experimental conditions. The detection accuracy results of the experiments are presented in Table 4.

Table 4. Average precision scores for different models and different datasets.

Dataset	YOLOv10n	YOLOv10s	YOLOv10m	YOLOv10b	YOLOv10l	YOLOv10x	Average
A	0.629	0.597	0.625	0.633	0.669	0.602	0.626
B	0.723	0.718	0.743	0.739	0.728	0.777	0.738
C	0.785	0.759	0.793	0.777	0.761	0.781	0.776
D	0.838	0.813	0.851	0.850	0.831	0.834	0.836
E	0.757	0.812	0.788	0.805	0.805	0.800	0.795
F	0.815	0.820	0.801	0.802	0.806	0.805	0.808

The results for dataset A reveal that even in the absence of any real images in the training set, the average precision of the six models is still high, 0.626, further confirming the high consistency between the generated and real images. With the incremental addition of generated images, the average precision could be enhanced by up to approximately 10%, significantly improving detection accuracy. However, when more than 800 generated images were incorporated into the training set, a slight decline in the model's average detection precision was observed, but this metric was still higher than that for dataset B. This phenomenon may be attributed to the inclusion of images located in the green area but not in the blue area of Figure 9, introducing some noise. Overall, incorporating the generated images into the real dataset effectively enhanced detection accuracy, especially when the volume of added images was 3.5 times that of the original images, yielding the most pronounced effect. Furthermore, we visualized the detection results based on YOLOv10n (as shown in Figure 10). The results in the second row of Figure 10 were caused by the training set being composed entirely of real images. Five targets were missed due to insufficient data. Notice that there is one missed detection instance in the first row, but its training set was entirely derived from the generated images. According to the results from dataset C to dataset F, the model performed better in reducing missed detection instances and false alarms, achieving better detection results.

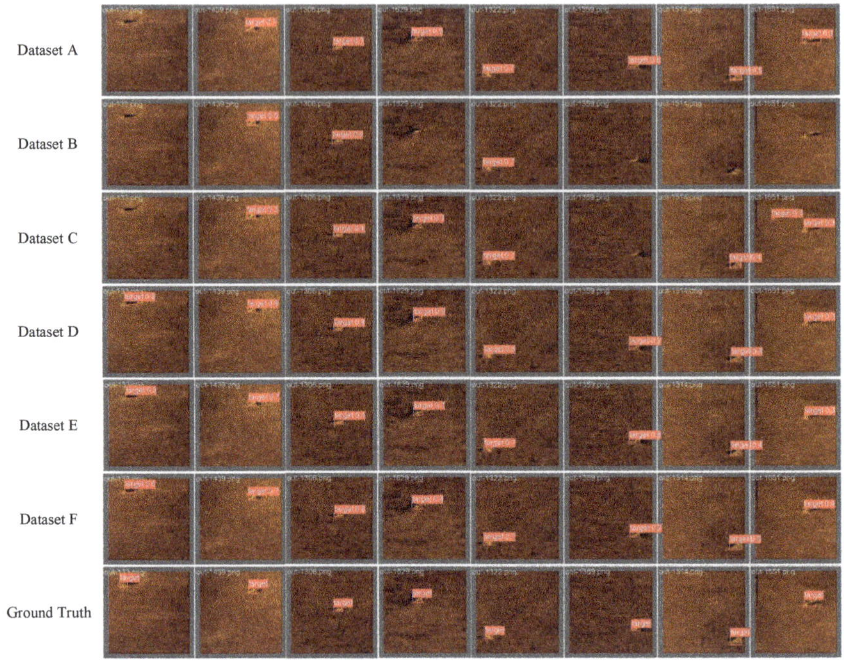

Figure 10. The detection results of the model trained on six datasets using YOLOv10n on the validation set.

5. Conclusions

In this study, we developed a controllable category SSS image generator capable of producing diverse and high-quality SSS images. To achieve this, we first established a data collection platform based on an AUV and gathered raw datasets from multiple marine areas. We also introduced a nonlinear gain enhancement algorithm for converting acoustic signals into image signals, effectively compensating for the signal strength at the far end of the sonar and enhancing the representation of target features. Based on diffusion models, we performed controllable five-category image generation without the need for additional classifiers and conducted comprehensive quantitative and qualitative evaluations. Furthermore, we validated the practical application of our generative model by using the YOLOv10 target detection algorithm. The experimental results indicate that the generated images can effectively complement the original images, improving target detection accuracy. Our work marks a significant milestone in the field of SSS data generation, providing data support for subsequent SSS-based developments.

However, it is worth noting that our current generative model only generates images and requires manual annotation when expanding the dataset. Therefore, our future research will focus on label-based generative models to achieve image–label matching generation.

Author Contributions: Conceptualization, F.Z. and X.H.; methodology, F.Z., X.H. and Z.W.; validation, Z.W., T.T. and X.H.; data collection, Z.W. and C.C.; writing—original draft preparation, X.H.; writing—review and editing, F.Z. All authors have read and agreed to the published version of the manuscript.

Funding: This research was funded by National Key R&D Program of China (2023QYXX).

Institutional Review Board Statement: Not applicable.

Informed Consent Statement: Not applicable.

Data Availability Statement: The datasets generated and analyzed during the current study are available from the corresponding author upon request.

Acknowledgments: We would like to acknowledge the Key Laboratory of Unmanned Underwater Transport Technology for the facilities and technical assistance provided.

Conflicts of Interest: The authors declare no conflicts of interest.

Abbreviations

The following abbreviations are used in this manuscript:

SSS	side-scan sonar
AUV	Autonomous Underwater Vehicle
FID	Fréchet Inception Distance
IS	Inception Score
Sonar	Sound Navigation and Ranging
USBL	Ultra-Short Baseline
VAEs	Variational Autoencoders
GANs	Generative Adversarial Networks
KL	Kullback–Leibler
VQVAE	Vector-Quantized VAE
LSGAN	Least Squares GAN
RNNs	Recurrent Neural Networks
LSTM	Long Short-Term Memory
MFSANet	multi-feature fusion self-attention network
SSAM	Simplified Self-Attention Module
GPS	Global Positioning System
DDPM	denoising diffusion probabilistic model
XTF	Extended Triton Format
GLCM	Gray-Level Co-occurrence Matrices
MDS	Multidimensional Scaling
SOTA	state-of-the-art
NMS	Non-Maximum Suppression

References

1. Wang, Y.; Chu, H.; Ma, R.; Bai, X.; Cheng, L.; Wang, S.; Tan, M. Learning-Based Discontinuous Path Following Control for a Biomimetic Underwater Vehicle. *Research* **2024**, *7*, 0299. [CrossRef]
2. Huy, D.Q.; Sadjoli, N.; Azam, A.B.; Elhadidi, B.; Cai, Y.; Seet, G. Object perception in underwater environments: A survey on sensors and sensing methodologies. *Ocean Eng.* **2023**, *267*, 113202. [CrossRef]
3. Xu, S.; Zhang, M.; Song, W.; Mei, H.; He, Q.; Liotta, A. A systematic review and analysis of deep learning-based underwater object detection. *Neurocomputing* **2023**, *527*, 204–232. [CrossRef]
4. Burguera, A.; Oliver, G. High-resolution underwater mapping using side-scan sonar. *PLoS ONE* **2016**, *11*, e0146396. [CrossRef]
5. Fallon, M.F.; Kaess, M.; Johannsson, H.; Leonard, J.J. Efficient AUV navigation fusing acoustic ranging and side-scan sonar. In Proceedings of the 2011 IEEE International Conference on Robotics and Automation, Shanghai, China, 9–13 May 2011; pp. 2398–2405.
6. Coiras, E.; Petillot, Y.; Lane, D.M. Multiresolution 3-D reconstruction from side-scan sonar images. *IEEE Trans. Image Process.* **2007**, *16*, 382–390. [CrossRef] [PubMed]
7. Rhinelander, J. Feature extraction and target classification of side-scan sonar images. In Proceedings of the 2016 IEEE Symposium Series on Computational Intelligence (SSCI), Athens, Greece, 6–9 December 2016; pp. 1–6.
8. de Souza, L.A.P.; Azevedo, A.A.; da Silva, M. Side Scan Sonar Applied to Water Reservoir. In Proceedings of the 2013 IEEE/OES Acoustics in Underwater Geosciences Symposium, Rio de Janeiro, Brazil, 24–26 July 2013; pp. 1–7.
9. Tang, Y.; Wang, L.; Jin, S.; Zhao, J.; Huang, C.; Yu, Y. AUV-based side-scan sonar real-time method for underwater-target detection. *J. Mar. Sci. Eng.* **2023**, *11*, 690. [CrossRef]
10. Yan, X.; Yang, J.; Sohn, K.; Lee, H. Attribute2image: Conditional image generation from visual attributes. In Proceedings of the Computer Vision–ECCV 2016: 14th European Conference, Amsterdam, The Netherlands, 11–14 October 2016; Proceedings, Part IV 14; Springer: Berlin/Heidelberg, Germany, 2016; pp. 776–791.
11. Xu, M.; Yoon, S.; Fuentes, A.; Park, D.S. A comprehensive survey of image augmentation techniques for deep learning. *Pattern Recognit.* **2023**, *137*, 109347. [CrossRef]

12. Ehrhardt, J.; Wilms, M. Autoencoders and variational autoencoders in medical image analysis. In *Biomedical Image Synthesis and Simulation*; Elsevier: Amsterdam, The Netherlands, 2022; pp. 129–162.
13. Goodfellow, I.; Pouget-Abadie, J.; Mirza, M.; Xu, B.; Warde-Farley, D.; Ozair, S.; Courville, A.; Bengio, Y. Generative adversarial networks. *Commun. ACM* **2020**, *63*, 139–144. [CrossRef]
14. Esser, P.; Rombach, R.; Blattmann, A.; Ommer, B. Imagebart: Bidirectional context with multinomial diffusion for autoregressive image synthesis. *Adv. Neural Inf. Process. Syst.* **2021**, *34*, 3518–3532.
15. Pinaya, W.H.L.; Vieira, S.; Garcia-Dias, R.; Mechelli, A. Autoencoders. In *Machine Learning*; Elsevier: Amsterdam, The Netherlands, 2020; pp. 193–208.
16. Oord, A.V.D.; Vinyals, O.; Kavukcuoglu, K. Neural Discrete Representation Learning. *Adv. Neural Inf. Process. Syst.* **2017**, *30*, 6307–6316.
17. Peng, J.; Liu, D.; Xu, S.; Li, H. Generating diverse structure for image inpainting with hierarchical VQ-VAE. In Proceedings of the IEEE/CVF Conference on Computer Vision and Pattern Recognition, Nashville, TN, USA, 20–25 June 2021; pp. 10775–10784.
18. Farnia, F.; Ozdaglar, A. Do GANs always have Nash equilibria? In Proceedings of the International Conference on Machine Learning. PMLR, Virtual Event, 13–18 July 2020; pp. 3029–3039.
19. Mao, X.; Li, Q.; Xie, H.; Lau, R.Y.; Wang, Z.; Paul Smolley, S. Least squares generative adversarial networks. In Proceedings of the IEEE International Conference on Computer Vision, Venice, Italy, 22–29 October 2017; pp. 2794–2802.
20. Arjovsky, M.; Chintala, S.; Bottou, L. Wasserstein generative adversarial networks. In Proceedings of the International Conference on Machine Learning, PMLR, Sydney, Australia, 6–11 August 2017; pp. 214–223.
21. Radford, A.; Metz, L.; Chintala, S. Unsupervised representation learning with deep convolutional generative adversarial networks. *arXiv* **2015**, arXiv:1511.06434.
22. Zhang, H.; Goodfellow, I.; Metaxas, D.; Odena, A. Self-attention generative adversarial networks. In Proceedings of the International Conference on Machine Learning, PMLR, Long Beach, CA, USA, 9–15 June 2019; pp. 7354–7363.
23. Karras, T.; Aila, T.; Laine, S.; Lehtinen, J. Progressive growing of gans for improved quality, stability, and variation. *arXiv* **2017**, arXiv:1710.10196.
24. Karras, T.; Laine, S.; Aila, T. A style-based generator architecture for generative adversarial networks. In Proceedings of the IEEE/CVF Conference on Computer Vision and Pattern Recognition, Long Beach, CA, USA, 15–20 June 2019; pp. 4401–4410.
25. Karras, T.; Laine, S.; Aittala, M.; Hellsten, J.; Lehtinen, J.; Aila, T. Analyzing and improving the image quality of stylegan. In Proceedings of the IEEE/CVF Conference on Computer Vision and Pattern Recognition, Seattle, WA, USA, 13–19 June 2020; pp. 8110–8119.
26. Van Den Oord, A.; Kalchbrenner, N.; Kavukcuoglu, K. Pixel recurrent neural networks. In Proceedings of the International Conference on Machine Learning, PMLR, New York, NY, USA, 19–24 June 2016; pp. 1747–1756.
27. Van den Oord, A.; Kalchbrenner, N.; Espeholt, L.; Vinyals, O.; Graves, A. Conditional image generation with pixelcnn decoders. *Adv. Neural Inf. Process. Syst.* **2016**, *29*, 4797–4805.
28. Salimans, T.; Karpathy, A.; Chen, X.; Kingma, D.P. Pixelcnn++: Improving the pixelcnn with discretized logistic mixture likelihood and other modifications. *arXiv* **2017**, arXiv:1701.05517.
29. Vaswani, A.; Shazeer, N.; Parmar, N.; Uszkoreit, J.; Jones, L.; Gomez, A.N.; Kaiser, Ł.; Polosukhin, I. Attention is all you need. *Adv. Neural Inf. Process. Syst.* **2017**, *30*, 5999–6009.
30. Ho, J.; Jain, A.; Abbeel, P. Denoising diffusion probabilistic models. *Adv. Neural Inf. Process. Syst.* **2020**, *33*, 6840–6851.
31. Nichol, A.Q.; Dhariwal, P. Improved denoising diffusion probabilistic models. In Proceedings of the International Conference on Machine Learning, PMLR, Virtual Event, 18–24 July 2021; pp. 8162–8171.
32. Song, J.; Meng, C.; Ermon, S. Denoising diffusion implicit models. *arXiv* **2020**, arXiv:2010.02502.
33. Dhariwal, P.; Nichol, A. Diffusion models beat gans on image synthesis. *Adv. Neural Inf. Process. Syst.* **2021**, *34*, 8780–8794.
34. Zhang, L.; Rao, A.; Agrawala, M. Adding conditional control to text-to-image diffusion models. In Proceedings of the IEEE/CVF International Conference on Computer Vision, Paris, France, 1–6 October 2023; pp. 3836–3847.
35. Jiang, Y.; Ku, B.; Kim, W.; Ko, H. Side-scan sonar image synthesis based on generative adversarial network for images in multiple frequencies. *IEEE Geosci. Remote Sens. Lett.* **2020**, *18*, 1505–1509. [CrossRef]
36. Bore, N.; Folkesson, J. Modeling and simulation of sidescan using conditional generative adversarial network. *IEEE J. Ocean. Eng.* **2020**, *46*, 195–205. [CrossRef]
37. Song, Y.; He, B.; Liu, P.; Yan, T. Side scan sonar image segmentation and synthesis based on extreme learning machine. *Appl. Acoust.* **2019**, *146*, 56–65. [CrossRef]
38. Wang, J.; Li, H.; Huo, G.; Li, C.; Wei, Y. Multi-modal multi-stage underwater side-scan sonar target recognition based on synthetic images. *Remote Sens.* **2023**, *15*, 1303. [CrossRef]
39. Ge, Q.; Ruan, F.; Qiao, B.; Zhang, Q.; Zuo, X.; Dang, L. Side-scan sonar image classification based on style transfer and pre-trained convolutional neural networks. *Electronics* **2021**, *10*, 1823. [CrossRef]
40. Xu, H.; Bai, Z.; Zhang, X.; Ding, Q. Mfsanet: Zero-shot side-scan sonar image recognition based on style transfer. *IEEE Geosci. Remote Sens. Lett.* **2023**, *20*, 1503105. [CrossRef]
41. Yang, Z.; Zhao, J.; Zhang, H.; Yu, Y.; Huang, C. A Side-Scan Sonar Image Synthesis Method Based on a Diffusion Model. *J. Mar. Sci. Eng.* **2023**, *11*, 1103. [CrossRef]

42. Zhang, F.; Zhang, W.; Cheng, C.; Hou, X.; Cao, C. Detection of Small Objects in Side-Scan Sonar Images Using an Enhanced YOLOv7-Based Approach. *J. Mar. Sci. Eng.* **2023**, *11*, 2155. [CrossRef]
43. Cheng, C.; Hou, X.; Wen, X.; Liu, W.; Zhang, F. Small-Sample Underwater Target Detection: A Joint Approach Utilizing Diffusion and YOLOv7 Model. *Remote Sens.* **2023**, *15*, 4772. [CrossRef]
44. Zhang, P.; Tang, J.; Zhong, H.; Ning, M.; Liu, D.; Wu, K. Self-trained target detection of radar and sonar images using automatic deep learning. *IEEE Trans. Geosci. Remote Sens.* **2021**, *60*, 1–14. [CrossRef]
45. Lee, B.; Ku, B.; Kim, W.; Kim, S.; Ko, H. Feature sparse coding with coordconv for side scan sonar image enhancement. *IEEE Geosci. Remote Sens. Lett.* **2020**, *19*, 1–5. [CrossRef]
46. Guo, Z.; Liu, J.; Wang, Y.; Chen, M.; Wang, D.; Xu, D.; Cheng, J. Diffusion models in bioinformatics and computational biology. *Nat. Rev. Bioeng.* **2024**, *2*, 136–154. [CrossRef]
47. Chen, S.; Sun, P.; Song, Y.; Luo, P. Diffusiondet: Diffusion model for object detection. In Proceedings of the IEEE/CVF International Conference on Computer Vision, Paris, France, 1–6 October 2023; pp. 19830–19843.
48. Xia, B.; Zhang, Y.; Wang, S.; Wang, Y.; Wu, X.; Tian, Y.; Yang, W.; Van Gool, L. Diffir: Efficient diffusion model for image restoration. In Proceedings of the IEEE/CVF International Conference on Computer Vision, Paris, France, 1–6 October 2023; pp. 13095–13105.
49. Song, Y.; Sohl-Dickstein, J.; Kingma, D.P.; Kumar, A.; Ermon, S.; Poole, B. Score-based generative modeling through stochastic differential equations. *arXiv* **2020**, arXiv:2011.13456.
50. Zhai, J.; Zhang, S.; Chen, J.; He, Q. Autoencoder and its various variants. In Proceedings of the 2018 IEEE International Conference on Systems, Man, and Cybernetics (SMC), Miyazaki, Japan, 7–10 October 2018; pp. 415–419.
51. Ronneberger, O.; Fischer, P.; Brox, T. U-net: Convolutional networks for biomedical image segmentation. In Proceedings of the Medical image computing and computer-assisted intervention–MICCAI 2015: 18th International Conference, Munich, Germany, 5–9 October 2015, Proceedings, Part III 18; Springer: Berlin/Heidelberg, Germany, 2015; pp. 234–241.
52. Zhang, B.; Sennrich, R. Root mean square layer normalization. *Adv. Neural Inf. Process. Syst.* **2019**, *32*.
53. Rodríguez, O.C.; Silva, A.J.; Hughes, A.P.; Moreira, A.C. Underwater Sonar as a Ray Tracing Problem. In *INCREaSE 2019: Proceedings of the 2nd International Congress on Engineering and Sustainability in the XXI Century, Faro, Portugal, 9–11 October 2019*; Springer: Cham, Switzerland, 2020; pp. 255–264.
54. Chong, M.J.; Forsyth, D. Effectively unbiased fid and inception score and where to find them. In Proceedings of the IEEE/CVF Conference on Computer Vision and Pattern Recognition, Seattle, WA, USA, 13–19 June 2020; pp. 6070–6079.
55. Szegedy, C.; Vanhoucke, V.; Ioffe, S.; Shlens, J.; Wojna, Z. Rethinking the inception architecture for computer vision. In Proceedings of the IEEE Conference on Computer Vision and Pattern Recognition, Las Vegas, NV, USA, 27–30 June 2016; pp. 2818–2826.
56. He, K.; Zhang, X.; Ren, S.; Sun, J. Deep residual learning for image recognition. In Proceedings of the IEEE Conference on Computer Vision and Pattern Recognition, Las Vegas, NV, USA, 27–30 June 2016; pp. 770–778.
57. Haralick, R.M.; Shanmugam, K.; Dinstein, I.H. Textural features for image classification. *IEEE Trans. Syst. Man Cybern.* **1973**, *SMC-3*, 610–621. [CrossRef]
58. Partio, M.; Cramariuc, B.; Gabbouj, M.; Visa, A. Rock texture retrieval using gray level co-occurrence matrix. In Proceedings of the 5th Nordic Signal Processing Symposium, Trondheim, Norway, 4–7 October 2002; Volume 75.
59. Hout, M.C.; Papesh, M.H.; Goldinger, S.D. Multidimensional scaling. *Wiley Interdiscip. Rev. Cogn. Sci.* **2013**, *4*, 93–103. [CrossRef] [PubMed]
60. Wang, A.; Chen, H.; Liu, L.; Chen, K.; Lin, Z.; Han, J.; Ding, G. Yolov10: Real-time end-to-end object detection. *arXiv* **2024**, arXiv:2405.14458.
61. Jiang, P.; Ergu, D.; Liu, F.; Cai, Y.; Ma, B. A Review of Yolo algorithm developments. *Procedia Comput. Sci.* **2022**, *199*, 1066–1073. [CrossRef]

Journal of
Marine Science and Engineering

MDPI

Article

A Method for Recognition and Coordinate Reference of Autonomous Underwater Vehicles to Inspected Objects of Industrial Subsea Structures Using Stereo Images

Valery Bobkov and Alexey Kudryashov *

Institute of Automation and Control Processes, Far Eastern Branch, Russian Academy of Sciences, 690041 Vladivostok, Russia; bobkov@dvo.ru

* Correspondence: alkud1981@mail.ru; Tel.: +7-(914)705-44-46

Abstract: To date, the development of unmanned technologies using autonomous underwater vehicles (AUVs) has become an urgent demand for solving the problem of inspecting industrial subsea structures. A key issue here is the precise localization of AUVs relative to underwater objects. However, the impossibility of using GPS and the presence of various interferences associated with the dynamics of the underwater environment do not allow high-precision navigation based solely on a standard suite of AUV navigation tools (sonars, etc.). An alternative technology involves the processing of optical images that, at short distances, can provide higher accuracy of AUV navigation compared to the technology of acoustic measurement processing. Although there have been results in this direction, further development of methods for extracting spatial information about objects from images recorded by a camera is necessary in the task of calculating the exact mutual position of the AUV and the object. In this study, in the context of the problem of subsea production system inspection, we propose a technology to recognize underwater objects and provide coordinate references to the AUV based on stereo-image processing. Its distinctive features are the use of a non-standard technique to generate a geometric model of an object from its views (foreshortening) taken from positions of a pre-made overview trajectory, the use of various characteristic geometric elements when recognizing objects, and the original algorithms for comparing visual data of the inspection trajectory with an a priori model of the object. The results of experiments on virtual scenes and with real data showed the effectiveness of the proposed technology.

Keywords: autonomous underwater vehicle (AUV); subsea production systems (SPS); computer vision; stereo; visual navigation

1. Introduction

The issue of inspection of underwater infrastructures (such as gas and oil pipelines, underwater mining systems, etc.) using autonomous underwater vehicles (AUV) [1–5], as an alternative to the previously deployed technology based on remotely operated underwater vehicles and divers' operations, has become particularly relevant recently. Accurate positioning of AUVs relative to such objects is a key problem of inspection.

However, the challenges in providing accurate AUV navigation in the underwater environment (the impossibility of using GPS, lack of terrain maps, insufficient accuracy of acoustic sensing, and limitations of optical sensing) prevent more extensive use of AUVs for solving inspection problems. Therefore, various tools and technologies are applied to improve navigation accuracy, depending on the tasks set for the AUV mission, characteristics of scenes, and availability of primary data. Traditional equipment includes sonar navigation systems, Doppler lags, navigation and piloting sensors, and Inertial Measurement Units (IMUs).

When solving problems of underwater object inspection, integrated navigation technologies for AUVs usually use sensors such as sonar, laser, and acoustic modems. A detailed description of the technologies with an example of applications for solving the problem of harbor inspection can be found in [6–8]. However, the data obtained are not always sufficient to ensure solutions to the problems of recognizing inspected objects and accurately localizing AUVs in relation to them. An alternative to the above-mentioned navigation equipment is technologies based on processing optical images taken with mono and stereo cameras. These methods can potentially provide higher accuracy of AUV navigation compared to the processing of acoustic measurements in local scenes at short distances to objects. Several groups of applications can be considered here.

A number of works are devoted to algorithms based on stereo visual odometry, which are used by AUVs for navigation. For example, in [9], it is shown that stereo vision-based odometry can be used by AUVs that navigate close to the seabed for velocity and incremental pose estimation in small areas. The article describes the integration of two different stereo visual odometry algorithms into an AUV and experiments carried out in laboratory and harbor conditions comparing vision-based pose estimates with ground truth. Since the inertial navigation system of an AUV suffers from drift, observing fixed objects in an inertial reference system can enhance the localization performance. For example, in [10], a method for localizing AUVs using visual measurements of underwater structures and artificial landmarks was described.

In a number of studies, e.g., in [11,12], methods to track the desired trajectory using visual measurements of points features and adaptive control, including neural networks, were considered.

To solve a typical problem of sending an AUV home to the dock station, it is advisable to apply a vision guidance method, such as in [13].

In [14], a method for monocular visual odometry with optical flow tracking was proposed, which, according to the authors, is more suitable for underwater imaging than the classic approaches based on descriptors.

The issue of determination of a vehicle's position relative to the objects being inspected is based on solving the problem of object recognition. Since primarily industrial subsea structures are considered here, it is advisable to take into account the geometric specifics of artificial objects (the evident presence of standard geometric elements in their shape).

A previous paper [15] addressed the problem of object recognition from colorless 3D point clouds in underwater environments. It compared the performances of the state-of-the-art methods for underwater object recognition (global descriptors). The studied methods were designed to assist AUVs in performing autonomous interventions in underwater Inspection, Maintenance, and Repair (IMR) applications. In that paper, a number of methods based on the analysis of surface characteristics of the object were considered [16–19].

In [20], the authors developed their previous study by proposing a 3D object recognition method for non-colored point clouds using point features. The method was designed for the application of scenarios such as the IMR of industrial subsea structures composed of pipes and connecting objects (such as valves, elbows, and R-Tee connectors). The recognition algorithm uses a database of partial views of the objects, stored as point clouds, which are available a priori.

When developing inspection technologies, preliminary information about the objects can be taken into account such as digital maps of the area [21] or the availability of coordinate measurements of characteristic point features of the objects in the external coordinate system (CS) [22].

Another paper [23] presented a brief review of the latest advancements in navigation and control technologies for AUVs. It discussed the challenges and solutions in AUV navigation, including the integration of various sensors and algorithms to ensure accurate and reliable operations in complex underwater environments.

As can be seen from the brief overview above, different methods are required to solve inspection tasks using AUVs depending on the sensor suite used, the underwater conditions, and the mission objectives. In particular, it is necessary to further develop methods for extracting spatial information about objects from images recorded by a camera (taking into account the dynamics of the underwater environment) to solve problems of object recognition and precise coordinate referencing of an AUV to objects.

In this article, we propose a new approach to addressing the problem of underwater object recognition and coordinate referencing of AUVs to them within the framework of the inspection of a subsea production system (SPS). The approach is based on the non-standard automated technique for forming a geometric model of the object based on images of an overview trajectory, the use of different types of characteristic geometric elements in object recognition, and the application of original algorithms for matching the obtained visual data with an a priori model of the object.

The article is structured as follows. The introduction provides a review of available information as regards the issues addressed in the article. Section 2 briefly considers the proposed approach to solving the problem of coordinate reference of the AUV to the SPS. Section 3.1 describes a technique to form a model of SPS objects on the basis of views of the overview trajectory. Section 3.2 describes algorithms for recognizing the characteristic geometric elements in a stereo pair of frames at the position of the AUV's working trajectory that provide identification of the SPS object and calculation of the AUV's movement in the coordinate space of the object. Section 4 presents the results of experiments with synthetic and actual data. Section 5 is devoted to the discussion of the results of the work.

2. Problem Statement and General Approach

The problem of regular IMR of various underwater engineering facilities, including the subsea production system (SPS), requires precise coordinate references of the AUV to underwater artificial objects during inspection missions. Referencing the AUV to the object's coordinate system (CS) is necessary for planning the AUV's trajectory and performing operations (photography of critical structural components, possible landing, and repair) directly in the coordinate space of the objects under inspection. It is assumed that the AUV, along with the standard suite of navigation sensors, is equipped with an optical stereo camera directed strictly down relative to the AUV longitudinal axis. In the context of this issue, the problem of calculating the exact coordinate reference of the AUV to the CS of a subsea artificial object is solved using video information. Structurally, the SPS is defined as a set of functionally integrated objects distributed over a limited area of the seabed and coordinated in the CS of the SPS. When the AUV is moving along the specified trajectory in the area of the SPS location, regular attempts are made to identify the SPS objects by processing the stereo video stream from a camera. Recognition of the object under inspection is performed by comparing the geometric information derived from stereo images to the a priori-specified 3D model of the object. Hereinafter, the term "object model" means a visual model, i.e., a set of spatial elements of the object visible through the camera. The visual model may not match the functional model of a large object that describes this object at the level of its certain functional units. Nevertheless, the visual model elements may have attributes indicating that they belong to certain functional units of the object, if necessary for inspection purposes. The model is built with the involvement of the operator. Unlike the traditional technique, where the model is formed based on in situ measurements or documentary information describing the object, the proposed technique uses images of a few views of the pre-made overview trajectory of the AUV and 3D information obtained on their basis. The overview trajectory is set in such a way that the camera could capture images of all objects (or parts of the object, if it does not completely fit into the camera's field of view due to its large size) of the distributed SPS. Then, we define the SPS model as a combination of models of all objects composing the SPS. Also, we define the model of a certain object as a combination of models of all of the used object's views from the overview trajectory. The model of a certain view is a set of 3D geometric elements of the object (visible from this aspect) that characterize the spatial structure of its visible part. Stereo pairs of frames of the views that were used to build the object model are linked to it. The characteristic 3D point features of the object and rectilinear segments and macro-elements composed of segments characterizing the shape of the object are considered geometric elements of the model. According to the technique, the operator, based on a visual analysis of the overview trajectory images used and the calculated 3D data, forms characteristic elements (CEs) for each view. CEs are coordinated both in the CS linked to a specific view and in the object's intrinsic CS built by the operator. For each view of the overview trajectory used, a coordinate transformation matrix is calculated, linking the CS of this view to the object's intrinsic CS. The matrix is formed by the well-known technique of specifying the unit vectors of the object's CS in the view's CS. One of the objects' CSs is fixed as a CS of the SPS. Accordingly, a coordinate transformation matrix is formed for each object, linking the object's CS with the CS of the SPS.

The use of several views of the object and CEs of different types increases the reliability of identification of certain objects and the accuracy of AUV navigation relative to the underwater object.

The object identification and coordinate referencing of the AUV to the object's intrinsic CS is generally performed by a sequential three-stage computational scheme.

First stage. Direct recognition of 3D point features of the object is made from a stereo pair of frames taken by the AUV camera at the analyzed position of the working (inspection) trajectory. This is achieved using the algorithm for matching the set of point features in the stereo pair of frames of the working trajectory with the set of point features in the stereo pair of the model for each view of the overview trajectory used (see the description of the algorithm in Section 3.2.1). The result of the matching algorithm is two matched sets of 3D points, of which one is determined in the model's CS and the other in the camera's CS at the position of the working trajectory (in the AUV's CS). The matrix for coordinate transformation between the AUV's CS and the CS of the object model is calculated using these two matched 3D sets. The resulting coordinate transformation matrix can be considered an acceptable solution to the problem of AUV referencing to the object or be used as an initial approximation when obtaining a more reliable and accurate solution at the second and third stages, taking into account all types of CEs.

Second stage. Rectilinear segments and macro-elements are recognized using the coordinate transformation matrix obtained in the first stage. Note that this approach does not involve the vectorized shapes of source images (which require significant computational costs); however, correlation estimates are used when comparing the segment line images in the model and the working stereo pair of frames. The recognition of elements in addition to points increases the reliability of object identification and the accuracy of coordinates in reference to them.

Third stage. An updated matrix of coordinate transformation from the CS of the AUV's working trajectory position to the object's CS is calculated with the point data for all views and CE types taken into account.

3. Proposed Methods

It is advisable to plan and perform the AUV inspection trajectory in the coordinate space of a subsea production system (SPS) or a separate object inspected. Therefore, when the AUV is moving along the working trajectory, SPS objects are searched/detected using the methods and algorithms based on the comparison of the geometric information derived from recorded stereo frames to the object models. Thus, the problems of detection of 2D characteristic elements in images, 3D reconstruction of CEs, and matching of spatial CEs with the object model are solved sequentially. Models of SPS objects are formed by an automated technique based on photographs of the overview trajectory using several views (aspects). Three types of CEs are considered spatial elements of an object model: point features, rectilinear segments, and, possibly, macro-elements built on a combination of segment lines. In this study, geometric elements of a "corner" type are considered macro-elements. Eventually, the problem of calculating the AUV's movement in the CS of a subsea object is solved.

Below, the following designations are used:

M_2D_POINT_V^kV^k_LR and M_2D_ POINT_V^kV^k_RL are matched sets of points in the left and right frames of the "model" stereo pair of the *k*th view (recall that the model is considered a set of several views of the AUV's overview trajectory). Here, the symbol "M" means that the set belongs to the model, and the symbol "S" will be used in the identifiers of sets belonging to the frames of the "working" stereo pair. V^k indicates that the *k*th view of the overview trajectory is used. The LR mnemonics means that the set in the left frame of the stereo pair is matched to the set in the right frame of the stereo pair. The RL mnemonics are used in a similar way. The designation V^kV^k below means that the matching refers to the stereo pair of frames of the *k*th view. Similarly, the mnemonics of the WW designation mean that the sets belonging to the frames of a "working" stereo-pair are matched. The designation V^kW means that the set in the frames of the working stereo pair is matched with the set of the model of the *k*th view (the *k*th view of the overview trajectory). Accordingly, the designation WV^k means that the set of the model of the *k*th view is matched with the set in the frames of the working stereo pair. The sets belonging to the object model are formed using an automated technique with the involvement of the operator. The abbreviations used in the following presentation are given in Abbreviations.

3.1. Forming the Model Object

Geometric models of SPS objects are formed at a preliminary stage, prior to launching inspection missions. These are used to visually recognize SPS objects and provide coordinate referencing of the AUV to them during each of the subsequent inspection missions. The model is usually formed using direct measurements of characteristics of the object/SPS with the involvement of documentary information and/or software tools. Point features are often used as characteristic features determining the spatial geometric structure: points generated and matched in the stereo pair of frames using the SURF detector or corner points [24] presumably recognized using the Harris Corner Detector on the basis of images taken by the AUV camera during inspection. The advantage of this technique for object identification is its versatility: it does not require the shape of the surface to be taken into account. However, using point features only (when matched with the model) may not be enough to reliably identify the object because (a) point features do not fully reflect the spatial geometric structure of the object and (b) the possibly small number of matched points (since photography from the overview trajectory and the working one is conducted under different conditions, i.e., at different altitudes, at different angles, and in different illuminations) may reduce the accuracy of referencing the AUV to the object. To increase the reliability of object identification and the accuracy of AUV's coordinate reference to the object, it is suggested, first, to use CEs of the "point" type from all views used, and then to use segment lines and macro-elements based on them, along with points, as characteristic features of the object. For example, elements such as the "corner" type (a vertex with adjacent sides and segment lines) are more reliably recognized. In particular, corners remain unchanged under varying illumination and external effects on the object (corrosion, algae fouling, etc.). Second, an automated method for forming an object model is proposed, aimed at improving the efficiency of object recognition when AUV performs a working mission. On the one hand, the method excludes preliminary direct in situ measurements, and, on the other hand, it contributes to a higher degree of object recognition at the stage of inspection. The essence of the automated technique is to involve the operator in constructing an object model based on processing the results of video records from the pre-made overview trajectory of the AUV. The operator can clearly indicate the CEs forming the object model, which, presumably, will be well recognized by the AUV moving along the inspection trajectory. When indicating CEs, the operator uses source images and auxiliary software and algorithmic tools in an interactive mode (the SURF detector, the ray triangulation method, the visual navigation method (VNM) [23], 3D visualization, and necessary interactive service). The overview trajectory is planned in such a way that the set of its used views provides a good overview of all the SPS objects to be inspected. The operator forms the object's intrinsic CS in which the spatial model of the object should be described, which is necessary for accurately setting the trajectory of the inspection mission at the planning stage. For each used view of the object, a matrix is calculated linking the CS of this view with the object's CS. By means of these matrices, the coordinates of all the CEs of all the views used are transformed into the object's CS and stored in the respective sets representing the complete 3D model of the object (with all the used views taken into account). Also, for each SPS object, the matrix to link the CS of this object with the CS of SPS is calculated, for which the CS of one of the objects is used.

The main steps of the method for object model formation are shown in Figure 1.

1. The operator selects one or more of the most informative object's views from the sequence of photographs of the overview trajectory (with respective stereo pairs of frames), which together can potentially provide recognition of the object for any position of the working (inspection) trajectory when the AUV is positioned above the object.
2. The operator fixes a rectangular area of the object's location in the seabed plane (using the VNM, which provides the calculation of points' coordinates in the external CS with some accuracy). This allows for a rough object search at the recognition stage (while the AUV is moving along the working trajectory) before activation of the object recognition algorithm based on the processing of images taken with the camera.
3. The operator creates a 3D model of the object that combines the 3D models of several used views of the overview trajectory. The spatial geometric elements making up the model are formed by processing the original stereo pairs of frames of the views used. The processing includes the application of algorithmic procedures with the explicit involvement of the operator in the process of CE formation. The types of elements used, as noted above, are as follows: points, rectilinear segments, and macro-elements based on segment lines (such as "corner" type or others).
   For each selected ***k***th view of the AUV's overview trajectory:
   3.1. The point features in the left and right frames of the stereo-pair are matched using the SURF detector. The points belonging to the object, specified by the operator, are filtered and selected. Manual generation of additional points is possible. It is also possible to include terrain points near the object since the scene is static. By matching the sets of points M_2D_POINT_V^kV^k_LR and M_2D_POINT_ V^kV^k_RL, a set of 3D points M_3D_POINT_ V^kV^k_CS^{view_k}, visible through the camera for this view, is constructed using the ray triangulation method. The points are coordinated in the CS^{view_k} of this view of the overview trajectory.
   3.2. The operator generates a set of 3D edge lines visible through the camera in CS^{view_k} of this view using the original frames of the stereo pair. Each spatial segment line is described by two 3D endpoints belonging to it with pointers at 2D images of these points in the 2D sets indicated above. The problem of 3D reconstruction of the matched 2D images of segment lines in the model stereo-pair of frames is solved in a traditional way: by matching the endpoints in the stereo-pair of frames with the calculation of correlation estimates when scanning epipolar lines and then calculating the 3D coordinates of the segment endpoints in CS^{view_k} by the ray triangulation method. The endpoints can also be matched by the operator clearly indicating the matched images. The formed segments are coordinated in CS^{view_k} and stored in the set M_3D_LINE_ V^kV^k_CS^{view_k}.
   3.3. Formation of "corner" type CEs based on the set of spatial segments obtained for this view. The formed "corners" CEs are coordinated in the CS^{view_k} and stored in the set M_3D_CORNER_ V^kV^k_CS^{view_k}.
4. To obtain a model description independent of the AUV's CS, the operator explicitly determines the object's intrinsic CS. Such a CS is determined on the basis of the segment line indicated by the operator (let it be referred to as the base segment line)

in such a way that the Z′-axis is oriented in the direction of the Z-axis of the external CS (see Figure 2). This can be achieved using the data received by the standard navigation system of the AUV. For each used *k*th view of the object, a coordinate transformation matrix is calculated that links the object's CS with the CS of the camera of this view of the overview trajectory (which is performed in a standard way by setting the unit vectors of one CS in the other CS). If the camera does not see the selected base segment line in one of the views, then the VNM method is applied, which provides a matrix for coordinate transformation from the CS of one trajectory position into the CS of the other position (i.e., the coordinates of the base segment line are calculated in the CS of the considered view implicitly, without matching in the stereo pair of frames). Afterward, the object's intrinsic CS is built, the same as for other views, on the base segment line, and the matrix of the link between the CS of this view and the object's CS is calculated. Thus, a single CS of the object is built using the same spatial base segment line in all the views used. For each *k*th view, its own matrix $H_{CS^{view\_k},CS^{object\_n}}$ of transformation into this object's CS is calculated (where n is the object's sequential number).

5. The coordinates of all three CE types of the used *k*th view are transformed using the calculated matrix into the object's intrinsic CS built. The obtained coordinate representations are stored, respectively, in the sets M_3D_POINT_V^{k}_CS^{object_n}, M_3D_LINE_V^{k}_CS^{object_n}, and M_3D_CORNER_V^{k}_CS^{object_n} specified in CS^{object_n}. Simultaneously, these coordinate representations are recorded to the respective accumulated sets representing the object's complete 3D model in CS^{object_n} with all processed views taken into account: M_3D_POINT_CS^{object_n}, M_3D_LINE_CS^{object_n}, and M_3D_CORNER_CS^{object_n}. Note that if CE is present in several views, then in the complete model it is represented by averaged coordinates.
6. The operator explicitly determines the CS of the SPS. As such a CS, the CS of one of the objects is used. The intrinsic CSs of all objects are coordinated in the CS of the SPS. Thus, the object model is formed in two representations:
    6.1. As a set of 3D models of several objects' views, where the model of a view is a combination of three CE sets (points, lines, and corners) specified in the CS of this view (CS^{view_k}). The matrix $H_{CS^{view\_k},CS^{object\_n}}$ for coordinate transformation from this view's CS into the objects' intrinsic CS (CS^{object_n}) is calculated for each view.
    6.2. As a combination of sets of three CE types (points, lines, and corners) specified in the object's intrinsic CS (CS^{object_n}), which represents the spatial structure of the object. Here, the set of each type is formed by summing the CEs from several views used. Note that a face plane is linked to each edge line, which belongs to it. The normal plane and its position relative to the segment line (the face on the right or left) are indicated. This information is necessary at the object recognition stage for th ecorrect calculation of the correlation estimate when matching the images of the segment line (two of its points) in the frames of the model's stereo-pair and the stereo pair of the working (inspection) trajectory (the rectangular area adjacent to the edge used to calculate the correlation coefficient is specified only on this plane).

Note an important positive feature of the approach proposed: the initial object model, formed by the method with the use of a single overview trajectory, can be extended with subsequent regular inspections of the AUV. In particular, the obtained views of each subsequent inspection trajectory can be used to process the following new inspection trajectory in the same way as the views of the initial overview trajectory in the above-considered method. This provides, as regular missions are launched, a greater total number of recognized points by increasing the number of views and, eventually, a higher accuracy of the coordinate reference of the AUV to the object.

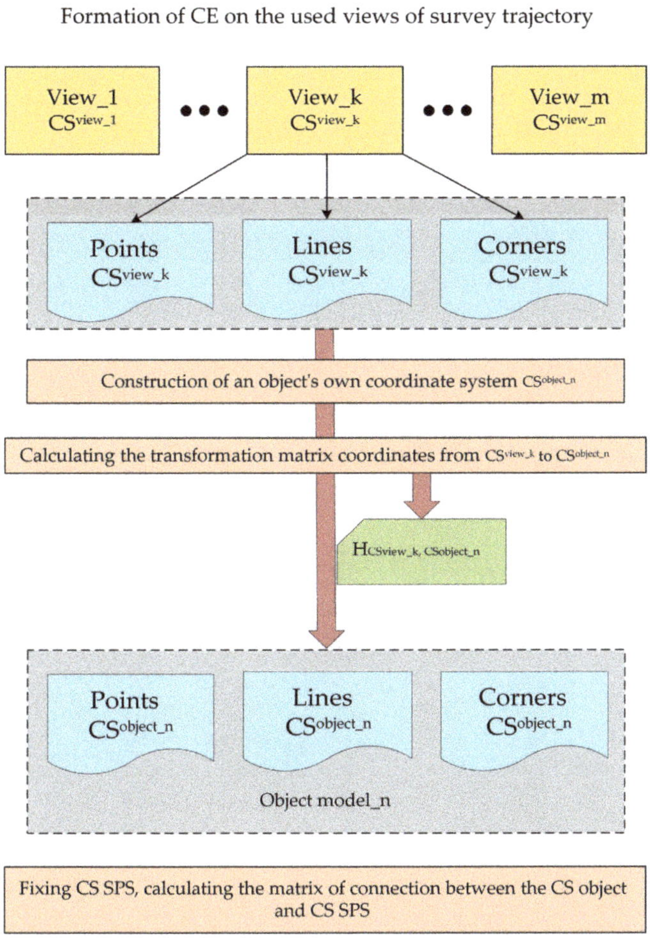

Figure 1. Forming an object model based on overview trajectory images.

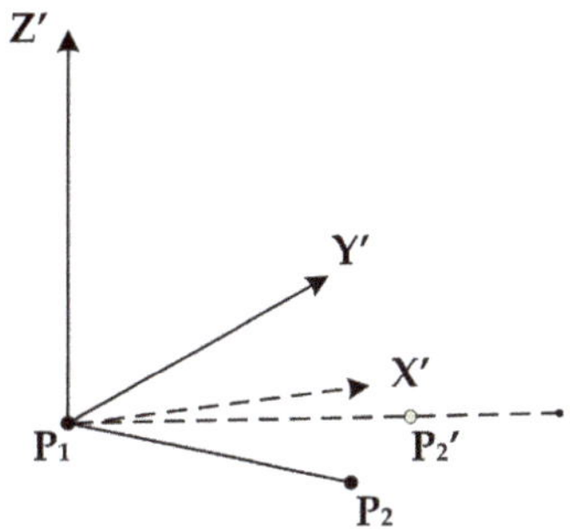

Figure 2. Building of the object's own CS (X′Y′Z′) based on the base segment line [P1, P2], specified in the camera's CS. Segment [P1, P2] is set in the camera's CS; point P1 is the origin of the CS being built; the direction of the Z′-axis coincides with the direction of the Z-axis of the external CS; [P1, P2′] is the projection of the segment [P1, P2] onto the plane perpendicular to the Z′-axis; the X′-axis is determined by the direction [P1, P2′]; the Y′-axis is completed by the right CS rule.

3.2. Recognition of SPS Objects

The recognition of SPS objects with referencing of the AUV to the object's CS is required for planning and performing the AUV's trajectory directly in the CS of the object under inspection. The problem of the coordinate reference of the AUV to the object is solved by comparing the visual data obtained at the position of the AUV's inspection trajectory with the object model. A three-stage computational scheme is implemented using several views (aspects) of the object that represent an object model (Figure 3). The choice of views for comparison is based on the criterion of proximity of the AUV's position in the external CS of the analyzed position of the AUV's working trajectory and the position of the overview trajectory of the selected view.

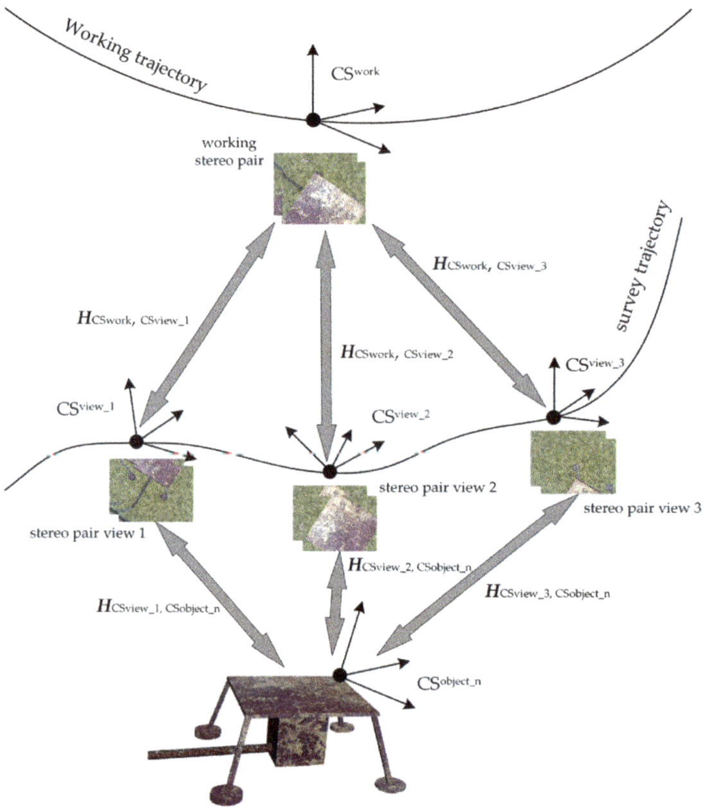

Figure 3. Stage 1: SPS object recognition by matching point features on stereo pairs of the overview and working (inspection) trajectory (three views are shown).

At the first stage, the desired object is recognized for each selected kth view by matching the point features in the working and model stereo pairs (see the description of the algorithm in Section 3.2.1). With successful recognition (a threshold criterion is used), the result is (a) a set of points S_3D_POINT_V^{k}_CS^{work} matched to the set M_3D_POINT_V^{k}_CS^{view_k} (belonging to the model of k-th view) and (b) the calculated matrix $H_{CS^{work}, CS^{view\_k}}$ for coordinate transformation from the CS of the inspection trajectory position into the CS of the model of the object's k-th view used.

Since for each view, there is a matrix $\boldsymbol{H}_{CSview\_k,\, CSobject\_n}$ for coordinate transformations from the CS of this view into the CS of the n-th object (obtained by forming a model of the k-th view), the transformation from CS^{work} into CS^{object_n} can be obtained by sequentially performing the two above-indicated transformations. Each resulting transformation (the number of which is equal to the number of views used) can be considered an acceptable solution to the problem set. The resulting matrix can also be used as an initial approximation in the optimization method when calculating the refined matrix for coordinate transformation from CS^{work} to CS^{object_n}, with point data for all views and all CE types taken into account (at the third stage). The obtained set of recognized "points" S_3D_POINT_V^kW_CS^{work} is entered into the set S_3D_ POINT_CS^{object_n} accumulated from all the views, which is coordinated in CS^{object_n} (the available matrix $H_{CSview\_k, CSobject\_n}$ is used). The number of matched points is input in Table 1. Table 1 records the relative number of recognized CEs (points, lines, and corners) in the images of the stereo pair (obtained at the position of the inspection trajectory) when compared with the model for each view k.

Table 1. Results of CEs (points, lines, and corners) recognition for all views.

	S_3D_POINT_CS_{work} (Number of Points Matched to Model/Number of Points in Model)	**S_3D_LINE_CS_{work} (Number of Lines Matched to Model/Number of Lines in Model)**	**S_3D_CORNER_CSwork (Number of Corners Matched to Model/Number of Corners in Model)**
View 1 in CS_1	np_1/mp	n_1_1/mL	$n^c{}_1$/mc
...	...	...	...
View k in CS_k	np_k/mp	$n^l{}_k$/mL	$n^c{}_k$/mc
...	...	...	...
View last in CS_{last}	np_last/mp	$n^l{}_{last}$/mL	$n^c{}_{last}$/mc

At the second stage, "lines" and "corners" in the working stereo pair, matched to the elements of the *k*-th view model, are recognized (using the resulting transformation $H_{CSwork, CSview\_k}$) (see Section 3.2.2). The result is the sets S_3D_LINE_V^kW_CS^{work} and S_3D_CORNER_V^kW_CS^{work}. These sets are entered into the respective sets S_3D_LINE_CS^{object_n} and S_CORNER_CS^{object_n}, coordinated in CS^{object_n} (the available matrix $H_{CS^{view\_k},CS^{object\_n}}$ is used). The matching results are input in Table 1. Thus, these sets, along with the set of points S_3D_ POINT_CS^{object_n}, are a full combination of recognized 3D CEs of all types of the object from all views in the object's intrinsic CS. The visualization of them can provide a visual representation of how fully and reliably the analyzed object was recognized.

In the third stage, the refined matrix $H_{CS^{work},CS^{object\_n}}$ is calculated. Refinement (increase in the accuracy of AUV coordination in the object's CS) is carried out by taking into account the recognized "points" in all the views used, and also by taking into account the points belonging to the recognized "lines" and "corners". Three approaches to calculating the coordinate reference of the AUV to the SPS object are considered:

1. The reference is, as mentioned above, based on the matrix for coordinate transformation from CS^{work} to CS^{object_n}, which is calculated using a model of the *k*-th view as a sequential execution of two available transformations:

$$H_{CS^{work},CS^{object\_n}} = H_{CS^{view\_k},CS^{object\_n}} \cdot H_{CS^{work},CS^{view\_k}} \tag{1}$$

In this case, we obtain *k* matrices (according to the number of object's views used). For each variant, the error of calculation accuracy is estimated on the basis of a test set of matched points.

2. The matrix $H_{CS^{work},CS^{object\_n}}$ for coordinate transformation from CS^{work} to CS^{object_n} is calculated on the basis of a combined set of matched points (for all views of the object). A standard method to minimize the target function is used, based on the estimate of the total discrepancy between the recognized points of the set S_3D_POINT_V^kW_CS^{work} (in the working stereo pair of frames) and the points of the set M_3D_POINT_V^k_CS^{view_k}

matched to them (a model of the *k*th view of the object). The total discrepancy is calculated in the coordinate space CS^{object_n}. Two variants of obtaining a combined point set are compared: (a) with only CE of the "point" type taken into account in all the views used and (b) with more points belonging to the recognized CEs of the "lines" and "corners" types added to the "point" type CE in each view.

To write the target function in a compact form, we introduce intermediate designations. We designate the set M_3D_POINT_V^k_CS^{view_k} as MP^k ($mp^k{}_1, \ldots mp^k{}_{np\_k}$) and designate the set S_3D_POINT_V^kW_CS^{wok} matched to the former set as SP^k ($sp^k{}_1, \ldots sp^k{}_{np\_k}$). Then the problem of minimizing the target function in variant (a) is formulated as follows:

$$\min\sum_1^k \sum_{i=1}^{np\_k} R1_i^k,\ \text{where } R1_i^k = \left\| mp_i^k \cdot H_{CS^{view\_k},CS^{object\_n}} - sp_i^k \cdot H_{CS^{work\_k},CS^{object\_n}} \right\| \quad (2)$$

where the matrix $H_{CS^{view\_k},CS^{object\_n}}$ is used, which was compiled at the stage of formation of the *k*-th view model.

Variant (b) additionally takes into account, as mentioned above, the points belonging to the "lines" and "corners" CEs (these points determine the coordinate position of the CE). Similarly, we simplify the writing by introducing intermediate designations. We designate the set of points M_3D_LINE_V^k_CS^{view_k} belonging to the "lines" type CEs of the *k*-th view model as $ML^k\left(ml_1^k, \ldots, ml_{nl\_k}^k\right)$, and designate the set of recognized lines S_3D_LINE_V^kW_CS^{work} matched to the former set as $SL^k\left(sl_1^k, \ldots, sl_{nl\_k}^k\right)$. Here, the elements ml^ki and $sl^k{}_i$ denote the start and end points of *i*-th segment line.

Then the points of the *i*-th "line" taken into account in the target function can be written as follows:

$$R2_i^k = \left\| ml_i^k \cdot H_{CS^{view\_k},CS^{object\_n}} - sl_i^k \cdot H_{CS^{work\_k},CS^{object\_n}} \right\| \quad (3)$$

Similarly, we introduce simplifying designations for points belonging to the CEs of the "corners" type. We designate the set of points M_3D_CORNER_V^k_CS^{view_k} belonging to the "corners" CEs of the *k*th view model as $MC^k\left(mc_1^k, \ldots, mc_{nc\_k}^k\right)$, and designate the set of recognized corners S_3D_CORNER_V^kW_CS^{work} matched to the former set as $SC^k\left(sc_1^k, \ldots, sc_{nl\_k}^k\right)$.

Accordingly, the points of the *i*-th "corner" taken into account in the target function can be written as follows:

$$R3_i^k = \left\| mc_i^k \cdot H_{CS^{view\_k},CS^{object\_n}} - sc_i^k \cdot H_{CS^{work\_k},CS^{object\_n}} \right\| \quad (4)$$

Then the problem of minimizing the target function in variant (b) is formulated as follows:

$$\min\sum_1^k \left(\sum_{i=1}^{np\_k} R1_i^k + \sum_{i=1}^{nl\_k} R2_i^k + \sum_{i=1}^{nc\_k} R3_i^k\right) \quad (5)$$

3. In this approach, unlike the previous ones, the AUV's position in CS^{object_n} is calculated without calculating the single-coordinate transformation matrix for all views. For each variant of $H_{CS^{work},CS^{object\_n}}$ (see approach 1), coordinates of the AUV are calculated (uniform coordinates of the AUV in CS^{work}—(0, 0, 0, 1)) in CS^{object_n}, i.e.,

$$P_{AUV^{object\_n}} = (0,0,0,1) \cdot H_{CS^{work\_k},CS^{object\_n}} \quad (6)$$

Thus, we have *k* variants of coordinate values for AUV in CS^{object_n}. As a solution is sought, we consider an averaged value of the coordinates. When averaging, the degree of confidence in each variant is taken into account in the form of weight coefficient α_k, where α_k is proportional to np_k (number of points matched).

Upon processing all views and inputting the results in Table 1, a decision is made on the reliability of the object identification according to the criteria set up previously.

The AUV's movement in the coordinate space of the object is calculated using the obtained matrix of reference to the object's CS.

3.2.1. Object Recognition by Point Features (Stage 1)

We consider the algorithm for matching the point features in stereo pairs of frames of the AUV's working trajectory position and a model of a certain object's view from the overview trajectory. As noted above, at the stage of object model formation:

(a) The object's intrinsic CS was built, for which the direction of the Z-axis of the external CS was used as the Z-axis direction.
(b) A set of 3D points M_3D_ V^kV^k_CS^{view_k} was obtained in the coordinate system CS^{view_k} from the matched 2D sets (see Section 3.1). The matrix $H_{CS^{view\_k},CS^{object\_n}}$ for coordinate transformation from CS^{view_k} to CS^{object_n} was calculated. Here, object_n is the identifier of *n*th object.

Then, the action of the Algorithm for object recognition on the basis of points and coordinate reference of the AUV to the object's CS is described by performing the following sequence of steps (Figure 4).

1. Selection of the *k*-th view in the model object for which the object's field of visibility best fits the potential object's field of visibility for the AUV's camera at the considered position of the inspection trajectory.
2. Generation and matching, using the SURF detector, of special points in the left frames of the model and working stereo pairs. The result of successful matching is the set M_2D_POINT_WV^k_LL in the image im_L_object_iew^k and the set S_2D_V^kW_LL in the image im_L_object_work.
3. If the matching in the previous step is successful, then the special points in the frames of the working stereo pair are matched using the SURF detector. The result is the set S_2D_POINT_WW_LR in the image im_L_object_work and the set S_2D_POINT_WW_RL in the image im_R_object_work.
4. The set of 3D points S_3D_POINT_WW_CS^{work} is built in the coordinate system CS^{work} from the 2D sets matched in the previous step.
5. The selection of a subset of points, also matched in the left frames of the "model" and "working" stereo pair, from the set of points matched in the frames of the "model" stereo pair (a model of this view): subM_2D_POINT_V^k = M_2D_POINT_V^kV^k_LR ∩ M_2D_POINT_ WV^k _LL (if there are fewer than three points, then we assume that the camera "does not see" the object). This operation is necessary to subsequently form a 3D representation of that subset of points, from the model of this object's view, for which matching with the recognized points of the object was obtained in the CS of the working trajectory position.
6. The selection, from the set of points matched in the frames of the working stereo pair, of a subset of points that was also matched with the respective subset of the model of this object's view (subM_2D_POINT_V^k):

subS_2D_point_W = S_2D_point_WW_LR ∩ S_2D_ point_ V^kW _LL.

7. The formation of the subset M_3D_POINT_V^k_CS^{view_k} (belonging to the set M_3D_POINT_V^k V^k_CS^{view_k}) corresponding to the subset subM_2D_POINT_V^k.
8. Building of the subset S_3D_POINT_V^kW_CS^{work} belonging to the set S_3D_POINT_WW_CS^{work} using the 2D subset subS_2D_point_W obtained in paragraph 6.
9. Since the subsets M_3D_POINT_V^k_CS^{view_k} and S_3D_POINT_W_CS^{work} were matched by performing the above operations, the matrix for coordinate transformation between the above-indicated CSs ($H_{CS^{work},CS^{view\_k}}$) is calculated from the set of these points. A standard method is used: minimizing the sum of discrepancies by the least squares method.

10. The matrix for coordinate transformation from CS^{AUV_work} into CS^{object_n} is calculated and provides the direct calculation of AUV's movement in the coordinate space of the object: $H_{CS^{work},CS^{object\_n}} = H_{CS^{view\_k},CS^{object\_n}} \cdot H_{CS^{work},CS^{view\_k}}$.
11. To calculate the AUV's movement in the coordinate space of the SPS, a matrix for coordinate transformation from CS^{object_n} to CS^{SPS} is calculated using the matrix $H_{CS^{object\_n},CS^{SPS}}$.

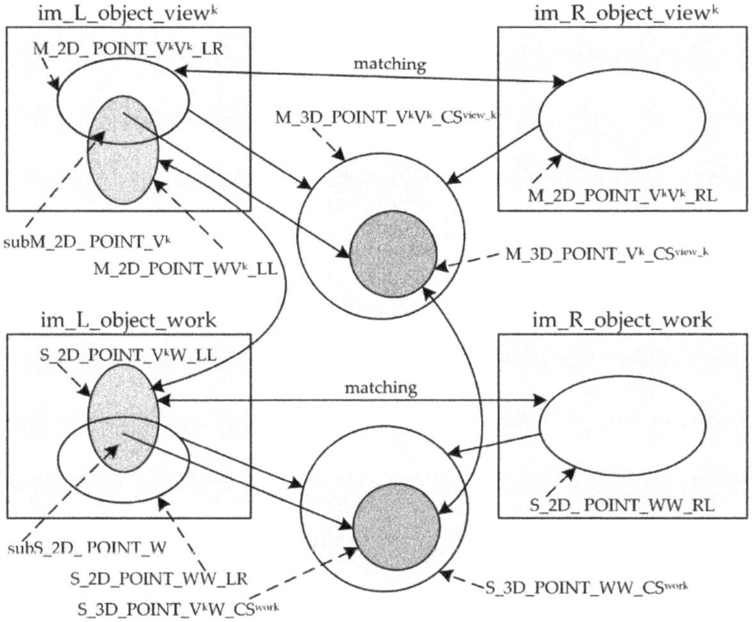

Figure 4. Object recognition on the basis of point features by comparing the left frames of the model and working stereo pairs. Dashed lines refer to the names of point sets. The solid one-way arrow links the original 2D set of points to the resulting 3D set. The solid two-way arrow links two matching sets.

3.2.2. Recognition and Calculation of 3D Coordinates of "Segment Line" and "Corner" Type CEs (Stage 2)

The coordinate transformation matrix $H_{CS^{work},CS^{view\_k}}$ obtained at the first stage, linking the coordinate system CS^{work} with the coordinate system of the *k*-th view of the overview trajectory CS^{view_k} (*k*-th view model), substantially simplifies the solution of the problem of recognizing the "segment line" type CEs in the images of the inspection trajectory since it does not require the direct selection of segment lines in vectorized images.

When recognizing segment lines belonging to the object in the input video stream of the inspection trajectory, two related problems are solved sequentially:

(1) Matching the segment images in the left (right) frame of the model stereo pair (of the k-th view) and the left (right) frame of the stereo pair of the inspection trajectory's position; (2) 3D reconstruction of the matched 2D segment images in the working stereo pair of frames.

The first problem is solved by the below-described algorithm for matching segments, which uses (a) the coordinate transformation matrix $H_{CS^{work},CS^{view\_k}}$, calculated from the points at the first stage, linking CS^{work} and CS^{view_k}; and (b) a set of 3D segments $M\_3D\_LINE\_V^kV^k\_CS^{view\_k}$ (the *k*-th view model). Each segment is defined by two spatial end points (in CS^{view_k}), and for each end point, there are pointers at its 2D images in the stereo pair of frames. The advantage of this technique for defining a segment is that, as mentioned above, no image vectorization is required to match the segment line image in the model with the segment line image selected in the working stereo pair of frames. In other words, the 3D problem of segment line matching is reduced to the 2D problem of matching the segment end points with the calculation of correlation estimates (Figure 5). It should be taken into account that the segment line is an edge of the object's face. Thus, when calculating the correlation of the images of the end points compared between the images, one uses the rectangular arrays of pixels on the side from the segment line where the object is located (this information is saved while forming the model). Eventually, the proposed matching technique without using the vectorized form of images substantially reduces computational costs.

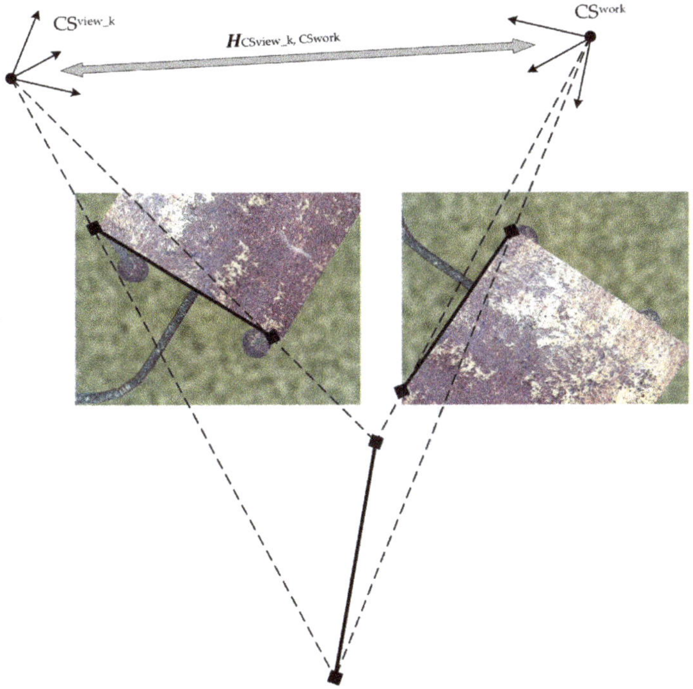

Figure 5. The matching of projections of an object edge on images of a stereo pair of object model and a stereo pair of the inspection trajectory position. Bold lines are projection images of the edges of the object in stereo pairs photographs in CS^{view_k} and in CS^{work}. The dashed lines indicate the end points of the projection of the object edge on the images of two stereo pairs.

Algorithm for recognizing and calculating 3D coordinates of "line" type CEs. Segment lines are sequentially selected from the 3D set M_3D_LINE_V^kV^k_CS^{view_k}; for each segment line, its 2D image is recognized in the stereo pair of frames (im_L_object_work, im_R_object_work) of the analyzed position of the working (inspection) trajectory. Thus, for each segment line l_i (defined by the end points $p1_i$ and $p2_i$) (Figure 5):

1. Pointers at 2D images of its two points in the stereo pair of frames of the model are extracted.
2. The segment line is placed in the coordinate space CS^{work} using a matrix $\left(H_{CS^{work},CS^{view\_k}}\right)^{-1}$ (calculated in step 1) and is projected onto the stereo pair of frames of the working trajectory. The expression "segment line is projected" means the projection of the segment's end points.

3. The coefficient of correlation between the compared images of the point in the model's stereo pair and the working stereo pair of frames is calculated for each end point (Figure 6). When comparing the end points of the segment lines, the orientation of the vehicle by the compass of both the model and the working trajectory is used to provide the same orientation of the surrounding terrain compared.
4. In the case of incomplete matching of the images, an attempt is made to find a match within a small radius, assuming that the projection (step 2) was performed with some error (step 3).

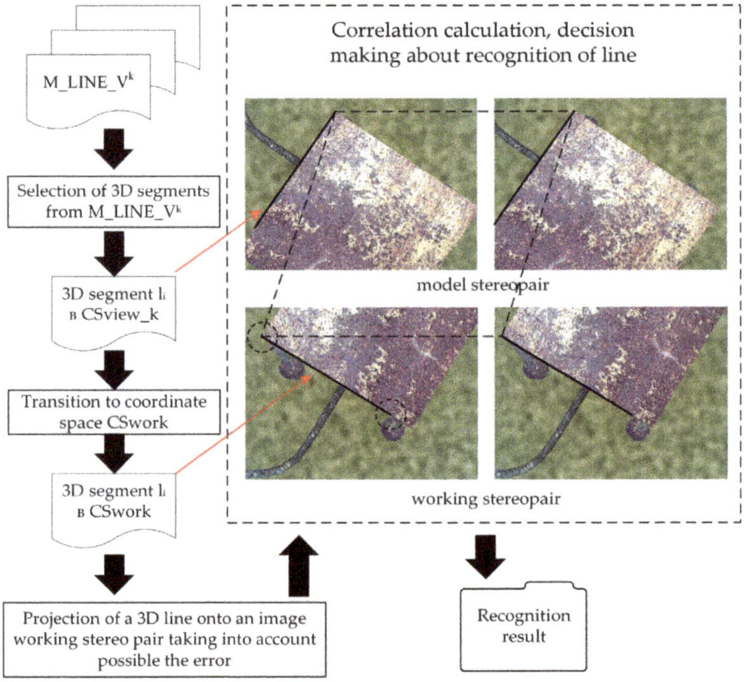

Figure 6. Recognition of rectilinear segments. The dashed and solid lines, as in Figure 5, refer to the images of the projections of the object's edge in the photographs of two stereo pairs, and to the end points of the edge projections.

A decision is made about the match or non-match of the images.

The second problem, 3D reconstruction of the matched 2D images of segment lines in the working stereo pair of frames, is solved in a traditional way: by matching the end points in the stereo pair of frames using epipolar constraints, then using the ray triangulation method to calculate the 3D coordinates of the segment line's end points in CS^{work}.

The result of the recognition of all segment lines is the set S_3D_LINE_W_CS^{work}_match matched to M_3D_LINE_V^k_CS^{view_k}_match. This result is input in the Table of Recognition Results, and the set is entered in S_3D_ LINE_CS^{object_n}.

Recognition and calculation of 3D coordinates of "corner" type CEs. CEs of the "corner" type are selected from the obtained set of "lines". The selection algorithm is quite simple since a "corner" CE is constructed from two segment lines lying in the same plane in space. By indicating whether the segment line belongs to a specific "corner", ambiguity is eliminated in case of similar geometries of corners.

The result of th erecognition of all corners (the number of corners in the set S_3D_CORNER_W_CS^{work}_match matched to M_3D_CORNER_V^k_CS^{view_k}_match, with pointers at these sets) is stored in the Table of Recognition Results, and the set is entered in S_3D_ CORNER_CS^{object_n}.

Object identification. Upon identifying the inspection trajectory of all three types of CEs at this position, corresponding to the object model, the data of the Table of Recognition Results are analyzed, and an algorithmic decision is made about the identification of the object.

Calculation of the refined coordinate transformation matrix $H_{CS^{work},CS^{object\_n}}$ is carried out by the standard technique with minimization of the total coordinate deviation when matching all points used. In this case, the matrix is calculated using all points belonging to CEs in the sets S_3D_ POINT_CS^{object_n}, S_3D_LINE_CS^{object_n}, and S_CORNER_CS^{object_n} since these points have a coordinate representation both in CS^{work} and in CS^{object_n}. As a result, the matrix $H_{CS^{work},CS^{object\_n}}$ (4 × 4) is calculated, which links the CS of the AUV's camera and the object's CS. Then, the AUV's coordinates at each position of the trajectory are recalculated from the CS of the AUV or its camera to the object's CS using this matrix $P^{CS^{object\_n}} = P(0,0,0,1) \cdot H_{AUV,object\_n}$, where $P(0,0,0,1)$ are the uniform AUV coordinates in the CS of the AUV's camera at this trajectory position(Figure 7). It should be noted that the CS of each position of the AUV's trajectory is linked to the CS of the initial trajectory position using the author's method for visual navigation [24].

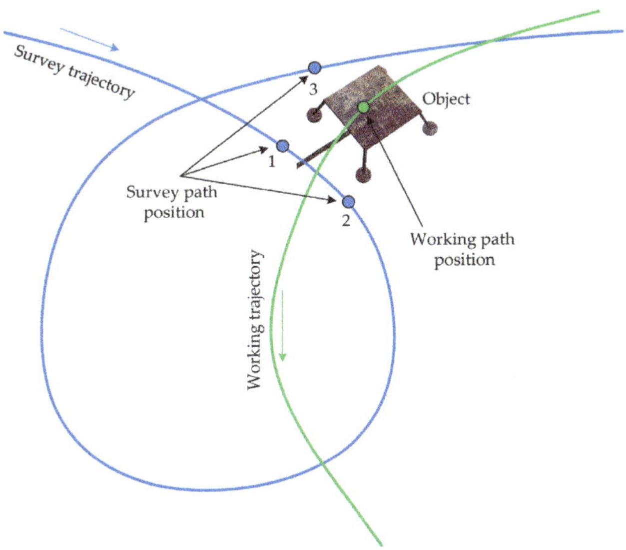

Figure 7. Overview and inspection trajectories on the seabed plane. The views close to the object are indicated on the overview trajectory; for the position indicated on the inspection trajectory, the coordinate reference of the AUV to the object is calculated.

The calculation of the AUV's movement in the object's coordinate space is performed using the matrix $H_{CS^{work},CS^{object\_n}}$ for transformation from CS^{work} into CS^{object_n}. As shown above, it is calculated as follows:

$$H_{CS^{work},CS^{object\_n}} = H_{CS^{view\_k},CS^{object\_n}} \cdot H_{CS^{work},CS^{view\_k}}$$

4. Experiments

We set up a series of experiments with a virtual scene and actual data used in order to evaluate the efficiency of the proposed method for coordinate referencing of AUV to SPS objects.

Model scene. To obtain a comparative estimate of the accuracy of the coordinate referencing of the AUV to the object on the basis of one view and several views of the object, we conducted a series of experiments using a virtual scene generated by a modeling system [25]. The AUV's altitude above the seabed varied from 6.4 to 7.1 m. The altitude above the well cover was 3.6 m; the frame rate was 10 fps; and the resolution of images was 1200 × 900 pixels. The AUV's coordinates were calculated at the trajectory positions every 10 frames of photography. The coordinate referencing was carried out by our method on the basis of the calculated matrix for coordinate transformation $H_{CS^{work},CS^{object\_n}}$ (see Section 3).

The results of one of the series of experiments are shown in Figure 8. With the use of a single view of the overview trajectory (view 1, with 26 matched points), the accuracy of the AUV coordinate reference to the object was 32 mm. After points of views 2 and 3 were added and duplicates removed, the number of matched points grew to 76, and the accuracy, accordingly, increased to 2.8 mm. Additionally, we set up three more similar experiments, but with three other views in each (having smaller general fields of visibility of the object by the AUV's camera at the considered position of the inspection trajectory). The resulting accuracies proved to be, as expected, slightly lower at 3.4, 4.2, and 12.3 mm, but still higher than those with the use of one view.

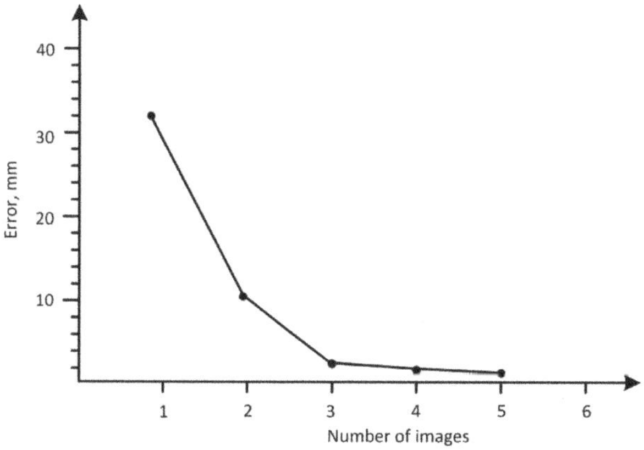

Figure 8. Virtual data: dependence of the accuracy of the coordinate referencing of the AUV to the object on the number of views used.

Thus, the use of several views of the overview trajectory to calculate the matrix $H_{CS^{work},CS^{object\_n}}$ in these experiments increased the accuracy of the coordinate reference of the AUV to the object 3–11-fold. The reference accuracy was also confirmed by the high value of the calculated coefficient of correlation (up to 94%) of end points when segment lines were re-projected.

Actual data. The experiment was conducted in conditions close to an actual underwater mission: a submerged mockup of the object was installed on the seabed at a depth of 1.5 m; photographs were taken with a camera (Karmin2 (Nerian's 3D Stereo Camera, 25 cm) Allied Vision Technologies GmbH, Stuttgart, Germany) housed in a sealed box and connected to a laptop aboard a boat. Photography (obtaining different views according to the above method) was carried out by an operator (diver).

The technique for evaluating the accuracy of the proposed method in this experiment was based on indirect estimates (unlike the virtual scene, where, when calculating the error, the value set in the model was considered as a true value of the parameter being estimated). The accuracy error of the method was estimated in two ways: (a) the operator measured the distance between a pair of characteristic points on the object before submerging the object mockup into the seabed (the distance calculated by our method was then compared to this value, which provided the accuracy error of the method evaluated); (b) the operator explicitly indicated the images of the selected pair of object points in the views of the overview trajectory and in the view of the working trajectory, the line calculated by our method between two 3D points of the object was projected onto the photograph of the working trajectory's view, and then the error was calculated as the difference between the calculated projection of the line in the photograph and the line built by the operator (in pixels). Figure 9 shows the results of the accuracy assessment obtained according to the first technique. Using one view (246 matched points), the accuracy error in 3D space was 1.7% of the segment line length, and when using three projections (the number of compared points increased to 473), the error of the method decreased to 1.3%.

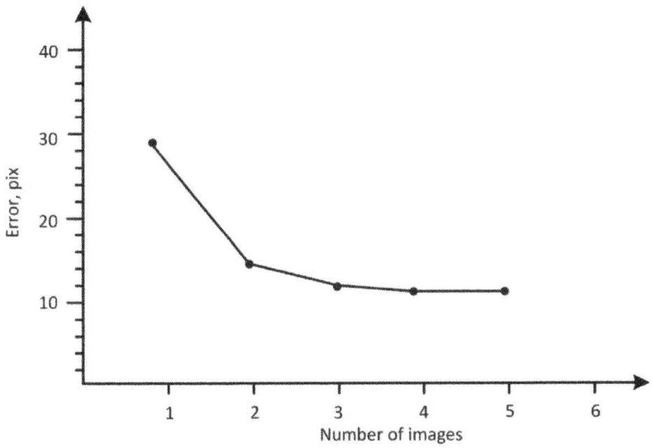

Figure 9. Actual data: dependence of the error of the coordinate referencing of the AUV to the object from the number of views used.

The calculation of the error using the second technique yielded the following results: when using one view, the error in reprojecting the end points of the segment lines onto the image was 28 pixels; and when using three types, the reprojection error decreased to 13 px (Figure 10).

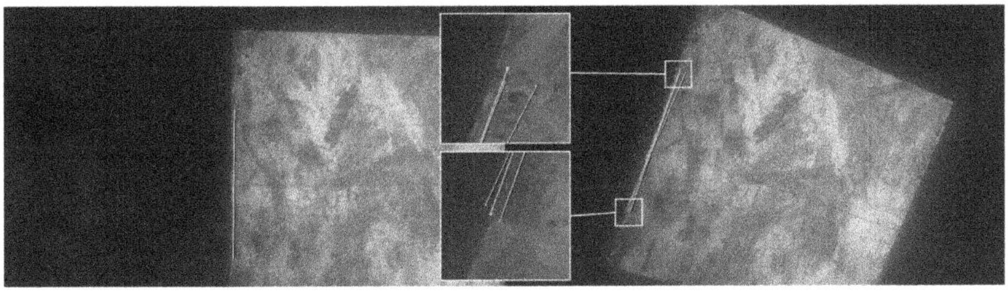

(a) (b)

Figure 10. Calculation of error of the coordinate reference method in a scene with a mockup of a subsea production system (SPS) object: (**a**) the left figure is a photograph of the object for a view on the overview trajectory, where the blue line was built by the operator; (**b**) the right figure is a photograph of the object for a view on the working trajectory. The blue line built by the operator is considered a true value to which the result of the method is compared (a projection onto the photograph of the spatial line calculated by the method). The red line is a projection of the line calculated on the basis of a single view; the green line is a projection of the line calculated on the basis of three views.

Thus, we confirmed the conclusion drawn in the case of the virtual scene about the increase in the accuracy of coordinate referencing of the AUV to an underwater object by using several views of the overview trajectory.

A separate experiment was also performed to obtain a comparative assessment of the efficiency of our algorithm for calculating the $H_{CS^{work},CS^{view\_k}}$ transformation matrix based on matching points of the one view (described in Section 3.2.1). The comparison was performed with the ICP algorithm from the Open3D library. The accuracy and speed of the programs were assessed under the same conditions. The results are shown in Table 2. It was also found that, unlike our algorithm, the ICP algorithm performs poorly (or does not work at all) with a small number of points.

Table 2. Comparative efficiency of algorithms.

	Image Size	Number of Matched Points/Number of Points in 3D Cloud	Our Algorithm, Time (s)	Our Algorithm, Error	ICP Algorithm, Time (s)	ICP Algorithm, Error
Model scene	1200 × 900	26/252	0.1	32 mm	0.14	41 mm
Real scene	1600 × 1200	246/584	0.15	28 px	0.21	52 px

5. Discussion

In this work, in the context of solving the problem of inspection of industrial subsea structures, a method for recognition and high-precision coordinate reference of AUVs to SPS objects is proposed and implemented. Accurate coordinate reference should provide the ability to plan the AUV inspection trajectory in the coordinate space of the SPS object to perform Maintenance and Repair operations in real time. The distinctive features of the method that realize its effectiveness are as follows:

- A model of an underwater object is formed on the basis of views (foreshortening) of a previously completed survey trajectory using an automated technique with the participation of an operator. The technique does not require much labor since it excludes preliminary direct in situ measurements and the use of documentation.
- The use of several views of the survey trajectory contributes to a higher degree of object recognition at the inspection stage and, as experiments on synthetic and real data have shown, several-fold increases the accuracy of the coordinate reference of the APR to the object.
- The use of several types of geometric elements when comparing data obtained at the position of the inspection trajectory with the object model increases the reliability of object recognition, despite the noisiness of the images, and also increases the accuracy of the coordinate reference due to the use of an expanded set of points when calculating the coordinate transformation matrix.
- Original and computationally low-cost algorithms for recognition and matching with the model of the characteristic geometric elements used (point features, segments, and angles) have been developed.

Comparison with other methods. An example of an approach to solving the problem of inspection of underwater structures, based on the integrated use of sensors of various types, with an emphasis on processing sonar images, is described in [6]. For obstacle and inspection target detection, a scanning sonar is used. A pan–tilt-enabled sensor head with a camera, laser measurement, and a Multi Beam Echo Sounder is used to inspect the detected objects. Here, the problem of AUV movement relative to inspected objects is solved in a different way and without the use of optical images. Namely, the sonar images are automatically processed with edge detection and line extraction algorithms to obtain a simplified environment description, which is used by the guidance methods. The inspection sensors also provide the vehicle guidance system (VGS) with distance values to the inspection object. A laser gyroscope is used for navigation and positioning supported by a Doppler velocity log and a pressure sensor. Additionally, a scanning sonar provides data for collision avoidance. This sonar is mainly used for the reactive vehicle control by the VGS and the detection of the inspection objects.

In terms of comparison with our approach, previous work [20] in which the problem of recognizing underwater objects was solved is also of interest. The method proposed by the authors aims to recognize objects of a specific type (pipes and connecting elements), using an a priori accessible database of partial views of objects stored in the form of 3D point clouds. The methods are tested using an experimental dataset containing laser scans and AUV navigation data.

Other authors [26] presented the integration of a stereo-vision Graph-SLAM system in the navigation and control architecture of the autonomous underwater vehicle (AUV) SPARUS II. Experiments were conducted with a virtual model of the SPARUS II in a simulated marine environment, all in the context of the UWSIM simulator. The vehicle navigated at a constant altitude of 2.5 m. The average error on a 180 m long trajectory was 0.64 m for Odometry and 0.15 m for SLAM.

In another study [27], different navigation technologies were considered from the point of view of their efficiency and limitations, including assessments of optical technologies. One of the test results demonstrated that the average localization error is approximately 5 cm and the average relative location error is approximately 2% in the range of 3.6 m. In principle, these values are consistent with the results obtained in our experiments with virtual data (3.2 cm for one view and 2.8 cm for three views) and real data (1.7% for one view and 1.3% for three views).

An interesting experiment on the use of stereo vision to eliminate the integrated error of AUV navigation (generated by the use of INS) was conducted in [28]. The experiment consisted of AUV navigation and returning to the docking station. Using the stereo-vision system, an error of about 100 mm was eliminated.

Another study [29] proposed a robust real-time AUV self-localization method based on a stereo camera and an inertial sensor, which merged point and diagonal features, as well as inertial measurements to overcome the challenges of poor visibility. The method also included an underwater loop detection algorithm based on the combination of points and diagonal segments, which can extract effective binary descriptors in low-texture underwater scenarios. Testing the method in a real underwater environment using a sensor suite yielded the following accuracy results:

Error for Stereo + IMU With Loop—39 cm; Error for Stereo + IMU Without Loop—66 cm

The differences between our approach and the above research are that (a) in our approach, the main emphasis is on solving the problem of coordinate reference of an AUV to an object using optical images, not limited to the problem of object recognition; (b) recognition is based on simple geometric elements characteristic of man-made objects (points, segments, and angles) while providing some universality in relation to types of objects, and without requiring significant computational costs; and (c) the use of several views of underwater objects during recognition allows us to increase accuracy of AUV positioning relative to the object.

6. Conclusions

This article presents a new approach to the accurate coordinate referencing of AUVs to SPS objects, based solely on stereo-image processing. It differs from other similar approaches in its use of a pre-made overview trajectory of the AUV to form models of SPS objects, its use of different types of geometric elements that increase the degree of object recognition, and its use of original algorithms to match the obtained visual data with an a priori model of the object. The results of experiments on virtual scenes and with real data showed the effectiveness of the proposed technology.

Future work. We plan to (a) extend experiments with the proposed method as part of the on-board AUV software to analyze the effectiveness of its use in different underwater conditions and (b) compare the effectiveness of the algorithmic approach (using our method as an example) to solve the problem of identifying underwater objects and the approach based on the use of neural networks.

Author Contributions: Conceptualization, V.B.; methodology, V.B. and A.K.; software, A.K.; validation, V.B. and A.K.; formal analysis, V.B.; investigation, V.B. and A.K.; resources, V.B.; data curation, A.K.; writing—original draft preparation, V.B.; writing—review and editing, V.B. and A.K.; visualization, A.K.; supervision, V.B.; project administration, V.B.; funding acquisition, V.B. All authors have read and agreed to the published version of the manuscript.

Funding: The study was funded by the Russian Science Foundation, grant No. 22-11-00032 (https://rscf.ru/en/project/22-11-00032/ (accessed on 19 August 2024)), and within the framework of the State budget topic No. 0202-2021-0004 of the Institute of Automation and Control Process, Far Eastern Branch, Russian Academy of Sciences (IACP FEB RAS). The method for tracking a subsea pipeline (SP) using stereo images taken by an autonomous underwater vehicle (AUV) was designed under the RSF grant No. 22-11-00032. The estimates of comparative efficiency of the methods for tracking an SP using stereo images taken by an AUV were obtained under the State budget topic No. 0202-2021-0004 of the IACP FEB RAS.

Institutional Review Board Statement: Not applicable.

Informed Consent Statement: Not applicable.

Data Availability Statement: The dataset are available at https://doi.org/10.6084/m9.figshare.26462257.v1 (accessed on 19 August 2024).

Conflicts of Interest: The authors declare no conflicts of interest.

Abbreviations

CEs—characteristic elements.

CS^{view_k}—the coordinate system of AUV's camera for the *k*th view of the overview trajectory.

CS^{work}—the coordinate system of AUV's camera at the considered position of the working (inspection) trajectory.

CS^{object_n}—the coordinate system of the *n*th object of SPS.

CS^{SPS}—the coordinate system of SPS (CS of one of the objects can be selected as CS of SPS).

M_3D_POINT_V^kV^k_CS^{view_k}—a set of 3D points in the coordinate system CS^{view_k} (belongs to the 3D model of the *k*th view of the overview trajectory) obtained from the matched sets M_2D_POINT_V^kV^k_LR and M_2D_POINT_V^kV^k_RL.

M_3D_LINE_V^kV^k_CS^{view_k}—a set of 3D "segment lines" in the coordinate system CS^{view_k} (part of a 3D model for the *k*th view of the overview trajectory).

M_3D_CORNER_ V^kV^k _CS^{view_k}—a set of 3D "corners" in the coordinate system CS^{view_k} (part of a 3D model for the *k*th view of the overview trajectory).

M_2_POINT_WV^k_LL and S_2D_POINT_V^kW_LL—2D sets of points matched (using the SURF detector). The first set is determined in the left frame of the "model" stereo-pair of the *k*th view; the second set is in the left frame of the "working" stereo-pair.

subM_2D_POINT_V^k V^k and subS_2D_ POINT_WW—2D subsets of points matched.

The first subset is the intersection of the sets M_2D_POINT_V^kV^k_LR and M_2D_POINT_WV^k_LL in the left frame of the "model" stereo-pair of the *k*th view; the second subset is the intersection of the sets S_2D_POINT_WW_LR and S_2D_POINT_V^kW_LL in the left frame of the "working" stereo-pair.

M_3D_POINT_V^k_CS^{view_k}—a subset of 3D points of the model of the *k*th view (in CS^{view_k}) to which the subset of recognized points S_3D_POINT_V^k_CS^{work} (being part of the set S_3D_POINT_WW_CS^{work}) is matched.

M_3D_LINE_V^k_CS^{view_k}—a subset of 3D segment lines of the model of the *k*th view (in CS^{view_k}) to which the subset of recognized segment lines S_3D_LINE_V^k_CS^{work} (being part of the set S_3D_LINE_WW_CS^{work}) is matched.

M_3D_CORNER_V^k_CS^{view_k}—a subset of 3D corners of the model of the *k*th view (in CS^{view_k}), to which the subset of recognized corners S_3D_CORNER_V^k_CS^{work} (being part of the set S_3D_CORNER_WW_CS^{work}) is matched.

M_3D_POINT_V^k_ CS^{object_n}, M_3D_LINE_V^k_ CS^{object_n}, M_3D_CORNER_V^k_ CS^{object_n}—the same above-listed sets of recognized CEs of three types belonging to the *k*th view, but coordinated in the object's CS (CS^{object_n}).

M_3D_POINT_CS^{object_n}—a combined set of objects' 3D points, recognized for all views used, at the considered position of the working (inspection) trajectory.

M_3D_LINE_CS^{object_n}—a combined set of objects' 3D segment lines, recognized for all views used, at the considered position of the working (inspection) trajectory.

M_3D_CORNER_CS^{object_n}—a combined set of objects' 3D corners, recognized for all views used, at the considered position of the working (inspection) trajectory.

S_2D_ POINT_ WW_LR and S_2D_POINT_WW_RL—matched sets of points in the left and right frames of the "working" stereo-pair. These sets are generated and matched using the SURF detector.

S_3D_POINT_WW_CS^{work}—a set of 3D points in the coordinate system CS^{work} obtained from the matched sets S_2D_POINT_WW_LR and S_2D_POINT_WW_RL.

S_3D_POINT_V^{k}W_CS^{work}—a subset of the 3D "points" recognized in the working stereo-pair of frames when compared to the *k*th view, coordinated in CS^{work} (np_k is the number of points in the subset for the *k*th view) (being part of the set S_3D_POINT_WW_CS^{work}). The subset is matched to the subset of the model M_3D_POINT_V^k_CS^{view_k}.

S_3D_LINE_V^{k}W_CS^{work}—a subset of the 3D "segment lines" recognized in the working stereo-pair of frames for the *k*th view, coordinated in CS^{work} (nl_k is the number of segment lines in the subset for the *k*th view).

S_3D_CORNER_V^{k}W_CS^{work}—a subset of the 3D "corners" recognized in the working stereo-pair of frames for the *k*th view, coordinated in CS^{work} (nc_k is the number of corners in the subset for the *k*th view).

S_3D_ POINT_CS^{object_n}—a set of "points" in CS^{object_n} obtained by combining the recognized "points" across all views.

S_3D_LINE_CS^{object_n}—a set of "segment lines" in CS^{object_n} obtained by combining the recognized "segment lines" across all views.

S_CORNER_CS^{object_n}—a set of "corners" in CS^{object_n} obtained by combining the recognized "corners" across all views.

$H_{CSwork, CSview\_k}$—a matrix for coordinate transformation from the CS of the AUV's inspection trajectory position into the CS of the *k*th view of the overview trajectory.

$H_{CSview\_k, CSobject\_n}$—a matrix for coordinate transformation from the CS of the *k*th view of the overview trajectory into the CS of the *n*th object.

$H_{CSwork, CSobject\_n}$—a matrix for coordinate transformation from the CS of the AUV's inspection trajectory position into the CS of *n*th object.

$H_{CSobject\_n, CS\_SPS}$—a matrix for coordinate transformation from the CS of *n*th object into the CS of SPS.

References

1. Mai, C.; Hansen, L.; Jepsen, K.; Yang, Z. Subsea Infrastructure Inspection: A Review Study. In Proceedings of the 6th International Conference on Underwater System Technology: Theory and Applications, Penang, Malaysia, 13–14 December 2016. [CrossRef]
2. Zhang, Y.; Zheng, M.; An, C.; Seo, J.K. A review of the integrity management of subsea production systems: Inspection and monitoring methods. *Ships Offshore Struct.* **2019**, *14*, 1–15. [CrossRef]
3. Manley, J.E.; Halpin, S.; Radford, N.; Ondler, M. Aquanaut: A New Tool for Subsea Inspection and Intervention. In Proceedings of the OCEANS 2018 MTS/IEEE Conference, Charleston, SC, USA, 22–25 October 2018. [CrossRef]
4. Albiez, J.; Cesar, D.; Gaudig, C.; Arnold, S.; Cerqueira, R.; Trocoli, T.; Mimoso, G.; Saback, R.; Neves, G. Repeated close-distance visual inspections with an AUV. In Proceedings of the OCEANS 2016 MTS/IEEE Monterey, San Diego, CA, USA, 19 September 2016; pp. 1–8. [CrossRef]
5. Terracciano, D.; Bazzarello, L.; Caiti, A.; Costanzi, R.; Manzari, V. Marine Robots for Underwater Surveillance. *Curr. Robot. Rep.* **2020**, *1*, 159–167. [CrossRef]
6. Jacobi, M. Autonomous inspection of underwater structures. Robot. *Auton. Syst.* **2015**, *67*, 80–86. [CrossRef]
7. Bao, J.; Li, D.; Qiao, X.; Rauschenbach, T. Integrated navigation for autonomous underwater vehicles in aquaculture: A review. *Inf. Process. Agric.* **2020**, *7*, 139–151. [CrossRef]
8. Sahoo, A.; Dwivedy, S.K.; Robi, P.S. Advancements in the field of autonomous underwater vehicle. *Ocean. Eng.* **2019**, *181*, 145–160. [CrossRef]
9. Wirth, S.; Carrasco, P.L.N.; Oliver-Codina, G. Visual odometry for autonomous underwater vehicles. In Proceedings of the 2013 MTS/IEEE OCEANS, Bergen, Norway, 10–13 June 2013; pp. 1–6. [CrossRef]
10. Jung, J.; Li, J.-H.; Choi, H.-T.; Myung, H. Localization of AUVs using visual information of underwater structures and artificial landmarks. *Intell. Serv. Robot.* **2016**, *10*, 67–76. [CrossRef]

11. Gao, J.; Wu, P.; Yang, B.; Xia, F. Adaptive neural network control for visual servoing of underwater vehicles with pose estimation. *J. Mar. Sci. Technol.* **2016**, *22*, 470–478. [CrossRef]
12. Xu, H.; Oliveira, P.; Soares, C.G. L1 adaptive backstepping control for path-following of underactuated marine surface ships. *Eur. J. Control.* **2020**, *58*, 357–372. [CrossRef]
13. Fan, S.; Liu, C.; Li, B.; Xu, Y.; Xu, W. AUV docking based on USBL navigation and vision guidance. *J. Mar. Sci. Technol.* **2018**, *24*, 673–685. [CrossRef]
14. Ferrera, M.; Moras, J.; Trouvé-Peloux, P.; Creuze, V. Real-Time Monocular Visual Odometry for Turbid and Dynamic Underwater Environments. *Sensors* **2019**, *19*, 687. [CrossRef] [PubMed]
15. Himri, K.; Ridao, P.; Gracias, N. 3D Object Recognition Based on Point Clouds in Underwater Environment with Global Descriptors: A Survey. *Sensors* **2019**, *19*, 4451. [CrossRef] [PubMed]
16. Kasaei, S.H.; Lopes, L.S.; Tomé, A.M.; Oliveira, M. An orthographic descriptor for 3D object learning and recognition. In Proceedings of the IEEE/RSJ International Conference on Intelligent Robots and Systems (IROS), Daejeon, Republic of Korea, 9–14 October 2016; pp. 4158–4163.
17. Osada, R.; Funkhouser, T.; Chazelle, B.; Dobkin, D. Matching 3D models with shape distributions. In Proceedings of the SMI 2001 International Conference On Shape Modeling and Applications, Genova, Italy, 7–11 May 2001; pp. 154–166.
18. Marton, Z.C.; Pangercic, D.; Rusu, R.B.; Holzbach, A.; Beetz, M. Hierarchical object geometric categorization and appearance classification for mobile manipulation. In Proceedings of the 10th IEEE-RAS International Conference on Humanoid Robots (Humanoids), Nashville, TN, USA, 6–8 December 2010; pp. 365–370.
19. Rusu, R.B.; Blodow, N.; Beetz, M. Fast point feature histograms (FPFH) for 3D registration. In Proceedings of the IEEE International Conference on Robotics and Automation, ICRA'09, Kobe, Japan, 12–17 May 2009; pp. 3212–3217.
20. Himri, K.; Ridao, P.; Gracias, N. Underwater Object Recognition Using Point-Features, Bayesian Estimation and Semantic Information. *Sensors* **2021**, *21*, 1807. [CrossRef] [PubMed]
21. Chemisky, B.; Menna, F.; Nocerino, E.; Drap, P. Underwater Survey for Oil and Gas Industry: A Review of Close Range Optical Methods. *Remote Sens.* **2021**, *13*, 2789. [CrossRef]
22. Bobkov, V.; Kudryashov, A.; Inzartsev, A. Method for the Coordination of Referencing of Autonomous Underwater Vehicles to Man-Made Objects Using Stereo Images. *J. Mar. Sci. Eng.* **2021**, *9*, 1038. [CrossRef]
23. Wen, L.; Youzhi, L. Autonomous Underwater Vehicle Navigation and Control: A Brief Review. *Robot Autom. Eng. J.* **2024**, *5*, 555672.
24. Bobkov, V.A.; Kudryashov, A.P.; Inzartsev, A.V. Technology of AUV High-Precision Referencing to Inspected Object. *Gyroscopy Navig.* **2019**, *10*, 322–329. [CrossRef]
25. Melman, S.; Bobkov, V.; Inzartsev, A.; Pavin, A. Distributed Simulation Framework for Investigation of Autonomous Underwater Vehicles' Real-Time Behavior. In Proceedings of the OCEANS'15 MTS/IEEE, Washington, DC, USA, 19–22 October 2015.
26. Carrasco, P.L.N.; Bonin-Font, F.; Campos, M.M.; Codina, G.O. Stereo-Vision Graph-SLAM for Robust Navigation of the AUV SPARUS II. *IFAC-PapersOnLine* **2015**, *48*, 200–205. [CrossRef]
27. Gómez-Espinosa, A.; Cuan-Urquizo, E.; González-García, J. Autonomous Underwater Vehicles: Localization, Navigation, and Communication for Collaborative Missions. *Appl. Sci.* **2020**, *10*, 1256. [CrossRef]
28. Hsu, H.Y.; Toda, Y.; Yamashita, K.; Watanabe, K.; Sasano, M.; Okamoto, A.; Inaba, S.; Minami, M. Stereo-vision-based AUV navigation system for resetting the inertial navigation system error. *Artif. Life Robot.* **2022**, *27*, 165–178. [CrossRef]
29. Wang, Y.; Gu, D.; Ma, X.; Wang, J.; Wang, H. Robust Real-Time AUV Self-Localization Based on Stereo Vision-Inertia. *IEEE Trans. Veh. Technol.* **2023**, *72*, 7160–7170. [CrossRef]

Journal of
*Marine Science
and Engineering*

MDPI

Article

Prescribed Performance Formation Tracking Control for Underactuated AUVs under Time-Varying Communication Delays

Haitian Zhang [1], Yanqing Jiang [1,2,*], Rui Gao [1], Hang Li [1] and Ao Li [1]

[1] Science and Technology on Underwater Vehicles Laboratory, Harbin Engineering University, Harbin 150001, China; 1120640488@hrbeu.edu.cn (H.Z.); gaorui16011103@hrbeu.edu.cn (R.G.); lihang2023@hrbeu.edu.cn (H.L.); myladyen@163.com (A.L.)

[2] Sanya Nanhai Innovation and Development Base, Harbin Engineering University, Sanya 572024, China

* Correspondence: jiangyanqing@hrbeu.edu.cn

Abstract: Achieving formation tracking control of underactuated autonomous underwater vehicles (AUVs) under communication delays presents a significant challenge. To address this challenge, a distributed prescribed performance control protocol based on a real-time state information online predictor (RSIOP) is proposed in this paper. First, we innovatively designed an RSIOP to achieve active compensation for the delayed state information of neighboring AUVs. Next, considering formation performance and safety, a low-complexity and practical nonlinear mapping function was used to implement prescribed performance tracking control for the AUV formation. Additionally, the adverse effects of external disturbance uncertainties and input saturation are also considered. Finally, the simulation tests demonstrated that the proposed formation control protocol can successfully achieve the predetermined formation tracking tasks in the presence of time-varying communication delays and external disturbances, while also enabling real-time changes in formation configuration during the process. Throughout, the protocol maintains input saturation limits, and the actual control inputs remain smooth, with no significant oscillations. Furthermore, comparative simulation tests verified the necessity of the RSIOP developed in this study and quantitatively demonstrated that the proposed control method exhibits superior performance in terms of formation control accuracy, error convergence speed, and transient-state constraints.

Keywords: underactuated autonomous underwater vehicles; formation control; time-varying communication delay; prescribed performance control

1. Introduction

Autonomous underwater vehicles (AUVs) constitute an emerging class of unmanned underwater platforms that have experienced rapid advancements in recent years, driven by significant progress in energy systems, sensing technologies, navigation, communication, and control technologies [1–5]. However, with the expansion of oceanic endeavors and the increasing complexity of underwater tasks, AUVs face significant challenges. The limited energy capacity, the susceptibility to ocean currents and marine life disturbances, and the relatively small operational coverage within a limited timeframe of single AUV platforms are inadequate to meet the requirements of more complex, real-time, large-scale marine missions [6,7]. These limitations highlight the urgent need for more robust and scalable solutions, driving the focus toward multi-AUV systems. Today, multi-AUV systems have been proven to collect more reliable data and exhibit higher efficiency, better fault tolerance, greater robustness, wider mission execution areas, and lower system costs compared to single AUVs [8–11]. Despite these advantages, the successful deployment of multi-AUV systems poses considerable challenges, especially with respect to formation control and communication reliability. The primary objective of AUV formation control is to manage the relative distance, speed, or orientation between AUVs, enabling them to move as a unified group. This facilitates tasks such as transitioning from initial positions to a designated location to form a specific formation or maintaining a certain formation during navigation [12].

The success of AUV formation control relies heavily on effective communication between vehicles. To achieve AUV formation, information transmission between AUVs must be conducted through wireless communication within a network control system. Acoustic communication has become the most widely used method due to its advantages of longer underwater propagation distances and lower attenuation. However, due to its relatively slow transmission speed and the time required for data parsing, underwater communication is subject to significant delays. This results in each AUV receiving the delayed state of neighboring AUVs rather than accurate, real-time states. These delays can severely impact the control performance and stability of the formation control system, making it a crucial challenge that must be addressed [13,14]. Recent research has made progress in addressing the challenges posed by communication delays, yet significant gaps remain. Li et al. [15,16] established the Lyapunov–Krasovskii equation and used linear matrix inequality (LMI) to solve the time-varying communication delay problem, obtaining boundary conditions for control system error convergence. Zhang et al. [17] modeled a fixed communication topology and, combined with a consensus algorithm, proposed a leader–follower control strategy designed to handle time-varying delays, establishing new stability criteria to ensure the consistency of multi-AUV systems. Zeng et al. [18], utilizing matrix theory and pertinent stability criteria, established sufficient conditions for the stability of the error system under both scenarios, with and without communication delays. Yan et al. [19] proposed a multi-UUV path-following controller based on a virtual leader to address communication delays, using Shuler theory to provide sufficient conditions for multi-UUV path-following convergence. While these approaches have advanced the field, they primarily rely on the delayed states of neighboring AUVs and require knowledge of the upper bound of communication delays. As a result, they address the communication delay problem between AUVs in a passive manner, which limits their effectiveness in more dynamic environments and presents certain limitations. Li et al. [20] studied fully actuated AUVs and used curve-fitting methods to predict the leader's state in a two-dimensional formation, but the fixed-order model used had difficulties ensuring prediction accuracy. Du et al. [21] pioneered an active communication delay compensation mechanism by designing a data-driven state predictor to estimate the current motion state of neighboring AUVs in real-time. However, the recursive step size of state estimation is dependent on the order of the prediction model, which can result in prediction failures due to data loss, making it potentially less suitable for practical AUV applications.

Moreover, another critical aspect of AUV formation control is the management of formation state errors, which has direct implications for system performance and safety. While several approaches have been proposed, challenges remain. Bechlioulis et al. [22] initially introduced an innovative approach known as prescribed performance control (PPC) for nonlinear systems. By transforming the constrained nonlinear system under the prescribed performance function into an unconstrained one, they designed a robust controller to maintain the error within predefined performance bounds. This method set a foundation, yet it involves intricate transformations and handling of constraints. Since then, numerous scholars have conducted research building upon this approach. For instance, Liu et al. [23] transformed tracking errors into virtual errors using a specified performance function and designed an adaptive region-tracking control scheme with transient performance using backstepping technology. Similarly, Sun et al. [24] created a PPC strategy for trajectory tracking control of fully actuated AUVs, utilizing performance functions and corresponding error transformations. While effective, these methods share a common drawback: the need for complex transformations to manage constrained tracking errors, which can complicate implementation and limit practical applicability. Recognizing these challenges, Huang et al. [25] used nonlinear mapping techniques to reduce the complexity of designing prescribed performance controllers, effectively addressing the PPC problem for underactuated single AUVs. However, even with these improvements, PPCs often require large control inputs at the initial stage. Due to the physical constraints of AUV equipment, actuator input saturation becomes unavoidable, leading to performance degradation [26]. This issue has been overlooked in the aforementioned studies, yet it is clearly a problem that urgently needs to be addressed in formation control. Thus, while significant progress has been made, these limitations underscore the need for further research to develop more practical and robust PPC solutions that can be effectively implemented in real-world AUV formations.

In response to these challenges and research gaps, this paper centers on the formation control of underactuated AUVs and develops a prescribed performance control protocol for formation tracking under time-varying communication conditions. Unlike previous studies, the proposed control protocol offers the following advantages:

(1) Combining time series analysis theory, an innovative real-time state information online predictor (RSIOP) is proposed. This approach does not necessitate prior knowledge of the upper bound of communication delays. Instead, it leverages the delayed motion state information of neighboring AUVs to recursively obtain their current state information. This approach not only ensures data validity but also achieves real-time active compensation for time-varying communication delays.

(2) To enhance both transient and steady-state performance, the AUV formation's motion states are constrained using prescribed performance functions, which significantly improves operational efficiency. Additionally, by incorporating nonlinear mapping techniques, the complex transformations of constrained errors are avoided, resulting in a more straightforward and intuitive control protocol design.

(3) The proposed approach addresses the adverse effects of input saturation and unknown external disturbances in AUV formation control without the need for observers, thereby simplifying the adjustment of control parameters.

(4) The formation is designed to be dynamic, enhancing its flexibility in practical applications and improving its ability to adapt to complex and evolving environments.

The rest of this paper is structured as follows. The preliminary and problem formulations are presented in Section 2. Section 3 describes the solution to time-varying communication delays. Section 4 presents the design of the prescribed performance formation tracking controller, along with a stability analysis of the closed-loop system. Section 5 offers the simulation results to validate the effectiveness of the control protocol. Finally, the conclusions are summarized in Section 6.

2. Preliminary and Problem Formulations

2.1. Graphy Theory

Communication topology is a critical component in the formation control of AUVs. It serves as a bridge connecting AUVs, establishing the mechanisms for information flow exchange. Graph theory can be effectively utilized to describe communication topology, providing robust theoretical support for formation control.

In this article, a weighted directed graph $\boldsymbol{G} = \{\boldsymbol{V}, \boldsymbol{E}, \boldsymbol{A}\}$ is utilized to represent the communication topology among AUVs within the network system. Assume there is a virtual leader and N AUVs in the formation, with each AUV considered as a node. $\boldsymbol{V} = \{v_0, v_1, \ldots, v_N\}$ represents the set of nodes in the formation, $\boldsymbol{E} \subseteq \boldsymbol{V} \times \boldsymbol{V}$ denotes the set of edges, and $\boldsymbol{A} = [a_{ij}] \in \mathbb{R}^{(N+1)\times(N+1)}$ is the weighted adjacency matrix. If an edge $(v_i, v_j) \in \boldsymbol{E}$, it indicates that information from the *jth* AUV can flow to the *ith* AUV, and $a_{ij} = 1$. The *jth* AUV can then be referred to as a neighbor of the *ith* AUV, and the neighbor set of ith vehicle can be denoted as $\boldsymbol{B}_i = \{j | (v_i, v_j) \in \boldsymbol{E}\}$. $b_i = \sum_{j=0}^{N} a_{ij}$ specifies the number of neighbors of the ith vehicle. The a_{ij} can be represented as follows:

$$a_{ij} = \begin{cases} 1, \; if\{(v_i, v_j) \in \boldsymbol{E}, i \neq j\} \\ 0, \; otherwise \end{cases}$$

Assumption 1. *Within the communication network system, at least one path exists between any two nodes, indicating that* **G** *is connected. Additionally, the virtual leader acts as the root node of the system.*

2.2. AUV Model

Consider a 5-DOF underactuated AUV mathematical model subject to external disturbances and saturation constraints. The north–east–down frame $\{NED\}$ and the body-fixed frame $\{B\}$ of the ith AUV are established separately, as illustrated in Figure 1.

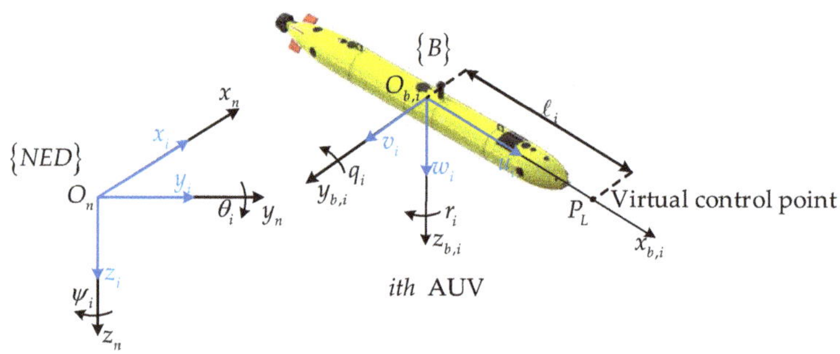

Figure 1. Coordinate frame system of the AUV.

According to reference [27], the three-dimensional motion model of the ith AUV can be derived.

$$\dot{\boldsymbol{\eta}}_i = \boldsymbol{J}(\boldsymbol{\eta}_i)\boldsymbol{v}_i \tag{1}$$

$$\boldsymbol{M}_i\dot{\boldsymbol{v}}_i + \boldsymbol{C}_i(\boldsymbol{v}_i)\boldsymbol{v}_i + \boldsymbol{D}_i(\boldsymbol{v}_i)\boldsymbol{v}_i + \boldsymbol{g}_i(\boldsymbol{\theta}_i) = \boldsymbol{T}_i + \boldsymbol{\Delta}_i \tag{2}$$

where $\boldsymbol{\eta}_i = [x_i, y_i, z_i, \theta_i, \psi_i]^T$ denotes the positions (i.e., surge, sway, and heave displacements) and orientations (i.e., pitch and yaw angles) in the north–east–down frame. $\boldsymbol{v}_i = [vu_i, v_i, w_i, p_i, r_i]^T$ represents the velocities in the body-fixed frame. $\boldsymbol{J}(\boldsymbol{\eta}_i)$ is the transformation matrix and can be described as

$$\boldsymbol{J}(\boldsymbol{\eta}_i) = \begin{bmatrix} \cos(\theta_i)\cos(\psi_i) & -\sin(\psi_i) & \sin(\theta_i)\cos(\psi_i) & 0 & 0 \\ \cos(\theta_i)\sin(\psi_i) & \cos(\psi_i) & \sin(\psi_i)\sin(\theta_i) & 0 & 0 \\ -\sin(\theta_i) & 0 & \cos(\theta_i) & 0 & 0 \\ 0 & 0 & 0 & 1 & 0 \\ 0 & 0 & 0 & 0 & \frac{1}{\cos(\theta_i)} \end{bmatrix}$$

$\boldsymbol{M}_i = diag[m_{11,i}, m_{22,i}, m_{33,i}, m_{55,i}, m_{66,i}]$ is the inertial matrix of the AUV.

$\boldsymbol{C}_i(\boldsymbol{v}_i) = \begin{bmatrix} 0 & 0 & 0 & m_{33,i}w_i & -m_{22,i}v_i \\ 0 & 0 & 0 & 0 & m_{11,i}u_i \\ 0 & 0 & 0 & -m_{11,i}u_i & 0 \\ -m_{33,i}w_i & m_{11,i}u_i & 0 & 0 & 0 \\ m_{22,i}v_i & -m_{11,i}u_i & 0 & 0 & 0 \end{bmatrix}$ is the Coriolis and centripetal matrix.

$\boldsymbol{D}(\boldsymbol{v}_i) = diag\{X_{u,i}, Y_{v,i}, Z_{w,i}, M_{q,i}, N_{r,i}\}$ represents the damping matrix.

$g_i(\boldsymbol{\theta_i}) = [0, 0, 0, \rho_i g_i \nabla_i GM_{L,i}\sin(\theta_i), 0]^T$ is the restoring moment matrix.
$T_i = [T_{1,i}, 0, 0, T_{4,i}, T_{5,i}]^T$ denotes the actual input control force.
And $\boldsymbol{\Delta_i} = [\Delta_{1,i}, \Delta_{2,i}, \Delta_{3,i}, \Delta_{4,i}, \Delta_{5,i}]^T$ is the uncertain external disturbance.

Assumption 2. *For the uncertain external disturbance $\boldsymbol{\Delta_i}$, there exists a positive constant $\overline{\boldsymbol{\Delta}}_i$ such that $\|\boldsymbol{\Delta_i}\| \le \overline{\boldsymbol{\Delta}}_i$ always holds.*

However, to address the underactuation problem, a coordinate transformation is required to integrate the dynamics in all directions and involve all control inputs [28].

$$\boldsymbol{y_i} = \boldsymbol{H_i}(\boldsymbol{\eta_i}) = [x_i + \ell_i\cos(\theta_i)\cos(\psi_i), y_i + \ell_i\cos(\theta_i)\sin(\psi_i), z_i - \ell_i\sin(\theta_i)]^T \tag{3}$$

where ℓ_i is a constant positive parameter that represents the distance between the virtual control point P_i and the center of mass (COM) along the $O_{b,i}x_{b,i}$ axis, as illustrated in Figure 1. Furthermore, by differentiating Equation (3) and incorporating Equations (1) and (2), the new kinematic and dynamic equations for the AUV can be obtained.

$$\dot{\boldsymbol{\eta}}_i^* = \boldsymbol{J}_{1,i}^*(\boldsymbol{\eta}_i)\boldsymbol{v}_i^* + \boldsymbol{J}_{2,i}^*(\boldsymbol{\eta}_i, \boldsymbol{v}_i) \tag{4}$$

$$\dot{\boldsymbol{v}}_i^* = (\boldsymbol{M}_i^*)^{-1}(\boldsymbol{T}_i^* + \boldsymbol{\Delta}_i^* - \boldsymbol{C}_i^*(\boldsymbol{v}_i)\boldsymbol{v}_i^* - \boldsymbol{D}_i^*(\boldsymbol{v}_i)\boldsymbol{v}_i^* - \boldsymbol{g}_i^*(\boldsymbol{\theta}_i)) \tag{5}$$

where

$$\boldsymbol{J}_{1,i}^*(\boldsymbol{\eta}_i) = \begin{bmatrix} \cos(\theta_i)\cos(\psi_i) & -\ell_i\sin(\theta_i)\cos(\psi_i) & -\ell_i\sin(\psi_i) \\ \cos(\theta_i)\sin(\psi_i) & -\ell_i\sin(\theta_i)\sin(\psi_i) & \ell_i\cos(\psi_i) \\ -\sin(\theta_i) & -\ell_i\cos(\theta_i) & 0 \end{bmatrix}$$

$$\boldsymbol{J}_{2,i}^*(\boldsymbol{\eta_i}, \boldsymbol{v_i}) = \begin{bmatrix} -v_i\sin(\psi_i) + w_i\sin(\theta_i)\cos(\psi_i) \\ v_i\cos(\psi_i) + w_i\sin(\theta_i)\sin(\psi_i) \\ w_i\cos(\psi_i) \end{bmatrix}$$

$$\boldsymbol{C}_i^*(\boldsymbol{v_i}) = \begin{bmatrix} 0 & m_{33,i}w_i & -m_{22,i}v_i \\ (m_{11,i} - m_{33,i})w_i & 0 & 0 \\ (m_{22,i} - m_{11,i})v_i & 0 & 0 \end{bmatrix}$$

$\boldsymbol{v}_i^* = [vu_i, q_i, r_i]^T$, $\boldsymbol{M}_i^* = diag(m_{11,i}, m_{44,i}, m_{55,i})$, $\boldsymbol{D}_i^* = -diag(X_{u,i}, M_{q,i}, N_{r,i})$, $\boldsymbol{g}_i^*(\boldsymbol{\theta}_i^*) = [0, \rho_i g_i \nabla_i GM_{L,i}\sin(\theta_i), 0]^T$, $\boldsymbol{T}_i^* = [T_{1,i}, T_{4,i}, T_{5,i}]^T$, $\boldsymbol{\Delta}_i^* = [\Delta_{1,i}, \Delta_{4,i}, \Delta_{5,i}]^T$.

2.3. Actuator Dynamics

In practical applications, due to actuator limitations, the actual output force cannot achieve the control command input force. The specific actuator model is shown in Equation (6).

$$T_{x,i} = sat(u_{x,i}) = \begin{cases} u_{x,i}, & if\ u_{x,i\min} \le u_i \le u_{x,i\max} \\ u_{x,i\min}, & if\ u_i < u_{x,i\min} \\ u_{x,i\max}, & if\ u_i > u_{x,i\max} \end{cases} \tag{6}$$

where $u_{x,i}$ is the command control input; $\boldsymbol{u_i} = [u_{1,i}, u_{5,i}, u_{6,i}]^T$ is the command control input vector; and $u_{x,i\min}$ and $u_{x,i\max}$ denote the saturation constraints.

The existence of such hard saturation limits introduces nonlinearity in the input, adding complexity to the subsequent controller design. Building on the ideas from reference [29], a dead-zone operator-based model (DOBM) is employed to reconfigure the saturation phenomenon, which ensures the continuity of the closed-loop control system.

$$h(u_{x,i}) = \alpha_{x,i}u_{x,i} - \int_0^{E_{x,i}} l_{x,i}(\varepsilon_{x,i})dz_{\varepsilon,x,i}(u_{x,i})d\varepsilon_{x,i} \tag{7}$$

where $l_{x,i}(\varepsilon_i)$ is a density function that vanishes at a finite horizon $E_{x,i}$ and satisfies $l_{x,i}(\varepsilon_i) \geq 0 \; \forall \varepsilon_i > 0$; $\alpha_{x,i}$ is a constant parameter, calculated from $\alpha_{x,i} = \int_0^{E_{x,i}} l_{x,i}(\varepsilon_{x,i}) d\varepsilon_{x,i}$; and $dz_{\varepsilon,x,i}(u_{x,i})$ denotes a dead-zone operator, which is defined as (8). To simplify notation, we denote $\overline{\Omega}_{x,i} = \int_0^{E_{x,i}} l_{x,i}(\varepsilon_{x,i}) dz_{\varepsilon,x,i}(u_{x,i}) d\varepsilon_{x,i}$.

$$dz_{\varepsilon,x,i}(u_{x,i}) = \max(u_{x,i} - \varepsilon_{x,i}, \min(0, u_{x,i} + \varepsilon_{x,i})) \tag{8}$$

The dynamic model of the actuator can be formulated by integrating Equations (6)–(8) as follows:

$$\boldsymbol{T}_i^* = \boldsymbol{\alpha}_i \boldsymbol{u}_i - \boldsymbol{\Omega}_i + \boldsymbol{o}_i \tag{9}$$

where $\boldsymbol{\alpha}_i = diag[\alpha_{1,i}, \alpha_{4,i}, \alpha_{5,i}]$, $\boldsymbol{\Omega}_i = [\overline{\Omega}_{1,i}, \overline{\Omega}_{4,i}, \overline{\Omega}_{5,i}]^T$ and $\boldsymbol{o}_i = \boldsymbol{sat}(\boldsymbol{u}_i) - \boldsymbol{h}(\boldsymbol{u}_i)$. Subsequently, the relevant lemma, *Lemma* 1, is presented. Specifically, Lemma 1 elucidates that $l_{x,i}(\varepsilon_{x,i})$ and $E_{x,i}$ can be appropriately chosen to approximate the limits of the saturation constraint, provided that $u_{x,i\min}$ and $u_{x,i\max}$ are known. Given that the dead-zone-based model $\boldsymbol{h}(\boldsymbol{u}_i)$ is continuous, it further implies that $\boldsymbol{h}(\boldsymbol{u}_i)$ and $\boldsymbol{o}_i$ are bounded.

Lemma 1 ([30]). *For any given input saturation limit $u_{x,i,\max}$, there always exists an appropriate density function $l_{x,i}(\varepsilon_{x,i})$ ensuring $u_{x,i,\max} = \lim_{u_{x,i} \to \infty} h(u_{x,i}) = \int_0^{E_{x,i}} \varepsilon_{x,i} l_{x,i}(\varepsilon_{x,i}) d\varepsilon_{x,i}$. The detailed approximation of saturation function effects are illustrated in Figure 2.*

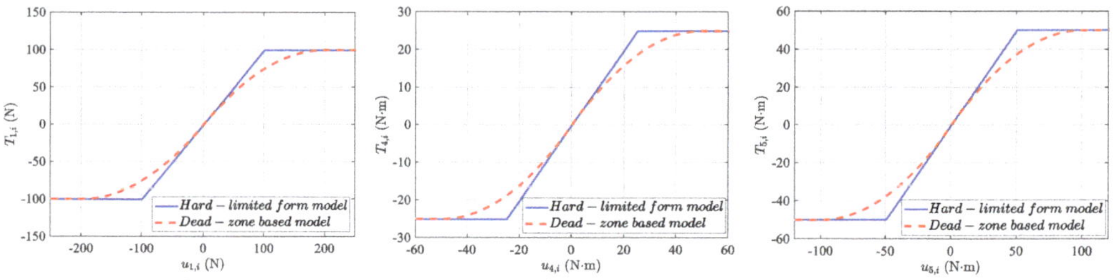

Figure 2. Saturation model; here, the density function of the Dead-zone based model is set to $l_{x,i}(\varepsilon_{x,i}) = \max(\mu_{x,i} sign(E_{x,i} - \varepsilon_{x,i}), 0)$. (**a**) Consider the scenario where $u_{1,i\max} = -u_{1,i\min} = 100$ N with $\mu_{1,i} = 0.005$, $E_{1,i} = 200$; (**b**) Consider the scenario where $u_{4,i\max} = -u_{4,i\min} = 50$ N·m with $\mu_{4,i} = 0.01$, $E_{4,i} = 100$. (**c**) Consider the scenario where $u_{5,i\max} = -u_{5,i\min} = 25$ N·m with $\mu_{5,i} = 0.02$, $E_{5,i} = 50$.

3. Approach to Address Time-Varying Underwater Acoustic Communication Delay

Due to the underwater acoustic communication delay τ_{ij}, when the leader *jth* AUV transmits its state information $\boldsymbol{\Phi}_j(t)$ to the *ith* AUV, the latter can only receive the delayed state information $\boldsymbol{\Phi}_j(t - \tau_{ij})$ at time t, as illustrated in the Figure 3, where the red arrows indicate data information flow transmission.

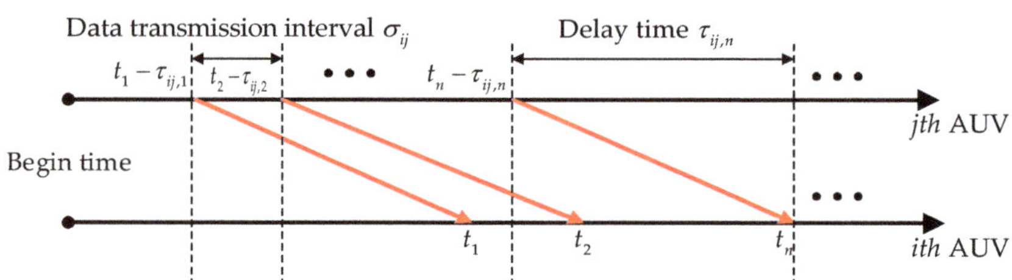

Figure 3. Data information flow transmission.

Where $\boldsymbol{\Phi}_j(t) = [\boldsymbol{\eta}_j(t); \boldsymbol{v}_j(t)] = [\Phi_{1,j}(t), \Phi_{2,j}(t), \cdots \Phi_{10,j}(t)]^T$ denotes the state information of j at the moment of data information transmission.

To address the formation-control accuracy issue caused by this delay, an autoregressive model in the form of Equation (10) is introduced to model the AUV state information, inspired by time series theory [31]. The current state information of the jth AUV can be inferred through time series analysis and multi-step prediction.

$$\Phi_{\rho,j}(t-\tau_{ij}) = \boldsymbol{\Phi}_{\rho,j}^T \boldsymbol{\beta}_{\rho,ij} + \delta_{\rho,ij} \tag{10}$$

where $\boldsymbol{\Phi}_{\rho,j} = [\Phi_{\rho,j}(t-\tau_{ij}(1+1/n_{ij})), \Phi_{\rho,j}(t-\tau_{ij}(1+2/n_{ij})), \cdots, \Phi_{\rho,j}(t-\tau_{ij}(1+p_{ij}/n_{ij}))]^T$ is the delayed state information of the jth AUV received by the ith AUV; $\sigma_{ij} = \tau_{ij}/n_{ij}$ is the time step of information transmission; p_{ij} is the order of the autoregression model, which can be determined using the Akaike Information Criterion (AIC) [32]; $\boldsymbol{\beta}_{\rho,ij} = \left[\beta_{\rho,ij,1}, \beta_{\rho,ij,2}, \cdots, \beta_{\rho,ij,p_{ij}}\right]^T$ is the model coefficient; and $\delta_{\rho,ij}$ is the bounded noise.

By using the least-squares method to estimate the coefficient $\boldsymbol{\beta}_{\rho,ij}$, the following equation can be obtained:

$$\hat{\boldsymbol{\beta}}_{\rho,ij} = \underset{\boldsymbol{\beta}_{\rho,ij}}{\operatorname{argmin}} \left\| \Phi_{\rho,j}(t-\tau_{ij}) - \boldsymbol{\Phi}_{\rho,j}^T \boldsymbol{\beta}_{\rho,ij} \right\|^2 \tag{11}$$

Subsequently, the update law for $\boldsymbol{\beta}_{\rho,ij}$ is formulated as

$$\dot{\hat{\boldsymbol{\beta}}}_{\rho,ij}(t) = \boldsymbol{Q}_{\rho,ij}(t) \tag{12}$$

$$\dot{\boldsymbol{Q}}_{\rho,ij}(t) = -c_{\rho,ij,1}\boldsymbol{Q}_{\rho,ij}(t) + c_{\rho,ij,2}\boldsymbol{\Phi}_{\rho,j}^T(\Phi_{\rho,j}(t-\tau_{ij}) - \boldsymbol{\Phi}_{\rho,j}\hat{\boldsymbol{\beta}}_{\rho,ij}(t)) \tag{13}$$

where $\boldsymbol{Q}_{\rho,ij}(t)$ is the transition matrix; $c_{\rho,ij,1}$ and $c_{\rho,ij,2}$ are the positive tuning constants.

Define

$$\tilde{\boldsymbol{\beta}}_{\rho,ij} = \boldsymbol{\beta}_{\rho,ij} - \hat{\boldsymbol{\beta}}_{\rho,ij} \tag{14}$$

Differentiating (14) with respect to Equations (12) and (13) results in

$$\dot{\tilde{\boldsymbol{\beta}}}_{\rho,ij} = \dot{\boldsymbol{\beta}}_{\rho,ij} - \int_0^t c_{\rho,ij,1}\boldsymbol{\Phi}_{\rho,j}^T\boldsymbol{\Phi}_{\rho,j}\tilde{\boldsymbol{\beta}}_{\rho,ij}(T)e^{-c_{\rho,ij,2}(t-T)}dT - \int_0^t c_{\rho,ij,1}\boldsymbol{\Phi}_{\rho,j}^T\delta_{\rho,ij}e^{-c_{\rho,ij,2}(t-T)}dT \tag{15}$$

The Lyapunov function is formulated as

$$V\left(\tilde{\boldsymbol{\beta}}_{\rho,ij}\right) = \frac{1}{2}\tilde{\boldsymbol{\beta}}_{\rho,ij}^T\tilde{\boldsymbol{\beta}}_{\rho,ij} \tag{16}$$

By taking the derivative of Equation (16) with respect to Equation (15), and choosing $0 < c_{\rho,ij,1} < 1$, $c_{\rho,ij,2} > 0$, we obtain

$$\dot{V}\left(\tilde{\boldsymbol{\beta}}_{\rho,ij}\right) \leq -E_{i,1}V\left(\tilde{\boldsymbol{\beta}}_{\rho,ij}\right) + E_{i,2} \tag{17}$$

with suitable positive constants $E_{i,1}$ and $E_{i,2}$, which demonstrates that the coefficient estimation error can converge to a vicinity near zero.

Subsequently, we can design an RSIOP in the following manner:

$$\hat{\Phi}_{\rho,j}\left(t-\tau_{ij}\left(1-\frac{\vartheta_{ij}}{n_{ij}}\right)\right)=\begin{cases}\sum\limits_{m=1}^{p_{ij}}\left(\hat{\beta}_{\rho,ij,m}\Phi_{\rho,j}\left(t-\tau_{ij}\left(1+\frac{m-1}{n_{ij}}\right)\right)\right) \\ \qquad \vartheta_{ij}=1 \\ \sum\limits_{m=1}^{\vartheta_{ij}-1}\left(\hat{\beta}_{\rho,ij,m}\hat{\Phi}_{\rho,j}\left(t-\tau_{ij}\left(1+\frac{m-\vartheta_{ij}}{n_{ij}}\right)\right)\right. \\ \left.+\sum\limits_{m=\vartheta_{ij}}^{p_{ij}}\hat{\beta}_{\rho,ij}\Phi_{\rho,j}\left(t-\tau_{ij}\left(1+\frac{m-\vartheta_{ij}}{n_{ij}}\right)\right)\right) & \text{when } \vartheta_{ij}\leq p_{ij} \\ \sum\limits_{m=1}^{p_{ij}}\left(\hat{\beta}_{\rho,ij,m}\hat{\Phi}_{\rho,j}\left(t-\tau_{ij}\left(1+\frac{m-\vartheta_{ij}}{n_{ij}}\right)\right)\right) & \text{when } \vartheta_{ij}>p_{ij} \\ \qquad \vartheta_{ij}=2,3,\cdots,n_{ij}\end{cases} \tag{18}$$

where $\hat{\Phi}_{\rho,j}$ denotes the predicted value of $\Phi_{\rho,j}$, ϑ_{ij} denotes the index value during the iteration process.

By performing n_{ij} iterations, the true value of the current state information of the *jth* AUV $\hat{\Phi}_{\rho,j}(t)$ can be calculated, effectively resolving the communication delay issue.

4. Formation Control Protocol Design

4.1. Error Transformation with Prescribed Performance

This study is primarily dedicated to designing an adaptive control protocol to enable the formation to achieve a time-varying formation configuration with prescribed performance under conditions of communication delays, external disturbances, and input saturation. Additionally, the formation is guided to follow the desired trajectory generated by the virtual leader. The block diagram of the proposed prescribed performance formation control protocol for the *ith* AUV is shown in Figure 4.

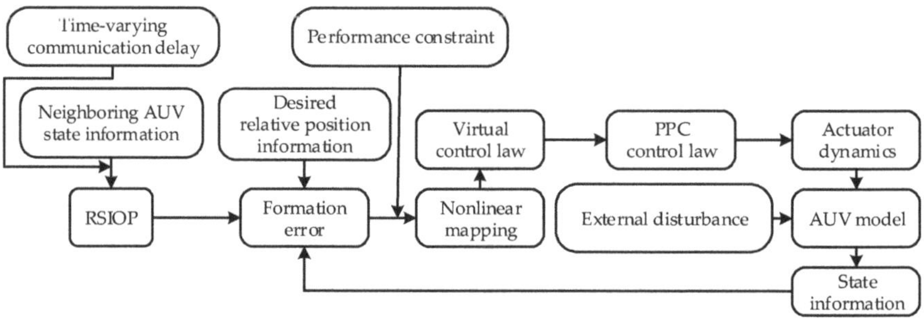

Figure 4. The block diagram of the proposed protocol for formation control.

First, the formation error for the *ith* AUV is characterized as

$$\boldsymbol{\Gamma}_{e,i}=\sum_{j\in\boldsymbol{B}_i}\left[a_{ij}\left(\boldsymbol{\eta}_i^*-\hat{\boldsymbol{\eta}}_j^*-\boldsymbol{d}_{ij}\right)\right] \tag{19}$$

where $\boldsymbol{\Gamma}_{e,i}=[x_{e,i},y_{e,i},z_{e,i}]^T$; $\boldsymbol{d}_{ij}=\left[d_{x,ij},d_{y,ij},d_{z,ij}\right]^T$ represents the distances between the *ith* AUV and its neighbors in the *x*, *y*, and *z* directions, respectively, defining the desired formation configuration to be preserved. In order to meet the performance requirements, the formation error must adhere to the following performance function constraints:

$$-\underline{\Gamma}_{e,i}(t)\leq\Gamma_{e,i}(t)\leq\overline{\Gamma}_{e,i}(t), t\geq 0 \tag{20}$$

$$\begin{cases}\overline{\Gamma}_{e,i}(t)=\left(\overline{\Gamma}_{e,0,i}-\overline{\Gamma}_{e,\infty,i}\right)\exp(-\sigma_{\Gamma,i}t)+\overline{\Gamma}_{e,\infty,i} \\ \underline{\Gamma}_{e,i}(t)=\left(\underline{\Gamma}_{e,0,i}-\underline{\Gamma}_{e,\infty,i}\right)\exp(-\sigma_{\Gamma,i}t)+\underline{\Gamma}_{e,\infty,i}\end{cases} \tag{21}$$

where $\overline{\Gamma}_{e,0,i}$, $\underline{\Gamma}_{e,0,i}$, $\overline{\Gamma}_{e,\infty,i}$, $\underline{\Gamma}_{e,\infty,i}$, and $\sigma_{\Gamma,i}$ are positive constants, with $\overline{\Gamma}_{e,\infty,i} \leq \overline{\Gamma}_{e,0,i}$ and $\underline{\Gamma}_{e,\infty,i} \leq \underline{\Gamma}_{e,0,i}$. As shown in Equations (20) and (21), as t approaches infinity, the formation error will stabilize within the predefined error tolerance $\left[-\underline{\Gamma}_{e,\infty,i}, \overline{\Gamma}_{e,\infty,i}\right]$.

To guarantee that the errors $\Gamma_{e,i}, \Gamma = x, y, z$ adhere to the prescribed performance defined in Equation (21), a mapping function was designed to convert the original constrained tracking errors into an unconstrained form.

$$W_{1,\Gamma_{e,i}} = f\left(\overline{\Gamma}_{e,i}, \underline{\Gamma}_{e,i}, \Gamma_{e,i}\right) = \frac{\overline{\Gamma}_{e,i}\underline{\Gamma}_{e,i}\Gamma_{e,i}}{\left(\overline{\Gamma}_{e,i} - \Gamma_{e,i}\right)\left(\underline{\Gamma}_{e,i} + \Gamma_{e,i}\right)} \tag{22}$$

We can observe that $W_{1,\Gamma_{e,i}}$ will tend toward infinity when the formation error $\Gamma_{e,i}$ reaches the prescribed bounds. Therefore, if the transformed state $W_{1,\Gamma_{e,i}}$ stays bounded, the formation tracking errors will adhere to the prescribed performance, as long as the initial states of the control system satisfy the performance constraint specified in Equation (20).

Subsequently, differentiating Equation (22) with respect to time results in

$$\dot{W}_{1,\Gamma_{e,i}} = P_{\Gamma_{e,i}}\dot{\Gamma}_{e,i} + Q_{\Gamma_{e,i}} \tag{23}$$

where $P_{\Gamma_{e,i}} = \frac{\left(\overline{\Gamma}_{e,i}^2\underline{\Gamma}_{e,i}^2 + \overline{\Gamma}_{e,i}\underline{\Gamma}_{e,i}\Gamma_{e,i}^2\right)}{\left[\left(\overline{\Gamma}_{e,i} - \Gamma_{e,i}\right)\left(\underline{\Gamma}_{e,i} + \Gamma_{e,i}\right)\right]^2}$, $Q_{\Gamma_{e,i}} = \frac{-\left(\underline{\Gamma}_{e,i}\dot{\overline{\Gamma}}_{e,i} + \dot{\underline{\Gamma}}_{e,i}\overline{\Gamma}_{e,i}\right)\Gamma_{e,i}^3 - \left(\underline{\Gamma}_{e,i}^2\dot{\overline{\Gamma}}_{e,i} - \overline{\Gamma}_{e,i}^2\dot{\underline{\Gamma}}_{e,i}\right)\Gamma_{e,i}^2}{\left[\left(\overline{\Gamma}_{e,i} - \Gamma_{e,i}\right)\left(\underline{\Gamma}_{e,i} + \Gamma_{e,i}\right)\right]^2}$.

4.2. Formation Control Law Design

Step 1. First, define $\boldsymbol{W}_{1,i} = \left[W_{1,x_{e,i}}, W_{1,y_{e,i}}, W_{1,z_{e,i}}\right]^T$, $\boldsymbol{P}_i = diag\left[P_{x_{e,i}}, P_{y_{e,i}}, P_{z_{e,i}}\right]^T$, and $\boldsymbol{Q}_i = \left[Q_{x_{e,i}}, Q_{y_{e,i}}, Q_{z_{e,i}}\right]^T$. Subsequently, by combining Equations (4), (19), and (23), the following relationship is obtained:

$$\dot{\boldsymbol{W}}_{1,i} = \boldsymbol{P}_i \sum_{j\in\boldsymbol{B}_i}\left[a_{ij}\left(\boldsymbol{J}_{1,i}^*(\boldsymbol{\eta}_i)\boldsymbol{v}_i^* + \boldsymbol{J}_{2,i}^*(\boldsymbol{\eta}_i, \boldsymbol{v}_i) - \boldsymbol{J}_{1,j}^*\left(\hat{\boldsymbol{\eta}}_j\right)\hat{\boldsymbol{v}}_j^* - \boldsymbol{J}_{2,j}^*\left(\hat{\boldsymbol{\eta}}_j, \hat{\boldsymbol{v}}_j\right) - \dot{\boldsymbol{d}}_{ij}\right)\right] - \boldsymbol{Q}_i \tag{24}$$

Following the backstepping procedure, the virtual control commands ς_i for $\boldsymbol{v}_i^*$ are designed as

$$\begin{aligned}\varsigma_i = \boldsymbol{J}_{1,i}^{*-1}(\boldsymbol{\eta}_i)\Big\{&\boldsymbol{P}_i^{-1}(-\boldsymbol{K}_{1,i}\boldsymbol{W}_{1,i} + \boldsymbol{Q}_i) + \\ &\sum_{j\in\boldsymbol{B}_i}\left[a_{ij}\left(-\boldsymbol{J}_{2,i}^*(\boldsymbol{\eta}_i, \boldsymbol{v}_i) + \boldsymbol{J}_{1,j}^*\left(\hat{\boldsymbol{\eta}}_j\right)\hat{\boldsymbol{v}}_j^* + \boldsymbol{J}_{2,j}^*\left(\hat{\boldsymbol{\eta}}_j, \hat{\boldsymbol{v}}_j\right) + \dot{\boldsymbol{d}}_{ij}\right)\right]\Big\}/b_i\end{aligned} \tag{25}$$

where $\boldsymbol{K}_{1,i} \in \mathbb{R}^{3\times3}$ is a positive-definite configuration matrix.

Step 2. To address the negative effect of the 'explosion of complexity' resulting from differentiating ς_i, a second-order robust exact differentiator with finite-time convergence capability is introduced [33]. Lemma 2 is given.

$$\begin{cases}\dot{\varsigma}_{0,i} = -s_{0,i}|\varsigma_{0,i} - \varsigma_i|^{\frac{1}{2}}sign(\varsigma_{0,i} - \varsigma_i) + \varsigma_{1,i} \\ \dot{\varsigma}_{1,i} = -s_{1,i}sign(\varsigma_{1,i} - \dot{\varsigma}_{0,i})\end{cases} \tag{26}$$

where $s_{0,i}$ and $s_{1,i}$ are positive design parameters. $\varsigma_{0,i} \in \mathbb{R}^{3\times3}$ and $\varsigma_{1,i} \in \mathbb{R}^{3\times3}$, respectively, denote the produce outputs and the outputs of this matrix; moreover, $\varsigma_{1,i}$ can be used as a substitute for $\dot{\varsigma}_i$.

Lemma 2 ([33]). *As long as the resulting system is homogeneous and appropriate parameters $s_{0,i}$ and $s_{1,i}$ are selected, this type of robust exact differentiator has finite-time convergence.*

Step 3. To achieve additional stability for the virtual control commands, the actual control inputs are crafted as follows:

$$W_{2,i} = v_i^* - \varsigma_{0,i} \tag{27}$$

$$u_i = -\alpha_i^{-1}\left(K_{2,i}W_{2,i} - b_i J_{1,i}^{*T}(\eta_i)P_i W_{1,i} + C_i^*(v_i)v_i^* + D_i^*(v_i)v_i^* + g_i^*(\theta_i) + M_i^*\dot{\varsigma}_{0,i}\right) \tag{28}$$

where $W_{2,i}$ denotes the velocity error. $K_{2,i} \in \mathbb{R}^{3\times 3}$ is a positive-definite designed matrix.

4.3. Stability Analysis

Theorem 1. *For the formation tracking control problem of the AUV model expressed by Equations (1), (2), and (9) obeying Assumption 1 and Assumption 2. If the real-time state information online predictor is designed as in Equation (18), with the virtual control input (25) and control input (28) applied, then under conditions of communication delay, input saturation, and unknown external disturbances, all signals within the closed-loop system will be uniformly ultimately bounded. Moreover, if the initial formation error satisfies condition (20), the prescribed performance can be achieved. This indicates that the AUV formation can follow the target trajectory while maintaining the specified time-varying formation.*

Proof of Theorem 1. The Lyapunov function is defined as follows to demonstrate the asymptotic stability of the closed-loop signals.

$$V = \frac{1}{2}\sum_{i=1}^{N}\left(V_{1,i}^2 + V_{2,i}^2\right) \tag{29}$$

$$V = \frac{1}{2}\sum_{i=1}^{N}\left(W_{1,i}^T W_{1,i} + W_{2,i}^T M_i^* W_{2,i}\right) \tag{30}$$

Differentiating V with respect to time gives

$$\dot{V} = \sum_{i=1}^{N}\left(W_{1,i}^T \dot{W}_{1,i} + W_{2,i}^T M_i^* \dot{W}_{2,i}\right) \tag{31}$$

Combining Equations (5), (9), (24), and (25), we further obtain

$$\begin{aligned}\dot{V} = \sum_{i=1}^{N}\Big[&W_{1,i}^T\left(-K_{1,i}W_{1,i} + b_i P_i J_{1,i}^*(\eta_i)W_{2,i}\right) + \\ &W_{2,i}^T\left(\alpha_i u_i - \Omega_i + o_i + \Delta_i^* - C_i^*(v_i)v_i^* - D_i^*(v_i)v_i^* - g_i^*(\theta_i) - M_i^*\dot{\varsigma}_{0,i}\right)\Big]\end{aligned} \tag{32}$$

Substituting the control input from Equation (28), we obtain:

$$\dot{V} = \sum_{i=1}^{N}\left(-K_{1,i}W_{1,i}^T W_{1,i} - K_{2,i}W_{2,i}^T W_{2,i} - W_{2,i}^T(-\Omega_i + o_i + \Delta_i^*)\right) \tag{33}$$

$$\dot{V} \le \sum_{i=1}^{N}\left(-K_{1,i}\|W_{1,i}\|^2 - K_{2,i}\|W_{2,i}\|^2 + \|W_{2,i}\|\|\Omega_i\| + \|W_{2,i}\|\|o_i + \Delta_i^*\|\right) \tag{34}$$

$$\dot{V} \le \sum_{i=1}^{N}\left(-K_{1,i}\|W_{1,i}\|^2 - \|W_{2,i}\|(K_{2,i}\|W_{2,i}\| - \lambda_i)\right) \tag{35}$$

$$\lambda_i = \|\Omega_i\| + \|o_i + \Delta_i^*\| \tag{36}$$

Based on Assumption 2 and the discussion below Equation (9), it can be concluded that

$$\sum_{i=1}^{N} \hbar_i = \sum_{i=1}^{N} (\|\boldsymbol{\Omega}_i\| + \|\boldsymbol{o}_i + \boldsymbol{\Delta}_i^*\|) < a_i \tag{37}$$

$$\dot{V} \le \sum_{i=1}^{N} \left(-\boldsymbol{K}_{1,i} \|\boldsymbol{W}_{1,i}\|^2 - \|\boldsymbol{W}_{2,i}\| \overline{a}_i \right) \tag{38}$$

where $\overline{a}_i = \boldsymbol{K}_{2,i} \|\boldsymbol{W}_{2,i}\| - a_i$; one further has $\dot{V} < 0 (or\ \overline{a}_i > 0)$ as long as $\|\boldsymbol{W}_{2,i}\| > \boldsymbol{\Gamma}_i$

$$\boldsymbol{\Gamma}_i = \frac{\overline{a}_i}{\boldsymbol{K}_{2,i}} \tag{39}$$

After calculations, the final result is obtained as

$$V \le \frac{1}{2} \sum_{i=1}^{N} \left(\boldsymbol{\Gamma}_{1,i}^2 + \boldsymbol{\Gamma}_{2,i}^2 \right) \tag{40}$$

$$\boldsymbol{\Gamma}_{1,i} = \max\left(\frac{b_i \|\boldsymbol{P}_i\| \left\| \boldsymbol{J}_{1,i}^*(\boldsymbol{\eta}_i) \right\| \|\boldsymbol{\Gamma}_{2,i}\|}{\boldsymbol{K}_{1,i}}, \sqrt{\frac{2V_{1,i}(0)}{\Upsilon_{1,i}}} \right), \ \boldsymbol{\Gamma}_{2,i} = \max\left(\frac{\overline{a}_i}{\boldsymbol{K}_{2,i}}, \sqrt{\frac{2V_{2,i}(0)}{\Upsilon_{2,i}}} \right) \tag{41}$$

$$\|\boldsymbol{W}_{1,i}\| \le \boldsymbol{\Gamma}_{1,i}, \ \|\boldsymbol{W}_{2,i}\| \le \boldsymbol{\Gamma}_{2,i} \tag{42}$$

where $\Upsilon_{1,i}$ and $\Upsilon_{2,i}$ are appropriate positive constants. From the above conclusions, we obtain that V is bounded as time approaches infinity, ensuring that all signals within the closed-loop system are kept bounded. More importantly, the boundedness of the mapped system states that $\boldsymbol{W}_{1,i}$ is also ensured. From the analysis following Equation (22), it can be concluded that the prescribed performance of the formation can be attained under the proposed protocol. Therefore, Theorem 1 is proven. □

Remark 1. *The proof demonstrates that by keeping the transformed errors $\boldsymbol{W}_{1,i}$ and $\boldsymbol{W}_{2,i}$ bounded, the closed-loop system can achieve the prescribed performance stability. This indicates that the proposed control protocol can achieve its objectives without reducing $\boldsymbol{W}_{1,i}$ and $\boldsymbol{W}_{2,i}$ to arbitrarily small regions. Following this logic, as described in reference [34], the uncertainty disturbance $\hbar_i$ can be compensated even without any auxiliary control signals, as it only affects the magnitude of $\boldsymbol{W}_{1,i}$ and $\boldsymbol{W}_{2,i}$ but does not alter the achieved stability properties. Separating the tracking performance from the uncertainty disturbances significantly enhances the robustness of the proposed control protocol.*

5. Simulation Results

In this section, the proposed control protocol's effectiveness is evaluated through a series of simulation tests. All simulations in this study were performed using MATLAB R2023a on a personal computer running Windows 10, equipped with an Intel Core i7-8750H processor (2.20 GHz) and 16 GB of RAM.

The study focuses on an AUV formation, which includes a virtual leader and four underactuated AUVs. Assume that the trajectory of the virtual leader is already loaded onto AUV1, meaning that there is no communication delay between them. The main physical parameters and partial hydrodynamic coefficients of the AUVs were measured experimentally and are tabulated in Tables 1 and 2. Additionally, the communication topology is shown in Figure 5, with the virtual leader designated as number 0.

Table 1. Main physical parameters of the AUVs.

Parameters	Value	Parameters	Value
m_i	183 Kg	L_i	2.55 M
$GM_{L,i}$	0.0217 M	ρ_i	1024 Kg/m^3
Maximum thrust	100 N	Maximum Horizontal/Vertical Rudder Force	$25/50\ \text{N}\cdot\text{m}$

Table 2. Partial hydrodynamic coefficients of the AUVs.

Parameters	Value	Parameters	Value
$m_{11,i}$	188 Kg	$m_{22,i}$	263 Kg
$m_{33,i}$	263 Kg	$m_{44,i}$	271 Kgm^2
$m_{55,i}$	271 Kgm^2	$X_{u,i}$	$16.65\lvert vu_i\rvert$ Kg/s
$Y_{v,i}$	$149.68\lvert vu_i\rvert$ Kg/s	$Z_{w,i}$	$142.23\lvert vu_i\rvert$ Kg/s
$M_{q,i}$	$220.92\lvert vu_i\rvert$ Kg/s	$N_{r,i}$	$252.83\lvert vu_i\rvert$ Kg/s

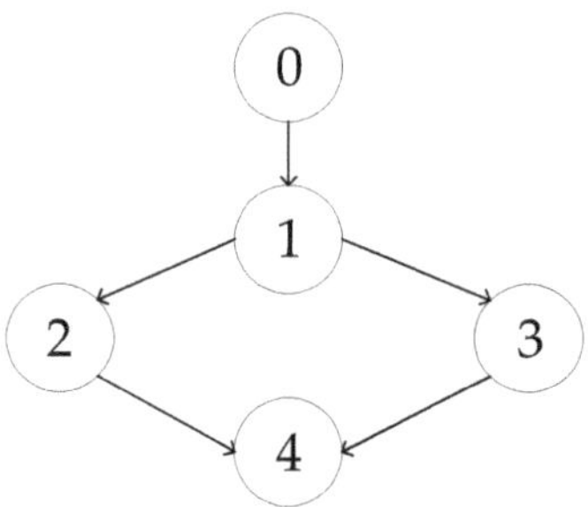

Figure 5. The communication topology graph.

The virtual leader's trajectory is defined as $[(50\cos(0.01t) - 50 + 50\sin(0.02t))$ m, $(50\sin(0.01\,t) + 50\cos(0.02\,t) - 50)$m,$(-0.1t)m]^T$. The initial position information, velocity information, and the desired relative position information between neighboring AUVs are shown in Table 3.

Table 3. Formation state information settings.

Parameters	Value
$\eta_1(0)$	$[-1.35\text{m}, 1\text{m}, -1\text{m}, 0\text{rad}, 0\text{rad}]^T$
$\eta_2(0)$	$[-8.75\text{m}, -7\text{m}, -2\text{m}, 0\text{rad}, 0\text{rad}]^T$
$\eta_3(0)$	$[6.75\text{m}, -7\text{m}, -3\text{m}, 0\text{rad}, 0\text{rad}]^T$
$\eta_4(0)$	$[-2.35\text{m}, -15.5\,\text{m}, -2\,\text{m}, 0\text{rad}, 0\text{rad}]^T$
$v_i(0)$	$[0\text{m/s}, 0\text{m/s}, 0\text{m/s}, 0\text{rad/s}, 0\text{rad/s}]^T$
$\boldsymbol{d}_{10}$	$[0\text{m}, 0\text{m}, 0\text{m}]^T$
$\boldsymbol{d}_{21}$	$[-10\text{m}, \left(-5\cos\left(\frac{\pi}{300}\text{t}\right) - 5\right)\text{m}, \left(-0.5\cos\left(\frac{\pi}{300}\text{t}\right) - 1.5\right)\text{m}]^T$
$\boldsymbol{d}_{31}$	$[10\text{m}, \left(-5\cos\left(\frac{\pi}{300}\text{t}\right) - 5\right)\text{m}, \left(-1.5\cos\left(\frac{\pi}{300}\text{t}\right) + 0.5\right)\text{m}]^T$
$\boldsymbol{d}_{42}$	$[8\text{m}, \left(-8\cos\left(\frac{\pi}{1200}\text{t}\right)\right)\text{m}, \left(-1.5\cos\left(\frac{\pi}{300}\text{t}\right) + 1\right)\text{m}]^T$
$\boldsymbol{d}_{43}$	$[-8\text{m}, \left(-8\cos\left(\frac{\pi}{1200}\text{t}\right)\right)\text{m}, \left(-1.5\cos\left(\frac{\pi}{300}\text{t}\right) + 2\right)\text{m}]^T$

The communication delays between AUVs are set as $\tau_{21} = \tau_{31} = \left(2 - 0.25\cos\left(\frac{\pi}{600}t\right)\right)$s, $\tau_{42} = \tau_{43} = \left(1.5 + 0.25\cos\left(\frac{\pi}{600}t\right)\right)$s. The external ocean current disturbances are defined as $\boldsymbol{\Delta}_i = \left[\left(1.3 + 2\sin\left(\frac{\pi}{50}\text{t}\right)\right)\text{N}, \left(-0.9 + 2\sin\left(\frac{\pi}{50}\text{t}\right)\right)\text{N}, \left(0.5\cos\left(\frac{\pi}{10}\text{t}\right)\right)\text{N}, \left(\cos\left(\frac{\pi}{10}\text{t}\right)\right)\text{N}\cdot\text{m}, \left(-\sin\left(\frac{\pi}{50}\text{t} + \frac{\pi}{3}\right) - 4\sin\left(\frac{\pi}{100}\text{t}\right)\right)\text{N}\cdot\text{m}\right]^T$. The values of the actuator dynamics model are set as described below Figure 2. Additionally, Table 4 presents the main control parameters used in the simulation experiment. All parameters were selected empirically within the range that ensures the stability of all closed-loop system signals, as determined by the stability analysis. Following this, a systematic tuning process was implemented, during which the control gains were iteratively adjusted to find the optimal balance between robustness and responsiveness, ultimately leading to a well-founded and effective determination of the parameters.

Table 4. Key control parameters of the AUVs.

Parameters	Value
$\left[\overline{\boldsymbol{\Gamma}}_{e,0,i}, \underline{\boldsymbol{\Gamma}}_{e,0,i}\right]^T$	$[5.1, 1.1]^T$, if $\boldsymbol{\Gamma}_{e,i}(0) > 0$ $[1.1, 5.1]^T$, else
$\left[\dot{\overline{\boldsymbol{\Gamma}}}_{e,\infty,i}, \underline{\boldsymbol{\Gamma}}_{e,\infty,i}\right]^T$	$[0.1, 0.1]^T$
$\left[c_{\rho,ij,1}, c_{\rho,ij,2}\right]^T$	$[0.8, 0.1]^T$
$\left[s_{0,i}, s_{1,i}\right]^T$	$[1, 0.1]^T$
$\left[\boldsymbol{K}_{1,i}, \boldsymbol{K}_{2,i}\right]^T$	$[0.2, 0.1, 0.1]^T$, $[200, 200, 400]^T$

The simulation outcomes of the proposed control protocol are presented in Figures 6–10. The trajectories of the AUV formation are shown in Figure 6, demonstrating that the AUVs successfully follow the desired trajectory. The formation configuration transitions from a quadrilateral to a triangle, achieving the predetermined formation transformation. The position state information prediction errors $\tilde{\boldsymbol{\eta}}_j(t)$ and the velocity state information prediction errors $\tilde{\boldsymbol{v}}_j(t)$ of the AUV formation are illustrated in Figures 7 and 8, respectively. In the legend, $\widetilde{\eta}_{ij}$ and $\widetilde{v}_{ij}$ represent the estimates of the state information of the ith AUV for the jth AUV. All state estimation errors converge within approximately 10 s. These results indicate that the RSIOP can accurately and quickly estimate the current state information $\boldsymbol{\Phi}_j(t)$ of neighboring AUVs considering communication delays, thereby ensuring the accuracy and stability of formation control. Observing the actual control inputs of the AUV formation in Figure 9, we can conclude that the saturation constraints are not violated under the dead-zone model. Additionally, all actual control input curves are smooth, with no significant oscillations, aligning with the actuator's executable capability in practical scenarios. The formation error response, shown in Figure 10, indicates that the AUV formation can achieve the prescribed performance with the designed control protocol. In other words, both the transient- and steady-state performance of the formation are ensured. The formation errors converge within 16 s and remain within the performance constraint bounds. Moreover, the steady-state formation control error for each AUV in the x, y, and z directions is less than 0.1 m Furthermore, it can be clearly observed that external uncertainty disturbances and input saturation have minimal impact on the formation control accuracy, indicating that the proposed control protocol exhibits good robustness. Thus, the validity of Theorem 1 is verified.

To validate the effectiveness of the designed control protocol, the simulation results of the proposed method were compared against those obtained from simulation tests using the traditional backstepping control protocol. Additionally, during the simulations, the designed RSIOP was not used to compensate for delayed state information. The simulation conditions were kept entirely consistent with those in the previous simulations. The results of the simulations are as follows:

Figure 11 shows the movement trajectories of the AUV formation under the traditional backstepping control method. It is evident that, under the condition of not considering the solution to communication delays, the control accuracy of the formation cannot be guaranteed, and the AUV formation cannot form the predetermined formation structure, thus failing to complete the mission objectives. Figure 12 shows the formation error response under the traditional backstepping control method. It can be concluded that the formation error needs to be reduced to within approximately 100 s and cannot be maintained within the performance constraint bounds proposed in this paper, resulting in a significant decrease in control performance. To further quantify the control accuracy differences between the two methods used in this study, we introduce integrated absolute error (IAE), which is expressed as $\mathrm{IAE}(\Gamma_{e,i}) = \sum_{i=1}^{N} \int_0^t \|\Gamma_{e,i}\|\, dt$, with $\boldsymbol{\Gamma}_{e,i} = x_{e,i}, y_{e,i}, z_{e,i}$. The maximum instantaneous formation error change is formulated as $\max(\Gamma_{e,i}) = \max\left(\sum_{i=1}^{N} \|\Gamma_{e,i}\|\right)$. The analysis results, as shown in Table 5, indicate that the control protocol proposed in this paper has superior control accuracy.

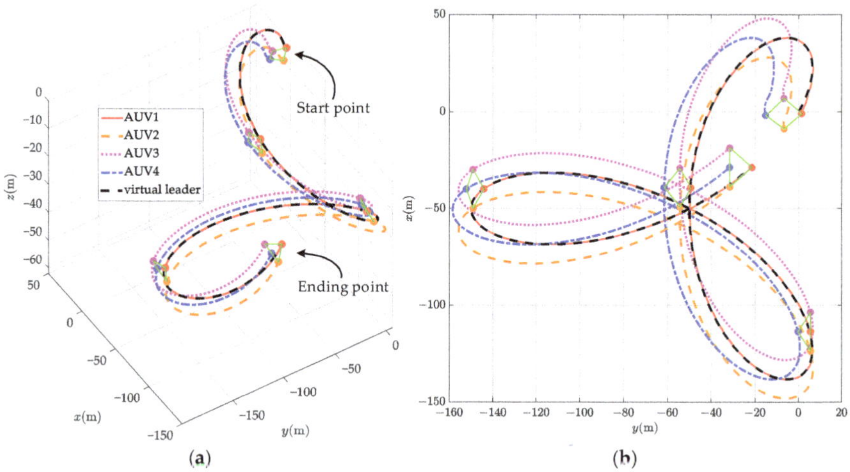

Figure 6. Motion trajectories of the AUV formation in (**a**) x–y–z plane; (**b**) x–y plane.

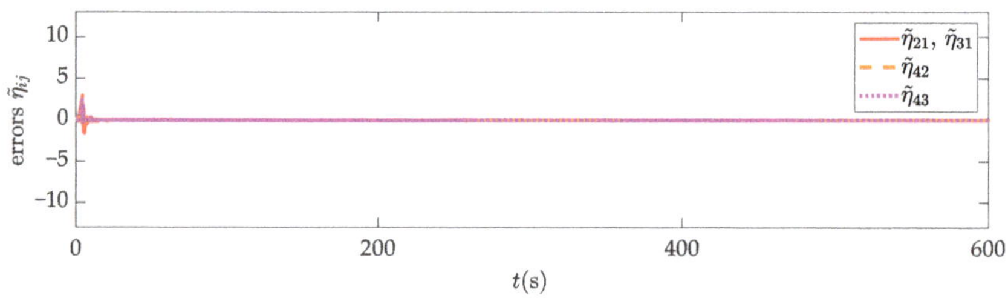

Figure 7. Position state information prediction error in AUV formation.

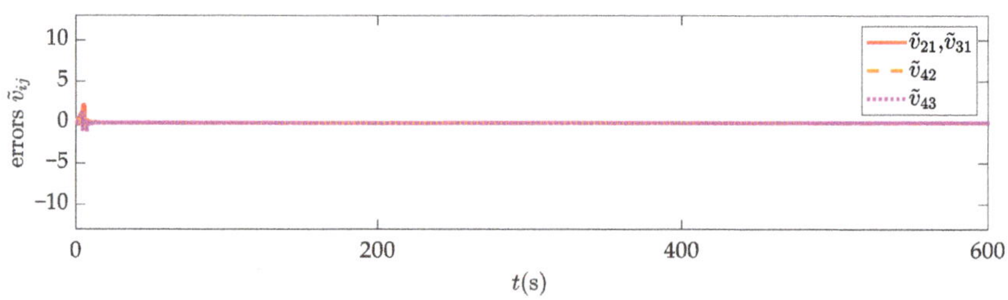

Figure 8. Velocity state information prediction error in AUV formation.

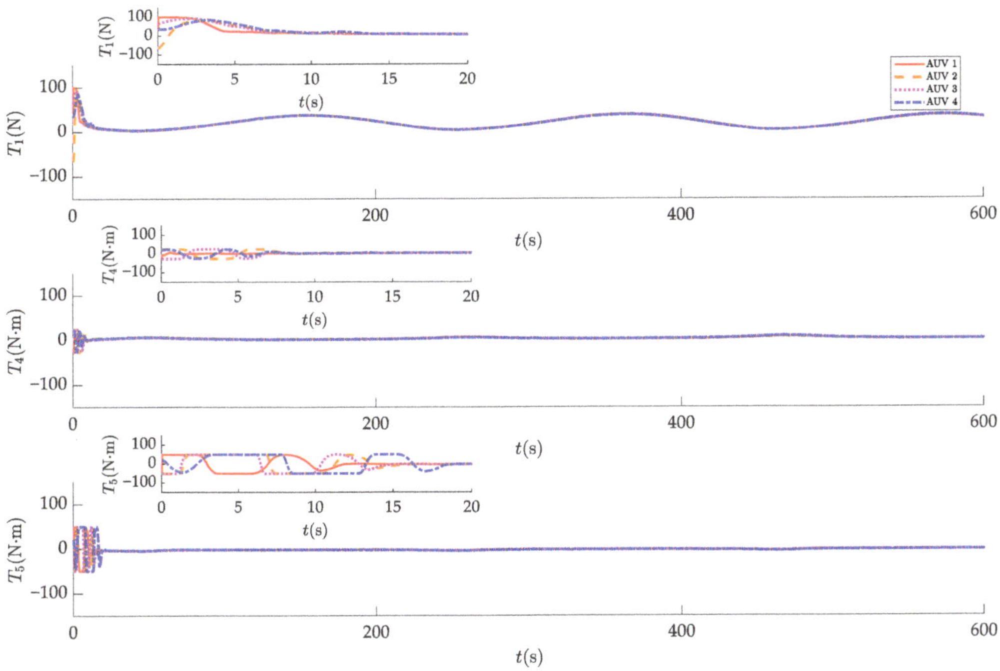

Figure 9. Actual control inputs of the AUV formation.

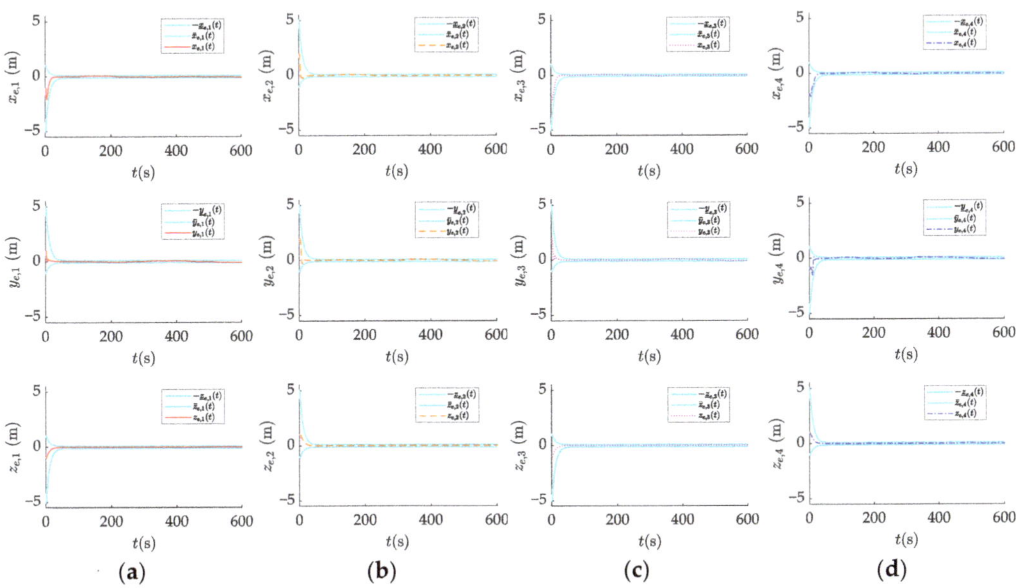

Figure 10. Formation error responses of (**a**) AUV1, (**b**) AUV2, (**c**) AUV3, and (**d**) AUV4.

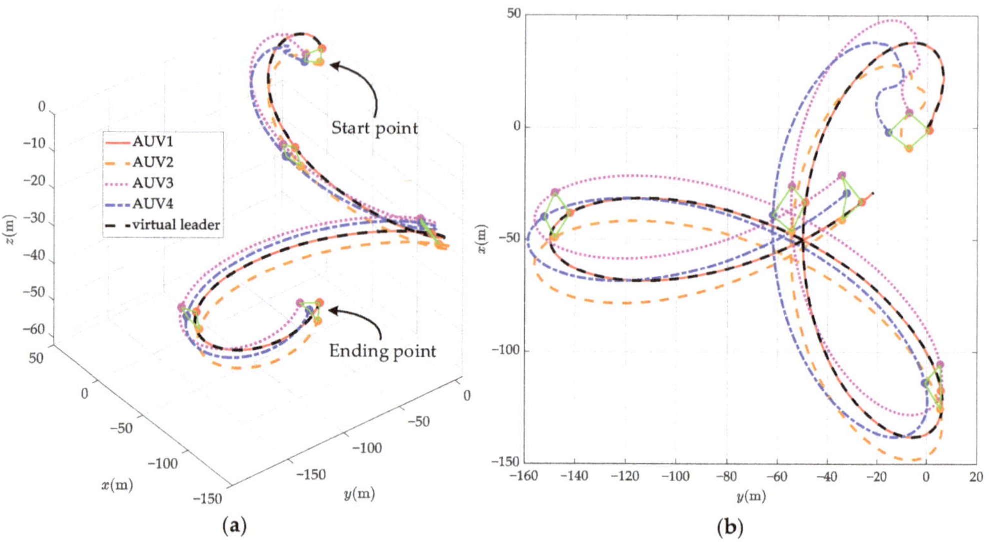

Figure 11. Motion trajectories of the AUV formation under the traditional backstepping control method in (**a**) x–y–z plane; (**b**) x–y plane.

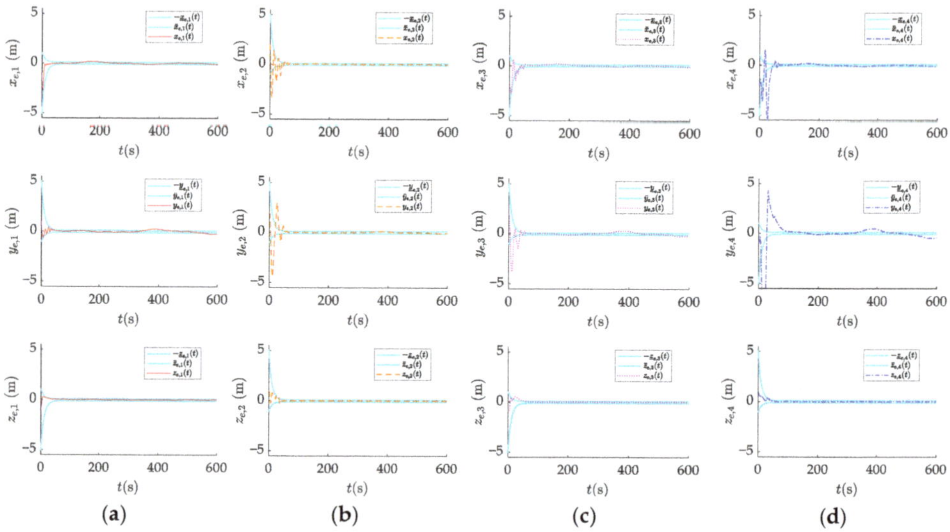

Figure 12. Formation error responses under the traditional backstepping control method of (**a**) AUV1, (**b**) AUV2, (**c**) AUV3, and (**d**) AUV4.

Table 5. The IAE values of the two control methods.

Parameters	$\mathrm{IAE}(x_{e,i})$	$\mathrm{IAE}(y_{e,i})$	$\mathrm{IAE}(z_{e,i})$	$\max(x_{e,i})$	$\max(y_{e,i})$	$\max(z_{e,i})$
The proposed protocol	1237.49	1252.75	612.64	7.43	6	4
Backstepping method	2989	5646.97	1056.65	9.66	15.61	4

Remark 2. *The traditional backstepping method was chosen for comparison in this section, rather than simply omitting the RSIOP, because without the state compensation provided by the RSIOP under conditions of communication delays, the proposed prescribed performance control protocol would not achieve convergence of the formation error. This phenomenon further underscores the necessity and effectiveness of developing the RSIOP.*

Although this paper has effectively addressed the issue of formation control under time-varying communication delays, there are still some limitations. One of the key limitations of the proposed formation control protocol with prescribed performance is that it requires the initial formation error of the AUVs to be within the performance constraint bounds. Moreover, the final value must be reasonably determined based on the AUVs' performance capabilities, as determined through experience. If these conditions are not met, the convergence of the formation control errors cannot be guaranteed. This is an area that requires further investigation, particularly in developing strategies for handling cases where the initial errors exceed the prescribed bounds. In addition, the application of the real-time state information online predictor (RSIOP) and the prescribed performance control (PPC) framework does involve certain computational complexities. In dynamic environments, the algorithm must quickly adapt to changes in communication delays and external disturbances. As the number of AUVs in the formation increases, this will inevitably increase the computational burden. Ensuring that these computations can be performed in real-time without causing delays in the control response is also a significant challenge. To address this, we have already made some trade-offs at the algorithm level. While the algorithm aims to maintain high accuracy in state prediction and control actions, some approximations have been made to reduce the computational load. Furthermore, we will explore the use of parallel computing techniques and hardware acceleration to enhance the real-time performance of the algorithm in future work, especially in more complex and dynamic scenarios.

6. Conclusions

In this brief, a distributed prescribed performance tracking control protocol is proposed for underactuated AUV formation systems under time-varying communication delays, which are constrained by external uncertainties and input saturation. The formation system is required to achieve the specified formation configuration transformation while also tracking the desired trajectory as a whole.

First, a delayed state information compensation method-based RSIOP is proposed to mitigate the adverse effects of acoustic communication delays on formation control. Second, a formation control protocol is developed by combining backstepping technology and nonlinear mapping methods, taking into account external uncertainties, input saturation, and predefined performance constraints during the design process. Furthermore, the stability of the closed-loop system is established using the Lyapunov method. Lastly, simulation results confirm the robustness and effectiveness of the proposed protocol.

In future work, we plan to consider more practical scenarios and further investigate the limitations of the proposed protocol, such as computational complexity and sensitivity to initial conditions. Additionally, we intend to validate the practicality of our algorithm through experimental testing.

Author Contributions: Conceptualization, H.Z. and Y.J.; methodology, H.Z.; software, H.Z. and R.G.; validation, H.Z. and R.G.; formal analysis, H.Z.; investigation, H.Z. and Y.J.; resources, Y.J.; data curation, H.L.; writing—original draft preparation, H.Z.; writing—review and editing, H.L. and Y.J.; visualization, H.Z. and A.L.; supervision, Y.J.; project administration, Y.J.; funding acquisition, Y.J. All authors have read and agreed to the published version of the manuscript.

Funding: This research was funded by the National Key R&D Program of China (Grant No. 2023YFC2810500), the National Key Laboratory of Autonomous Marine Vehicle Technology Fund (Grant No. 2022JCJQ-SYSJJ-LB06907), and the State Administration of Science, Technology and Industry Fund (Grant No. JCKYS2022SXJQR-08).

Institutional Review Board Statement: Not applicable.

Informed Consent Statement: Not applicable.

Data Availability Statement: Data is contained within the article.

Conflicts of Interest: The authors declare no conflicts of interest.

References

1. Xie, T.; Li, Y.; Jiang, Y.; Pang, S.; Wu, H. Turning Circle Based Trajectory Planning Method of an Underactuated AUV for the Mobile Docking Mission. *Ocean Eng.* **2021**, *236*, 109546. [CrossRef]
2. Wang, C.; Cai, W.; Lu, J.; Ding, X.; Yang, J. Design, Modeling, Control, and Experiments for Multiple AUVs Formation. *IEEE Trans. Autom. Sci. Eng.* **2022**, *19*, 2776–2787. [CrossRef]
3. Li, B.; Wen, G.; Peng, Z.; Huang, T.; Rahmani, A. Fully Distributed Consensus Tracking of Stochastic Nonlinear Multiagent Systems With Markovian Switching Topologies via Intermittent Control. *IEEE Trans. Syst. Man Cybern. Syst.* **2022**, *52*, 3200–3209. [CrossRef]
4. Qiao, L.; Zhang, W. Trajectory Tracking Control of AUVs via Adaptive Fast Nonsingular Integral Terminal Sliding Mode Control. *IEEE Trans. Ind. Inform.* **2020**, *16*, 1248–1258. [CrossRef]
5. Vu, M.T.; Le Thanh, H.N.N.; Huynh, T.-T.; Thang, Q.; Duc, T.; Hoang, Q.-D.; Le, T.-H. Station-Keeping Control of a Hovering Over-Actuated Autonomous Underwater Vehicle Under Ocean Current Effects and Model Uncertainties in Horizontal Plane. *IEEE Access* **2021**, *9*, 6855–6867. [CrossRef]
6. Zhou, Z.; Liu, J.; Yu, J. A Survey of Underwater Multi-Robot Systems. *IEEE/CAA J. Autom. Sin.* **2022**, *9*, 1–18. [CrossRef]
7. Liu, Y.; Liu, J.; He, Z.; Li, Z.; Zhang, Q.; Ding, Z. A Survey of Multi-Agent Systems on Distributed Formation Control. *Unmanned Syst.* **2023**, *12*, 913–926. [CrossRef]
8. Li, D.; Du, L. AUV Trajectory Tracking Models and Control Strategies: A Review. *J. Mar. Sci. Eng.* **2021**, *9*, 1020. [CrossRef]
9. Wang, J.; Dong, H.; Chen, F.; Vu, M.T.; Shakibjoo, A.D.; Mohammadzadeh, A. Formation Control of Non-Holonomic Mobile Robots: Predictive Data-Driven Fuzzy Compensator. *Mathematics* **2023**, *11*, 1804. [CrossRef]
10. Yan, T.; Xu, Z.; Yang, S.X. Consensus Formation Tracking for Multiple AUV Systems Using Distributed Bioinspired Sliding Mode Control. *IEEE Trans. Intell. Veh.* **2023**, *8*, 1081–1092. [CrossRef]
11. Chen, B.; Hu, J.; Zhao, Y.; Ghosh, B.K. Finite-Time Velocity-Free Rendezvous Control of Multiple AUV Systems With Intermittent Communication. *IEEE Trans. Syst. Man Cybern. Syst.* **2022**, *52*, 6618–6629. [CrossRef]
12. Hadi, B.; Khosravi, A.; Sarhadi, P. A Review of the Path Planning and Formation Control for Multiple Autonomous Underwater Vehicles. *J. Intell. Robot. Syst.* **2021**, *101*, 67. [CrossRef]
13. Yang, Y.; Xiao, Y.; Li, T. A Survey of Autonomous Underwater Vehicle Formation: Performance, Formation Control, and Communication Capability. *IEEE Commun. Surv. Tutor.* **2021**, *23*, 815–841. [CrossRef]
14. Yan, Z.; Zhang, C.; Tian, W.; Zhang, M. Formation Trajectory Tracking Control of Discrete-Time Multi-AUV in a Weak Communication Environment. *Ocean Eng.* **2022**, *245*, 110495. [CrossRef]
15. Li, J.; Zhang, H.; Chen, T.; Wang, J. AUV Formation Coordination Control Based on Transformed Topology under Time-Varying Delay and Communication Interruption. *J. Mar. Sci. Eng.* **2022**, *10*, 950. [CrossRef]
16. Li, J.; Tian, Z.; Zhang, H. Discrete-Time AUV Formation Control with Leader-Following Consensus under Time-Varying Delays. *Ocean Eng.* **2023**, *286*, 115678. [CrossRef]
17. Zhang, W. Leader-Following Consensus of Discrete-Time Multi-AUV Recovery System with Time-Varying Delay. *Ocean Eng.* **2021**, *219*, 108258. [CrossRef]
18. Zeng, Z.; Yu, H.; Guo, C.; Yan, Z. Finite-Time Coordinated Formation Control of Discrete-Time Multi-AUV with Input Saturation under Alterable Weighted Topology and Time-Varying Delay. *Ocean Eng.* **2022**, *266*, 112881. [CrossRef]
19. Yan, Z.; Yang, Z.; Pan, X.; Zhou, J.; Wu, D. Virtual Leader Based Path Tracking Control for Multi-UUV Considering Sampled-Data Delays and Packet Losses. *Ocean Eng.* **2020**, *216*, 108065. [CrossRef]
20. Li, L.; Li, Y.; Zhang, Y.; Xu, G.; Zeng, J.; Feng, X. Formation Control of Multiple Autonomous Underwater Vehicles under Communication Delay, Packet Discreteness and Dropout. *J. Mar. Sci. Eng.* **2022**, *10*, 920. [CrossRef]
21. Du, J.; Li, J.; Lewis, F.L. Distributed 3-D Time-Varying Formation Control of Underactuated AUVs With Communication Delays Based on Data-Driven State Predictor. *IEEE Trans. Ind. Inform.* **2023**, *19*, 6963–6971. [CrossRef]
22. Bechlioulis, C.P.; Rovithakis, G.A. Robust Adaptive Control of Feedback Linearizable MIMO Nonlinear Systems With Prescribed Performance. *IEEE Trans. Autom. Control* **2008**, *53*, 2090–2099. [CrossRef]
23. Liu, X.; Zhang, M.; Wang, S. Adaptive Region Tracking Control with Prescribed Transient Performance for Autonomous Underwater Vehicle with Thruster Fault. *Ocean Eng.* **2020**, *196*, 106804. [CrossRef]
24. Sun, Y.; Liu, M.; Qin, H.; Wang, H.; Ding, Z. Full Prescribed Performance Trajectory Tracking Control Strategy of Autonomous Underwater Vehicle with Disturbance Observer. *ISA Trans.* **2024**, *151*, 117–130. [CrossRef]
25. Huang, B.; Zhou, B.; Zhang, S.; Zhu, C. Adaptive Prescribed Performance Tracking Control for Underactuated Autonomous Underwater Vehicles with Input Quantization. *Ocean Eng.* **2021**, *221*, 108549. [CrossRef]

26. Wei, C.; Chen, Q.; Liu, J.; Yin, Z.; Luo, J. An Overview of Prescribed Performance Control and Its Application to Spacecraft Attitude System. *Proc. Inst. Mech. Eng. Part I J. Syst. Control Eng.* **2021**, *235*, 435–447. [CrossRef]
27. Li, Y.; He, J.; Shen, H.; Zhang, W.; Li, Y. Adaptive Practical Prescribed-Time Fault-Tolerant Control for Autonomous Underwater Vehicles Trajectory Tracking. *Ocean Eng.* **2023**, *277*, 114263. [CrossRef]
28. Shojaei, K.; Arefi, M.M. On the Neuro-adaptive Feedback Linearising Control of Underactuated Autonomous Underwater Vehicles in Three-dimensional Space. *IET Control Theory Appl.* **2015**, *9*, 1264–1273. [CrossRef]
29. Huang, B.; Zhu, C.; Xu, Y.; Zhu, G.; Su, Y. Energy Tradeoff-Oriented Quasi-Optimal Distributed Affine Formation Maneuver Control for Electric Marine Surface Vehicles. *IEEE Trans. Transp. Electrif.* **2024**, 1. [CrossRef]
30. Wang, Y.; Ji, H. Input-to-state Stability-based Adaptive Control for Spacecraft Fly-around with Input Saturation. *IET Control Theory Appl.* **2020**, *14*, 1365–1374. [CrossRef]
31. Akaike, H. A New Look at the Statistical Model Identification. *IEEE Trans. Autom. Control* **1974**, *19*, 716–723. [CrossRef]
32. Shibata, R. Selection of the Order of an Autoregressive Model by Akaike's Information Criterion. *Biometrika* **1976**, *63*, 117–126. [CrossRef]
33. Levant, A. Higher-Order Sliding Modes, Differentiation and Output-Feedback Control. *Int. J. Control* **2003**, *76*, 924–941. [CrossRef]
34. Zhu, C. Approximation-Free Appointed-Time Tracking Control for Marine Surface Vessel with Actuator Faults and Input Saturation. *Ocean Eng.* **2022**, *245*, 110468. [CrossRef]

Journal of
Marine Science and Engineering

MDPI

Article

A Bio-Inspired Sliding Mode Method for Autonomous Cooperative Formation Control of Underactuated USVs with Ocean Environment Disturbances

Zaopeng Dong [1,2], Fei Tan [1,2], Min Yu [1,2,*], Yuyang Xiong [1,2] and Zhihao Li [1,2]

1 Key Laboratory of High Performance Ship Technology (Wuhan University of Technology), Ministry of Education, Wuhan University of Technology, Wuhan 430063, China; dongzaopeng@whut.edu.cn (Z.D.); tan_f@whut.edu.cn (F.T.); 19520513036@163.com (Y.X.); lizhihao@whut.edu.cn (Z.L.)
2 School of Naval Architecture, Ocean and Energy Power Engineering, Wuhan University of Technology, Wuhan 430063, China
* Correspondence: yumin@whut.edu.cn

Abstract: In this paper, a bio-inspired sliding mode control (bio-SMC) and minimal learning parameter (MLP) are proposed to achieve the cooperative formation control of underactuated unmanned surface vehicles (USVs) with external environmental disturbances and model uncertainties. Firstly, the desired trajectory of the follower USV is generated by the leader USV's position information based on the leader–follower framework, and the problem of cooperative formation control is transformed into a trajectory tracking error stabilization problem. Besides, the USV position errors are stabilized by a backstepping approach, then the virtual longitudinal and virtual lateral velocities can be designed. To alleviate the system oscillation and reduce the computational complexity of the controller, a sliding mode control with a bio-inspired model is designed to avoid the problem of differential explosion caused by repeated derivation. A radial basis function neural network (RBFNN) is adopted for estimating and compensating for the environmental disturbances and model uncertainties, where the MLP algorithm is utilized to substitute for online weight learning in a single-parameter form. Finally, the proposed method is proved to be uniformly and ultimately bounded through the Lyapunov stability theory, and the validity of the method is also verified by simulation experiments.

Keywords: unmanned surface vehicle; leader-follower; bio-inspired model; MLP-RBF neural network; sliding mode control; cooperative formation control

1. Introduction

Recently, due to significant advancements in marine science and technology, such as marine resource exploration, maritime casualty and ocean mapping [1–4], the intelligent control of unmanned surface vehicles (USVs) has become a hot topic. As small intelligent unmanned maritime carrier platforms, USVs can replace humans to perform various complex tasks, particularly in harsh marine environments [5,6]. However, with the increasing complexity and diversity of maritime operations, a single USV often struggles to efficiently accomplish these tasks [7,8]. In response to this challenge, the cooperative formation of multiple USVs has gained widespread attention [9,10]. One significant problem concerned with multiple USVs is formation tracking control, which requires the USVs to keep a prescribed formation while USVs are navigating along a given desired trajectory [11,12]. A number of related research works on the autonomous cooperative formation control of USVs have been carried out, and many valuable research results have been published.

A novel terminal sliding surface control (TSMC) was designed by [13] to solve the cooperative formation control problem of unmanned surface vehicle-remotely operated vehicles (USV-ROVs) subject to uncertainties under deceptive attacks. The USV's dynamic model was established by [14] for the leader–follower formation in polar coordinates, where a sliding mode control is employed to design the controller. Based on the leader–follower framework, USV formation control was designed by utilizing the model predictive control (MPC) algorithm [15]. Combined with the leader–follower formation strategy, a deep reinforcement learning (DRL) algorithm was introduced in [16], where the follower USVs can self-adjust even through deviating from the formation. A hierarchical sliding mode control strategy was designed to solve the formation control problem of USVs based on the sampling communication by [17], where a sign function was developed for the sliding mode control. In [18], an adaptive sliding mode control was used to reduce system oscillations, and the hyperbolic tangent function was employed to enhance the robustness of the control system. A distributed control law was designed by [19] based on a novel sliding surface in which the constructed sliding surface is capable of being reached in a finite time.

On account of the time-varying external disturbances and mode uncertainties, two novel distributed coordinated finite time-tracking controllers were proposed based on the sliding mode control method in [20], and the radial basis function neural networks (RBFNNs) combined with the minimal learning parameter (MLP) was applied in [21]. Combined with a lateral velocity tracking differentiator (LVTD), a novel dynamic surface sliding mode control (DSSMC) method was designed by [22] for the cooperative formation control of underactuated USVs with complex disturbances. The OAST-SMC with backstepping method was employed to solve the external disturbances and enhance the precision of formation control in [23]. An accurate disturbance observer (ADO) and a novel fixed-time fast terminal sliding mode (FTFTSM) were designed for parametric uncertainties and complex disturbances [24]. To investigate the formation tracking control problem of networked USVs under a directed graph, a prescribed-time distributed cooperative formation tracking control scheme was provided by [25], where the prescribed-time sliding surface was designed based on the state observer.

Through the analysis of the above existing research results, it should be pointed out that, in studies [14–16], the problem of cooperative formation control was addressed based on the leader–follower framework. The cooperative formation control algorithms of USVs were designed for fully actuated USVs in [19,24,25]; however, due to factors such as system dynamic characteristics and physical constraints in the actual engineering practice, the control system may become underactuated. Meanwhile, autonomous cooperative formation control problems were solved by using the sliding mode method in [13,17,18,20,21], but the problem of differential explosion was not discussed in these paper. This problem is caused by repeated derivation during the sliding mode design process and may increase the computational complexity of the controller and the system oscillation. Besides, although the complex ocean environment disturbances were addressed in [22,23,25], the problem of model uncertainties was not considered.

Motivated by the above analysis, to address the control problem of USV formation with ocean environment disturbances and model uncertainties, a bio-inspired sliding mode control (bio-SMC) method is designed in this paper. Based on the leader–follower framework, the problem of cooperative formation control is transformed into a trajectory tracking error stabilization problem for each USV, and then the virtual control laws of the longitudinal and lateral velocities are designed by using a backstepping method to stabilize the position error. A sliding mode control with a bio-inspired model is designed to alleviate the system oscillation and reduce the computational complexity of the controller, and then the problem of differential explosion caused by repeated derivation could be avoided. Besides, a radial basis function neural network (RBFNN) with the MLP algorithm is adopted for estimating and compensating for the environmental disturbances and model uncertainties. Compared with the existing research results, the main contributions of this paper can be summarized as follows: (1) Based on the leader–follower framework, the virtual longitudinal and lateral velocities are designed by using the backstepping method, and then the controller design can become more convenient. (2) The second-order sliding surface with an integral is constructed for reducing steady-state error, and a hyperbolic tangent function is employed in place of the sign function to reduce system oscillations. (3) A bio-inspired model is designed to avoid the problem of differential explosion caused by repeated derivation, and then the virtual velocities can be smoothed. (4) A radial basis function neural network (RBFNN) with the minimal learning parameter (MLP) algorithm is adopted for approximating the environmental disturbances and model uncertainties.

This paper is organized as follows: the mathematical model of USV and error model of USV formation are described in Section 2. In Section 3, the control law including the surge force and the yaw moment are described. The stability of the controller is proved in Section 4. The simulation experiments are presented to verify the accuracy and effectiveness of the proposed method in Section 5. Finally, the conclusion is given in Section 6.

2. Problem Description

A coordinated formation control algorithm of USVs is designed in this paper, through which the formation trajectory tracking issue can be addressed effectively. The formation trajectory tracking problem can be described as shown in Figure 1.

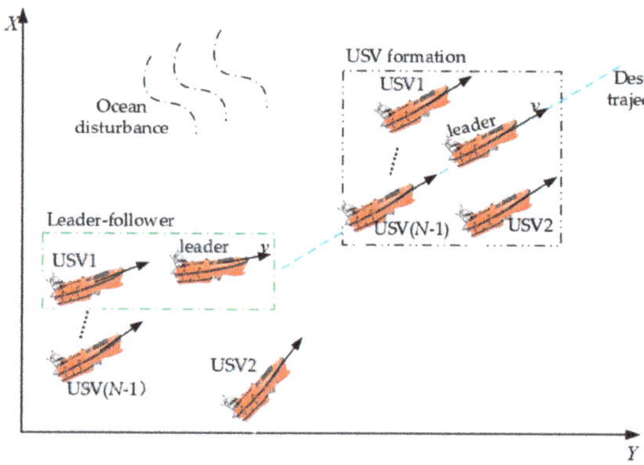

Figure 1. Formation trajectory tracking diagram.

The formation trajectory tracking problem can be considered as how the USV formation should navigate along the desired trajectory under the influence of ocean disturbances, where each USV is required to maintain the predefined formation relative to other USVs. To achieve the problem mentioned above, the model of USVs and the tracking error model of USV formation are built, as shown in Sections 2.1 and 2.2.

2.1. Mathematical Model of USV

As shown in Figure 1, assume that there exist N USVs in formation, which includes one leader USV and $N-1$ follower USVs. According to the literature [26], the mathematical model of the $i^{th}(i=1,2,\cdots,N)$ USV under the influence of ocean disturbances and model uncertainties can be expressed as follows:

$$\begin{cases} \dot{\boldsymbol{\eta}}_i = \boldsymbol{R}(\varphi_i)\boldsymbol{v}_i \\ \boldsymbol{M}_i\dot{\boldsymbol{v}}_i + \boldsymbol{C}_i(\boldsymbol{v}_i)\boldsymbol{v}_i + \boldsymbol{D}_i(\boldsymbol{v}_i)\boldsymbol{v}_i = \boldsymbol{\tau}_i + \boldsymbol{d}_i + \Delta\boldsymbol{f}_i \end{cases} \tag{1}$$

where $\boldsymbol{\eta}_i = [x_i, y_i, \varphi_i]^T$ is the position and heading angle vector of the ith USV, in which x_i is the longitudinal position, y_i is the lateral position and φ_i is the heading angle. $\boldsymbol{v}_i = [u_i, v_i, r_i]^T$ represents the velocity vector of the ith USV, in which u_i, v_i and r_i are the longitudinal velocity, lateral velocity and heading angle speed. $\boldsymbol{\tau}_{ci} = [\tau_{ui}, 0, \tau_{ri}]^T$ is the control input vector of the ith USV. $\boldsymbol{d}_i = [d_{ui}, d_{vi}, d_{ri}]^T$ represents the external disturbances vector of the ith USV. $\Delta\boldsymbol{f}_i = [\Delta f_{u_i}, \Delta f_{v_i}, \Delta f_{r_i}]^T$ denotes the model uncertainties of the ith USV. $\boldsymbol{M}_i = diag(m_{1i}, m_{2i}, m_{3i})$ is the inertia matrix including added mass of the ith USV.

$D_i(v_i) = diag(d_{1i}, d_{2i}, d_{3i})$ represents the hydrodynamic damping coefficient matrix of the ith USV. $R(\varphi_i)$,$C_i(v_i)$ are the rotation matrix and the Coriolis-centripetal matrix of the ith USV, specifically defined as follows:

$$R(\varphi_i) = \begin{bmatrix} \cos\varphi_i & -\sin\varphi_i & 0 \\ \sin\varphi_i & \cos\varphi_i & 0 \\ 0 & 0 & 1 \end{bmatrix}, C(v_i) = \begin{bmatrix} 0 & 0 & -m_{2i}v_i \\ 0 & 0 & m_{1i}u_i \\ m_{2i}v_i & m_{1i}u_i & 0 \end{bmatrix}$$

According to Equation (1), the model of the ith USV with three degrees of mathematical freedom can be expanded as follows:

$$\begin{cases} \dot{x}_i = u_i \cos\varphi_i - v_i \sin\varphi_i \\ \dot{y}_i = u_i \sin\varphi_i + v_i \cos\varphi_i \\ \dot{\varphi}_i = r_i \\ \dot{u}_i = f_{u_i}(u_i, v_i, r_i) + \tau_{ui}/m_{1i} \\ \dot{v}_i = f_{v_i}(u_i, v_i, r_i) \\ \dot{r}_i = f_{r_i}(u_i, v_i, r_i) + \tau_{ri}/m_{3i} \end{cases} \tag{2}$$

where the specific expressions of the $f_{u_i}(u_i, v_i, r_i), f_{v_i}(u_i, v_i, r_i), f_{r_i}(u_i, v_i, r_i)$ are as follows:

$$\begin{cases} f_{u_i}(u_i, v_i, r_i) = (m_{2i}/m_{1i})v_i r_i - (d_{1i}/m_{1i})u_i + d_{ui}/m_{1i} + \Delta f_{u_i} \\ f_{v_i}(u_i, v_i, r_i) = -(m_{1i}/m_{2i})u_i r_i - (d_{2i}/m_{2i})v_i + d_{vj}/m_{2i} + \Delta f_{v_i} \\ f_{r_i}(u_i, v_i, r_i) = ((m_{1i} - m_{2i})/m_{3i})u_i v_i - (d_{3i}/m_{3i})r_i + d_{ri}/m_{3i} + \Delta f_{r_i} \end{cases} \tag{3}$$

2.2. Error Model of USV Formation

The leader–follower framework proposed in Figure 1 is designed to maintain a predefined formation while USVs are navigating, as specifically described in Figure 2 below:

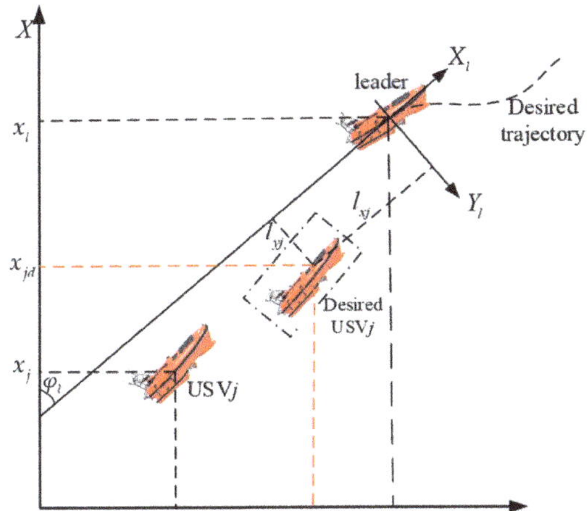

Figure 2. Leader–follower framework.

As shown in Figure 2, the $j^{\text{th}} (j = 1, 2, \ldots, N-1)$ USV serves as the follower USV for the leader USV. $Y_l = [x_l, y_l]^T$ is the position vector of the leader USV. $Y_j = [x_j, y_j]^T$ is the position vector of the jth follower USV. φ_l is the heading angle of the leader USV; $L_j = [l_{xj}, l_{yj}]^T$ is the formation position vector from the jth follower USV to the leader USV, where l_{xj} is the desired relative longitudinal position in the coordinate system of the leader USV and l_{yj} is the desired relative lateral position in the coordinate system of the leader USV. $Y_{jd} = [x_{jd}, y_{jd}]^T$ is the desired position vector of the jth follower USV, where x_{jd} is the desired longitudinal position of the jth follower USV and y_{jd} is the desired lateral position of the jth follower USV. The desired position of the jth follower USV can be expressed as follows:

$$\boldsymbol{Y}_{jd} = \boldsymbol{Y}_l + \boldsymbol{J}(\varphi_l) \cdot \boldsymbol{L}_j \tag{4}$$

where $\boldsymbol{J}(\varphi_l) = \begin{bmatrix} \cos(\varphi_l) & -\sin(\varphi_l) \\ \sin(\varphi_l) & \cos(\varphi_l) \end{bmatrix}$ is the standard rotation matrix.

Based on the above analysis, the trajectory tracking problem of the jth follower USV could be transformed to design a suitable control $\boldsymbol{\tau}_i$ that ensures tracking error $\boldsymbol{Y}_{je} = \boldsymbol{Y}_j - \boldsymbol{Y}_{jd}$ approaches zero.

3. Controller Design

As shown in Figure 1, to enable USVs to track a desired trajectory in a specified formation, it is crucial to design an appropriate control algorithm. The flowchart of cooperative formation control of USVs is designed as shown in Figure 3.

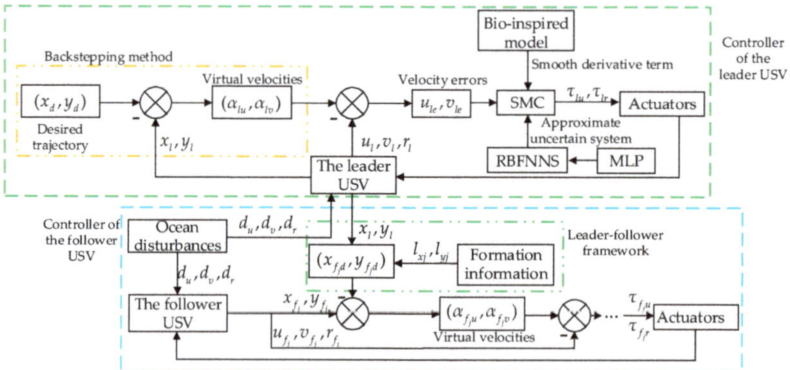

Figure 3. Flowchart of cooperative formation control for underactuated USVs.

As seen in Figure 3, the cooperative formation control process of USVs can be divided into two main parts: the controller of the lead USV would be designed by a backstepping method and the bio-inspired sliding model method. Then, the actual trajectory of the leader USV and formation information are used to generate the desired trajectories for each follower USV, and the controllers of the follower USVs would be designed the same as for the leader USV, which includes the design of the virtual velocity control law in Section 3.1, the design of the surge force controller in Section 3.2 and the design of the yaw moment controller in Section 3.3.

3.1. Design of Virtual Velocities

As seen in Figure 3, it is crucial for the controller to design virtual velocities based on the position errors. The position errors of the jth follower USV could be proposed as follows:

$$\begin{bmatrix} x_{f_je} \\ y_{f_je} \end{bmatrix} = \begin{bmatrix} x_{f_j} - x_{f_jd} \\ y_{f_j} - y_{f_jd} \end{bmatrix} \tag{5}$$

where x_{f_j}, y_{f_j} are the actual longitudinal position and lateral position of the jth follower USV and x_{f_jd}, y_{f_jd} are the desired longitudinal position and lateral position of the jth follower USV. x_{f_je}, y_{f_je} are longitudinal position error and lateral position error of the jth follower USV.

Combining this with Equation (2), the derivation of Equation (5) could be written as follows:

$$\begin{bmatrix} \dot{x}_{f_je} \\ \dot{y}_{f_je} \end{bmatrix} = \begin{bmatrix} \cos(\varphi_{f_j}) & -\sin(\varphi_{f_j}) \\ \sin(\varphi_{f_j}) & \cos(\varphi_{f_j}) \end{bmatrix} \begin{bmatrix} u_{f_j} \\ v_{f_j} \end{bmatrix} - \begin{bmatrix} \dot{x}_{f_jd} \\ \dot{y}_{f_jd} \end{bmatrix} \tag{6}$$

where u_{f_j}, v_{f_j} are the actual longitudinal velocity and lateral velocity of the jth follower USV. φ_{f_j} is the actual heading angle of the jth follower USV.

To stabilize the position errors of the jth follower, the Lyapunov function could be designed as follows:

$$V_{1f_i} = x_{f_ie}^2/2 + y_{f_ie}^2/2 \tag{7}$$

Combining this with Equations (2) and (6), the derivation of Equation (7) can be obtained as follows:

$$\begin{aligned} \dot{V}_{1f_i} &= \frac{\partial V_{1f_i}}{\partial x_{f_ie}}\frac{\partial x_e}{\partial t} + \frac{\partial V_{1f_i}}{\partial y_{f_ie}}\frac{\partial y_{f_ie}}{\partial t} \\ &= x_{f_ie}\dot{x}_{f_ie} + y_{f_ie}\dot{y}_{f_ie} \\ &= x_{f_ie}(u_{f_j}\cos(\varphi_{f_j}) - v_{f_j}\sin(\varphi_{f_j}) - \dot{x}_{f_jd}) + y_{f_ie}(u_{f_j}\sin(\varphi_{f_j}) + v_{f_j}\cos(\varphi_{f_j}) - \dot{y}_{f_jd}) \end{aligned} \tag{8}$$

Generally, the values of virtual velocities α_{f_ju} and α_{f_jv} are related to the position errors, and based on the backstepping method, the virtual velocities α_{f_ju} and α_{f_jv} of the jth follower USV could be designed as follows:

$$\begin{bmatrix} \alpha_{f_ju} \\ \alpha_{f_jv} \end{bmatrix} = \begin{bmatrix} \cos(\varphi_{f_j}) & \sin(\varphi_{f_j}) \\ -\sin(\varphi_{f_j}) & \cos(\varphi_{f_j}) \end{bmatrix} \begin{bmatrix} \dot{x}_{f_jd} - k_{1f_j}x_{f_je}/\sqrt{x_{f_je}^2 + y_{f_je}^2 + C_{f_j}} \\ \dot{y}_{f_jd} - k_{2f_j}y_{f_je}/\sqrt{x_{f_je}^2 + y_{f_je}^2 + C_{f_j}} \end{bmatrix} \tag{9}$$

where $k_{1f_j}, k_{2f_j}, C_{f_j}$ are positive numbers.

Based on the above analysis, the trajectory tracking problem of the jth follower USV could be transformed to design a suitable control $\boldsymbol{\tau}_j$ that ensures errors of the virtual velocities and actual velocities approach zero.

3.2. Design of the Surge Force

Utilizing the virtual longitudinal velocity α_{f_ju} designed in Section 3.1, the surge force of the jth follower USV could be designed based on the bio-inspired sliding mode method. Define the longitudinal velocity error u_{f_je} as follows:

$$u_{f_je} = u_{f_j} - \alpha_{f_ju} \tag{10}$$

To reduce the steady-state error, the integral sliding surface s_{1f_j} with u_{f_je} could be designed as follows:

$$s_{1f_j} = u_{f_je} + \lambda_{1f_j}\int_0^t u_{f_je}(s)ds \tag{11}$$

where λ_{1f_j} is a positive number.

Combining this with the Equation (2), the derivation of Equation (11) could be obtained as follows:

$$\begin{aligned} \dot{s}_{1f_j} &= \dot{u}_{f_j} - \dot{\alpha}_{f_ju} + \lambda_{1f_j}u_{f_je} \\ &= f_{f_ju}(u_{f_j}, v_{f_j}, r_{f_j}) + \tau_{uj}/m_{1j} - \dot{\alpha}_{f_ju} + \lambda_{1f_j}u_{f_je} \end{aligned} \tag{12}$$

In order to avoid differential explosion caused by repeated derivation and smooth the virtual velocity α_{f_ju}, the bio-inspired model could be designed [27], where the virtual longitudinal velocity derivative $\dot{\alpha}_{f_iu}$ in Equation (12) is transformed into $\dot{\beta}_{f_ju}$, shown as follows:

$$\begin{aligned} &\dot{\beta}_{f_ju} = -B_{1f_j}\beta_{f_ju} + (C_{1f_j} - \beta_{f_ju})f(\alpha_{f_ju}) - (D_{1f_j} + \beta_{f_ju})g(\alpha_{f_ju}) \\ &f(\alpha_{f_ju}) = \begin{cases} \alpha_{f_ju}, \alpha_{f_ju} \geq 0 \\ 0, \alpha_{f_ju} < 0 \end{cases}, g(\alpha_{f_ju}) = \begin{cases} 0, \alpha_{f_ju} > 0 \\ -\alpha_{f_ju}, \alpha_{f_ju} \leq 0 \end{cases} \end{aligned} \tag{13}$$

where β_{f_ju} is the output of the bio-inspired model. B_{1f_j} represents the model of the attenuation rate and is the positive constant. C_{1f_j} and D_{1f_j} denote the upper and lower bounds of the β_{f_iu}. $f(\alpha_{f_ju})$ and $g(\alpha_{f_ju})$ are linear threshold functions.

To prevent the system oscillation caused by using the sign function, Equation (14) could be designed with a hyperbolic tangent function, listed as follows:

$$\dot{s}_{1f_j} = -\varepsilon_{1f_j}\tanh(s_{1f_j}) - \sigma_{1f_j}s_{1f_j} \tag{14}$$

where ε_{1f_j} and σ_{1f_j} are positive numbers.

Combining this with the Equations (12)–(14), the surge force of the jth follower USV could be obtained as follows:

$$\tau_{f_ju} = m_{1j}[-f_{f_ju}(u_{f_j}, v_{f_j}, r_{f_j}) + \dot{\beta}_{f_ju} - \lambda_{1f_j}u_{f_je} - \varepsilon_{1f_j}\tanh(s_{1f_j}) - \sigma_{1f_j}s_{1f_j}] \tag{15}$$

3.3. Design of the Yaw Moment

Utilizing the virtual lateral velocity α_{f_jv} designed in Section 3.1, the yaw moment of the jth follower USV could be designed based on the bio-inspired sliding mode method. Define the lateral velocity error v_{f_je} as follows:

$$v_{f_je} = v_{f_j} - \alpha_{f_jv} \tag{16}$$

where v_{f_j} is the actual lateral velocity of the jth follower USV.

Due to the lack of a controller in the lateral direction for the underactuated USV, the second-order integral sliding surface with the lateral velocity error v_{f_je} could be designed as follows:

$$s_{2f_j} = \dot{v}_{f_je} + \lambda_{2f_j}v_{f_je} + \lambda_{3_j}\int_0^t v_{f_je}(s)ds \tag{17}$$

where λ_{2f_j} and λ_{3f_j} are positive numbers.

Combining this with the Equation (2), the derivation of Equation (17) could be obtained as follows:

$$\dot{s}_{2f_j} = \ddot{v}_{f_j} - \ddot{\alpha}_{f_jv} + \lambda_{2f_j}(\dot{v}_{f_j} - \dot{\alpha}_{f_jv}) + \lambda_{3_j}(v_{f_j} - \alpha_{f_jv}) \tag{18}$$

To obtain $\ddot{\alpha}_{f_jv}$ in Equation (18), the derivation of Equation (9) is proposed as follows:

$$\begin{aligned} \begin{bmatrix} \dot{\alpha}_{f_ju} \\ \dot{\alpha}_{f_jv} \end{bmatrix} &= r_{f_j}\begin{bmatrix} -\sin(\varphi_{f_j}) & \cos(\varphi_{f_j}) \\ -\cos(\varphi_{f_j}) & -\sin(\varphi_{f_j}) \end{bmatrix}\begin{bmatrix} \dot{x}_{f_jd} - k_{1f_j}w_{f_j}^{-1}x_{f_je} \\ \dot{y}_{f_jd} - k_{2f_j}w_{f_j}^{-1}y_{f_je} \end{bmatrix} + \\ &\begin{bmatrix} \cos(\varphi_{f_j}) & \sin(\varphi_{f_j}) \\ -\sin(\varphi_{f_j}) & \cos(\varphi_{f_j}) \end{bmatrix}\begin{bmatrix} \ddot{x}_{f_jd} - k_{1f_j}(w_{f_j}^{-1} - w_{f_j}^{-3}x_{f_je}^2)\dot{x}_{f_je} + k_{1f_j}w_{f_j}^{-3}x_{f_je}y_{f_je}\dot{y}_{f_je} \\ \ddot{y}_{f_jd} - k_{2f_j}(w_{f_j}^{-1} - w_{f_j}^{-3}y_{f_je}^2)\dot{y}_{f_je} + k_{2f_j}w_{f_j}^{-3}x_{f_je}y_{f_je}\dot{x}_{f_je} \end{bmatrix} \\ &= r_{f_j}\begin{bmatrix} \alpha_{f_jv} \\ -\alpha_{f_ju} \end{bmatrix} + \begin{bmatrix} \cos(\varphi_{f_j}) & \sin(\varphi_{f_j}) \\ -\sin(\varphi_{f_j}) & \cos(\varphi_{f_j}) \end{bmatrix}\begin{bmatrix} \ddot{x}_{f_jd} - k_{1f_j}(w_{f_j}^{-1} - w_{f_j}^{-3}x_{f_je}^2\dot{x}_{f_je} + k_{1f_j}w_{f_j}^{-3}x_{f_je}y_{f_je}\dot{y}_{f_je} \\ \ddot{y}_{f_jd} - k_{2f_j}(w_{f_j}^{-1} - w_{f_j}^{-3}y_{f_je}^2)\dot{y}_{f_je} + k_{2f_j}w_{f_j}^{-3}x_{f_je}y_{f_je}\dot{x}_{f_je} \end{bmatrix} \end{aligned} \tag{19}$$

where $w_{f_j} = \sqrt{x_{f_je}^2 + y_{f_je}^2 + C_{f_j}}$.

Thus, combining this with Equation (19), $\ddot{\alpha}_{f_jv}$ could be obtained as follows:

$$\begin{cases} \ddot{\alpha}_{f_jv} = -\dot{r}_{f_j}\alpha_{f_ju} - r_{f_j}\dot{\alpha}_{f_ju} + \dot{g}_{f_j} \\ g_{f_j} = -[\ddot{x}_{f_jd} - k_{1f_j}(w_{f_j}^{-1} - w_{f_j}^{-3}x_{f_je}^2)\dot{x}_{f_je} + k_{1f_j}w_{f_j}^{-3}x_{f_je}y_{f_je}\dot{y}_{f_je}]\sin(\varphi_{f_j}) + \\ [\ddot{y}_{f_jd} - k_{2f_j}(w_{f_j}^{-1} - w_{f_j}^{-3}y_{f_je}^2)\dot{y}_{f_je} + k_{2f_j}w_{f_j}^{-3}x_{f_je}y_{f_je}\dot{x}_{f_je}]\cos(\varphi_{f_j}) \end{cases} \tag{20}$$

According to Equation (20), Equation (18) could be rewritten as follows:

$$\begin{aligned} \dot{s}_{2f_j} &= \ddot{v}_{f_j} - (-\dot{r}_{f_j}\alpha_{f_ju} - r_{f_j}\dot{\alpha}_{f_ju} + \dot{g}_{f_j}) + \lambda_{2f_j}(\dot{v}_{f_j} - \dot{\alpha}_{f_jv}) + \lambda_{3_j}(v_{f_j} - \alpha_{f_jv}) \\ &= \ddot{v}_{f_j} - [-(f_{f_jr}(u_{f_j}, v_{f_j}, r_{f_j}) + \tfrac{1}{m_{3j}}\tau_{r_j})\alpha_{f_ju} - r_{f_j}\dot{\alpha}_{f_ju} + \dot{g}_{f_j}) \\ &\quad + \lambda_{2f_j}(\dot{v}_{f_j} - \dot{\alpha}_{f_jv}) + \lambda_{3_j}(v_{f_j} - \alpha_{f_jv}) \end{aligned} \tag{21}$$

By utilizing the bio-inspired model, the virtual lateral velocity derivative $\dot{\alpha}_{f_jv}$ in Equation (21) is transformed into $\dot{\beta}_{f_jv}$, shown as follows:

$$\begin{aligned} \dot{\beta}_{f_jv} &= -B_{2f_j}\beta_{f_jv} + (C_{2f_j} - \beta_{f_jv})f(\alpha_{f_jv}) - (D_{2f_j} + \beta_{f_jv})g(\alpha_{f_jv}) \\ f(\alpha_{f_jv}) &= \begin{cases} \alpha_{f_jv}, \alpha_{f_jv} \geq 0 \\ 0, \alpha_{f_jv} < 0 \end{cases}, g(\alpha_{f_jv}) = \begin{cases} 0, \alpha_{f_jv} > 0 \\ -\alpha_{f_jv}, \alpha_{f_jv} \leq 0 \end{cases} \end{aligned} \tag{22}$$

where β_{f_jv} is the output of the bio-inspired model. B_{2f_j} represents the model of attenuation rate and is the positive constant. C_{2f_j} and D_{2f_j} denote the upper and lower bounds of β_{f_iv}. $f(\alpha_{f_ju})$ and $g(\alpha_{f_ju})$ are linear threshold functions.

The sliding mode reaching law with a hyperbolic tangent function could be designed as follows:

$$\dot{s}_{2f_j} = -\varepsilon_{2f_j}\tan\text{h}(s_{2f_j}) - \sigma_{2f_j}s_{2f_j} \tag{23}$$

where ε_{2f_j} and σ_{2f_j} are positive numbers.

Combining this with Equations (21)–(23), the yaw moment of the jth follower USV can be obtained as follows:

$$\begin{aligned} \tau_{r_j} = &\tfrac{m_{3j}}{\alpha_{f_ju}}[-\ddot{v}_{f_j} - \lambda_{2f_j}(\dot{v}_{f_j} - \dot{\beta}_{f_jv}) - \lambda_{3_j}(v_{f_j} - \beta_{f_jv}) - r_{f_j}\dot{\beta}_{f_ju} \\ &+ \dot{g}_{f_j} - \varepsilon_{2f_j}\tan\text{h}(s_{2f_j}) - \sigma_{2f_j}s_{2f_j}] - m_{3j}f_{f_jr}(u_{f_j}, v_{f_j}, r_{f_j}) \end{aligned} \tag{24}$$

3.4. Design of RBF Neural Network

According to Equations (15) and (24), the functions $f_{f_ju}(u_{f_j}, v_{f_j}, r_{f_j})$ and $f_{f_jr}(u_{f_j}, v_{f_j}, r_{f_j})$ are defined as unknown functions, which represent the ocean disturbances and model uncertainties. As shown in Figure 3, the RBF neural network is proposed to approximate the functions $f_{f_ju}(u_{f_j}, v_{f_j}, r_{f_j})$ and $f_{f_jr}(u_{f_j}, v_{f_j}, r_{f_j})$. The structure diagram of the RBF neural network with multiple inputs and a single output is described in Figure 4.

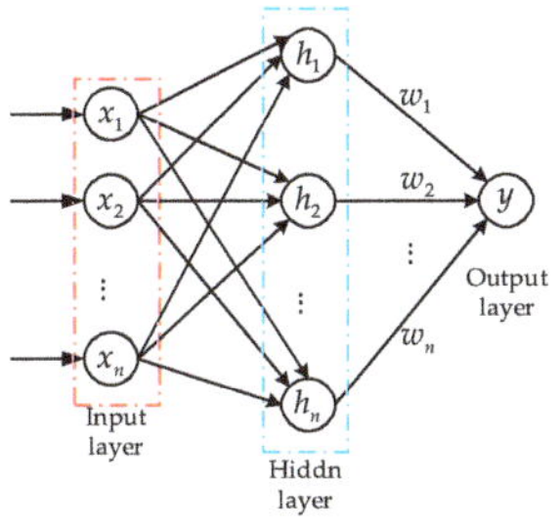

Figure 4. The structure diagram of the RBF neural network.

As seen in Figure 4, the RBF neural network is a three-layer feedforward network, which includes an input layer, a hidden layer and an output layer. Based on the Gaussian basis function, the RBF neural network can approximate the uncertain terms without the need for complex mathematical theoretical analysis. The expression of the RBF neural network could be designed as follows:

$$\begin{cases} f_{f_ju}(u_{f_j}, v_{f_j}, r_{f_j}) = \boldsymbol{W}_{f_ju}^{*T}\boldsymbol{h}(\boldsymbol{z}_{f_j}) + \delta_{1f_j}, \delta_{1f_j} < \delta_{f_jU} \\ f_{f_jr}(u_{f_j}, v_{f_j}, r_{f_j}) = \boldsymbol{W}_{f_jr}^{*T}\boldsymbol{h}(\boldsymbol{z}_{f_j}) + \delta_{2f_j}, \delta_{2f_j} < \delta_{f_jR} \\ \boldsymbol{h}(\boldsymbol{z}_{f_j}) = \exp\left(- \left\| \boldsymbol{z}_{f_j} - \boldsymbol{c}_k \right\|^2 / 2\boldsymbol{b}_k^2 \right), \boldsymbol{b}_k > 0,\ (k = 1, 2, \cdots, n) \end{cases} \tag{25}$$

where δ_{1f_j} and δ_{2f_j} are the approximation errors of the neural network. $\delta_{f_jU} > 0$ and $\delta_{f_jR} > 0$ are the upper bounds of δ_{1f_j} and δ_{2f_j}. $\boldsymbol{z}_{f_j} = [u_{f_j}, v_{f_j}, r_{f_j}]^T$ is the input vector of the RBF neural network, $\boldsymbol{h}(\boldsymbol{z}_{f_j}) = [h_1(\boldsymbol{z}_{f_j}), h_2(\boldsymbol{z}_{f_j}), \cdots, h_n(\boldsymbol{z}_{f_j})]^T$ represents the radial basis function vector of the neural network. $\boldsymbol{b}_k$ is the width of the Gaussian basis function. The positive integer k represents the kth node in the hidden layer of the neural network.$\boldsymbol{c}_k = [c_1, c_2, \cdots, c_n]^T \in \boldsymbol{R}^n$ denotes the center value of the kth hidden neuron. $\boldsymbol{W}_{f_ju}^* = [w_{f_ju,1}^*, w_{f_ju,2}^*, \cdots, w_{f_ju,n}^*]^T \in \boldsymbol{R}^n$ and $\boldsymbol{W}_{f_jr}^* = [w_{f_jr,1}^*, w_{f_jr,2}^*, \cdots, w_{f_jr,n}^*]^T \in \boldsymbol{R}^n$ are the weights associated with the kth hidden neuron, which are designed as follows:

$$\begin{cases} \boldsymbol{W}_{f_ju}^* = \underset{\boldsymbol{W} \in \boldsymbol{R}^n}{\text{argmin}} \left\{ \sup_{z_{f_j} \in \Omega} \left| f_{f_ju}(u_{f_j}, v_{f_j}, r_{f_j}) - \boldsymbol{W}_{f_ju}^T \boldsymbol{h}(\boldsymbol{z}_{f_j}) \right| \right\} \\ \boldsymbol{W}_{f_jr}^* = \underset{\boldsymbol{W} \in \boldsymbol{R}^n}{\text{argmin}} \left\{ \sup_{z_{f_j} \in \Omega} \left| f_{f_jr}(u_{f_j}, v_{f_j}, r_{f_j}) - \boldsymbol{W}_{f_jr}^T \boldsymbol{h}(\boldsymbol{z}_{f_j}) \right| \right\} \end{cases} \tag{26}$$

Since $\boldsymbol{W}_{f_ju}^*$ and $\boldsymbol{W}_{f_jr}^*$ are often unattainable in practical applications, the estimated values $\hat{\boldsymbol{W}}_{f_ju}$ and $\hat{\boldsymbol{W}}_{f_jr}$ are designed to replace $\boldsymbol{W}_{f_ju}^*$ and $\boldsymbol{W}_{f_jr}^*$. Combining this with Equations (15) and (24), the surge force and the yaw moment of the jth follower USV could be rewritten as follows:

$$\begin{cases} \tau_{f_ju} = m_{1j}[-\hat{f}_{f_ju}(u_{f_j}, v_{f_j}, r_{f_j}) + \dot{\beta}_{f_ju} - \lambda_{1f_j} u_{f_je} - \varepsilon_{1f_j} \tanh(s_{1f_j}) - \sigma_{1f_j} s_{1f_j}] \\ \tau_{r_j} = \frac{m_{3j}}{\alpha_{f_ju}}[-\ddot{v}_{f_j} - \lambda_{2f_j}(\dot{v}_{f_j} - \dot{\beta}_{f_jv}) - \lambda_{3_j}(v_{f_j} - \beta_{f_jv}) - r_{f_j}\dot{\beta}_{f_ju} \\ \quad + \dot{g}_{f_j} - \varepsilon_{2f_j} \tanh(s_{2f_j}) - \sigma_{2f_j} s_{2f_j}] - m_{3j}\hat{f}_{f_jr}(u_{f_j}, v_{f_j}, r_{f_j}) \end{cases} \tag{27}$$

where $\hat{f}_{f_ju}(u_{f_j}, v_{f_j}, r_{f_j}) = \hat{\boldsymbol{W}}_{f_ju}^T \boldsymbol{h}(\boldsymbol{z}_{f_j})$ is the estimated value of $f_{f_ju}(u_{f_j}, v_{f_j}, r_{f_j})$ and $\hat{f}_{f_jr}(u_{f_j}, v_{f_j}, r_{f_j}) = \hat{\boldsymbol{W}}_{f_jr}^T \boldsymbol{h}(\boldsymbol{z}_{f_j})$ is the estimated value of $f_{f_jr}(u_{f_j}, v_{f_j}, r_{f_j})$.

Meanwhile, to simplify the calculation, the MLP algorithm is utilized to substitute for online weight in a single-parameter form. Specifically, $\left\| \boldsymbol{W}_{f_ju}^* \right\|^2$ and $\left\| \boldsymbol{W}_{f_jr}^* \right\|^2$ could be designed as follows:

$$\begin{cases} \phi_{1f_j} = \left\| \boldsymbol{W}_{f_ju}^* \right\|^2 \\ \phi_{2f_j} = \left\| \boldsymbol{W}_{f_jr}^* \right\|^2 \end{cases} \tag{28}$$

where ϕ_{1f_j} and ϕ_{2f_j} are positive constants. And by defining $\hat{\phi}_{1f_j}$ and $\hat{\phi}_{2f_j}$ as the estimated value of ϕ_{1f_j} and ϕ_{2f_j}, the adaptive parameters could be designed as follows:

$$\begin{cases} \dot{\hat{\phi}}_{1f_j} = \zeta_{1f_j}(s_{1f_j}^2 \boldsymbol{h}_{f_j}^T \boldsymbol{h}_{f_j}/2 - \gamma_{1f_j}\hat{\phi}_{1f_j}) \\ \dot{\hat{\phi}}_{2f_j} = \zeta_{2f_j}(s_{2f_j}^2 \boldsymbol{h}_{f_j}^T \boldsymbol{h}_{f_j}/2 - \gamma_{2f_j}\hat{\phi}_{2f_j}) \end{cases} \tag{29}$$

where $\zeta_{1f_j}, \zeta_{2f_j}, \gamma_{1f_j}, \gamma_{2f_j}$ are positive numbers.

Combining this with Equations (27) and (29), the surge force and the yaw moment of the jth follower USV could be designed as follows:

$$\begin{cases} \tau_{f_ju} = m_{1j}[-s_{1f_j}\hat{\phi}_{1f_j}\boldsymbol{h}_{f_j}^T\boldsymbol{h}_{f_j}/2 + \dot{\beta}_{f_ju} - \lambda_{1f_j}u_{f_je} - \varepsilon_{1f_j}\tanh(s_{1f_j}) - \sigma_{1f_j}s_{1f_j}] \\ \tau_{r_j} = \frac{m_{3j}}{\alpha_{f_ju}}[-\ddot{v}_{f_j} - \lambda_{2f_j}(\dot{v}_{f_j} - \dot{\beta}_{f_jv}) - \lambda_{3_j}(v_{f_j} - \beta_{f_jv}) - r_{f_j}\dot{\beta}_{f_ju} \\ \quad + \dot{g}_{f_j} - \varepsilon_{2f_j}\tanh(s_{2f_j}) - \sigma_{2f_j}s_{2f_j}] - m_{3j}s_{2f_j}\hat{\phi}_{2f_j}\boldsymbol{h}_{f_j}^T\boldsymbol{h}_{f_j}/2 \end{cases} \tag{30}$$

4. Stability Analysis

To verify the stability of the designed controller, combining with the errors in sliding mode controller design and estimation error of the RBF neural network, the Lyapunov function could be designed as follows:

$$\begin{cases} V_{2f_j} = s_{1f_j}^2/2 + s_{2f_j}^2/2 + y_{f_ju}^2/2 + y_{f_jv}^2/2 + \widetilde{\phi}_{1f_j}^2/(2\zeta_{1f_j}) + \widetilde{\phi}_{2f_j}^2/(2\zeta_{2f_j}) \\ y_{f_ju} = \beta_{f_ju} - \alpha_{f_ju}, y_{f_jv} = \beta_{f_jv} - \alpha_{f_jv} \\ \widetilde{\phi}_{1f_j} = \phi_{1f_j} - \hat{\phi}_{1f_j}, \widetilde{\phi}_{2f_j} = \phi_{2f_j} - \hat{\phi}_{2f_j} \end{cases} \tag{31}$$

Combining this with Equations (13), (14), (22), (24) and (31), the derivation of the Lyapunov function V_{2f_j} could be presented as follows:

$$\begin{aligned} \dot{V}_{2f_j} &= s_{1f_j}\dot{s}_{1f_j} + s_{2f_j}\dot{s}_{2f_j} + y_{f_ju}\dot{y}_{f_ju} + y_{f_jv}\dot{y}_{f_jv} - \widetilde{\phi}_{1f_j}\dot{\hat{\phi}}_{1f_j}/\zeta_{1f_j} - \widetilde{\phi}_{2f_j}\dot{\hat{\phi}}_{2f_j}/\zeta_{2f_j} \\ &= s_{1f_j}[\boldsymbol{W}_{f_ju}^{*}{}^T\boldsymbol{h}(\boldsymbol{z}_{f_j}) + \delta_{1f_j} - s_{1f_j}\hat{\phi}_{1f_j}\boldsymbol{h}_{f_j}^T\boldsymbol{h}_{f_j}/2 - \varepsilon_{1f_j}\tanh(s_{1f_j}) - \sigma_{1f_j}s_{1f_j}] + s_2[\boldsymbol{W}_{f_jr}^{*}{}^T\boldsymbol{h}(\boldsymbol{z}_{f_k}) + \delta_{2f_j} \\ &\quad - s_{2f_j}\hat{\phi}_{2f_j}\boldsymbol{h}_{f_j}^T\boldsymbol{h}_{f_j}/2 - \varepsilon_{2f_j}\tanh(s_{2f_j}) - \sigma_{2f_j}s_{2f_j}] + y_{f_ju}(\dot{\beta}_{f_ju} - \dot{\alpha}_{f_ju}) + y_{f_jv}(\dot{\beta}_{f_jv} - \dot{\alpha}_{f_jv}) \\ &\quad - \widetilde{\phi}_{1f_j}\dot{\hat{\phi}}_{1f_j}/\zeta_{1f_j} - \widetilde{\phi}_{2f_j}\dot{\hat{\phi}}_{2f_j}/\zeta_{2f_j} \\ &= -\sigma_{1f_j}s_{1f_j}^2 - \sigma_{2f_j}s_{2f_j}^2 + s_{1f_j}\delta_{1f_j} + s_{2f_j}\delta_{2f_j} - s_{1f_j}\varepsilon_{1f_j}\tanh(s_{1f_j}) - s_{2f_j}\varepsilon_{2f_j}\tanh(s_{2f_j}) \\ &\quad + s_{1f_j}\boldsymbol{W}_{f_jr}^{*}{}^T\boldsymbol{h}(\boldsymbol{z}_{f_j}) - s_{1f_j}^2\hat{\phi}_{1f_j}\boldsymbol{h}_{f_j}^T\boldsymbol{h}_{f_j}/2 + s_{2f_j}\boldsymbol{W}_{f_jr}^{*}{}^T\boldsymbol{h}(\boldsymbol{z}_{f_j}) - s_{2f_j}^2\hat{\phi}_{2f_j}\boldsymbol{h}_{f_j}^T\boldsymbol{h}_{f_j}/2 + y_{f_ju}[-B_{1f_j}\beta_{f_ju} \\ &\quad + (C_{1f_j} - \beta_{f_ju})f(\alpha_{f_ju}) - (D_{1f_j} + \beta_{f_ju})g(\alpha_{f_ju}) - \dot{\alpha}_{f_ju}] + y_{f_jv}[-B_{2f_j}\beta_{f_jv} + (C_{2f_j} - \beta_{f_jv})f(\alpha_{f_jv}) \\ &\quad - (D_{2f_j} + \beta_{f_jv})g(\alpha_{f_jv}) - \dot{\alpha}_{f_jv}] - \widetilde{\phi}_{1f_j}\dot{\hat{\phi}}_{1f_j}/\zeta_{1f_j} - \widetilde{\phi}_{2f_j}\dot{\hat{\phi}}_{2f_j}/\zeta_{2f_j} \\ &\leq -\sigma_{1f_j}s_{1f_j}^2 - \sigma_{2f_j}s_{2f_j}^2 + s_{1f_j}\delta_{1f_j} + s_{2f_j}\delta_{2f_j} - \varepsilon_{1f_j}\left|s_{1f_j}\right| - \varepsilon_{2f_j}\left|s_{2f_j}\right| + s_{1f_j}\boldsymbol{W}_{f_jr}^{*}{}^T\boldsymbol{h}(\boldsymbol{z}_{f_j}) \\ &\quad - s_{1f_j}^2\phi_{1f_j}\boldsymbol{h}_{f_j}^T\boldsymbol{h}_{f_j}/2 + s_{2f_j}\boldsymbol{W}_{f_jr}^{*}{}^T\boldsymbol{h}(\boldsymbol{z}_{f_j}) - s_{2f_j}^2\phi_{2f_j}\boldsymbol{h}_{f_j}^T\boldsymbol{h}_{f_j}/2 + y_{f_ju}[-B_{1f_j}\beta_{f_ju} + (C_{1f_j} - \beta_{f_ju})f(\alpha_{f_ju}) \\ &\quad - (D_{1f_j} + \beta_{f_ju})g(\alpha_{f_ju}) - \dot{\alpha}_{f_ju}] + y_{f_jv}[-B_{2f_j}\beta_{f_jv} + (C_{2f_j} - \beta_{f_jv})f(\alpha_{f_jv}) - (D_{2f_j} + \beta_{f_jv})g(\alpha_{f_jv}) \\ &\quad - \dot{\alpha}_{f_jv}] - \widetilde{\phi}_{1f_j}\dot{\hat{\phi}}_{1f_j}/\zeta_{1f_j} - \widetilde{\phi}_{2f_j}\dot{\hat{\phi}}_{2f_j}/\zeta_{2f_j} \end{aligned} \tag{32}$$

According to Equations (25) and (28), the following inequalities could be presented as follows:

$$\begin{cases} s_{1f_j}\delta_{1f_j} \leq s_{1f_j}^2/2 + \delta_{1f_j}^2/2 \leq s_{1f_j}^2/2 + \delta_{f_jU}^2/2 \\ s_{2f_j}\delta_{2f_j} \leq s_{2f_j}^2/2 + \delta_{2f_j}^2/2 \leq s_{2f_j}^2/2 + \delta_{f_jR}^2/2 \\ 2s_{1f_j}\boldsymbol{W}_{f_ju}^{*}{}^T\boldsymbol{h}_{f_j} \leq s_{1f_j}^2\left\|\boldsymbol{W}_{f_ju}^{*}\right\|^2\left\|\boldsymbol{h}_{f_j}\right\|^2 + 1 = s_{1f_j}^2\phi_{1f_j}\boldsymbol{h}_{f_j}^T\boldsymbol{h}_{f_j} + 1 \\ 2s_{2f_j}\boldsymbol{W}_{f_ju}^{*}{}^T\boldsymbol{h}_{f_j} \leq s_{2f_j}^2\left\|\boldsymbol{W}_{f_jr}^{*}\right\|^2\left\|\boldsymbol{h}_{f_j}\right\|^2 + 1 = s_{2f_j}^2\phi_{2f_j}\boldsymbol{h}_{f_j}^T\boldsymbol{h}_{f_j} + 1 \end{cases} \tag{33}$$

As seen from Equations (13) and (22), there exists $f(\alpha_{f_ju}) \geq 0, f(\alpha_{f_jv}) \geq 0$ and $g(\alpha_{f_ju}) \geq 0, g(\alpha_{f_jv}) \geq 0$, and then Equation (32) could be simplified as follows:

$$\begin{aligned}&\dot{V}_{2f_j} \leq -\sigma_{1f_j}s_{1f_j}^2 - \sigma_{2f_j}s_{2f_j}^2 + s_{1f_j}^2/2 + \delta_{f_jU}^2/2 + s_{2f_j}^2/2 + \delta_{f_jR}^2/2 - \varepsilon_{1f_j}\left|s_{1f_j}\right| - \varepsilon_{2f_j}\left|s_{2f_j}\right| \\ &+s_{1f_j}\widetilde{\phi}_{1f_j}\boldsymbol{h}_{f_j}^T\boldsymbol{h}_{f_j}/2 + s_{2f_j}\widetilde{\phi}_{2f_j}\boldsymbol{h}_{f_j}^T\boldsymbol{h}_{f_j}/2 + 1 + y_{f_ju}(-B_{f_ju}\beta_{f_ju} + C_{1f_j}\alpha_{f_ju} - \dot{\alpha}_{f_ju}) \\ &+y_{f_jv}(-B_{f_jv}\beta_{f_jv} + C_{2f_j}\alpha_{f_jv} - \dot{\alpha}_{f_jv}) - \widetilde{\phi}_{1f_j}\dot{\hat{\phi}}_{1f_j}/\zeta_{1f_j} - \widetilde{\phi}_{2f_j}\dot{\hat{\phi}}_{2f_j}/\zeta_{2f_j} \\ &\text{s.t.}\begin{cases} C_{1f_j} = D_{1f_j}, C_{2f_j} = D_{2f_j} \\ B_{f_ju} = B_{1f_j} + f(\alpha_{f_ju}) + g(\alpha_{f_ju}) > 0 \\ B_{f_jv} = B_{2f_j} + f(\alpha_{f_jv}) + g(\alpha_{f_jv}) > 0 \end{cases}\end{aligned} \tag{34}$$

Combining this with Equation (29), and assuming $B_{f_ju} = C_{1f_j}$ and $B_{f_jv} = C_{2f_j}$, then the Equation (34) can be written as follows:

$$\begin{aligned}\dot{V}_{2f_j} &\leq -(2\sigma_{1f_j} - 1)s_{1f_j}^2/2 - (2\sigma_{2f_j} - 1)s_{2f_j}^2/2 + \delta_{f_jU}^2/2 + \delta_{f_jR}^2/2 + s_{1f_j}\widetilde{\phi}_{1f_j}\boldsymbol{h}_{f_j}^T\boldsymbol{h}_{f_j}/2 \\ &+s_{2f_j}\widetilde{\phi}_{2f_j}\boldsymbol{h}_{f_j}^T\boldsymbol{h}_{f_j}/2 + 1 - B_{f_ju}y_{f_ju}^2 - y_{f_ju}\dot{\alpha}_{f_ju} - B_{f_jv}y_{f_jv}^2 - y_{f_jv}\dot{\alpha}_{f_jv} \\ &-\widetilde{\phi}_{1f_j}(s_{1f_j}\boldsymbol{h}_{f_j}^T\boldsymbol{h}_{f_j}/2 - \gamma_{1f_j}\hat{\phi}_{1f_j}) - \widetilde{\phi}_{2f_j}(s_{2f_j}\boldsymbol{h}_{f_j}^T\boldsymbol{h}_{f_j}/2 - \gamma_{2f_j}\hat{\phi}_{2f_j}) \\ &\leq -(2\sigma_{1f_j} - 1)s_{1f_j}^2/2 - (2\sigma_{2f_j} - 1)s_{2f_j}^2/2 + \delta_{f_jU}^2/2 + \delta_{f_jR}^2/2 - B_{f_ju}y_{f_ju}^2 - y_{f_ju}\dot{\alpha}_{f_ju} - B_{f_jv}y_{f_jv}^2 \\ &-y_{f_jv}\dot{\alpha}_{f_jv} + \gamma_{1f_j}\widetilde{\phi}_{1f_j}\hat{\phi}_{1f_j} + \gamma_{2f_j}\widetilde{\phi}_{2f_j}\hat{\phi}_{2f_j}\end{aligned} \tag{35}$$

Since the desired trajectories of USVs are smooth and bounded, the control inputs and velocities of USVs are bounded, it leads to:

$$\begin{cases} -y_{f_ju}\dot{\alpha}_{f_ju} \leq \frac{\sigma_{f_ju}}{2}y_{f_ju}^2 + \frac{1}{2\sigma_{f_ju}}\dot{\alpha}_{f_ju}^2 \leq \frac{\sigma_{f_ju}}{2}y_{f_ju}^2 + \frac{1}{2\sigma_{f_ju}}\dot{U}_{f_j\max}^2 \\ -y_{f_jv}\dot{\alpha}_{f_jv} \leq \frac{\sigma_{f_jv}}{2}y_{f_jv}^2 + \frac{1}{2\sigma_{f_jv}}\dot{\alpha}_{f_jv}^2 \leq \frac{\sigma_{f_jv}}{2}y_{f_jv}^2 + \frac{1}{2\sigma_{f_jv}}\dot{V}_{f_j\max}^2 \\ 2\widetilde{\phi}_{1f_j}\hat{\phi}_{1f_j} = (\phi_{1f_j} - \hat{\phi}_{1f_j})\hat{\phi}_{1f_j} + \widetilde{\phi}_{1f_j}(\phi_{1f_j} - \widetilde{\phi}_{1f_j}) = \phi_{1f_j}^2 - \hat{\phi}_{1f_j}^2 - \widetilde{\phi}_{1f_j}^2 \leq \phi_{1f_j}^2 - \widetilde{\phi}_{1f_j}^2 \\ 2\widetilde{\phi}_{2f_j}\hat{\phi}_{2f_j} = (\phi_{2f_j} - \hat{\phi}_{2f_j})\hat{\phi}_{2f_j} + \widetilde{\phi}_{2f_j}(\phi_{2f_j} - \widetilde{\phi}_{2f_j}) = \phi_{2f_j}^2 - \hat{\phi}_{2f_j}^2 - \widetilde{\phi}_{2f_j}^2 \leq \phi_{2f_j}^2 - \widetilde{\phi}_{2f_j}^2 \end{cases} \tag{36}$$

where $\dot{U}_{f_j\max}$ is the maximum value of $\dot{\alpha}_{f_ju}$ and $\dot{V}_{f_j\max}$ is the maximum value of $\dot{\alpha}_{f_jv}$.

Combining this with Equation (36), Equation (35) could be simplified as follows:

$$\begin{aligned}\dot{V}_{2f_j} &\leq -(2\sigma_{1f_j} - 1)s_{1f_j}^2/2 - (2\sigma_{2f_j} - 1)s_{2f_j}^2/2 - (B_{f_ju} - \sigma_{f_ju}/2)y_{f_ju}^2 - (B_{f_jv} - \sigma_{f_jv}/2)y_{f_jv}^2 \\ &-\gamma_{1f_j}\widetilde{\phi}_{1f_j}^2/2 - \gamma_{2f_j}\widetilde{\phi}_{2f_j}^2/2 + \delta_{f_jU}^2/2 + \delta_{f_jR}^2/2 + \dot{U}_{f_j\max}^2/(2\sigma_{f_ju}) + \dot{V}_{f_j\max}^2/(2\sigma_{f_jv}) \\ &+\gamma_{1f_j}\phi_{1f_j}^2/2 + \gamma_{2f_j}\phi_{2f_j}^2/2 \\ &\leq \mu V_{2f_j} + C_3 \\ \text{s.t.} &\begin{cases} \mu = \min\left\{(2\sigma_{1f_j} - 1)/2, (2\sigma_{2f_j} - 1)/2, B_{f_ju} - \sigma_{f_ju}/2, B_{f_jv} - \sigma_{f_jv}/2, \gamma_{1f_j}/2, \gamma_{2f_j}/2\right\} \\ C_3 = \delta_{f_jU}^2/2 + \delta_{f_jR}^2/2 + \dot{U}_{f_j\max}^2/(2\sigma_{f_ju}) + \dot{V}_{f_j\max}^2/(2\sigma_{f_jv}) + \gamma_{1f_j}\phi_{1f_j}^2/2 + \gamma_{2f_j}\phi_{2f_j}^2/2 + 1 \\ B_{f_ju} - \sigma_{f_ju}/2 > 0, B_{f_jv} - \sigma_{f_jv}/2 > 0 \end{cases}\end{aligned} \tag{37}$$

By solving Equation (37), it leads to

$$0 \leq V_{2f_j}(t) \leq C_3/\mu + [V_{2f_j}(0) - C_3/\mu]e^{-\mu t} \tag{38}$$

According to Equation (38), $V_{2f_i}(t)$ converges within a ball of radius C_3/μ, so that it is ultimately uniformly bounded. By choosing appropriate control parameters, the convergent limit of $V_{2f_i}(t)$ is inclined to zero. And it can be demonstrated that the designed sliding mode controller finally tends to be stable at $t \to \infty$ with $x_{f_je} \to 0$, $y_{f_je} \to 0$, $u_{f_je} \to 0$ and $v_{f_je} \to 0$.

5. Computer Simulation

To highlight the advantages of the proposed bio-inspired sliding mode control (bio-SMC) method and the radial basis function neural network (RBF), four experiment cases including circular trajectory, straight line trajectory, sinusoidal trajectory and combination trajectory of straight line and circle are set in this section, and three USVs are taken as an example, where one is the leader USV, and the other two are the follower USVs. The USV model parameters are presented by [28], and the specific parameters are shown in Table 1.

Table 1. USVs model parameters.

Parameter	Value (Unit)	Parameter	Value (Unit)
m_{1j}	25.8 kg	d_{1j}	12 kg/s
m_{2j}	33.8 kg	d_{2j}	17 kg/s
m_{3j}	2.76 kg.m^2	d_{3j}	0.5 kg.m^2/s

The simulation time interval is 0.1 s and the external environmental disturbance is $d_j = [2\sin(0.1t), 0.2\sin(0.01t), \sin(0.1t)]^T$. The model uncertainties of USV are considered as $\Delta f_j = [-0.2d_{1j}u_j^2/m_{1j} - 0.1d_{1j}u_j^3/m_{1j}, -0.2d_{2j}v_j^2/m_{2j} - 0.1d_{2j}v_j^3/m_{2j}, -0.2d_{3j}r_j^2/m_{3j} -0.1d_{3j}r_j^3/m_{3j}]^T$.

Considering the constraints of the actuators on USVs in practical engineering, limitations are set on the control input, especially described as follows:

$$\begin{cases} \tau_{u\max} = 50\ \mathrm{N}, \tau_{r\max} = 30\ \mathrm{N\cdot m} \\ d\tau_{u\max} = 50\ \mathrm{N/s}, d\tau_{r\max} = 50\ \mathrm{N\cdot m/s} \end{cases} \tag{39}$$

In order to intuitively describe the tracking effect of the USV, the longitudinal position error of the USV is defined as $x_{ej} = \left|x_j - x_{dj}\right|$ and lateral position error of the USV is defined as $y_{ej} = \left|y_j - y_{dj}\right|$. And the longitudinal velocity error and lateral velocity error are $u_{je} = u_j - \alpha_{ju}$ and $v_{je} = v_j - \alpha_{jv}$.

Case 1. Circular trajectory

The desired circular trajectory of the leader USV is described as: $x_d = 30\sin 0.05t$, $y_d = -30\cos 0.05t$. The initial state information for each USV in the formation system is presented in Table 2.

Table 2. USVs initial state (Case 1).

Parameter	Leader USV	Follower USV1	Follower USV2
η	$[-5(m), -35(m), 45(°)]^T$	$[-6(m), -32(m), 45(°)]^T$	$[-6(m), -42(m), 45(°)]^T$
v	$[0, 0, 0]^T$	$[0, 0, 0]^T$	$[0, 0, 0]^T$
L		$[0(m), 5(m)]^T$	$[0(m), -5(m)]^T$

The input of the RBF neural network $\hat{f}_{f_ju}(u_{f_j}, v_{f_j}, r_{f_j})$ is 3; the number of nodes in the hidden layer is 9; the width of the Gaussian basis function is 2 and the vector values for the center points are normally distributed in the range of [−15, 15]. The input of the RBF neural network $\hat{f}_{f_jr}(u_{f_j}, v_{f_j}, r_{f_j})$ is 3; the number of nodes in the hidden layer is 9; the width of the Gaussian basis function is 4 and the vector values for the center points are normally distributed in the range of [−0.3, 0.3]. The controller parameters are presented in Table 3.

Table 3. USVs controller parameters (Case 1).

Parameter	Leader USV	Follower USV1	Follower USV2
C	50	40	50
k_1	3	2	3
k_2	5	2	6.5
λ_1	0.002	0.002	0.002
λ_2	6	4.5	5
λ_3	0.05	0.05	0.05
ε_1	0.01	0.001	0.01
ε_2	0.01	0.01	0.01
σ_1	0.5	0.5	0.5
σ_2	0.01	0.01	0.01
B_1	2	1.5	2
B_2	4	4	4
C_1, D_1	3.4	2.5	3.75
C_2, D_2	4	4	4
ζ_1	20	60	10
ζ_2	0.01	0.015	0.015
γ_1	0.0005	0	0
γ_2	0.001	0.001	0.001

The simulation results of Case 1 are shown in Figures 5–11.

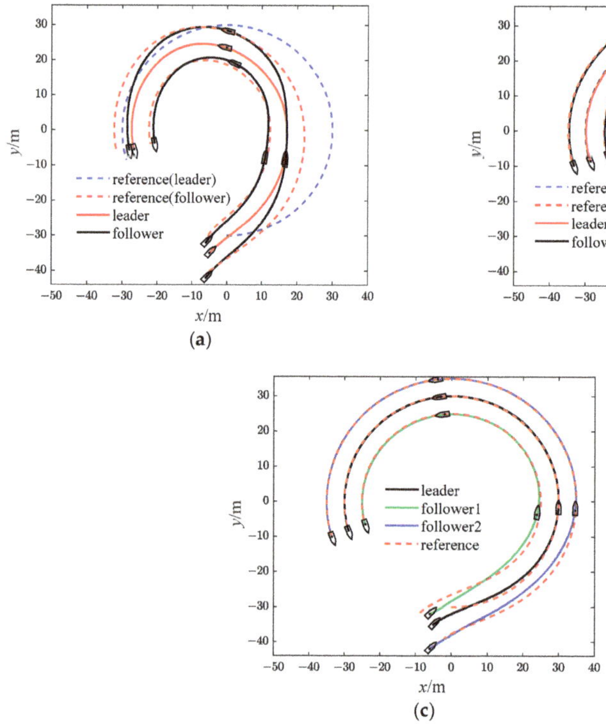

Figure 5. Diagrams of USVs trajectory (Case 1). (**a**)with SMC; (**b**) with SMC and RBF; (**c**) with bio-SMC and RBF.

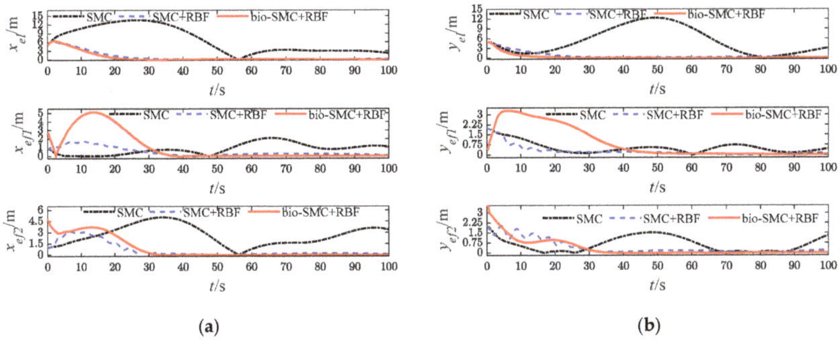

Figure 6. Diagrams of USVs' tracking error (Case 1). (**a**) longitudinal position error; (**b**) lateral position error.

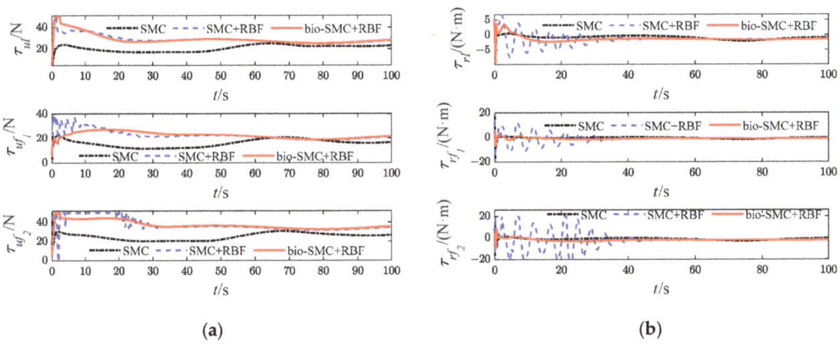

Figure 7. Diagrams of USVs' control input signals (Case 1). (**a**) surge force; (**b**) yaw moment.

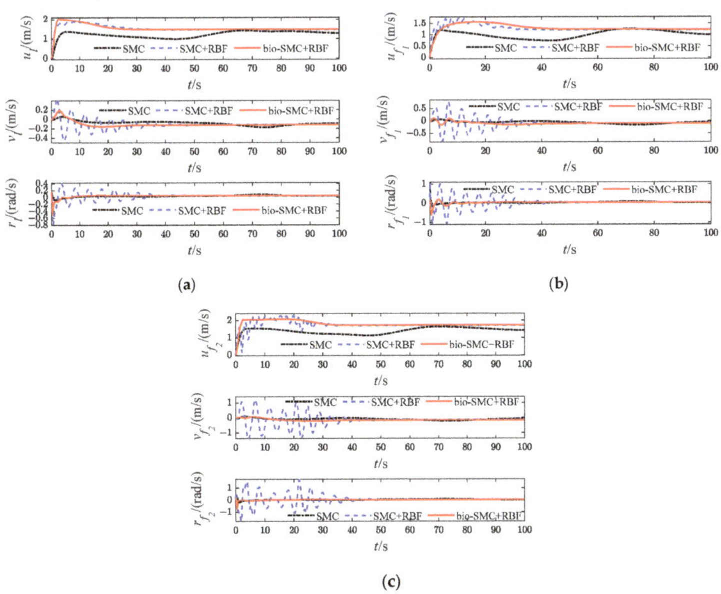

Figure 8. Diagrams of USVs' velocity variables (Case 1). (**a**) Leader USV; (**b**) Follower USV1; (**c**) Follower USV2.

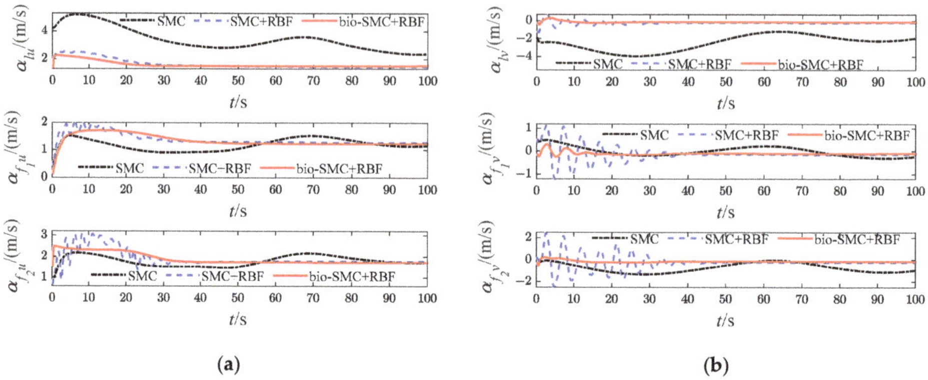

Figure 9. Diagrams of USVs' virtual velocity variables (Case 1). (a) longitudinal virtual velocity; (b) lateral virtual velocity.

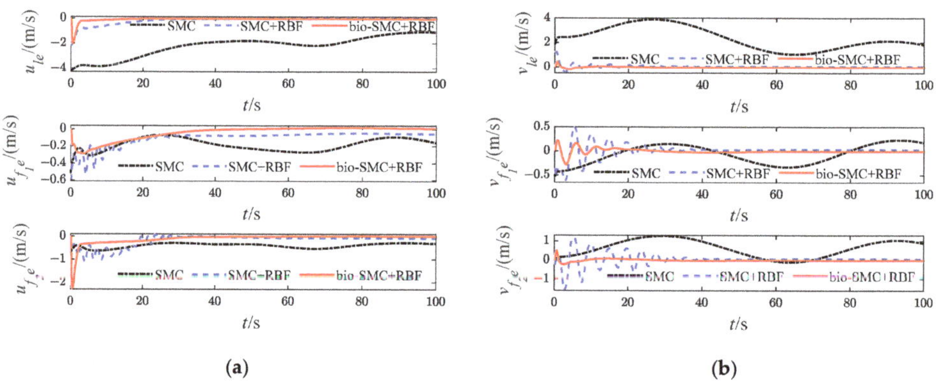

Figure 10. Diagrams of USVs' velocity error (Case 1). (a) longitudinal velocity error; (b) lateral velocity error.

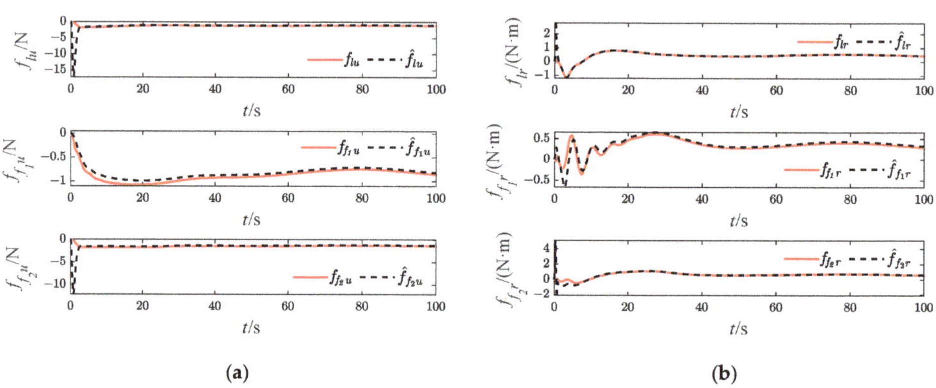

Figure 11. Approximation results of USVs (Case 1). (a) surge dynamic damping; (b) yaw dynamic damping.

Case 2. Straight line trajectory

The desired straight line trajectory of the leader USV is described as: $x_d = t, y_d = t$. The initial state information for each USV in the formation system is presented in Table 4.

Table 4. USVs initial state (Case 2).

Parameter	Leader USV	Follower USV1	Follower USV2
η	$[-5(m), -2(m), 0]^T$	$[-12(m), -15(m), 90(°)]^T$	$[-4(m), -20(m), -90(°)]^T$
v	$[0,0,0]^T$	$[0,0,0]^T$	$[0,0,0]^T$
L		$[-5(m), 10(m)]^T$	$[-5(m), -10(m)]^T$

The parameters of the RBF neural network are same as for Case 1, besides the width of the Gaussian basis function, and the width of the Gaussian basis function is 2. The controller parameters are presented in Table 5.

Table 5. USVs' controller parameters (Case 2).

Parameter	Leader USV	Follower USV1	Follower USV2
C	50	50	40
k_1	3.5	3.5	2
k_2	6	6	3
λ_1	0.012	0.012	0.001
λ_2	6.5	5	5
λ_3	0.001	0.05	0.05
ε_1	0.05	0.05	0.05
ε_2	0.01	0.01	0.01
σ_1	0.05	0.1	0.1
σ_2	0.01	0.01	0.01
B_1	2.5	5.5	5.5
B_2	4	4	4
C_1, D_1	4	7	7
C_2, D_2	4	4	4
ζ_1	3	8	8
ζ_2	0.01	0.025	0.025
γ_1	0.008	0.08	0.001
γ_2	0.1	0.1	0.1

The simulation results of Case 2 are shown in Figures 12–18.

Case 3. Sinusoidal trajectory

The desired sinusoidal trajectory of the leader USV is described as $x_d = t$, $y_d = 25\sin(0.05t) - 35$. The initial state information for each USV in the formation system is presented in Table 6.

Table 6. USVs' initial state (Case 3).

Parameter	Leader USV	Follower USV1	Follower USV2
η	$[-5(m), -35(m), 45(°)]^T$	$[-6(m), -32(m), 45(°)]^T$	$[-6(m), -42(m), 45(°)]^T$
v	$[0,0,0]^T$	$[0,0,0]^T$	$[0,0,0]^T$
L		$[0(m), 5(m)]^T$	$[0(m), -5(m)]^T$

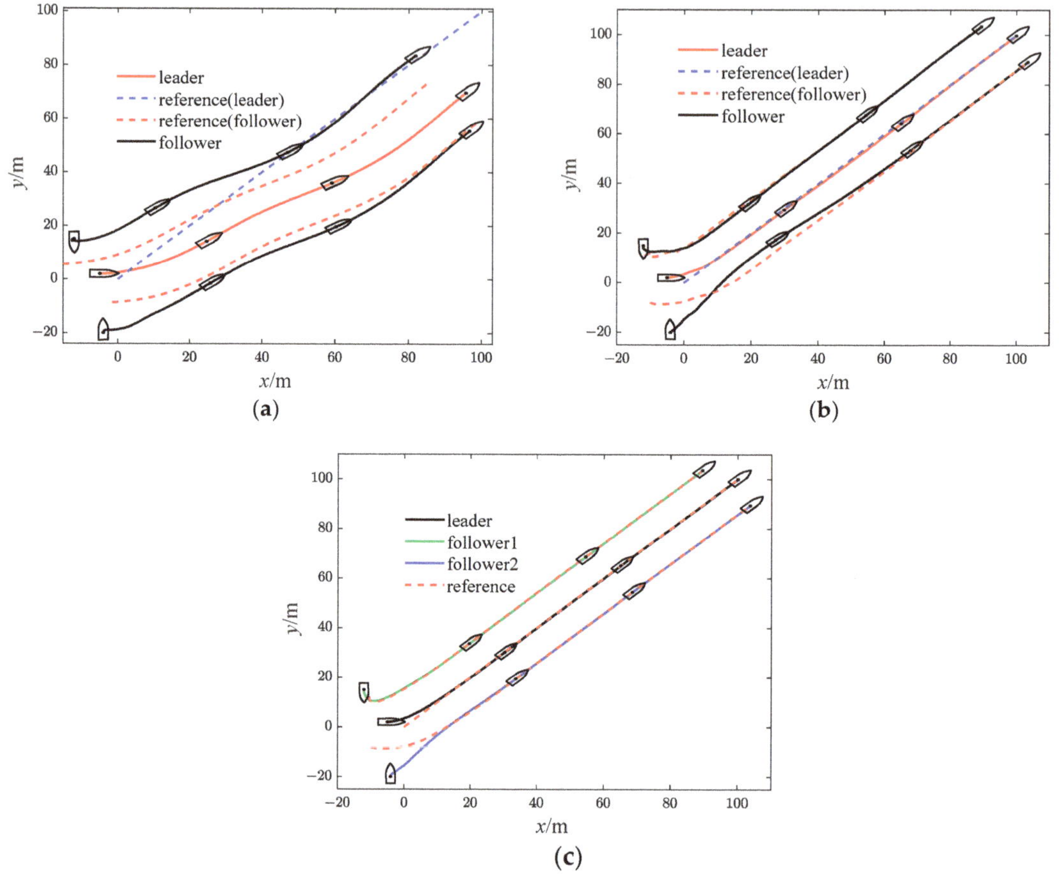

Figure 12. Diagrams of USVs' trajectory (Case 2). (**a**)with SMC; (**b**) with SMC and RBF; (**c**) with bio-SMC and RBF.

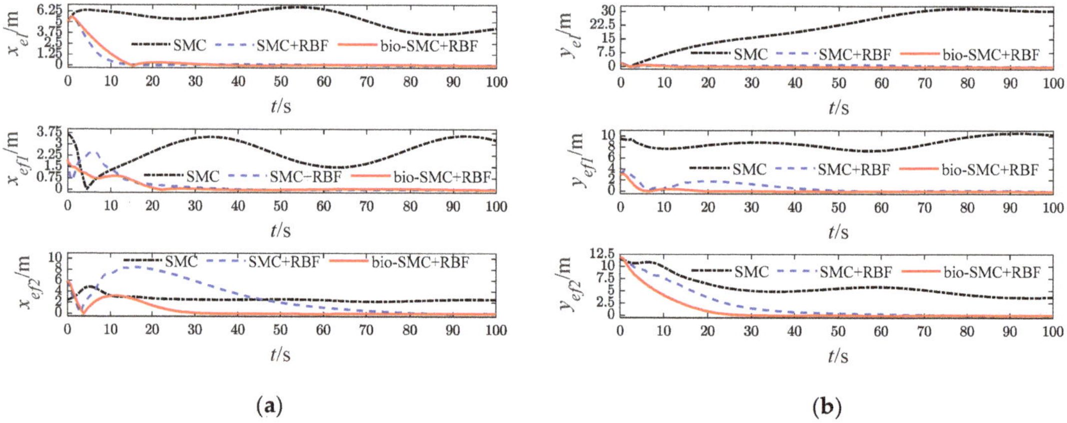

Figure 13. Diagrams of USVs' tracking error (Case 2). (**a**) longitudinal position error; (**b**) lateral position error.

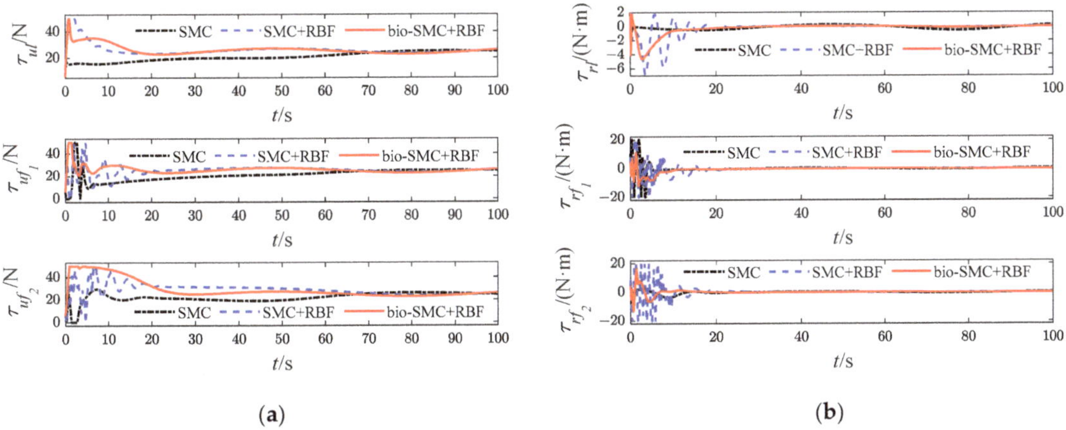

Figure 14. Diagrams of USVs' control input signals (Case 2). (**a**) surge force; (**b**) yaw moment.

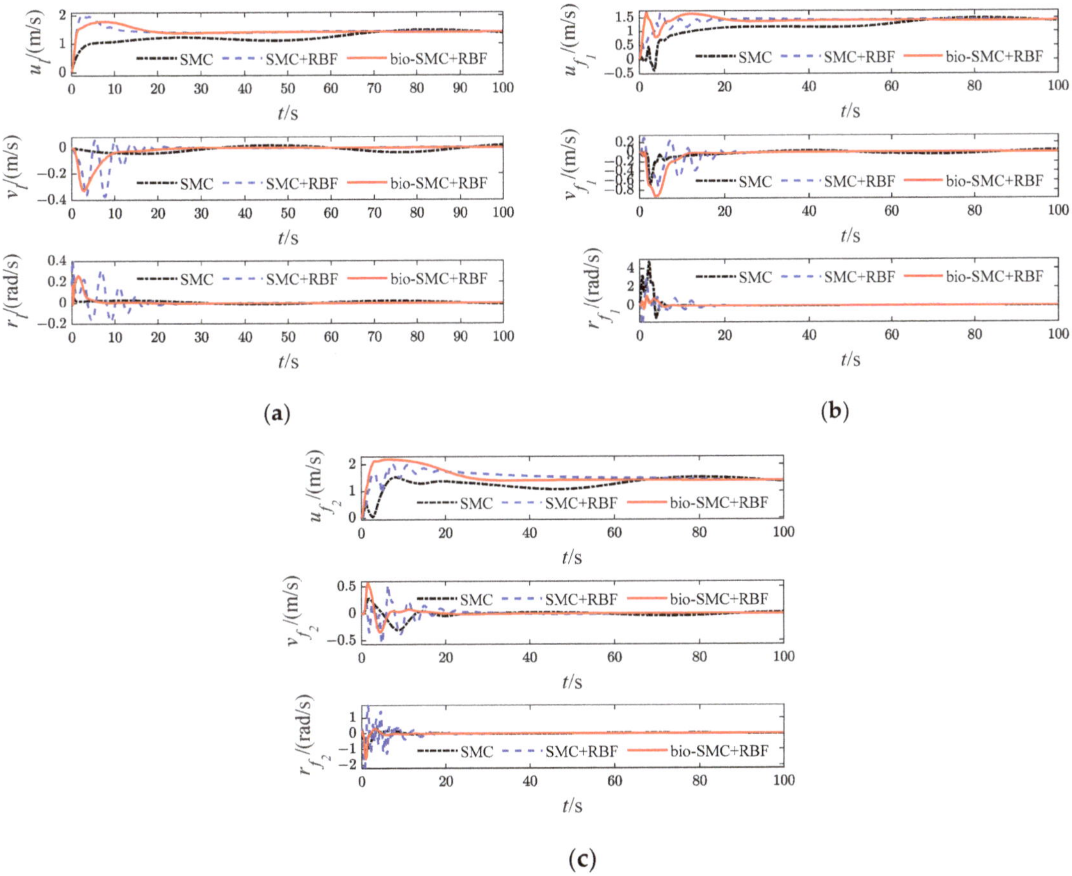

Figure 15. Diagrams of USVs' velocity variables (Case 2). (**a**) Leader USV; (**b**) Follower USV1; (**c**) Follower USV2.

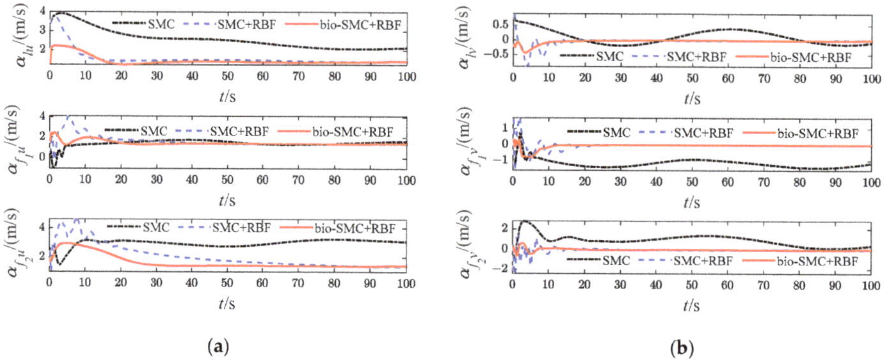

Figure 16. Diagrams of USVs' virtual velocity variables (Case 2). (**a**) longitudinal virtual velocity; (**b**) lateral virtual velocity.

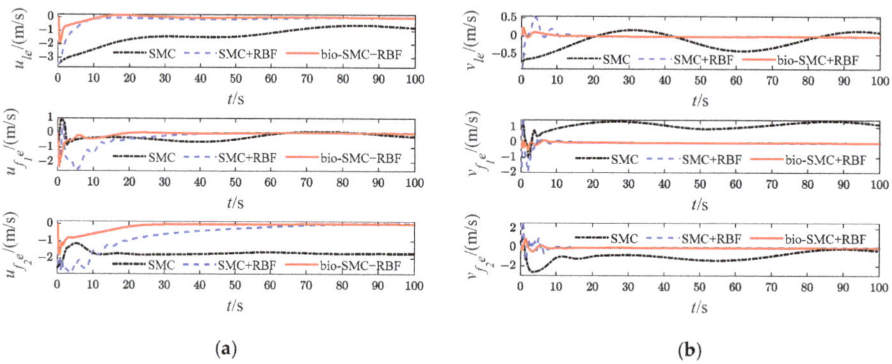

Figure 17. Diagrams of USVs' velocity error (Case 2). (**a**) longitudinal velocity error; (**b**) lateral velocity error.

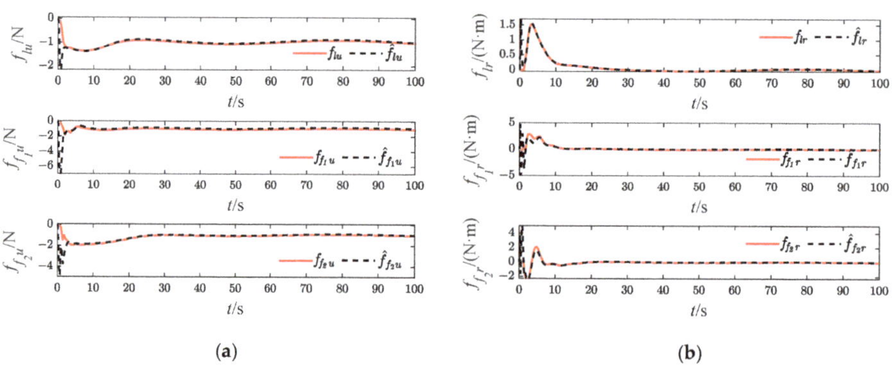

Figure 18. Approximation results of USVs' (Case 2). (**a**) surge dynamic damping; (**b**) yaw dynamic damping.

The parameters of the RBF neural network are same as for Case 2, and the controller parameters are presented in Table 7.

Table 7. USVs' controller parameters (Case 3).

Parameter	Leader USV	Follower USV1	Follower USV2
C	30	30	40
k_1	2	9.5	9
k_2	3.5	4	5
λ_1	0.0001	0.0001	0.0001
λ_2	1.2	1.2	1.2
λ_3	0.001	0.01	0.001
ε_1	0.01	0.05	0.05
ε_2	0.01	0.01	0.01
σ_1	0.02	0.01	0.01
σ_2	0.1	0.01	0.01
B_1	4	4	4
B_2	4	6.7	7.75
C_1, D_1	4	4	4
C_2, D_2	4	4	4
ζ_1	1	0.01	0.1
ζ_2	0.015	0.01	0.006
γ_1	0.0005	0.0005	0.0005
γ_2	0.001	0.001	0.001

The simulation results of Case 3 are shown in Figures 19–25.

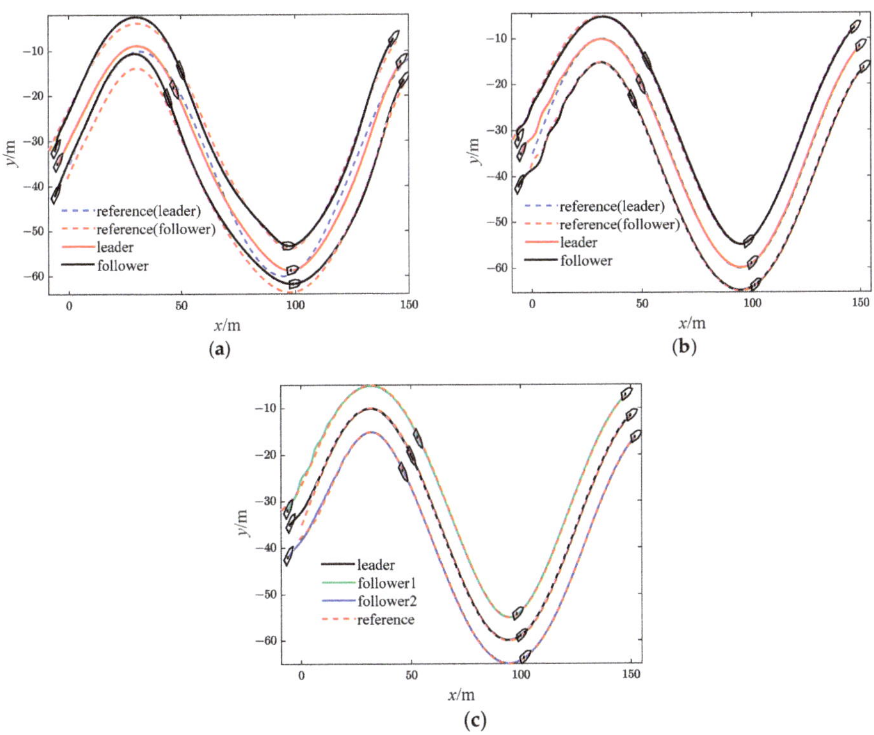

Figure 19. Diagrams of USVs' trajectory (Case 3). (**a**)with SMC; (**b**) with SMC and RBF; (**c**) with bio-SMC and RBF.

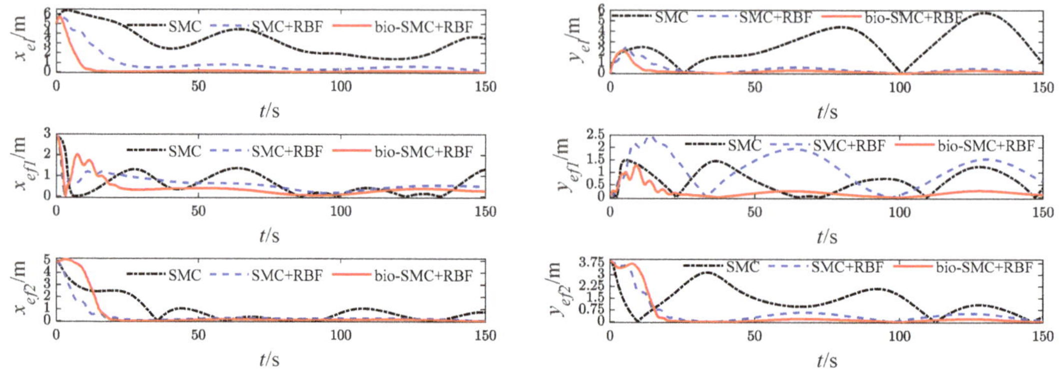

(a) (b)

Figure 20. Diagrams of USVs' tracking error (Case 3). (**a**) longitudinal position error; (**b**) lateral position error.

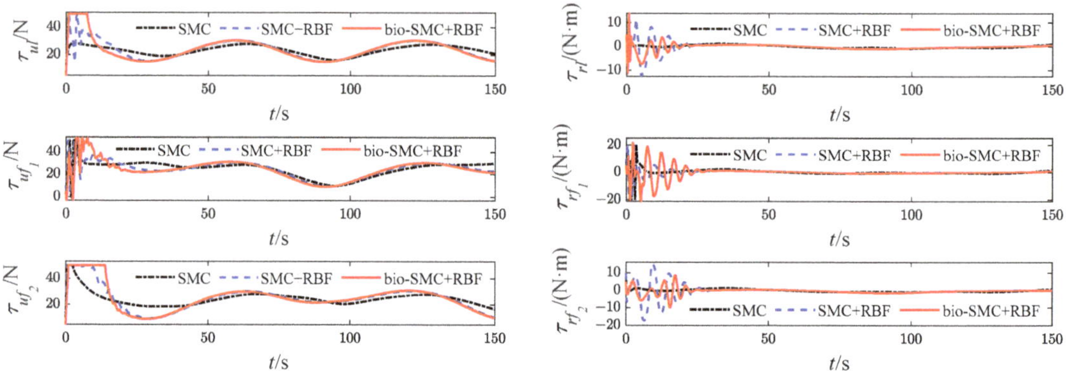

(a) (b)

Figure 21. Diagrams of USVs' control input signals (Case 3). (**a**) surge force; (**b**) yaw moment.

Case 4. Combination trajectory of straight line and circle

The desired sinusoidal trajectory of the leader USV is described as follows:

$$\begin{cases} x_d = t - 90, y_d = -25, 0 \le t < 90 \text{ s} \\ x_d = 25\sin(0.05(t-90)), y_d = 25\cos(0.05(t-90)), t \ge 90 \text{ s} \end{cases}$$

The initial state information for each USV is presented in Table 8.

Table 8. USVs' initial state (Case 4).

Parameter	Leader USV	Follower USV1	Follower USV2
η	$[-90(m), -30(m), 0(°)]^T$	$[-95(m), -20(m), 0(°)]^T$	$[-95(m), -40(m), 0(°)]^T$
v	$[0,0,0]^T$	$[0,0,0]^T$	$[0,0,0]^T$
L		$[0(m), 5(m)]^T$	$[0(m), -5(m)]^T$

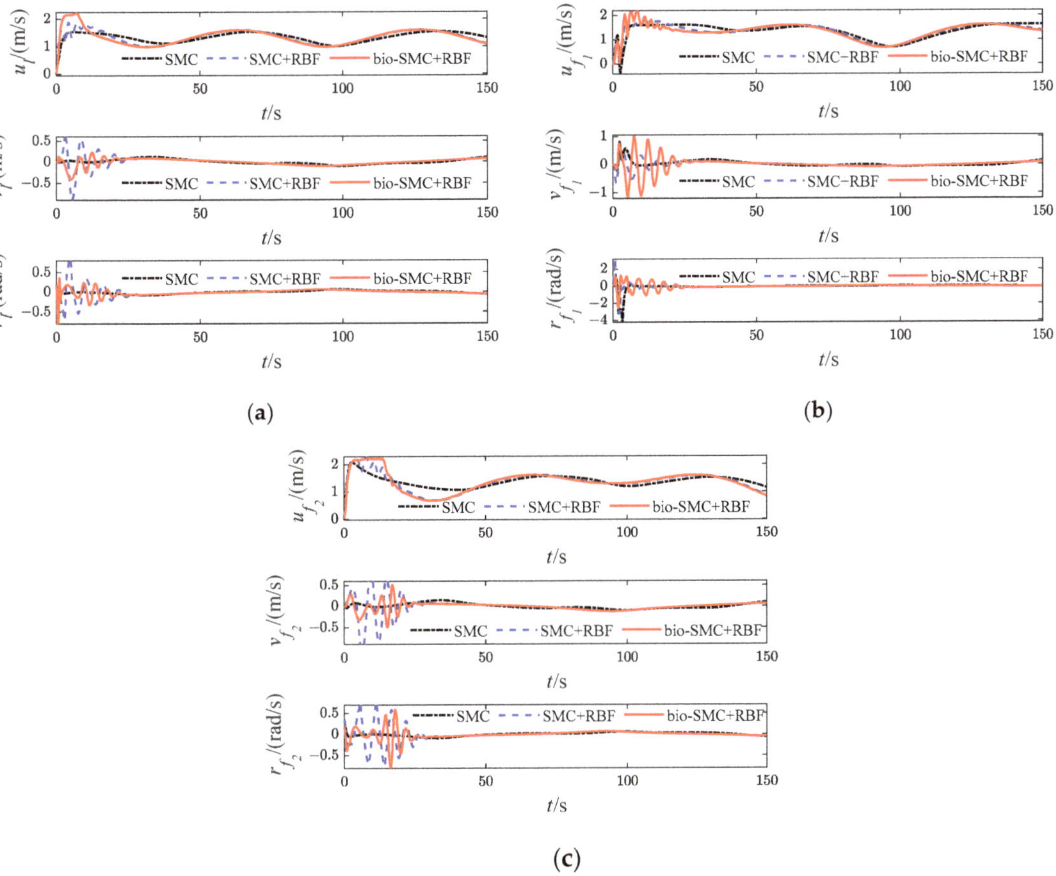

Figure 22. Diagrams of USVs' velocity variables (Case 3). (**a**) Leader USV; (**b**) Follower USV1; (**c**) Follower USV2.

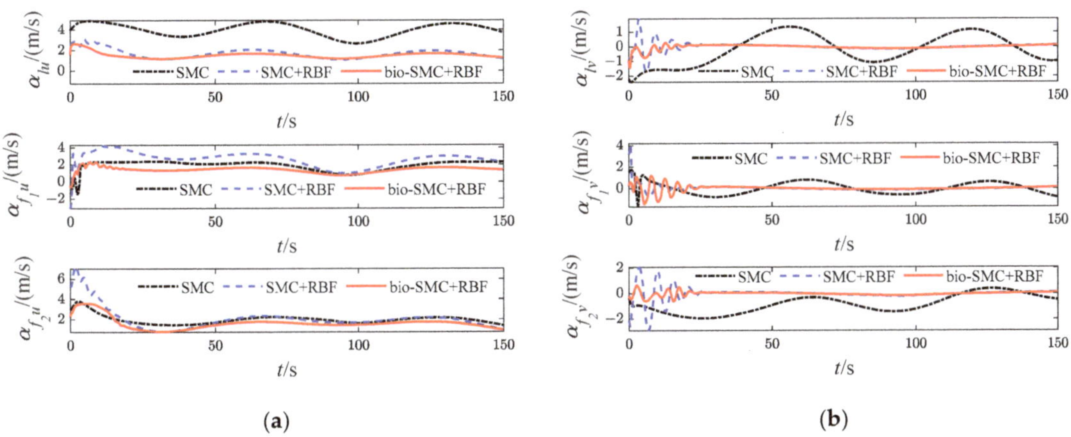

Figure 23. Diagrams of USVs' virtual velocity variables (Case 3). (**a**) longitudinal virtual velocity; (**b**) lateral virtual velocity.

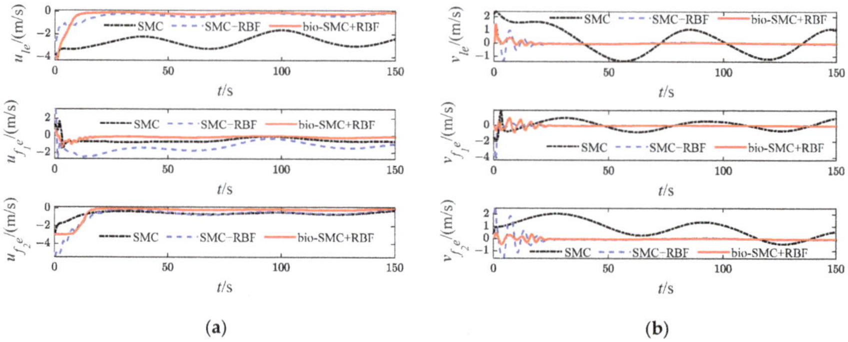

Figure 24. Diagrams of USVs' velocity error (Case 3). (**a**) longitudinal velocity error; (**b**) lateral velocity error.

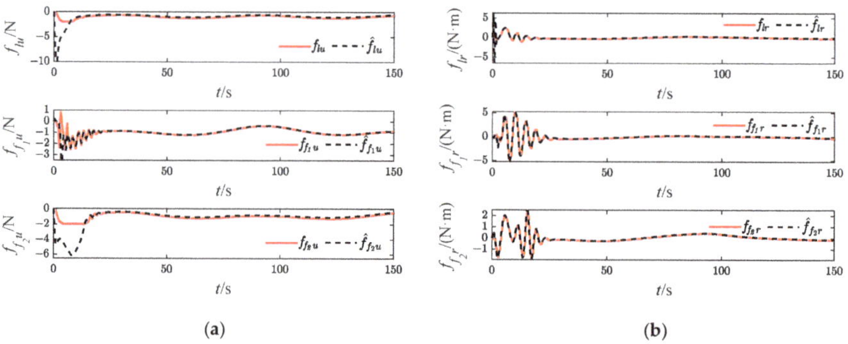

Figure 25. Approximation results of USVs' (Case 3). (**a**) surge dynamic damping; (**b**) yaw dynamic damping.

The parameters of the RBF neural network are same as for Case 1, and the controller parameters are presented in Table 9.

Table 9. USVs' controller parameters (Case 4).

Parameter	Leader USV	Follower USV1	Follower USV2
C	30	30	30
k_1	5	0.8	1.4
k_2	3	2	1.94
λ_1	0.002	0.002	0.002
λ_2	2.5	3.1	2
λ_3	0.001	0.001	0.001
ε_1	0.01	0.01	0.01
ε_2	0.05	0.05	0.05
σ_1	0.1	0.01	0.01
σ_2	0.01	0.01	0.01
B_1	2.5	5.5	6
B_2	4	4	4
C_1, D_1	4	6.7	7.75
C_2, D_2	4	4	4
ζ_1	1	60	60
ζ_2	0.015	0.01	0.05
γ_1	0.0005	0.0005	0.0005
γ_2	0.001	0.0001	0.001

The simulation results of Case 4 are shown in Figures 26–32.

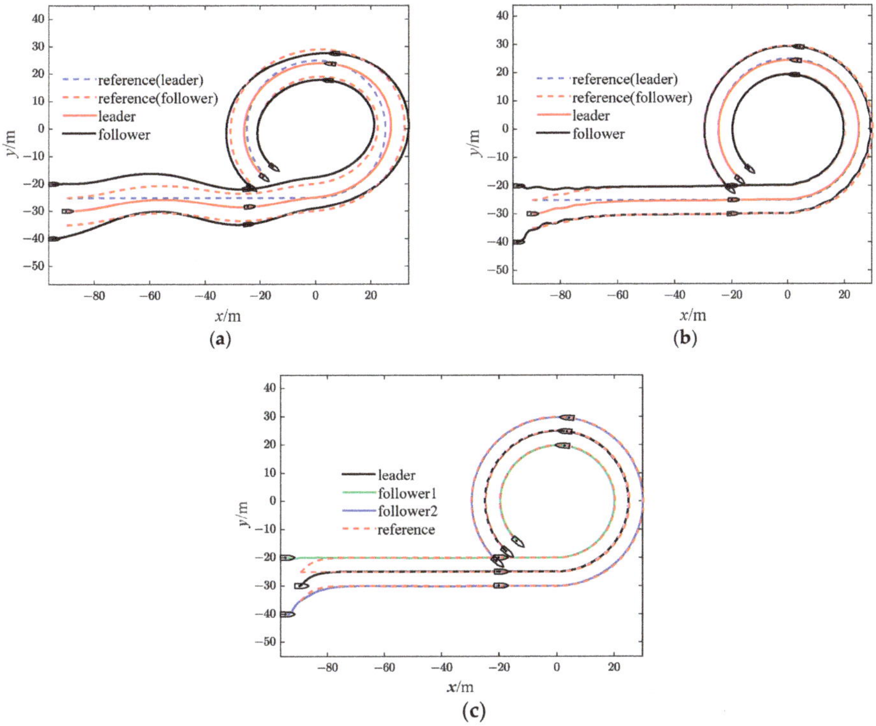

Figure 26. Diagrams of USVs' trajectory (Case 4). (**a**) with SMC; (**b**) with SMC and RBF; (**c**) with bio-SMC and RBF.

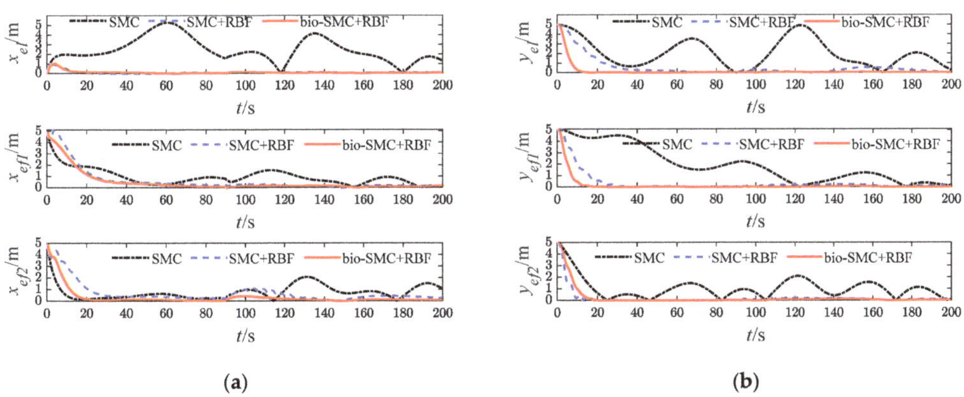

Figure 27. Diagrams of USVs' tracking error (Case 4). (**a**) longitudinal position error; (**b**) lateral position error.

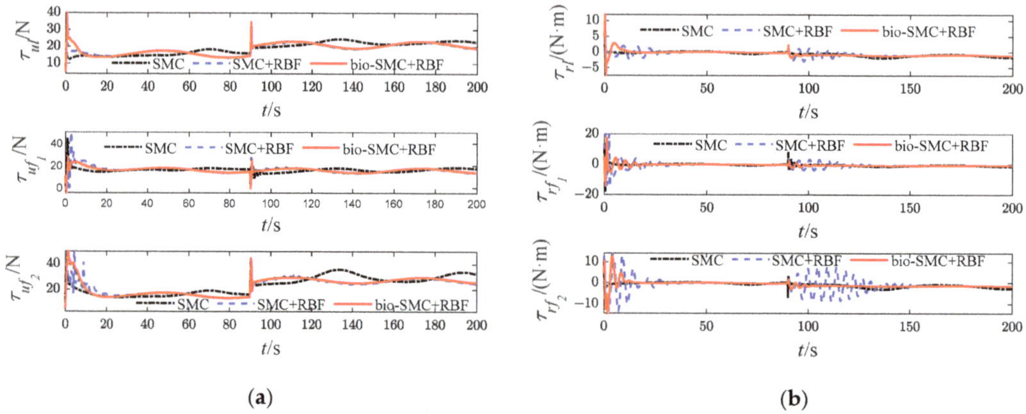

Figure 28. Diagrams of USVs' control input signals (Case 4). (**a**) surge force; (**b**) yaw moment.

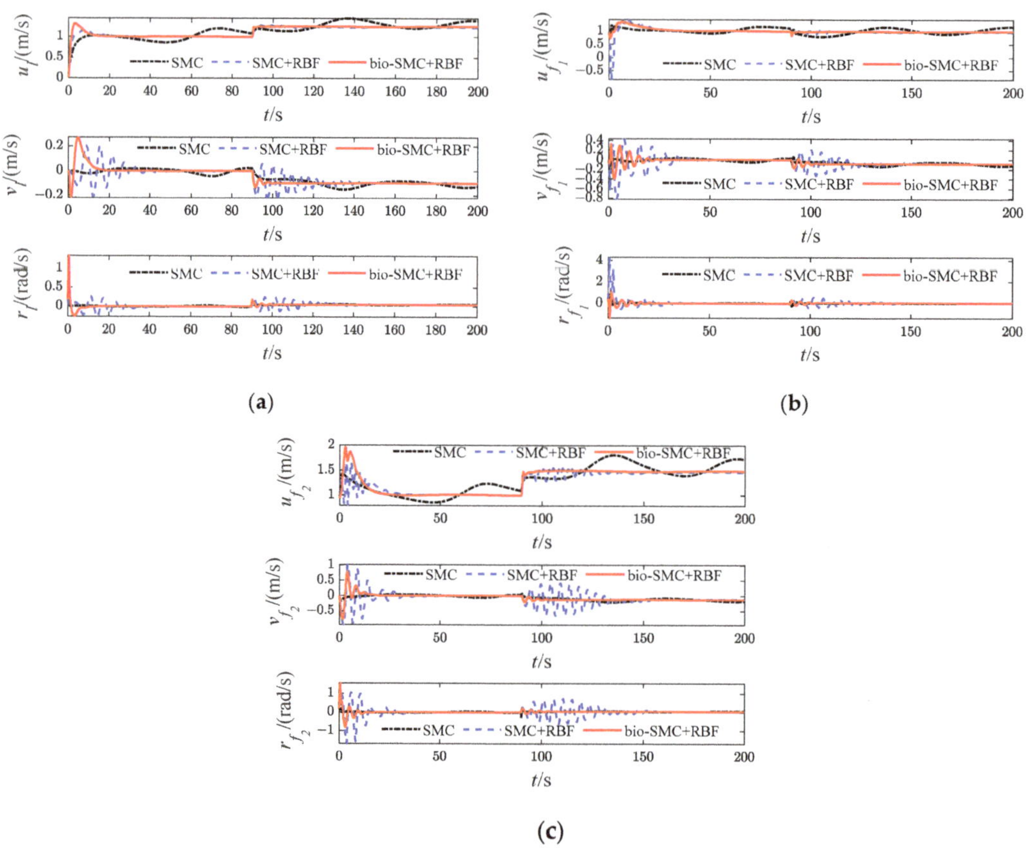

Figure 29. Diagrams of USVs' velocity variables (Case 4). (**a**) Leader USV; (**b**) Follower USV1; (**c**) Follower USV2.

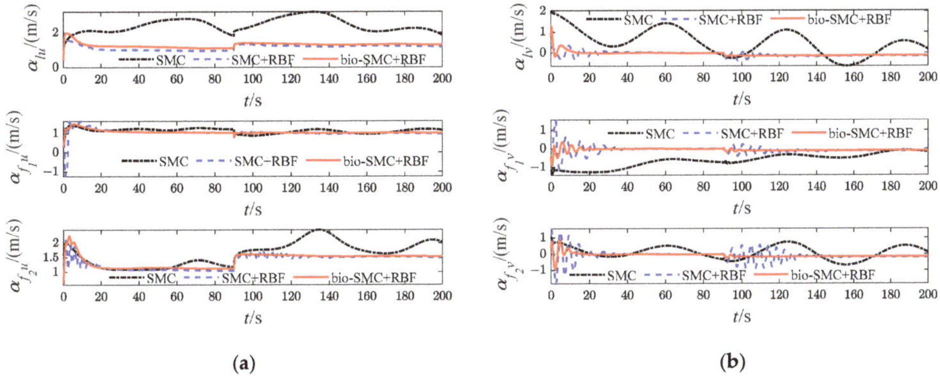

Figure 30. Diagrams of USVs' virtual velocity variables (Case 4). (**a**) longitudinal virtual velocity; (**b**) lateral virtual velocity.

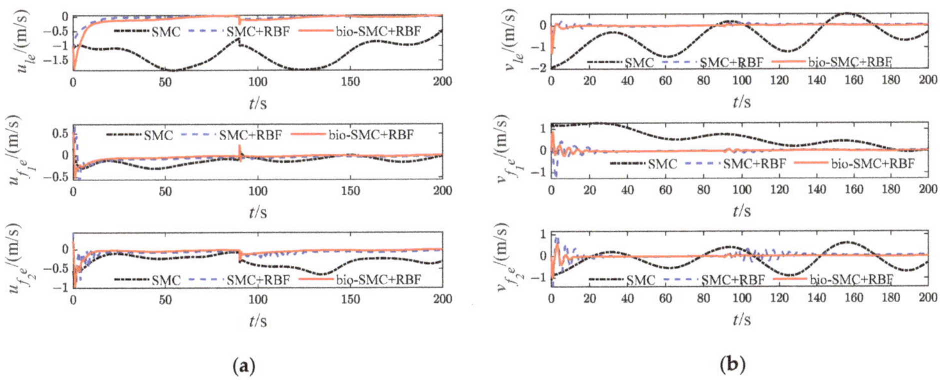

Figure 31. Diagrams of USVs' velocity error (Case 4). (**a**) longitudinal velocity error; (**b**) lateral velocity error.

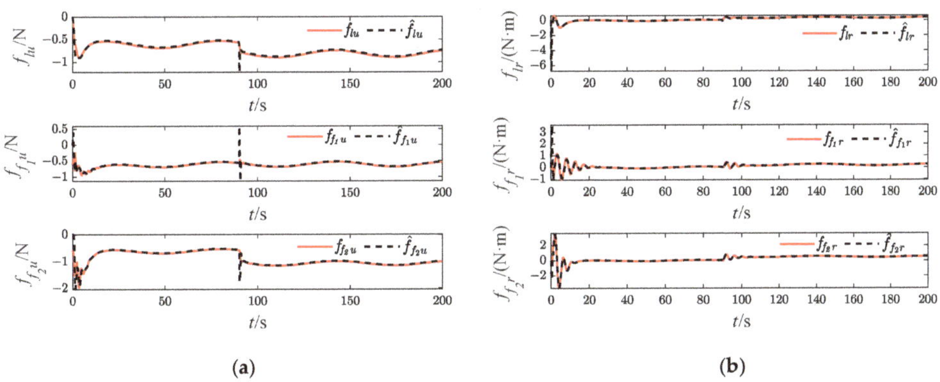

Figure 32. Approximation results of USVs' (Case 4). (**a**) surge dynamic damping; (**b**) yaw dynamic damping.

The effectiveness of the RBF and the bio-inspired sliding mode control (bio-SMC) are analyzed through the above four experiment cases, in which Figures 5a, 12a, 19a and 26a show the trajectory tracking results of the traditional sliding mode control (SMC) without RBF, while Figures 5b, 12b, 19b and 26b indicate the formation trajectories with RBF. Through comparative analysis of these figures, USVs are able to track the desired trajectory with RBF under the influence caused by ocean disturbances and model uncertainties. Combining Figures 6, 13, 20 and 27, after the formation system is stable, the tracking errors of USVs are generally greater than 1 m without RBF, but the tracking errors are far less than 0.5 m without RBF. Similarly, as seen from the velocity error plots, which include Figures 10, 17, 24 and 31, the velocity errors of USVs are smaller with RBF. Meanwhile, according to Figures 11, 18, 25 and 32, the RBF can accurately observe ocean disturbances and model uncertainties of USVs, and the effectiveness of RBF can be intuitively verified.

Then, as shown in Figures 5c, 12c, 19c and 26c, which indicate the results of formation trajectories with bio-SMC and RBF, it is found that the fluctuation of the bio-SMC trajectory is smaller compared to Figures 5b, 12b, 19b and 26b in the early stage of trajectory tracking. Meanwhile, as seen from Figures 9, 16, 23 and 30, where the virtual velocity variables can be obtained, by utilizing bio-SMC, the virtual velocity changes become smoother within the first 30 s. Also, by analyzing Figures 8, 15, 22 and 29, the bio-SMC has a significant effect on stabilizing these velocity variables within the first 30 s.

According to Figures 6, 13, 20 and 27, the bio-SMC has no adverse effect on USVs' trajectory tracking accuracy when the formation system is stable. Figures 7, 14, 21 and 28 illustrate that within the first 30 s, the controllers of USVs designed by bio-SMC can obtain smoother control inputs. Therefore, based on the above analysis, the bio-SMC has better control capability compared to the traditional sliding mode control.

6. Conclusions

To address the cooperative formation control problem of underactuated USVs with external environmental disturbances and model uncertainties, a bio-inspired sliding mode control (bio-SMC) and the RBF neural network with minimal learning parameter (MLP) have been designed in this paper. The virtual control laws of the longitudinal and lateral velocities are designed by using a backstepping approach, and the controller design has become more convenient. The sliding surface has been designed with a hyperbolic tangent function to avoid excessive oscillations. Meanwhile, a bio-inspired model has been proposed to smooth the virtual velocities successfully and to avoid the problem of differential explosion caused by repeated derivation. Furthermore, the RBF neural network with MLP algorithm has been adopted to compensate for the ocean disturbances and uncertain system. The effectiveness and reliability of the proposed method have been verified by simulation experiments, and all the conclusions are verified by simulation experiment results. This paper is highly focused on the theoretical aspects with less emphasis on practical deployment, and future research will be devoted towards extending the current framework into real-world applicability.

Author Contributions: Conceptualization, Z.D. and F.T.; methodology, Z.D. and F.T.; software, F.T.; validation, Z.D., F.T. and M.Y.; formal analysis, Z.D.; investigation, F.T.; resources, Z.D.; data curation, Z.D. and F.T.; writing—original draft preparation, Z.D. and F.T. and M.Y.; writing—review and editing, Z.D., F.T., M.Y., Y.X. and Z.L.; visualization, Z.D., F.T. and M.Y.; supervision, M.Y.; project administration, M.Y.; funding acquisition, M.Y. All authors have read and agreed to the published version of the manuscript.

Funding: This research was funded by the National Natural Science Foundation of China under Grant No. 51709214 and 51779052, the China Postdoctoral Science Foundation funded project (No. 2018M642939, 2019T120693).

Institutional Review Board Statement: Not applicable.

Informed Consent Statement: Not applicable.

Data Availability Statement: Data are contained within the article.

Acknowledgments: The authors gratefully acknowledge the support provided by the National Natural Science Foundation of China and the National Natural Science Foundation of China.

Conflicts of Interest: The authors declare no conflicts of interest.

References

1. Thanh, P.N.N.; Tom, P.M.; Anh, H.P.H. A new approach for three-dimensional trajectory tracking control of under-actuated AUVs with model uncertainties. *Ocean Eng.* **2021**, *228*, 108951. [CrossRef]
2. Tong, H.Y.; Fu, X.L.; Wang, H.B.; Shang, Z.; Wang, J.J. Global finite-time guidance algorithm and constrained formation control using novel nonlinear mapping function for underactuated multiple unmanned surface vehicles. *Ocean Eng.* **2024**, *293*, 116756. [CrossRef]
3. Hao, Y.; Hu, K.; Liu, L.; Li, J.X. Distributed dynamic event-triggered flocking control for multiple unmanned surface vehicles. *Ocean Eng.* **2024**, *309*, 118307. [CrossRef]
4. Krell, E.; King, S.A.; Carrillo, L.R.G. Autonomous surface vehicle energy-efficient and reward-based path planning using particle swarm optimization and visibility graphs. *Appl. Ocean Res.* **2022**, *122*, 103125. [CrossRef]
5. Chen, D.; Zhang, J.; Li, Z. A novel fixed-time trajectory tracking strategy of unmanned surface vessel based on the fractional sliding mode control method. *Electronics* **2022**, *11*, 726. [CrossRef]
6. Ghommam, J.; Saad, M.; Mnif, F.; Zhu, Q.M. Guaranteed performance design for formation tracking and collision avoidance of multiple USVs with disturbances and unmodeled dynamics. *IEEE Syst. J.* **2021**, *15*, 4346–4357. [CrossRef]
7. Rowinski, L.; Kaczmarczyk, M. Evaluation of effectiveness of waterjet propulsor for a small underwater vehicle. *Pol. Marit. Res.* **2022**, *28*, 30–41. [CrossRef]
8. Miao, R.; Wang, L.; Pang, S. Coordination of distributed unmanned surface vehicles via model-based reinforcement learning methods. *Appl. Ocean Res.* **2022**, *122*, 103106. [CrossRef]
9. Azarbahram, A.; Pariz, N.; Naghibi-Sistani, M.B.; Moghaddam, R.K. Platoon of uncertain unmanned surface vehicle teams subject to stochastic environmental loads. *Int. J. Adapt. Control Signal Process.* **2022**, *36*, 729–750. [CrossRef]
10. Duan, H.; Yuan, Y.; Zeng, Z. Distributed robust learning control for multiple unmanned surface vessels with fixed-time prescribed performance. *IEEE Trans. Syst. Man Cybern.-Syst* **2024**, *54*, 787–799. [CrossRef]
11. Jiang, X.; Xia, G. Sliding mode formation control of leaderless unmanned surface vehicles with environmental disturbances. *Ocean Eng.* **2022**, *244*, 110301. [CrossRef]
12. Peng, Z.H.; Wang, J.; Wang, D.; Han, Q.L. An overview of recent advances in coordinated control of multiple autonomous surface vehicles. *IEEE Trans. Ind. Inform.* **2021**, *17*, 732–745. [CrossRef]
13. Zhang, Q.; Zhang, S.H.; Liu, Y.; Zhang, Y.; Hu, Y.C. Adaptive terminal sliding mode control for USV-ROVs formation under deceptive attacks. *Front. Mar. Sci.* **2024**, *11*, 1320361. [CrossRef]
14. Fahimi, F. Sliding-mode formation control for underactuated surface vessels. *IEEE Trans. Robot.* **2007**, *23*, 617–622. [CrossRef]
15. Dong, Z.P.; Zhang, Z.Q.; Qi, S.J.; Zhang, H.S.; Li, J.K.; Liu, Y.C. Autonomous cooperative formation control of underactuated USVs based on improved MPC in complex ocean environment. *Ocean Eng.* **2023**, *270*, 113633. [CrossRef]
16. Zhao, Y.J.; Ma, Y.; Hu, S.L. USV formation and path-following control via deep reinforcement learning with random braking. *IEEE Trans. Neural Netw. Learn. Syst.* **2021**, *32*, 5468–5478. [CrossRef]
17. Liu, Z.W.; Hou, H.; Wang, Y.W. Formation-containment control of multiple underactuated surface vessels with sampling communication via hierarchical sliding mode approach. *ISA Trans.* **2022**, *124*, 458–467. [CrossRef]
18. Zhang, L.; Zheng, Y.; Huang, B. Finite-time trajectory tracking control for under-actuated unmanned surface vessels with saturation constraint. *Ocean Eng.* **2022**, *249*, 110745. [CrossRef]
19. Jiang, X.L.; Xia, G.H.; Feng, Z.G.; Wu, Z.G. Nonfragile formation seeking of unmanned surface vehicles: A sliding mode control approach. *IEEE Trans. Netw. Sci. Eng.* **2022**, *9*, 431–444. [CrossRef]
20. Zhu, Y.; Bai, J.; Li, S.; Guo, G. Selection strategies and finite-time target tracking of multiple unmanned surface vehicles with mode uncertainty and disturbances. *Ocean Eng.* **2023**, *283*, 115088. [CrossRef]
21. Huang, B.; Song, S.; Zhu, C.; Li, J.; Zhou, B. Finite-time distributed formation control for multiple unmanned surface vehicles with input saturation. *Ocean Eng.* **2021**, *233*, 109158. [CrossRef]
22. Dong, Z.P.; Qi, S.J.; Yu, M.; Zhang, Z.Q.; Zhang, H.S.; Li, J.K.; Liu, Y. An improved dynamic surface sliding mode method for autonomous cooperative formation control of underactuated USVs with complex marine environment disturbances. *Pol. Marit. Res.* **2022**, *29*, 47–60. [CrossRef]
23. Zou, Y.H.; Kun, W.; Wei, S.; Zhou, L.; Zheng, Z.Z.; Hao, J.G. Back-stepping formation control of unmanned surface vehicles with input saturation based on adaptive super-twisting algorithm. *IEEE Access* **2022**, *10*, 114885–114896.
24. Shen, H.L.; Yin, Y.; Qian, X.B. Fixed-time formation control for unmanned surface vehicles with parametric uncertainties and complex disturbance. *J. Mar. Sci. Eng* **2022**, *10*, 1246. [CrossRef]
25. Sui, B.W.; Zhang, J.Q.; Liu, Z.; Wei, J.B. Distributed prescribed-time cooperative formation tracking control of networked unmanned surface vessels under directed graph. *Ocean Eng.* **2024**, *305*, 117993. [CrossRef]

26. Perez, T.; Fossen, T. Kinematic models for maneuvering and sea keeping of marine vessels. *Model. Identif. Control* **2007**, *28*, 19–30. [CrossRef]
27. Fu, M.; Yu, L. Finite-time extended state observer-based distributed formation control for marine surface vehicles with input saturation and disturbances. *Ocean Eng.* **2018**, *159*, 219–227. [CrossRef]
28. Do, K.D.; Pan, J. Global robust adaptive path following of underactuated ships. *Automatic* **2006**, *42*, 1713–1722. [CrossRef]

Journal of
Marine Science
and Engineering

MDPI

Article

Graph Matching for Underwater Simultaneous Localization and Mapping Using Multibeam Sonar Imaging

Lingfei Zhuang, Xiaofeng Chen, Wenjie Lu * and Yiting Yan

Guangdong Provincial Key Laboratory of Intelligent Morphing Mechanisms and Adaptive Robotics, the School of Mechanical Engineering and Automation, Harbin Institute of Technology (Shenzhen), Shenzhen 518055, China; 13600939547@163.com (L.Z.); 15156058248@163.com (X.C.); yitingyan0805@163.com (Y.Y.)
* Correspondence: luwenjie@hit.edu.cn

Abstract: This paper addresses the challenges of underwater Simultaneous Localization and Mapping (SLAM) using multibeam sonar imaging. The widely used Iterative Closest Point (ICP) often falls into local optima due to non-convexity and the lack of features for correct registration. To overcome this, we propose a novel registration algorithm based on Gaussian clustering and Graph Matching with maximal cliques. The proposed approach enhances feature-matching accuracy and robustness in complex underwater environments. Inertial measurements and velocity estimates are also fused for global state estimation. Comprehensive tests in simulated and real-world underwater environments have demonstrated that the proposed registration method effectively addresses the issue of the ICP algorithm easily falling into local optima while also exhibiting excellent inter-frame registration performance and robustness.

Keywords: underwater SLAM; multibeam sonar; registration; Graph Matching

1. Introduction

In recent years, autonomous underwater robots have been used in extensive applications, such as pipeline inspection, geological structure mapping, rescue operations, and resource exploration [1]. These robots' autonomous positioning and navigation control in complex environments necessitate environmental modeling and robot state estimation as inputs. Particularly in ocean environments that still need to be mapped, the limitations of communication infrastructure and remote control operations make autonomous exploration capabilities even more critical. Simultaneous Localization and Mapping (SLAM) technology is vital in this context. The Global Navigation Satellite System (GNSS) and other communication systems, such as Wi-Fi or electromagnetic waves, are significantly attenuated and thus unusable in underwater settings. Autonomous Underwater Vehicles (AUVs) primarily rely on dead reckoning for navigation; however, this method tends to drift over time. While cameras can capture detailed underwater scenes, their effectiveness is constrained by water turbidity and manual control or data processing limitations in such environments. Sonar is a commonly used alternative sensor. However, challenges such as low signal-to-noise ratios, low resolution, uneven sound waves, and multi-path reflections hinder the accurate extraction and registration of target features. Therefore, we address these challenges and focus on solving positioning and mapping issues underwater robots face when navigating complex ocean environments.

Sonar-based underwater SLAM has been a subject of research for many years. Earlier studies have described sonar inertia-based SLAM systems for underwater reconstruction and navigation [2–4]. Corke et al. [5] compared acoustic and visual underwater positioning methods, demonstrating their applicability in specific scenarios. Subsequently, research on sonar-based underwater SLAM gradually progressed, with Folkesson et al. [6] utilizing blazed array sonar sensors for real-time feature tracking. A system employing resampled features from sub-sonar is detailed in [7]. The sunfish model launching robot described in [8] employs multibeam sonar, a Fiber Optic Gyroscope (FOG) Inertial Measurement Unit (IMU), an Acoustic Doppler Velocimeter (ADV), and a pressure depth sensor. While it has been successfully applied in cave exploration, its inter-frame registration performance remains less than optimal, relying primarily on trajectory estimation for position determination.

In this paper, to enhance the performance of inter-frame registration and address the issue of local optima due to the non-convexity of the Iterative Closest Point (ICP) algorithm, we have augmented the application of the Gaussian clustering and maximal clique algorithms, which have been previously studied, to adapt to the unique challenges faced by multibeam imaging sonar-equipped robots navigating cluttered underwater environments. Our approach encompasses the following:

- We introduce a novel registration algorithm to effectively address the issue of local optima in inter-frame registration due to the non-convex nature of the Iterative Closest Point (ICP) algorithm.
- We employed an enhanced Accelerated-KAZE (AKAZE) algorithm to improve feature-matching accuracy in sparse environments and along extended straight lines, thereby enhancing the robustness of SLAM.
- We integrated an Inertial Measurement Unit and Doppler velocimeter data for the global state estimation of the robot, incorporating various pose design factors into a factor graph for back-end optimization.

2. Related Work

2.1. Point Cloud Registration

Point cloud registration algorithms can be categorized into two main approaches based on probabilistic fields and point-to-point matching. The probabilistic field approach often employs Gaussian Mixture Models (GMMs) to represent point clouds. In contrast, the point-to-point matching approach encompasses the well-established Iterative Closest Point (ICP) algorithm and its various extensions, as detailed by Segal et al. [9]. Besl and McKay conducted pioneering work in the registration of 3D shapes [10], while Chen and Medioni [11] concentrated on aligning range data for object modeling. Since then, numerous ICP variants have emerged. Due to the extensive body of literature on ICP algorithms, a complete and exhaustive introduction is beyond the scope of this paper. Although registration methods that ensure optimality have been published, they are relatively few and primarily rely on the Branch and Bound (BnB) algorithms. For instance, Li and Hartley [12,13] have proposed a 3D registration method that utilizes a Lipschitz-regularized L2 error function, which is minimized through BnB.

The widely recognized technique for field-based scan matching is the Normal Distribution Transform (NDT), which was first introduced by Biber and Strasser for the purpose of aligning two-dimensional point clouds [14]. This method has since been expanded to three dimensions by Magnusson [15]. Saarinen and colleagues developed the Normal Distribution Transform Occupancy Maps (NDT-OM), integrating NDT with the OctoMap framework crafted by Hornung and his team [16]. In response to the challenge of accurately aligning misaligned scans, Bouraine and others [17] have introduced a Particle Swarm Optimization approach for the Direct-to-Direct (D2D) method, known as NDT-P2D-PSO, which ensures a globally optimal solution and is capable of operating in real time for SLAM applications. However, in underwater sonar SLAM, NDT has some limitations. It employs a Cartesian grid to sample point clouds, which may result in an inadequate capture of the underlying structure of the point cloud, particularly when the scanned surface has irregular shapes. Therefore, we propose a new registration algorithm based on ICP, which provides a robust prediction matrix and effectively addresses the local optimal solution problem associated with ICP.

2.2. Underwater Sonar SLAM

In underwater settings, sonar images typically exhibit low resolution, low signal-to-noise ratios, and high ambiguity, characteristics that pose significant challenges to the crucial processes of automatic feature extraction and data association for underwater SLAM. To address these challenges, various methods have been introduced. Santos et al. [18] introduced a method based on topological Graph Matching, which utilizes Gaussian functions and the maximal clique algorithm to describe and match image features, demonstrating good robustness and rotational invariance. Eustice and colleagues proposed the first SLAM implementation using the Exact Sparse Extended Information Filter (ESEIF) [19], with features manually selected from sonar images, focusing on constructing a feature map to ensure real-time performance. Johannsson, Kaess, and Englot introduced the Normal Distribution Transform (NDT) to detect and match FLS image features [20], assuming a planar surface to address the ambiguity in elevation measurements. Similarly, Shin, Lee, and Choi engaged in seabed mapping, utilizing accelerated KAZE (AKAZE) features to process sonar images and demonstrated effective data association [21].

Huang and Kaess introduced the Acoustic Sensor-based Feature Motion (ASFM), which reconstructs 3D structures and sensor motion from multiple sonar viewpoints without making specific assumptions about the scene [22]. In their work, statistical analysis was conducted to identify and discuss degenerate cases that lead to ASFM failures, particularly analyzing which motion patterns result in poor constraints for the optimization problem. In 2018, Jie Li proposed a machine-learning-based loop closure method that accounts for the sparsity of sonar image features and the false data association, developing a reliable mathematical association method and demonstrating the robustness of the system [23]. In 2022, the Stevens Institute of Technology proposed an exploration framework based on virtual maps, which demonstrated favorable performance in simulated experiments [24]. However, practical applications necessitate the consideration of the costs associated with the construction and updating of virtual maps, as well as the resolution and noise levels of the sonar equipment.

Our research significantly enhances the navigation capabilities of multibeam imaging sonar-equipped robots in complex underwater environments, the robustness of the SLAM system, and the accuracy of global state estimation by introducing a novel registration algorithm to address the local optima issue of the Iterative Closest Point (ICP) algorithm, improving feature matching accuracy with an enhanced Accelerated-KAZE (AKAZE) algorithm, and integrating Inertial Measurement Unit and Doppler velocimeter data for factor graph optimization.

3. Inter-Frame Registration

To address the issue of ICP falling into local optima, we employ an inter-frame registration method based on the parallel Accelerated-KAZE (AKAZE) algorithm [25] and a newly proposed Graph Matching (GM) registration algorithm. This approach effectively performs sonar feature matching, adapts to rotational and long straight-line scenes, and accurately estimates the robot's motion state. The two algorithms complement each other's weaknesses.

3.1. Graph Matching ICP

The Iterative Closest Point (ICP) is the most widely used inter-frame registration algorithm, which is implemented by minimizing the following value function:

$$J(x_{ts}) = \sum_{\mathbf{p}_t \in \mathbf{P}_t, \mathbf{p}_s \in \mathbf{P}_s} \|\mathbf{T}_{ts}\mathbf{p}_s - \mathbf{p}_t\|^2 = \sum_{\mathbf{p}_t \in \mathbf{P}_t, \mathbf{p}_s \in \mathbf{P}_s} \|\mathbf{R}_{ts}\mathbf{p}_s + \mathbf{t}_{ts} - \mathbf{p}_t\|^2, \tag{1}$$

where $\mathbf{p}_t$ represents a point in the target point cloud; $\mathbf{p}_s$ represents a point in the source point cloud; $\mathbf{P}s$ and $\mathbf{P}t$ represent the point sets of the source point cloud and the target point cloud, respectively; $\mathbf{T}_{ts}$ denotes the transformation matrix from the source point cloud to the target point cloud; $\mathbf{R}_{ts}$ indicates the rotation matrix that aligns the source point cloud with the target point cloud; and $\mathbf{t}_{ts}$ signifies the translation vector that positions the source point cloud relative to the target point cloud.

However, ICP can fall into local optimum issues due to its non-convexity [26]. The application of ICP matching for sonar images is often hindered by the low angular resolution of the sonar, the high noise levels of the imaging sonar, and the low accuracy of trajectory calculations. To solve the problem of ICP falling into a local optimal solution, we propose a new Graph Matching (GM) registration algorithm. This algorithm aims to solve the following equation:

$$\tilde{\mathbf{x}}_s, \mathbf{T}_\theta = \arg\max_{\mathbf{x}_s, \mathbf{T}_\theta} \sum_{\mathbf{p}_s \in \mathbf{P}_s} \text{II}(d(\mathbf{p}_s)/n \leq \varepsilon), \tag{2}$$

where $\tilde{\mathbf{x}}_s$ represents the globally initialized source point cloud pose and $\mathbf{x}_s$ denotes the source point cloud pose that is being optimized. $\mathbf{T}_\theta$ denotes the predicted transformation matrix of ICP. ε is the reprojection error convergence threshold, and n represents the total number of points in the source point cloud. d is the sum of distances and represents the cumulative measure used to search for the nearest point through the course of the last iterative transformation. II is an indicator function that equals one if the solution is successful and zero if the solution fails. Simultaneously, the definition of $d(\mathbf{p}_s)$ is as follows:

$$d(\mathbf{p}_s) = \min_{\mathbf{p}_t \in \mathbf{P}_t} \|\mathbf{R}_{ts}\mathbf{p}_s + \mathbf{t}_{ts} - \mathbf{p}_t\|, \tag{3}$$

where $\mathbf{p}_s$ refers to an individual point within the source point cloud, and $\mathbf{p}_t$ refers to a point within the target point cloud. The rotation matrix $\mathbf{R}_{ts}$ and the translation vector $\mathbf{t}_{ts}$ together form the transformation that maps points from the source coordinate system to the target coordinate system, facilitating the alignment and registration process between the two point clouds.

3.1.1. Sonar Image Processing and Feature Extraction

Initially, we employ a Constant False Alarm Rate (CFAR) algorithm [27] for initial filtering of the sonar imagery, followed by the removal of discrete points, and finally apply voxel filtering to downsample the point cloud for feature extraction, as shown in Figure 1. Subsequently, we apply the DBSCAN (Density-Based Spatial Clustering of Applications with Noise) algorithm [28] for clustering. DBSCAN is distinguished by its ability to identify dense regions within a dataset and effectively partition them into distinct clusters, even in the presence of noise.

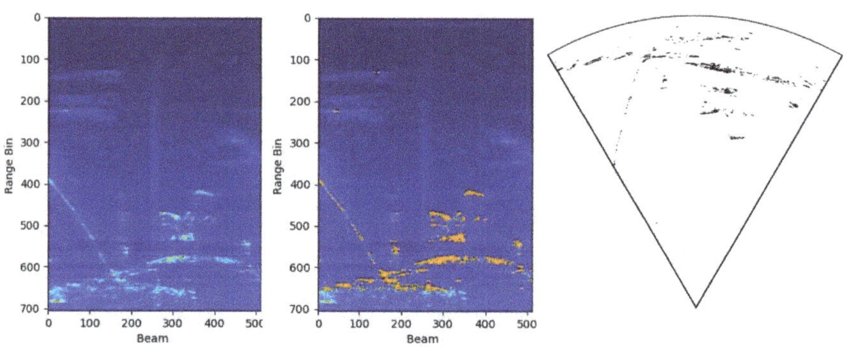

Figure 1. Filtering and feature extraction effects: The image on the **left** shows the raw data, the image in the **center** depicts the detected target features in orange, and the image on the **right** illustrates the target features interpolated from polar to Cartesian coordinates.

After downsampling the feature point cloud, clustering is performed using DBSCAN. When the search radius $\varepsilon = 0.06$ and the minimum number of points in a cluster $points_{min} = 4$, the results are depicted in Figure 2. After clustering, we filter the clusters based on the number of point clouds they contain to eliminate discrete noise points. Clusters with several point clouds greater than $Pts_{threshold}$ are retained for subsequent Gaussian generation. Here, $Pts_{threshold}$ represents the threshold for the number of point clouds. Next, a Gaussian block is generated for each cluster by traversing the clusters' labels. Gaussian blocks are used to characterize the shape of each cluster, and the parameters of the Gaussian are defined as follows:

$$GA = \{\mu X, \mu Y, \mu I, \sigma B, \sigma S, \sigma I, \theta, N\}, \quad (4)$$

where $\mu X, \mu Y$ denote the positional center point of the Gaussian block in the image, μI is the average intensity of the sonar image within the Gaussian block, σB represents the length of major axis of the Gaussian block, σS is the length of the minor axis of the Gaussian block, σI is the standard deviation of sonar image intensity within the Gaussian block, N is the number of points in the cluster, and θ is the angle measured clockwise from the positive underwater x-axis to the major axis of the Gaussian block. We restrict θ to the range of $(-\frac{\pi}{2}, \frac{\pi}{2})$. Consistent with the standard Gaussian block approach, we employ the eigenvalues and eigenvectors derived from the covariance matrix $\mathbf{\Sigma}_{xy}$ to calculate σB and σS.

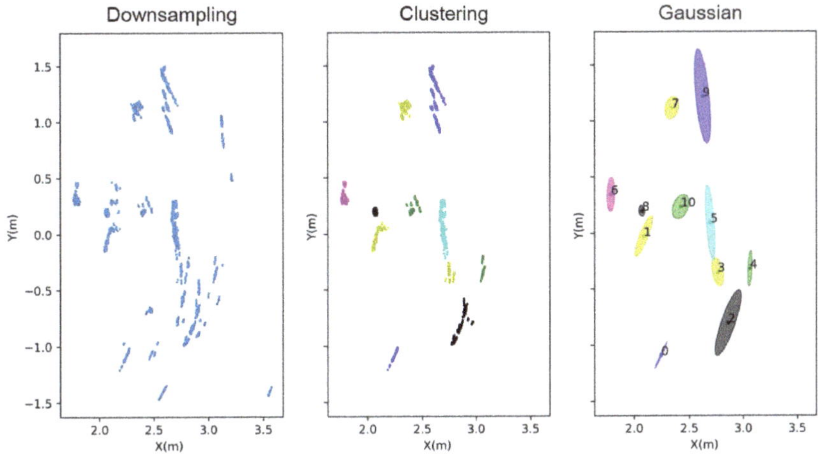

Figure 2. Point cloud clustered using DBSCAN and segmented into Gaussian blocks: The **left** column presents the point cloud after downsampling, the **center** shows the clustering effect, and the **right** depicts the resulting Gaussian blocks, color-coded for clarity.

Compute the center $(\mu X, \mu Y)$ of the point cloud for the cluster and the covariance matrix as follows:

$$(\mu X, \mu Y) = \left(\frac{1}{N} \sum_{i=1}^{N} x_i, \frac{1}{N} \sum_{i=1}^{N} y_i \right), \quad (5)$$

$$\mathbf{\Sigma}_{xy} = \begin{bmatrix} cov(X,X) & cov(X,Y) \\ cov(Y,X) & cov(Y,Y) \end{bmatrix}, \quad (6)$$

where X represents the set of values derived from the X coordinates of the points in the cluster minus μX, and Y represents the set of values derived from the Y coordinates of the points in the cluster minus μY.

The eigenvalues and eigenvectors of the covariance matrix $\mathbf{\Sigma}_{xy}$ can be determined using QR decomposition, the Jacobi method, etc. Once solved, the eigenvector $\boldsymbol{v}_{max}$ corresponding to the largest eigenvalue λ_{max} represents the direction of the major axis of the Gaussian block, while the eigenvector $\boldsymbol{v}_{min}$ associated with the smallest eigenvalue λ_{min} indicates the direction of the minor axis, as shown in Figure 3.

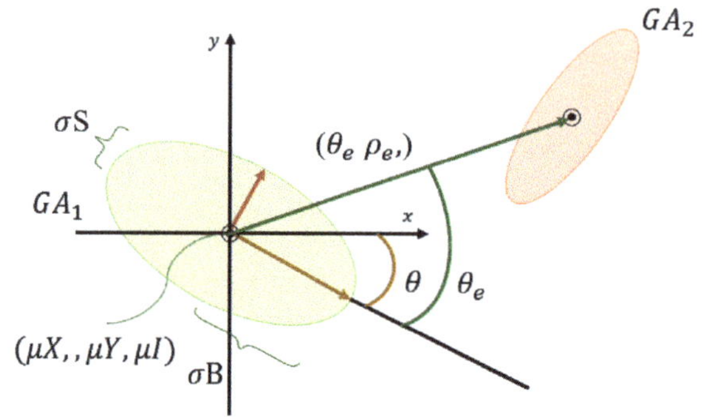

Figure 3. Gaussian block parameters and adjacency relationships.

Solving for angles θ and intercepts b

$$k = \frac{v_{max_y}}{v_{max_x}}, b = \mu Y - k\mu X, \theta = arctan(k). \tag{7}$$

Eventually, the clusters derived from the clustering process are further refined into Gaussian blocks, each endowed with relevant parameter attributes. This facilitates the establishment of relationships between neighbors. Figure 2 illustrates the shapes of Gaussian blocks generated for each cluster. It demonstrates that our algorithm is adaptable to various feature shapes for generating Gaussian blocks.

3.1.2. Gaussian Descriptors and Matching

In the pursuit of robust feature representation in image motion, especially with the challenge of rotation invariance, we introduce a novel approach leveraging Gaussian descriptors and graph theory. Our method aims to encapsulate the spatial relationships between feature blocks through the construction of an undirected graph G. Maximal cliques within this graph serve as a stable structural framework to describe the complex relationships among adjacent feature blocks. These cliques, and their associated descriptors, are crucial for our method as they allow for a detailed characterization of the spatial configuration of Gaussian blocks.

After generating the Gaussian blocks, we construct an undirected graph G and define $G = (V, E)$, where $V = \{GA_1, GA_2, GA_3, ..., GA_n\}$ is a set of vertices representing the centers of the Gaussian blocks. $E = \{E_1, E_2, E_3, ...E_m\}$ is a set of edges in the graph, with each Gaussian vertex connected by edges to describe the direct relationships between Gaussian blocks. The attributes associated with the edges are defined as $E_i = \{\theta_{e1}, \theta_{e2}, \rho_e, GA_{source}, GA_{target}\}$, where θ_{e1} is the angle between the positive direction of the major axis of the target Gaussian cloud and that of the source Gaussian cloud, and θ_{e2} is the angle between the positive direction of the major axis of the source Gaussian cloud and that of the target Gaussian cloud. These angles are obtained by substituting characteristic information of the two Gaussian blocks into the following equation, with the range of θ_e limited to $(0, 2\pi]$:

$$\theta_{e1} = arctan(\mu Y_{\text{target}} - \mu Y_{\text{source}}, \mu X_{\text{target}} - \mu X_{\text{source}}) - \theta_{source}, \tag{8}$$

$$\theta_{e2} = arctan(\mu Y_{\text{source}} - \mu Y_{\text{target}}, \mu X_{\text{source}} - \mu X_{\text{target}}) - \theta_{target}. \tag{9}$$

To address the issue of quadrant unification during the actual solution process, we use the coordinate transformation method, as illustrated in the following formulas:

$$X_{q1} = \mu X_{\text{target}} - \mu X_{\text{source}}, Y_{q1} = \mu Y_{\text{target}} - \mu Y_{\text{source}}, \tag{10}$$

$$\begin{bmatrix} X_{p1} \\ Y_{p1} \end{bmatrix} = \begin{bmatrix} \cos(\theta_{source}) & -\sin(\theta_{source}) \\ \sin(\theta_{source}) & \cos(\theta_{source}) \end{bmatrix}^T \begin{bmatrix} X_{q1} \\ Y_{q1} \end{bmatrix}. \tag{11}$$

Furthermore, we obtain θ_{e1} from Equation (12). If $\theta_{e1} < 0$, we adjust $\theta_{e1} = \theta_{e1} + 2\pi$, ensuring that the coordinates are unified into the interval $(0, 2\pi]$. The method for obtaining θ_{e2} is analogous to that for θ_{e1}, with details omitted for the sake of brevity.

$$\theta_{e1} = arctan(\frac{Y_{p1}}{X_{p1}}). \tag{12}$$

In addition, in the edge constraints, ρ_e represents the Euclidean distance between two Gaussian center points, calculated using the following equation:

$$\rho_e = \sqrt{(\mu X_{source} - \mu X_{target})^2 + (\mu Y_{source} - \mu Y_{target})^2}. \tag{13}$$

To construct an effective graph, we implement a criterion for establishing constraint edges: an edge is created between two Gaussian blocks if the Euclidean distance ρ_e between their centers is below a predefined threshold R_{max}. An increase in the value of R_{max} results in a more densely connected graph. This setting is conducive to establishing a Gaussian descriptor that quantifies the adjacency relationships between Gaussian blocks, as illustrated in Figure 4.

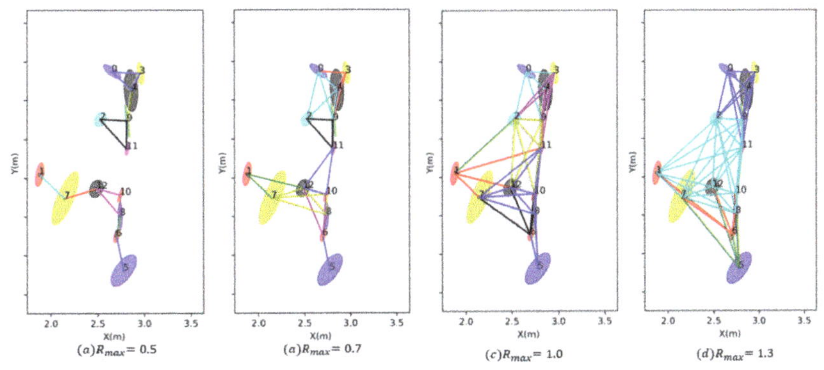

Figure 4. The impact of R_{max} on draph density.

Our method is designed to describe the relationship between adjacent feature blocks, addressing the challenge of rotation invariance during image motion. This is achieved by leveraging the angle θ_e stored in the edge between two vertices, which depends on the relationship between the Gaussian block and the edge. When the vertex corresponds to the source Gaussian block (GA_{source}), the angle is denoted as θ_{e1}, and when it corresponds to the target Gaussian block (GA_{target}), the angle is denoted as θ_{e2}. Stated differently, provided the surrounding environment remains static and maintains its shape, the values of θ_e and ρ_e for adjacent Gaussian blocks will remain constant, irrespective of changes in time and the robot's observational perspective. Figure 4 illustrates the effect of different R_{max} values on graph density, demonstrating that an increase results in denser Gaussian descriptor generation. However, this increased density corresponds to a longer relative computation time. Empirical tests have indicated that a setting of $R_{max} = 1.0$ yields optimal results.

We constructed a graph containing an image's Gaussian information and adjacency relationships through the above process. Then, we further optimize the filtering descriptors by searching for the maximal clique [29], which aids in establishing subsequent matching constraints. Given an undirected graph $G = (V, E)$, a subset $C = (V', E')$ of G is considered a clique. A maximal clique is a clique that is not a proper subset of any larger clique within the graph. This means no additional vertex from G can be incorporated to form a larger clique. Specifically, when a maximal clique contains the largest number of vertices among all maximal cliques in the graph, it is referred to as a maximum clique.

We employ the recursive Bron–Kerbosch algorithm [30] to identify maximal cliques. Our experiments reveal that a clique with three nodes yields the most stable cluster structure. When the number of nodes in a maximal clique is too small, such as two, the clique is insufficient to adequately describe the adjacency relationships among Gaussian blocks, leading to potential similarities that cannot be effectively distinguished. Conversely, identifying maximal cliques may become difficult when the number of nodes reaches four. This is attributed to the disappearance of some obstacles from the robot's field of view during its movement, making it unsuitable for environments with sparse features. We seek a stable structure to address this issue by targeting a maximal clique with exactly three nodes. Figure 4 illustrates a schematic of the identified maximal clique.

Following the search for the maximal clique, we utilize it as a reference to extract the vertices it contains and to retrieve the associated edge information. Both the edge length ρ_e and the angle θ_e are encapsulated within the descriptor to characterize the maximal clique. The descriptor for the i-th vertex within the maximal clique is formalized in the following equation, where the data in descriptor $\mathbf{descriptor}_i$ are sorted in ascending order of angles.

$$\mathbf{descriptor}_i = \{[\rho_1, \theta_1], [\rho_2, \theta_2], ..., [\rho_m, \theta_m]\}. \tag{14}$$

The obtained $\mathbf{descriptor}_1, \mathbf{descriptor}_2, ..., \mathbf{descriptor}_m$ are then assigned to the corresponding attributes of $Vertex_i$ to facilitate indexing the Gaussian center point and stored in **descriptors**. We propose an algorithm to align two sets of descriptors, which we call the GM (Graph Matching) algorithm.

Through the process above, we obtain the descriptors of the corresponding Gaussian blocks for the two frames. Next, we apply matching constraints to these descriptors: during the traversal of the i-th descriptor $\mathbf{descriptor}_{1,i}$ of **descriptors1** in $frame1$ and the j-th descriptor $\mathbf{descriptor}_{2,j}$ of descriptors **descriptors2** in $frame2$ [31], we define the similarity score, the index of the best matching vertex $best\_match$, and the optimal distance d_{best}. Using a brute-force descriptor matching approach [32], the corresponding vertex $Vertex_j$ of the $Vertex_i$ is found in turn. The specific algorithm procedure is depicted in Algorithm 1.

Initially, we employ Algorithm 2, a similarity algorithm, to calculate the similarity score between the descriptors $\mathbf{descriptor}_i$ and $\mathbf{descriptor}_j$ of the two vertices using Formula (15), where m and n represent the m-th and n-th pieces of data in $\mathbf{descriptor}_i$ and $\mathbf{descriptor}_j$, respectively. By adjusting the given angular threshold θ_{diff} and distance threshold ρ_{diff}, we effectively control the adjacency relationship between Gaussian blocks and mitigate the uncertainties associated with them. After calculating the similarity score, the optimal pairing vertex is identified using Algorithm 1. d_{pos} is defined as the Euclidean distance threshold of the matched vertices. To enhance robustness, we adjust d_{pos} by the robot's movement velocity, which influences the keyframe distance generation conditions on the trajectory calculation side, as depicted in Formula (16). This adjustment ensures consistency, preventing matched points from diverging beyond the distance of the dead reckoning keyframe motion. Through these steps, a set of preliminary matching points is obtained. Subsequently, the ICP algorithm [33] is employed to refine this set, resulting in the predictive transformation matrix and addressing the issue of the ICP algorithm converging to a local optimum.

$$|\theta_{i,m} - \theta_{j,n}| \le \theta_{diff}, |\rho_{i,m} - \rho_{j,n}| < \rho_{diff}, \tag{15}$$

$$\sqrt{(x_i - x_j)^2 + (y_i - y_j)^2} \le \xi_{pas}. \tag{16}$$

Algorithm 1 Matching constraint algorithm

Require: Descriptor set for two pictures **descriptors1** and **descriptors2** and abnormal distance threshold d_{pos}

```
descriptors1 = {descriptor1_1, descriptor1_2, ..., descriptor1_m}, descriptors2 = {descriptor2_1, descriptor2_2, ..., descriptor2_n}
connections = ∅
for every descriptor1_i in descriptors1 /*i ∈ [1, m]*/
    max_score ← −1
    best_match ← −1
    d_best ← −1
    for every descriptor2_j in descriptors2 /*j ∈ [1, n]*/
    /*Calculate the similarity between two descriptors*/
        score ← calculate_score
            if score > max_score then
                /*Calculate the similarity between two descriptors*/
                d* ← distance(descriptor1_i, descriptor2_j)
                if d* ≤ d_pos then
                    max_score ← score
                    best_match ← j
                    d_best ← d*
                end if
            else if score == max_score then
                if d* < d_best and d* < d_pos then
                    best_match ← j
                    d_best ← d*
                end if
            end if
    end for
    if best_match ≠ −1 then
        connections ← connections ∪ (best_match)
    end if
end for
connections ← remove_duplicate_point_pairs
return connections
```

Algorithm 2 Similarity scoring algorithm

Require: Two descriptors **descriptor1_i** $= \{[\rho_{i,1}, \theta_{i,1}], [\rho_{i,2}, \theta_{i,2}], \ldots, [\rho_{i,p}, \theta_{i,p}]\}$ and **descriptor2_j** $= \{[\rho_{j,1}, \theta_{j,1}], [\rho_{j,2}, \theta_{j,2}], \ldots, [\rho_{j,q}, \theta_{j,q}]\}$, angle threshold θ_{diff}, score distance threshold ρ_{diff}

```
score ← 0
m, n ← 0
/*m ∈ [1, p], n ∈ [1, q]*/
while m < len(descriptor1_i) and n < len(descriptor2_j) do
    /*calculate score*/
    θ* ← |θ_{i,m} − θ_{j,n}|
    ρ* ← |ρ_{i,m} − ρ_{j,n}|
    if θ* ≤ θ_diff then
        if ρ* < ρ_diff then
            score ← score + 1
        end if
        m ← m + 1, n ← n + 1
    else if θ_{i,m} < θ_{j,n} then
        m ← m + 1
    else
        n ← n + 1
    end if
end while
return score
```

3.2. Parallel Inter-Frame Registration Algorithm Using AKAZE and GM Methods

When the environmental feature comprises a long straight line and the robot's trajectory is parallel to this line, both the GM and ICP algorithms may degenerate and fail. In such scenarios, salient echo points typically emerge at the endpoints of the linear feature, which can be effectively tracked by the AKAZE (Accelerated-KAZE) algorithm [25]. The AKAZE algorithm is proficient in feature detection and description, leveraging the FED (Fast Explicit Diffusion) method to accelerate scale space construction and employing the M-LDB (Modified Local Difference Binary) paradigm to generate descriptors. When the robot's trajectory is perpendicular to the environmental feature line, the AKAZE algorithm exhibits robust performance and high precision in feature tracking and matching. However, it encounters difficulties matching when the motion involves rotation [34].

The final inter-frame registration process is as follows: after the front end obtains the critical frame information, denoted as $frame$, we initialize the waiting point cloud and set the current frame's point cloud to $\mathbf{P}_s$ and the previous frame's point cloud to $\mathbf{P}_t$. Following initialization, we employ the GM and AKAZE algorithms to extract effective sparse points. We then apply the ICP algorithm for registration estimation on these sparse points and calculate the corresponding reprojection errors ε_{GM} and ε_{AKAZE} for the transformation matrices $\mathbf{T}_{GM}$ and $\mathbf{T}_{AKAZE}$, respectively. We select the transformation matrix associated with the smaller error and integrate it into the dense point clouds $\mathbf{P}_s$ and $\mathbf{P}_t$ for further ICP refinement, obtaining an accurate solution $\hat{\mathbf{x}}_{ts}$.

Descriptor matching in the traditional AKAZE algorithm is performed using the brute force method, with the Hamming distance as a similarity metric, and a global search is conducted to identify the closest descriptor. However, the effectiveness of this approach is diminished when dealing with underwater sonar imagery, which is prone to multipath echoes and higher noise levels. We propose a method to eliminate false matching points to further improve the accuracy of matching. We sort the distance d_i between the matching points from small to large, record it as $d_i = [d_1, d_2, ..., d_n]$, and calculate it using the following formula, thereby eliminating mismatches caused by similar features due to sonar reverberation. In the formula, k is the proportion coefficient, used to eliminate data with low confidence. The scaling factor ξ controls the extent of the culling space.

$$\bar{d} = \frac{1}{(1-2k)n} \sum_{kn+1}^{(1-k)n} d_i, \tag{17}$$

$$d_i < \xi \bar{d}. \tag{18}$$

4. SLAM System

Let $\mathbf{x}_i$ represent the state of the robot at the i-th timestamp, where $i \in \{0, \cdots, N\}$. The robot state $\mathbf{x}_i \triangleq [\mathbf{R}_i \quad \mathbf{t}_i] \in \mathrm{SE}(2)$, where $\mathbf{R}_i \in \mathrm{SO}(2)$ represents the direction of the robot at the i-th timestamp and $\mathbf{t}_i \in \mathbb{R}^2$ represents its position [35]. To embed the environmental map into the state space, we depict an entity, such as a point cloud or salient image features, as a two-dimensional point feature. This feature is designated as $\mathbf{l}_j \in \mathbb{R}^2$, where $j \in \{1, \ldots, M\}$. Then, the measurement result can be represented by $\mathbf{z}_k$, where $k \in \{1, \ldots, K\}$, and the landmark $\mathbf{l}_{jk}$ observed in the state of $\mathbf{x}_{ik}$ during the measurement process usually corresponds to a set of paired associations $(\mathbf{i}_j, \mathbf{i}_k)$.

The objective of SLAM is to find the robot's trajectory $\mathbf{X} = \mathbf{x}_i$, the position of the landmark map, and the landmark map $\mathbf{L} = \{\mathbf{l}_j\}$ based on a set of data from the measurement set $\mathbf{Z} = \{\mathbf{z}_k\}$.

4.1. Sonar's Field of View

The multibeam sonar M1200d, as utilized in this research, encompasses a three-dimensional spatial field of view for the sonar sensor. Given that the horizontal beam angle is θ, the elevation angle is φ, and the effective detection radius is r, as shown in Figure 5a. Imaging sonar transforms the observed three-dimensional space into a two-dimensional plane for visualization, as shown in Figure 5b.

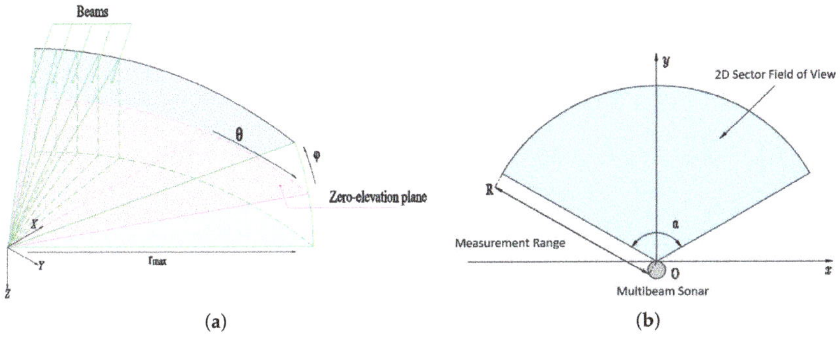

Figure 5. The geometric model of the multibeam sonar M1200d. (**a**) Sonar coordinate system. (**b**) Projection effects on sonar's 2D plane.

4.2. Underwater Robot Motion Estimation

We harness an Inertial Measurement Unit (IMU), a Doppler Velocity Log (DVL), and a depth meter—three pivotal sensors that contribute to estimating the robot's underwater motion posture. These three sensors are used to estimate the motion posture of the underwater robot, providing initial value constraints, which serve as tight coupling.

4.3. Factor Graph and Mapping

In the front-end calculations, pose estimation and mapping use the relationship between adjacent frames. However, relying solely on local constraints in a chained algorithm can lead to error accumulation and drift. Back-end optimization selects keyframes globally to mitigate this issue, establishing constraints across a larger spatial and temporal scale. This enables us to utilize the factor graph [31] from GTSAM for back-end optimization [36].

As shown in Figure 6, each keyframe's pose is denoted by a black square and linked to feature points in the sonar imagery. The factor graph includes Sequential Scan Matching (SSM) and Non-Sequential Scan Matching (NSSM) factors. Sequential factors represent the incremental transformation relative to the previous pose, leading to error accumulation. On the contrary, non-sequential factors act as loop closures, connecting two separate poses to correct accumulated drift.

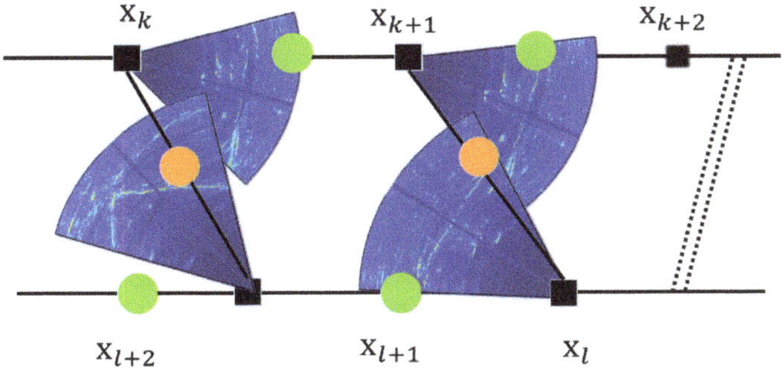

Figure 6. Composition of the factor graph for SLAM back-end optimization: The two horizontal lines represent two different time periods. The black squares on the lines indicate the poses of the corresponding keyframes. The green spheres represent Sequential Scan Matching (SSM), while the orange spheres represent Non-Sequential Scan Matching (NSSM).

The fundamental principle of the factor graph is to estimate the maximum a posteriori (MAP) probability of the robot's trajectory and the state of its surrounding environment based on dead reckoning and various sensor measurements [37]. This estimation is typically formulated as follows:

$$\mathbf{X}^* = \arg\max_{\mathbf{X},\mathbf{M}} P(\mathbf{X}, \mathbf{M} | \mathbf{U}, \mathbf{Z}), \tag{19}$$

where **U** signifies the robot's control input, **Z** represents the measurement information, **X** denotes the robot's posture, and **M** indicates the position of the observation point.

We assume that the SLAM system's system model and observation model follow a Gaussian noise distribution and take into account the scaling factor, while also enhancing the fault tolerance of the SLAM system by incorporating switch variables in the nonlinear iterative process. Furthermore, by assigning weight values to loop closures, the system's maximum a posteriori (MAP) probability estimation can be formulated as follows:

$$\mathbf{X}^*, \mathbf{S}^* = \arg\min_{\mathbf{X},\mathbf{S}} \sum_i \|f(\mathbf{x}_i, \mathbf{u}_i) - \mathbf{x}_{i+1}\|^2_{\boldsymbol{\Sigma}_i} + \sum_{ij} \|\psi(\mathbf{S}_{ij}) \cdot [f(\mathbf{x}_i, \mathbf{u}_{ij}) - \mathbf{x}_j]\|^2_{\boldsymbol{\Lambda}_{ij}} + \sum_{ij} \|\mathbf{r}_{ij} - \mathbf{s}_{ij}\|^2_{\boldsymbol{\Sigma}_{ij}}, \tag{20}$$

where $\mathbf{X}^*$ denotes the optimal estimate of the robot's trajectory; $\mathbf{S}^*$ represents the optimal estimate for the switch variables; **X** refers to the trajectory of the robot; **S** symbolizes the switch variables that determine whether specific loop closure constraints are activated; $f(\mathbf{x}_i; \mathbf{u}_i)$ is the motion model function that predicts the next state of the robot based on its current state $\mathbf{x}_i$ and control input $\mathbf{u}_i$ at timestamp i; $\mathbf{x}_{i+1}$ denotes the actual state of the robot at the subsequent timestamp; $\boldsymbol{\Sigma}_i$ represents the covariance matrix of the motion model, which characterizes the uncertainty in the predictions made by the motion model; $\psi(\mathbf{S}_{ij})$ is the function of the switch variable $\mathbf{S}_{ij}$ that adjusts the weight of the loop closure constraint; $\mathbf{u}_{ij}$ is the sequence of control inputs from timestamp i to timestamp j; $\boldsymbol{\Lambda}_{ij}$ represents the covariance matrix associated with the loop closure constraint, which describes the uncertainty of the loop closure; $\mathbf{r}_{ij}$ refers to the residual of the loop closure constraint; $\mathbf{s}_{ij}$ denotes the actual residual corresponding to the switch variable for the loop closure constraint; and $\boldsymbol{\Sigma}_{ij}$ represents the covariance matrix for the loop closure constraint.

5. Results

5.1. Underwater Platform and Datasets

We employ the BlueRov2 robot as the primary motion carrier, integrating the Oculus M1200d multibeam sonar, the DVL-a50 Doppler Velocity Log, and an IMU into its hardware suite, as shown in Figure 7a. Communication is enabled through a mini switch and a Raspberry Pi, with signals relayed to a distant PC via a carrier module. The multibeam sonar functions in a high-frequency mode at 2.1 MHz, offering a maximum field of view of 10 m, a horizontal aperture of 80°, a vertical aperture of 12°, and a maximum update frequency of 40 Hz. It boasts a range resolution of 0.007 m and an angular resolution of 0.4°. These sensor characteristics are detailed in Table 1. Sensor data are aggregated for seamless communication using the Robot Operating System (ROS). This enables the visualization of sonar raw data and image data, as well as the DVL data visualization module, as shown in Figure 7b. This setup is crucial for subsequent validation with the SLAM system.

Table 1. Sensor parameter characteristics.

Sensor Type	Model	Frequency	Features
Sonar	Oculus M1200d	12.6 Hz	High resolution, high update rate
Doppler Velocimeter	A50	10 Hz	Minimal size, max speed 2 m/s, ±0.1 cm/s accuracy
Inertial Measurement Unit	BMI055	9.5 Hz	High stability
Motion Capture System	NOKOV	171 Hz	2048 × 2048 resolution, 4.2 M pixels, 5.2 ms latency, 52° × 52° FOV

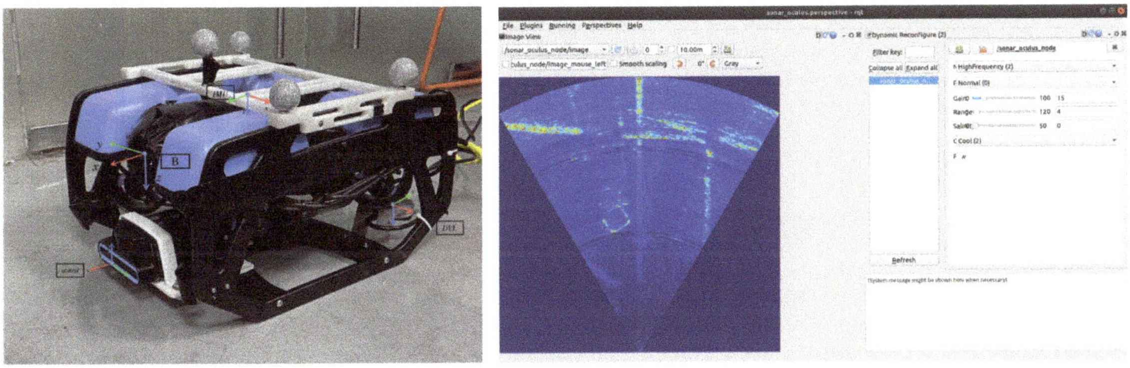

(a) (b)

Figure 7. Underwater robot and sonar image example. (**a**) BlueRov2 Robot with Oculus M1200d sonar, DVL-a50, and IMU installations. (**b**) Visualization collection of sonar raw data and images.

As shown in Figure 8a, the indoor environmental dataset was collected within a pool, which measures approximately 21 square meters with dimensions of 3.5 m by 8 m by 1.5 m. To enhance the environmental complexity, large water buckets were strategically positioned as underwater obstacles, each measuring 0.31 m by 0.295 m by 0.45 m. An optical positioning tracking system, equipped with 11 optical cameras, was deployed in the experimental setup to accurately capture the robot's motion trajectory. This setup facilitates error analysis by comparing the captured trajectory with that estimated by the SLAM algorithm. Additionally, a wide-angle camera was installed to monitor the robot's motion, which is crucial for investigating the factors influencing the quality of sonar images for future enhancements. The sonar's detection range was configured to 4 m with a gain compensation set at 15. A PID algorithm was implemented to govern the robot's low-speed movement. To ensure planar motion, the robot received z-depth data from the motion capture system and utilized real-time feedback to maintain its movement at a constant height of 0.55 m above the water surface, enabling it to complete a full circuit.

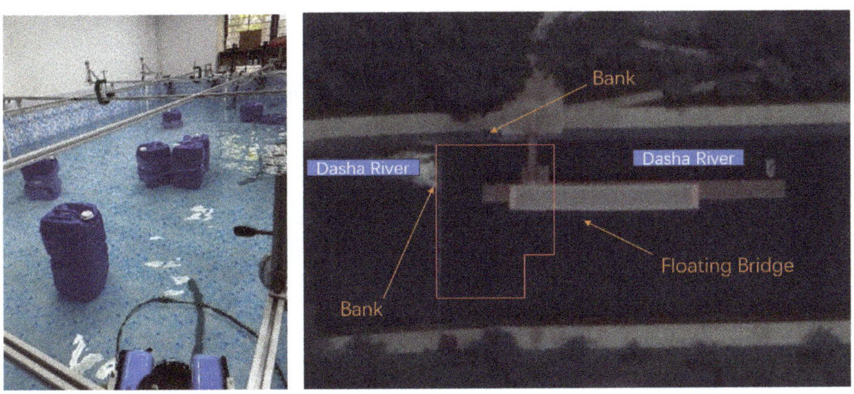

(a) (b)

Figure 8. Data collection environments. (**a**) The indoor environment. (**b**) The outdoor environment.

As shown in Figure 8b, the outdoor environmental dataset was collected in the Dasha River, located within the University Town of Nanshan District, Shenzhen City. The experimental site featured a water depth of 2 m, abundant silt and sand, aquatic life, small harbors along the riverbanks, and various obstacles, including bridge piers, effectively simulating the natural environment for underwater robots. The robot's buoyancy was adjusted using weight blocks throughout the experiment, and its movement was carefully controlled to maintain a consistent depth. The sonar angle was fine-tuned to face slightly downward to enhance detection capabilities. The robot was maneuvered in a counterclockwise circuit within the river to acquire the dataset.

5.2. Inter-Frame Registration Results

5.2.1. GM Results

After obtaining multiple clusters through clustering, the clusters are further generated as Gaussian blocks, each endowed with relevant parameter attributes to facilitate the establishment of adjacency relationships in the subsequent section. These Gaussian blocks are designed to capture the local geometric properties of the point cloud data, which are crucial for accurate feature matching and pose estimation. Figure 9 shows the shapes of the Gaussian blocks generated by each cluster, which can be seen to adapt to various shapes of features for generating Gaussian blocks. The adaptability of these blocks is a key aspect of our method, as it allows for a more robust representation of the underlying geometry, which is essential for handling complex scenes and varying environmental conditions.

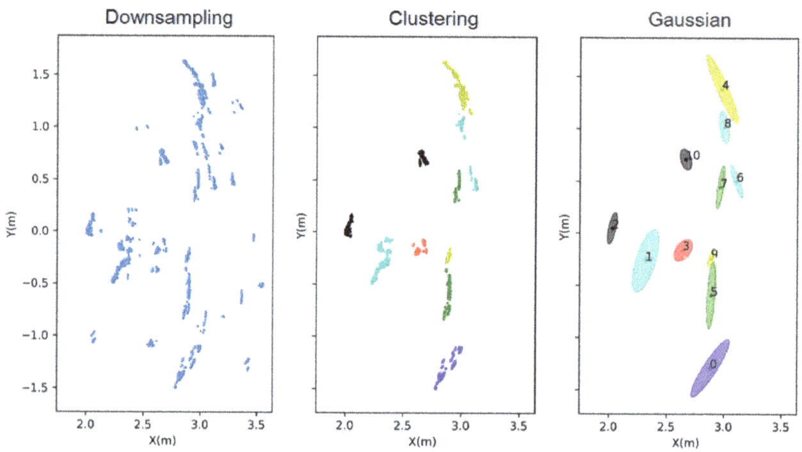

Figure 9. Point cloud segmentation into Gaussian blocks: The **left** column presents the point cloud after downsampling, the **center** shows the clustering effect, and the **right** depicts the resulting Gaussian blocks, color-coded for clarity.

After acquiring the descriptors of the two images **descriptors2** and **descriptors1** [31], the GM algorithm is employed for descriptor registration. Figure 10 illustrates the matching outcome between the source and target point cloud of the fifth frame, demonstrating high accuracy. This high accuracy is a result of the robust Gaussian block representation and the effective descriptor-matching strategy employed by our GM algorithm. The Gaussian blocks are effectively aligned, providing initial points for calculating the predicted transformation matrix T_0. The dense point cloud and the expected matrix are then substituted into the ICP for resolution. Figure 11a indicates that the traditional ICP algorithm encounters convergence issues by the 26th iteration, settling into a local minimum with a final error of 0.0623 m. This error is specifically related to the registration accuracy between the two point clouds. Figure 11b displays the result after ICP iterations, where the point clouds are nearly superimposed with an error of 0.036 m, achieving convergence around the 29th iteration. This suggests that the enhanced algorithm effectively addresses the issue of ICP getting stuck in local optima and failing to converge, thereby strengthening the system's pose estimation accuracy.

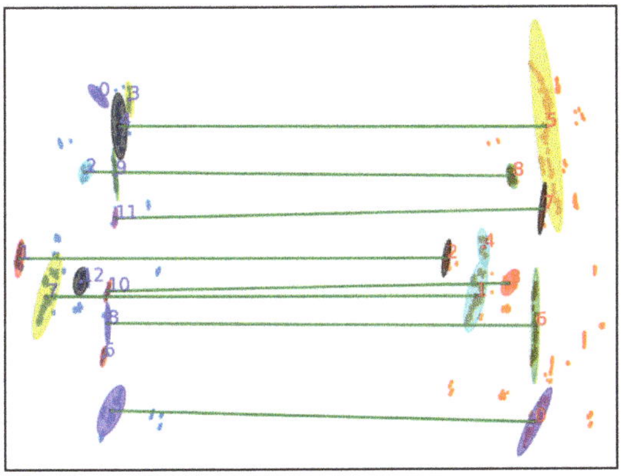

Figure 10. Matching Results of the Source and Target Point Clouds in the 5th Frame: The source point cloud is depicted in blue, the target point cloud is in orange, and the green lines indicate the successfully matched pairs. Additionally, the generated Gaussian blocks are color-coded for clarity.

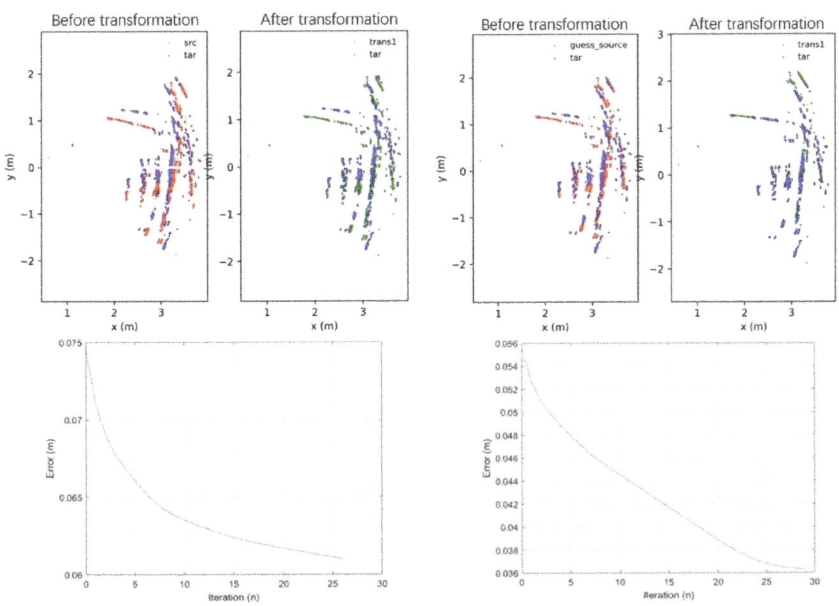

(a) (b)

Figure 11. Comparison of ICP solution results before and after improvement. (**a**) Point cloud matching effects and error analysis pre-improvement: The first column illustrates the initial state of the source and target point clouds before ICP matching, and the second column in green demonstrates the transformed source point cloud after the initial alignment. (**b**) Point cloud matching effects and error analysis post-improvement: The first column presents the initial transformation effect of the source point cloud when applied with the predicted matrix, and the second column displays the refined alignment outcome after subsequent ICP iterations.

5.2.2. AKAZE Results

In a monotonous environment with sparse sonar target features, the quality of adjacency relationship information is compromised, resulting in the generation of descriptor information that is not viable for use, as illustrated in Figure 12. This, consequently, causes the failure of the ICP solution. Moreover, the direct application of the ICP algorithm is typically ineffective in linear environments. Therefore, we employ the AKAZE algorithm to extract features from long, straight-line segments. As depicted in Figure 13b, target point clouds and registered source point clouds are nearly superimposed, indicating successful feature matching. Consequently, employing a frame-by-frame matching approach that leverages both the AKAZE and GM algorithms in tandem effectively addresses sonar feature matching. This method is adept at handling scenarios involving rotation and long, straight lines, facilitating the precise estimation of the robot's motion state. The complementary nature of these algorithms compensates for the limitations inherent in each.

Figure 14 presents a comparative analysis of the traditional versus the improved culling method at the 37th frame. The prevalent engineering approach for eliminating incorrect matches is two times the Hamming distance method, typically utilized in terrestrial visual matching. However, underwater sonar imagery is subject to multipath echoes and higher noise levels, which render this method less effective, as shown in Figure 14a. Building upon this, our approach further refines the elimination of mismatches induced by echoes, thereby enhancing the registration precision, as shown in Figure 14b. The improved method is observed to effectively eliminate mismatches due to sonar echo characteristics, leading to a marked enhancement in registration accuracy and robustness.

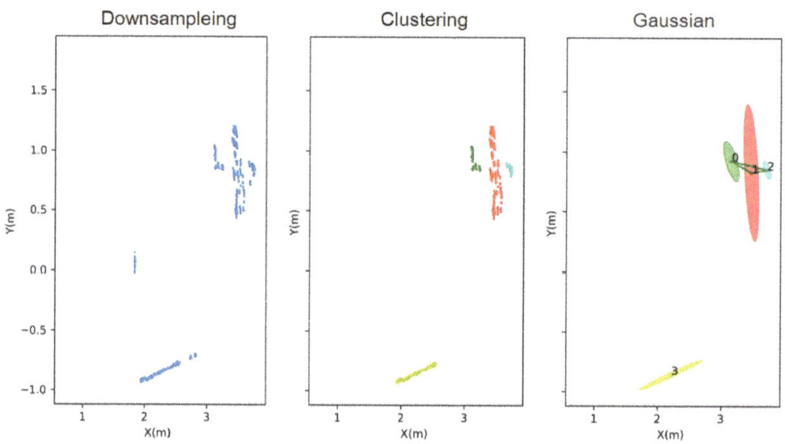

Figure 12. Failure of sonar feature matching for long straight lines.

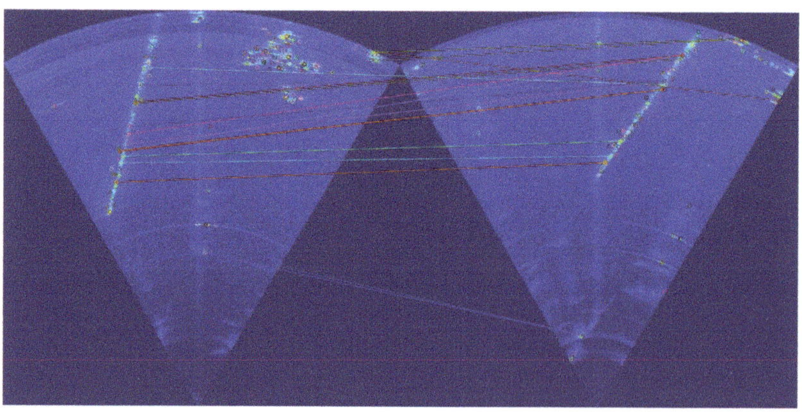

(a)

Figure 13. *Cont.*

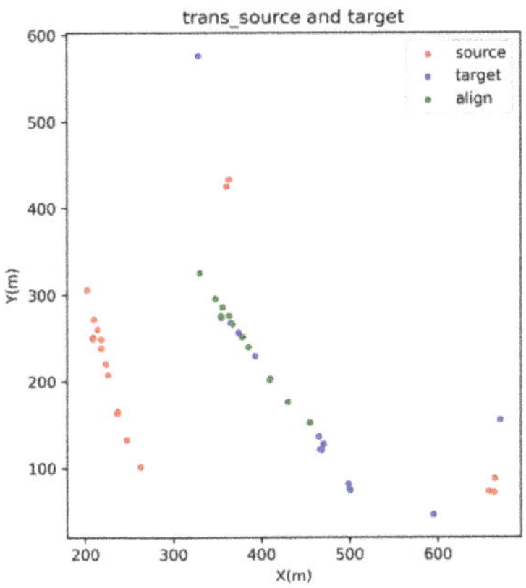

(**b**)

Figure 13. AKAZE algorithm for long straight line features. (**a**) The matching result of the AKAZE algorithm for long straight lines. (**b**) Solution effects of the AKAZE algorithm for long straight line features: the source point cloud is represented in red, the target point cloud in blue, and the registered source point cloud in green.

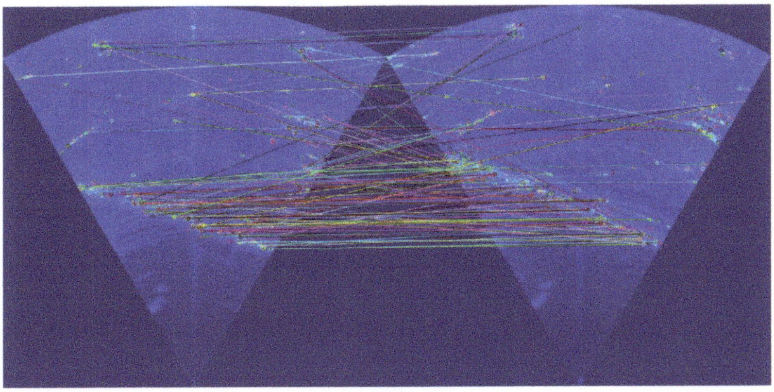

(**a**)

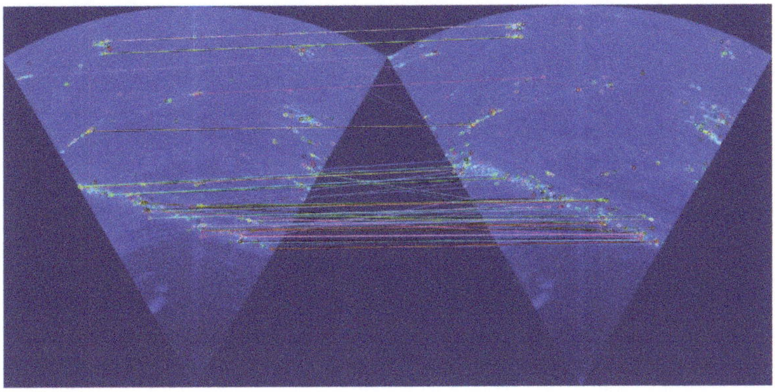

(**b**)

Figure 14. Comparison of traditional culling method and improved method used in the 37th frame. (**a**) Matching results using the 2-times Hamming distance culling method. (**b**) Matching results using the improved culling method.

5.3. SLAM Results

5.3.1. Indoor Environment Test Results

The SLAM algorithm presented in this work is denoted as GM SLAM. Additionally, we compare it with other state-of-the-art methods, e.g., Bruce SLAM [38]. Figure 15a illustrates the indoor positioning and mapping outcomes achieved by the GM SLAM algorithm. Here, colored markers represent mapped obstacles, including the pool walls and water tank obstacles. The gray grid delineates the area recognized as unobstructed space. In the indoor test, as the robot navigates two circles around the pool, the mapping results closely mirror the actual environmental layout, despite some observable positional drift. In contrast, as shown in Figure 15b, the Bruce SLAM algorithm's performance in similar indoor environments is not as optimal, particularly during limited-range movements within complex or feature-rich settings. The GM SLAM demonstrates a superior ability to delineate the environment, offering enhanced positioning accuracy and robustness. Conversely, the Bruce SLAM's frame-matching capabilities tend to be suboptimal, often failing to converge effectively and exhibiting a propensity for positional drift, leading to less precise mapping and localization outcomes.

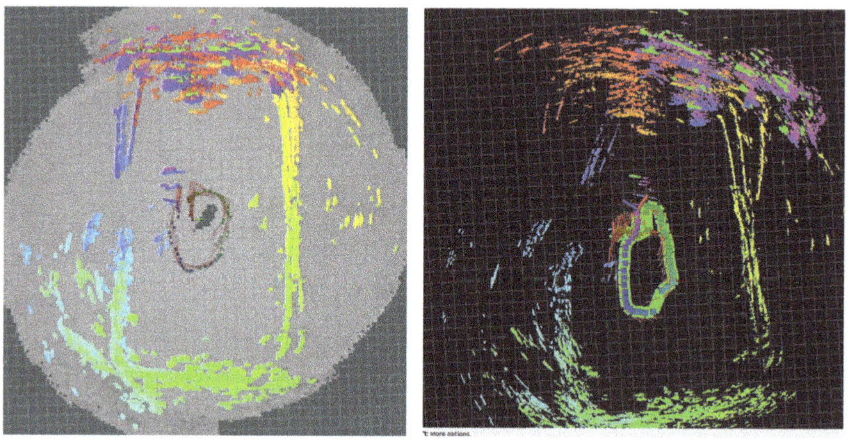

(**a**) (**b**)

Figure 15. The mapping effect performed by GM SLAM and Bruce SLAM, with colored point cloud maps and trajectories. (**a**) The mapping effect performed by GM SLAM. (**b**) The mapping effect performed by Bruce SLAM.

Figure 16a presents a comparison between the predicted trajectory generated by GM SLAM, the trajectory produced by the existing Bruce SLAM algorithm, and the actual trajectory obtained from motion capture data. Utilizing the evo tool [39], the trajectories were aligned for a thorough analysis. The analysis reveals that GM SLAM experiences fewer frame-matching failures and exhibits superior robustness in complex environments. Bruce SLAM, however, shows more drift, indicating feature tracking issues and a higher rate of incorrect frame matches, which underscores the higher positioning accuracy of GM SLAM. Figure 16b presents an error analysis for the x, y, and z axes. The operational height of the robot is set at approximately 0.58 m, and the motion capture system boasts a measurement accuracy of 0.02 m. Given the assumption of horizontal plane movement, maintaining the robot at a consistent depth is challenging, resulting in deviations along the z-axis. These deviations, in turn, impact the accuracy of the x and y directions in SLAM.

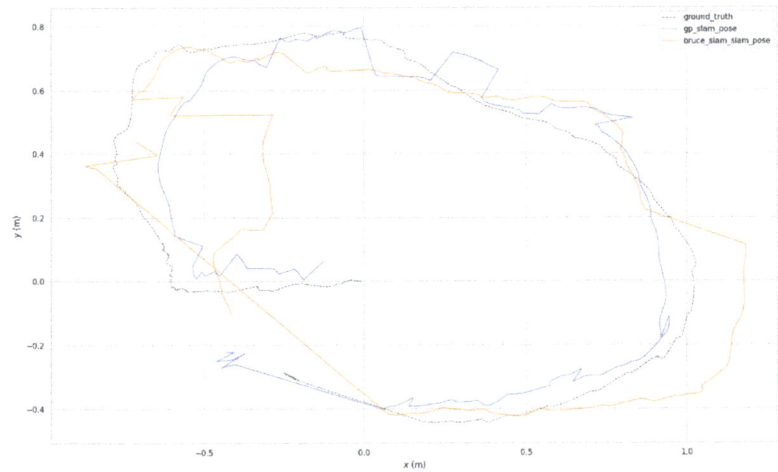

(**a**)

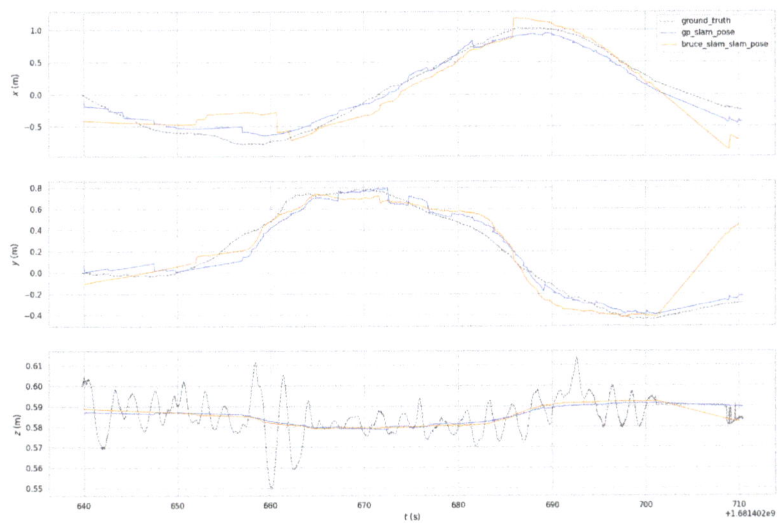

(**b**)

Figure 16. Evo trajectory analysis results: The blue line represents the trajectory generated by the GM SLAM algorithm, the orange line indicates the trajectory generated by the Bruce SLAM algorithm, and the dashed line corresponds to the actual robot trajectory captured through motion capture technology. (**a**) Comparison of GM SLAM trajectory, Bruce SLAM trajectory, and true trajectory. (**b**) The error analysis for the x, y, and z axes.

We compared the positioning trajectories of GM SLAM and Bruce SLAM with the true trajectory, using Equation (21) to calculate the Absolute Pose Error (APE) and Equation (22) to calculate the Root Mean Square Error (RMSE) for analysis [40]. The smaller the RMSE value, the better the fit between the observation model and the actual observations, indicating that the SLAM algorithm's estimated pose is more accurate.

$$APE = \sqrt{1/N \sum_{i=1}^{N} \| \mathbf{trans}\left(\mathbf{T}_{gt,i}^{-1}\mathbf{T}_{\text{est},i}\right) \|_2^2}, \tag{21}$$

where N represents the total number of observation points, $\mathbf{T}_{gt,i}$ denotes the true pose transformation matrix for the i-th observation point, and $\mathbf{T}_{\mathrm{est}\,,i}$ refers to its estimated pose transformation matrix.

$$RMSE = \sqrt{\frac{1}{n}\sum_{i=1}^{n}(y_i - \hat{y}_i)^2}, \tag{22}$$

where y_i represents the observed value for each sample and $\hat{y}_i$ represents the actual true value.

The Absolute Pose Error (APE) calculation results are detailed in Table 2, with Figure 17 illustrating the corresponding results for the estimated motion trajectories compared to the true values. Figure 17a illustrates that the main robot errors in GM SLAM occur during turning sections, which require further improvement. By comparing the trajectory errors of the three, it can be seen that the average error (mean) of the predicted trajectory of GM SLAM compared to the true trajectory is 0.118 m, with a root mean square error (rmse) of 0.132 m. In contrast, as shown in Figure 17b, the average error (mean) of the trajectory predicted by the Bruce SLAM algorithm is 0.189 m, with a root mean square error (rmse) of 0.247 m, both of which are greater than the errors of GM SLAM, indicating that the positioning accuracy of the GM SLAM algorithm is superior to that of the Bruce SLAM algorithm. Moreover, the standard deviation (std) of the trajectory of GM SLAM is 0.059 m, while that of the Bruce SLAM algorithm is 0.159 m. The other error items of GM SLAM are also generally smaller than those of Bruce SLAM, further demonstrating that the GM SLAM positioning algorithm has higher robustness and is superior to the Bruce SLAM algorithm.

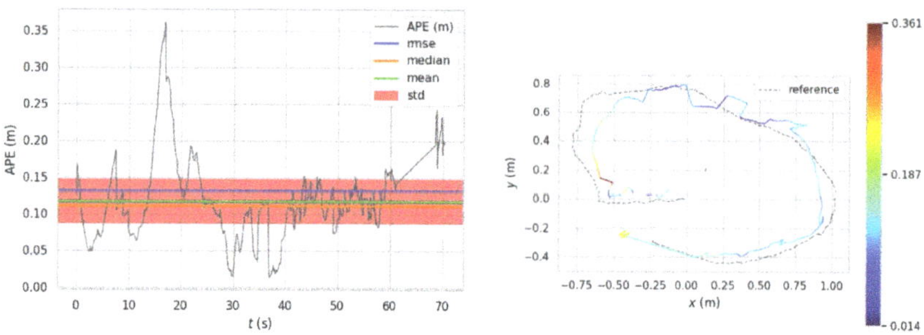

(**a**)

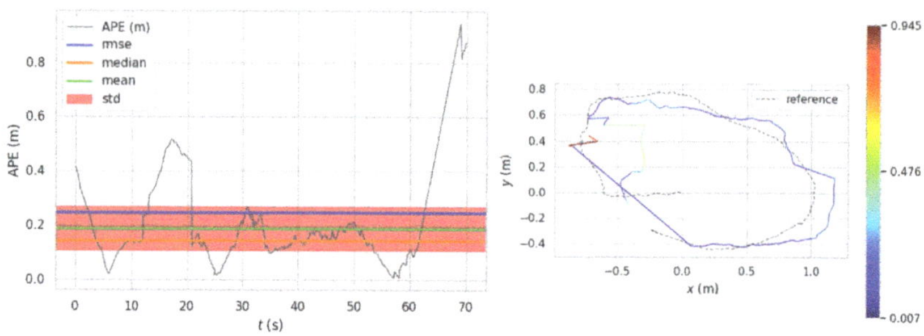

(**b**)

Figure 17. Comparison of absolute pose error (APE) calculation results between GP SLAM and Bruce SLAM. (**a**) Absolute pose error (APE) calculation results for GP SLAM: The left plot shows the APE calculation curve, and the right plot illustrates the error intensity. (**b**) Absolute pose error (APE) calculation results for Bruce SLAM: the left plot shows the APE calculation curve, and the right plot illustrates the error intensity.

Table 2. APE calculation results.

	GM SLAM/m	Bruce SLAM/m
max	0.360	0.944
mean	0.118	0.189
median	0.112	0.143
min	0.014	0.007
rmse	0.132	0.247
sse	9.559	33.796
std	0.059	0.159

5.3.2. Outdoor Environment Test Results

The sonar-observed objects from the outdoor experiment are depicted in Figure 18a, with straight lines representing the walls along the shore, circular objects representing underwater bridge piers, and other bright spots representing sand and stones on the seabed. We controlled the robot to navigate counterclockwise within the pool to collect the dataset, verifying the positioning and mapping feasibility using the GM SLAM algorithm. Figure 18b illustrates the mapping process, demonstrating that the algorithm can effectively capture and represent the shape and position of actual walls and bridge piers in the map.

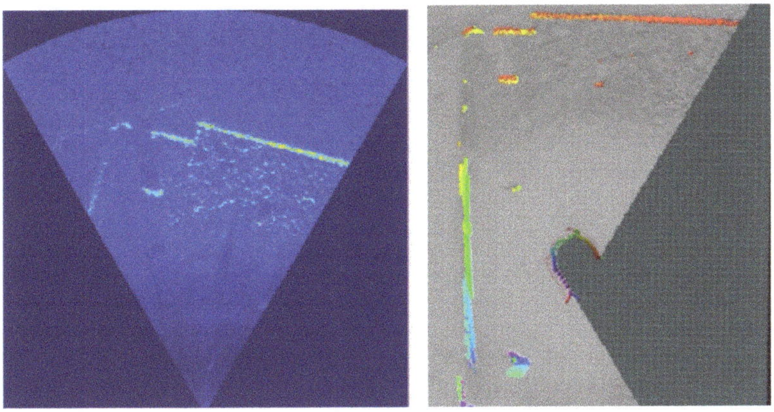

(a) (b)

Figure 18. Outdoor SLAM Process. (**a**) A schematic diagram of objects observed by sonar. (**b**) Sonar-observed outdoor sand and stone wall environment.

As an outdoor environment lacks a ground truth trajectory, this experiment focuses solely on comparing the mapping effects. Figure 19 presents a comparison between the point cloud mapping results of the GM SLAM and Bruce SLAM algorithms. Figure 19a illustrates that the mapping of GM SLAM closely approximates the actual environment, and the robot's positioning trajectory aligns closely with the expected controlled path. The trajectory is optimized by incorporating non-sequential factors into the factor graph and is further refined during loop closure detection when the robot revisits known locations. Figure 19b depicts the Bruce SLAM's mapping, revealing certain discrepancies after a circular traversal, attributed to error accumulation and erroneous frame matching. Comparative analysis indicates that the GM SLAM provides superior mapping quality and greater accuracy than the Bruce SLAM.

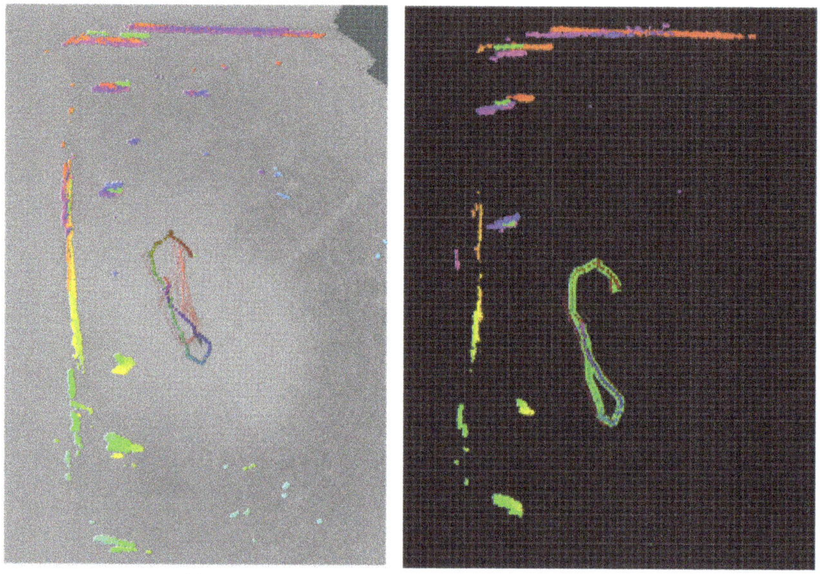

(**a**) (**b**)

Figure 19. The mappingeffect performed by GM SLAM and Bruce SLAM, with colored point cloud maps and trajectories. (**a**) Point cloud mapping and robot positioning trajectory for GM SLAM. (**b**) Point cloud mapping and robot positioning trajectory for Bruce SLAM.

6. Discussion and Conclusions

In this paper, we present a robust keyframe-based SLAM framework specifically designed for shallow water environments, employing a multibeam sonar. To combat the propensity of ICP matching to fall into local optima during sonar feature matching, we have introduced a GM algorithm that utilizes Gaussian clustering and maximal clique algorithms, designed to calculate a predictive initial transformation matrix for ICP, effectively reducing the risk of being ensnared in local optima. Furthermore, for cases in which features are sparse, we have implemented an enhanced AKAZE algorithm that filters out erroneous match points and enhances matching precision. Experimental validation has confirmed that our algorithm effectively tackles the ICP local optima issue and exhibits robust performance against rotational motion. A comparative analysis with the Bruce SLAM algorithm highlights the superiority of our approach in terms of both positioning and mapping accuracy, achieving impressive results.

Future endeavors will focus on fully leveraging the textural information of sonar by transforming it into semantic data, as the current method only utilizes edge point information and overlooks the sonar's textural content.

Author Contributions: Conceptualization, W.L.; Methodology, L.Z., X.C. and W.L.; Software, L.Z.; Validation, L.Z. and X.C.; Formal analysis, L.Z.; Investigation, L.Z. and X.C.; Resources, W.L.; Data curation, X.C.; Writing—original draft, L.Z., X.C. and Y.Y.; Writing—review & editing, L.Z., W.L. and Y.Y.; Visualization, L.Z., X.C. and Y.Y.; Supervision, W.L.; Funding acquisition, W.L. All authors have read and agreed to the published version of the manuscript.

Funding: This work is supported in part by the National Natural Science Foundation of China #62003110, the Shenzhen Science and Technology Innovation Foundations #JCYJ20210324132 607018, #JSGG20210420091804012, and #GXWD2022081 1163649003, the Science Center Program of the National Natural Science Foundation of China under Grant 62188101, and the open research fund of Guangdong Key Laboratory of Modern Control Technology.

Institutional Review Board Statement: Not applicable.

Data Availability Statement: Data is contained within the article.

Conflicts of Interest: The authors declare no conflict of interest.

References

1. Wang, Y.; Ji, Y.; Woo, H.; Tamura, Y.; Tsuchiya, H.; Yamashita, A.; Asama, H. Acoustic camera-based pose graph slam for dense 3-d mapping in underwater environments. *IEEE J. Ocean. Eng.* **2020**, *46*, 829–847. [CrossRef]
2. Salvi, J.; Petillo, Y.; Thomas, S.; Aulinas, J. Visual slam for underwater vehicles using video velocity log and natural landmarks. In Proceedings of the OCEANS 2008, Quebec City, QC, Canada, 15–18 September 2008; pp. 1–6.
3. Beall, C.; Dellaert, F.; Mahon, I.; Williams, S.B. Bundle adjustment in large-scale 3d reconstructions based on underwater robotic surveys. In Proceedings of the OCEANS 2011 IEEE, Santander, Spainm, 6–9 June 2011; pp. 1–6.
4. Shkurti, F.; Rekleitis, I.; Scaccia, M.; Dudek, G. State estimation of an underwater robot using visual and inertial information. In Proceedings of the 2011 IEEE/RSJ International Conference on Intelligent Robots and Systems, San Francisco, CA, USA, 25–30 September 2011; pp. 5054–5060.
5. Corke, P.; Detweiler, C.; Dunbabin, M.; Hamilton, M.; Rus, D.; Vasilescu, I. Experiments with underwater robot localization and tracking. In Proceedings of the 2007 IEEE International Conference on Robotics and Automation, Roma, Italy, 10–14 April 2007; pp. 4556–4561.
6. Folkesson, J.; Leonard, J.; Leederkerken, J.; Williams, R. Feature tracking for underwater navigation using sonar. In Proceedings of the 2007 IEEE/RSJ International Conference on Intelligent Robots and Systems, San Diego, CA, USA, 29 October–2 November 2007; pp. 3678–3684.
7. Fallon, M.F.; Folkesson, J.; McClelland, H.; Leonard, J.J. Relocating underwater features autonomously using sonar-based SLAM. *IEEE J. Ocean. Eng.* **2013**, *38*, 500–513. [CrossRef]
8. Richmond, K.; Flesher, C.; Lindzey, L.; Tanner, N.; Stone, W.C. SUNFISH®: A human-portable exploration AUV for complex 3D environments. In Proceedings of the OCEANS 2018 MTS/IEEE, Charleston, CA, USA, 22–25 October 2018; pp. 1–9.
9. Segal, A.; Haehnel, D.; Thrun, S. Generalized-icp. In Proceedings of the Robotics: Science and Systems, Seattle, WA, USA, 28 June–1 July 2009; Volume 2, p. 435.
10. Besl, P.; McKay, N. A Method for Registration of 3-D Shapes. *IEEE Trans. Pattern Anal. Mach. Intell.* **1992**, *14*, 239–256. [CrossRef]
11. Chen, Y.; Medioni, G. Object Modeling by Registration of Multiple Range Images. In Proceedings of the 1992 IEEE International Conference on Robotics and Automation, Nice, France, 12–14 May 1992; pp. 2724–2729.
12. Li, H.; Hartley, R. The 3D-3D registration problem revisited. In Proceedings of the International Conference on Computer Vision, Rio De Janeiro, Brazil, 14–21 October 2007; pp. 1–8.
13. Bazin, J.C.; Seo, Y.; Pollefeys, M. Globally optimal consensus set maximization through rotation search. In Proceedings of the Asian Conference on Computer Vision, Daejeon, Korea, 5–9 November 2012; pp. 539–551.
14. Biber, P.; Strasser, W. The normal distributions transform: A new approach to laser scan matching. In Proceedings of the 2003 IEEE/RSJ International Conference on Intelligent Robots and Systems (IROS 2003) (Cat. No. 03CH37453), Las Vegas, NA, USA, 27 October–1 November 2003; Volume 3, pp. 2743–2748.
15. Magnusson, M. The Three-Dimensional Normal-Distributions Transform, an Efficient Representation for Registration, Surface Analysis, and Loop Detection. Ph.D. Thesis, Örebro University, Örebro, Sweden, 2009.
16. Hornung, A.; Wurm, K.; Bennewitz, M.; Stachniss, C.; Burgard, W. OctoMap: An efficient probabilistic 3D mapping framework based on Octrees. *Auton. Robot.* **2013**, *34*, 189–206. [CrossRef]
17. Bouraine, S.; Bougouffa, A.; Azouaoui, O. Particle swarm optimization for solving a scan matching problem based on the normal distributions transform. *Evol. Intell.* **2021**, *15*, 1–12. [CrossRef]
18. Santos, M.M.; Zaffari, G.B.; Ribeiro, P.O.; Drews-Jr, P.L.; Botelho, S.S. Underwater place recognition using forward-looking sonar images: A topological approach. *J. Field Robot.* **2019**, *36*, 355–369. [CrossRef]
19. Eustice, R.; Walter, M.; Leonard, J. Sparse extended information filters: Insights into sparsification. In Proceedings of the 2005 IEEE/RSJ International Conference on Intelligent Robots and Systems, Edmonton, AB, Canada, 2–6 August 2005; pp. 3281–3288.
20. Johannsson, H.; Kaess, M.; Englot, B.; Hover, F.; Leonard, J. Imaging sonar-aided navigation for autonomous underwater harbor surveillance. In Proceedings of the 2010 IEEE/RSJ International Conference on Intelligent Robots and Systems, Taipei, Taiwan, 18–22 October 2010; pp. 4396–4403.
21. Shin, Y.S.; Lee, Y.; Choi, H.T.; Kim, A. Bundle adjustment from sonar images and SLAM application for seafloor mapping. In Proceedings of the OCEANS 2015-MTS/IEEE, Washington, DC, USA, 19–22 October 2015; pp. 1–6.
22. Jiang, M.; Song, S.; Li, Y.; Jin, W.; Liu, J.; Feng, X. A survey of underwater acoustic SLAM system. In Proceedings of the Intelligent Robotics and Applications: 12th International Conference, ICIRA 2019, Shenyang, China, 8–11 August 2019; Part II 12; Springer: Berlin/Heidelberg, Germany, 2019; pp. 159–170.
23. Li, J.; Kaess, M.; Eustice, R.M.; Johnson-Roberson, M. Pose-graph SLAM using forward-looking sonar. *IEEE Robot. Autom. Lett.* **2018**, *3*, 2330–2337. [CrossRef]
24. Joshi, B.; Xanthidis, M.; Roznere, M.; Burgdorfer, N.J.; Mordohai, P.; Li, A.Q.; Rekleitis, I. Underwater exploration and mapping. In Proceedings of the 2022 IEEE/OES Autonomous Underwater Vehicles Symposium (AUV), Singapore, 19–21 September 2022; pp. 1–7.

25. Alcantarilla, P.F.; Solutions, T. Fast explicit diffusion for accelerated features in nonlinear scale spaces. *IEEE Trans. Patt. Anal. Mach. Intell* **2011**, *34*, 1281–1298.
26. Yang, J.; Li, H.; Campbell, D.; Jia, Y. Go-ICP: A globally optimal solution to 3D ICP point-set registration. *IEEE Trans. Pattern Anal. Mach. Intell.* **2015**, *38*, 2241–2254. [CrossRef]
27. Nitzberg, R. Constant-false-alarm-rate signal processors for several types of interference. *IEEE Trans. Aerosp. Electron. Syst.* **1972**, *AES-8*, 27–34. [CrossRef]
28. Ester, M.; Kriegel, H.P.; Sander, J.; Xu, X. Density-based spatial clustering of applications with noise. In Proceedings of the International Conference on Knowledge Discovery and Data Mining (KDD-96), Portland, OR, USA, 2–4 August 1996; Volume 240.
29. Leordeanu, M.; Hebert, M.; Sukthankar, R. An integer projected fixed point method for Graph Matching and map inference. *Adv. Neural Inf. Process. Syst.* **2009**, *22*, 1114–1122.
30. Johnston, H. Cliques of a graph-variations on the Bron-Kerbosch algorithm. *Int. J. Comput. Inf. Sci.* **1976**, *5*, 209–238. [CrossRef]
31. Zass, R.; Shashua, A. Probabilistic graph and hyperGraph Matching. In Proceedings of the 2008 IEEE Conference on Computer Vision and Pattern Recognition, Anchorage, AK, USA, 23–28 June 2008; pp. 1–8.
32. Zhou, F.; De la Torre, F. Factorized Graph Matching. *IEEE Trans. Pattern Anal. Mach. Intell.* **2015**, *38*, 1774–1789. [CrossRef]
33. Zhou, F.; De la Torre, F. Deformable Graph Matching. In Proceedings of the IEEE Conference on Computer Vision and Pattern Recognition, Portland, OR, USA, 23–28 June 2013; pp. 2922–2929.
34. Tareen, S.A.K.; Saleem, Z. A comparative analysis of sift, surf, kaze, akaze, orb, and brisk. In Proceedings of the 2018 International Conference on Computing, Mathematics and Engineering Technologies (iCoMET), Sukkur, Pakistan, 3–4 March 2018; pp. 1–10.
35. Censi, A. An accurate closed-form estimate of ICP's covariance. In Proceedings of the 2007 IEEE International Conference on Robotics and Automation, Roma, Italy, 10–14 April 2007; pp. 3167–3172.
36. Nabiyev, V.V.; Yılmaz, S.; Günay, A.; Muzaffer, G.; Ulutaş, G. Shredded banknotes reconstruction using AKAZE points. *Forensic Sci. Int.* **2017**, *278*, 280–295. [CrossRef]
37. Dellaert, F.; Kaess, M. Factor graphs for robot perception. *Found. Trends® Robot.* **2017**, *6*, 1–139. [CrossRef]
38. Wang, J.; Chen, F.; Huang, Y.; McConnell, J.; Shan, T.; Englot, B. Virtual maps for autonomous exploration of cluttered underwater environments. *IEEE J. Ocean. Eng.* **2022**, *47*, 916–935. [CrossRef]
39. Olson, E.B. Real-time correlative scan matching. In Proceedings of the 2009 IEEE International Conference on Robotics and Automation, Kobe, Japan, 12–17 May 2009; pp. 4387–4393.
40. Carroll, S.B. Evo-devo and an expanding evolutionary synthesis: A genetic theory of morphological evolution. *Cell* **2008**, *134*, 25–36. [CrossRef]

Journal of
*Marine Science
and Engineering*

MDPI

Article

Underwater Gyros Denoising Net (UGDN): A Learning-Based Gyros Denoising Method for Underwater Navigation

Chun Cao [1], Can Wang [1], Shaoping Zhao [1], Tingfeng Tan [1], Liang Zhao [2] and Feihu Zhang [1,*]

[1] School of Marine Science and Technology, Northwestern Polytechnical University, Xi'an 710072, China; caochun@mail.nwpu.edu.cn (C.C.); zhaoshaoping@mail.nwpu.edu.cn (S.Z.); ttf@mail.nwpu.edu.cn (T.T.)
[2] College of Electrical and Power Engineering, Taiyuan University of Technology, Taiyuan 030000, China; zhaoliang163yx@163.com
* Correspondence: feihu.zhang@nwpu.edu.cn; Tel.: +86-029-88492611

Abstract: Autonomous Underwater Vehicles (AUVs) are widely used for hydrological monitoring, underwater exploration, and geological surveys. However, AUVs face limitations in underwater navigation due to the high costs associated with Strapdown Inertial Navigation System (SINS) and Doppler Velocity Log (DVL), hindering the development of low-cost vehicles. Micro Electro Mechanical System Inertial Measurement Units (MEMS IMUs) are widely used in industry due to their low cost and can output acceleration and angular velocity, making them suitable as an Attitude Heading Reference System (AHRS) for low-cost vehicles. However, poorly calibrated MEMS IMUs provide an inaccurate angular velocity, leading to rapid drift in orientation. In underwater environments where AUVs cannot use GPS for position correction, this drift can have severe consequences. To address this issue, this paper proposes Underwater Gyros Denoising Net (UGDN), a method based on dilated convolutions and LSTM that learns and extracts the spatiotemporal features of IMU sequences to dynamically compensate for the gyroscope's angular velocity measurements, reducing attitude and heading errors. In the experimental section of this paper, we deployed this method on a dataset collected from field trials and achieved significant results. The experimental results show that the accuracy of MEMS IMU data denoised by UGDN approaches that of fiber-optic SINS, and when integrated with DVL, it can serve as a low-cost underwater navigation solution.

Keywords: SINS; MEMS; IMU; AHRS; deep learning; gyroscope

1. Introduction

The advancement of science and technology has led to increased resource consumption and a growing scarcity of land resources, making the resource-rich oceans the target of competition among nations [1]. AUVs, with their outstanding underwater operational capabilities, have become one of the most widely used devices in ocean exploration [2].

Since the beginning of the new century, with the acceleration of ocean development, underwater robots, particularly AUVs, have entered the public eye. The excellent underwater mobility and high level of autonomous control of AUVs have significantly enhanced human capabilities in exploring and developing the deep sea, showing great potential in marine exploration, surveying, and mapping [3]. Precise navigation of AUVs is a prerequisite for performing various underwater missions, and its accuracy directly affects the effectiveness of underwater detection, especially in hydrographic data collection and seabed mapping. Due to the inability to obtain GNSS signals underwater, AUVs cannot correct their position in real-time [4]; thus, inertial navigation systems have been widely applied in the field of underwater navigation. Currently, the intergration of SINS and DVL remains the mainstream solution in underwater navigation. Although the cooperative navigation of inertial and acoustic velocity measuring devices generally meets requirements and keeps navigation errors within a controllable range, its limitation lies in the dependency of navigation accuracy on the precision of the equipment. The high cost of SINS and DVL limits the application and data collection of small AUVs [5]. Specifically, the cost of fiber optic inertial navigation systems can reach tens of thousands of dollars, making it unaffordable for small teams. For AUVs that cannot frequently surface to obtain GNSS signals, acoustic navigation is a suitable alternative. In fact, numerous studies have been conducted by scholars based on USBL/LBL to correct the cumulative errors of dead reckoning, but this method inherently relies on external sensors, which undoubtedly increases costs. Additionally, the USBL/LBL base stations need to be pre-deployed, restricting the activity range of AUVs and making them unsuitable for long-distance missions [6].

In recent years, Visual-Inertial Navigation Systems (VINS) have demonstrated significant potential in applications such as autonomous driving and drone navigation, providing accurate and robust navigation solutions [7]. However, despite the many advantages of VINS, its application in AUV navigation faces substantial limitations in underwater environments characterized by weak textures and low lighting conditions [8]. Additionally, underwater SLAM (Simultaneous Localization and Mapping) technology based on imaging sonar has also made significant progress, but so far, most SLAM algorithms designed for AUVs remain prohibitively expensive, making the real-time requirements for practical implementation in AUVs quite challenging [9,10].

MEMS inertial sensors have dominated the market due to their low cost and relatively reliable performance. Thanks to advanced MEMS manufacturing processes, these sensors can achieve precise measurements and have significant advantages in terms of size, weight, and power consumption. However, the accuracy of IMUs is susceptible to calibration parameters, including scale factors and axis misalignment [11,12]. Inaccurate calibration can lead to imprecise measurements of angular velocity and linear acceleration, causing errors to accumulate rapidly during integration, which could lead to severe consequences [13].

To limit IMU error drift and address the issue of cumulative error fundamentally, scholars have conducted extensive research. By improving traditional calibration methods, such as gravity-based accelerometer correction and continuous rotation-based gyroscope correction, Rohac proposed a calibration method based on a sensor error model for accurately correcting MEMS sensor errors [12]. Additionally, by extending the open-source Kalibr toolkit [14], Rehder and colleagues achieved simultaneous calibration of multiple IMU parameters [15]. Over the past decade, with the tremendous success of neural networks in various fields, many researchers have started applying them to dead reckoning [16], indoor pedestrian navigation [17], gyroscope noise reduction [18], and attitude estimation [19]. Despite the relatively shallow layers of neural networks designed in these methods, the results obtained have been remarkably effective [20]. Additionally, most of these studies focus on datasets collected by UAVs/rovers, and have not yet conducted research related to surface or underwater environments.

In this paper, we propose a lightweight network based on CNN–LSTM for the calibration of low-cost IMU gyroscopes. This network is based on the mathematical model of gyroscope calibration and extracts spatiotemporal features from historical IMU sequences to dynamically calibrate IMU measurements. By using the calibrated IMU measurements, we can achieve orientation estimation accuracy approaching that of fiber-optic SINS. Integrated with DVL (instruments that use the Doppler effect to measure underwater velocity), this method provides a new approach for low-cost underwater navigation solutions. The methodology framework can be found in Figure 1.

Our contributions can be summarized as follows:

- We propose a lightweight CNN–LSTM model based on dilated convolutions to dynamically compensate for IMU measurement errors through learning.
- We introduce a gyroscope calibration matrix (including additive and multiplicative noise) to construct a training and calibration framework, optimizing the calibration matrix and hyperparameters within the network through learning.
- We conduct qualitative and quantitative evaluations of the proposed method using a self-made dataset. Current research mostly focuses on vehicles and drones, to fill the gap in waterborne or underwater navigation systems, we created a waterway

navigation dataset using a motorboat and the experimental results show that the denoised MEMS IMU data achieve accuracy close to that of fiber-optic SINS. When integrated with DVL, this method provides a reliable reference for low-cost underwater navigation solutions.

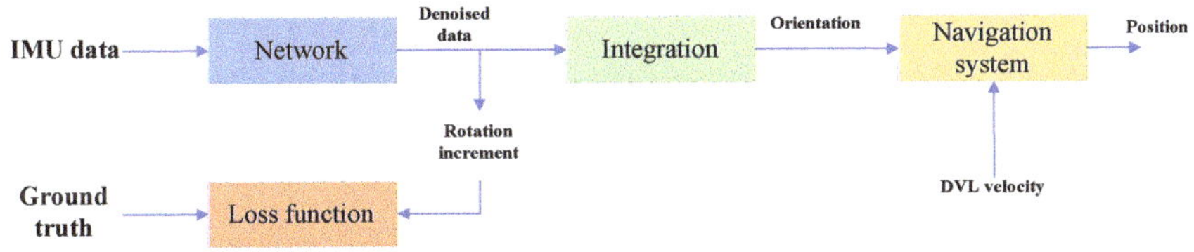

Figure 1. Overview of the proposed method.The network denoises the raw IMU data to provide real-time compensation for the gyroscope output. During training, the network output is used to obtain the rotation increments for loss function. During testing, the denoised data are used as the input for the intergration to estimate the orientation, which is then integrated with DVL measurements to estimate the position of the navigation system.

This paper will be organized as follows: Section 2 introduces related work, Section 3 presents the mathematical theory and our method, Section 4 qualitatively and quantitatively tests the proposed method through experiments, and Section 5 concludes with future works.

2. Related Works

To improve the accuracy of IMUs, their parameters need to be calibrated. Tedaldi et al. [21] proposed an IMU calibration method that does not require external equipment. This method involves placing the IMU in different static orientations at multiple positions. To avoid unobservability in the estimation of the calibration parameters, data from at least nine different orientations of the IMU must be collected. The more orientations used, the more accurate the calibration results. Cheuk et al. [22] proposed an automatic IMU calibration method that obtains the IMU's scale factors, biases, and angle alignment errors by rotating all axes of the IMU and holding it stationary in 12 positions. Zhang et al. [23] proposed an IMU calibration method based on a three-dimensional turntable, which not only removes biases but also focuses on correcting the angular correlations between different sensors. To address on-site IMU calibration issues, Qureshi [24] proposed a method using the Earth's gravitational field as a reference. This method does not require external equipment and only involves a few simple rotations of the IMU within 20 min. These methods often rely on mathematical models and cannot achieve online calibration, making it difficult to meet the needs for dynamic calibration.

Aimed at VINS, Furgale et al. proposed a method that calculates the calibration parameters of the IMU offline by pre-calibrating the external parameters between the camera and the IMU, known as the Kalibr library [14]. This method has been widely used in visual–inertial odometry and was extended in 2016 to achieve axis calibration for multiple IMUs [15]. For visual–inertial odometry, Qin et al. proposed an online calibration method for dynamically optimizing model parameters [25], and also introduced VINS-Mono, which is one of the most advanced visual–inertial odometry systems [26]. However, such methods rely on additional equipment, and due to the difficulty of obtaining clear optical images under low-light conditions underwater, these methods are challenging to apply in underwater navigation.

Additionally, given the tremendous success of deep learning in various fields, researchers have also started to focus on using deep learning techniques to calibrate IMU errors. Herath et al. proposed RoNIN, which uses three network architectures based on ResNet, LSTM, and TCN to regress true velocities [27]. Chen et al. introduced IONet, a network structure based on LSTM that learns positional transformations in polar coordinates from raw IMU data and constructs an inertial odometry system, reducing errors by segmenting inertial data into independent windows [28]. Esfahani et al. proposed OriNet, a deep learning framework that achieves accurate orientation estimation for drones by learning IMU sequences [29]. Liu et al. proposed TLIO, which combines ResNet with EKF to predict displacement and uncertainty in error propagation by learning from data collected by head-mounted IMUs [30]. Nobre et al. introduced a reinforcement learning-based framework that models the IMU calibration process as a Markov decision process and uses reinforcement learning to regress optimal calibration parameters [31]. Brossard et al. proposed a method based on dilated convolutions [18], using only five layers of dilated convolutions to extract spatiotemporal features from past IMU sequences to regress the true values of orientation. This method has been validated on datasets such as EuRoC [32] and performs comparably to top visual–inertial odometry systems like OpenVINS [33].

Russo et al. developed an intelligent deep denoising autoencoder to enhance Kalman filter outputs, with its key advantage being a comprehensive noise compensation model that eliminates the need to handle each influencing factor separately [34]. Di Ciaccio et al. introduced DOES, a deep-learning-based method tailored for maritime navigation, designed to improve roll and pitch estimations from conventional AHRS [35]. Zhang et al. proposed a multi-learning fusion model for denoising low-cost IMU data, integrating convolutional autoencoders, LSTM, and Transformer multi-head attention mechanisms, which is roven to be more effective than traditional signal processing techniques for the complex motion characteristics of ships [36]. Wang et al. presented a denoising approach combining the nonlinear generalization of SVM with the multi-resolution capabilities of wavelet transform, merging uncertain data from INS and GPS [37].

In summary, although research on deep-learning-based IMU calibration is still in its infancy, existing studies indicate that deep-learning-based MEMS IMU calibration can be performed online without relying on external equipment. Its application in underwater navigation is feasible, providing a new approach to reducing the cost of underwater navigation.

3. Mathematical Preliminaries and Methods

3.1. Low-Cost IMU and Modeling

IMUs typically include 3 gyroscopes and 3 accelerometers, and sometimes magnetometers to detect the Earth's magnetic field for carrier's heading. These sensors measure angular velocity and linear acceleration, allowing for calculating the carrier's attitude, velocity, and position.

$$\mathrm{u}_n^{imu} = \begin{bmatrix} \omega_n^{imu} \\ \mathrm{a}_n^{imu} \end{bmatrix} = \mathrm{C} \begin{bmatrix} \omega_n \\ \mathrm{a}_n \end{bmatrix} + b_n + \eta_n \tag{1}$$

Because it suffers from axis misalignment, scale factors, bias, temperature drift, and random walk, there is a discrepancy between IMU measurements and ground truth. Equation (1) describes the sensor measurement model of the IMU, where u_n^{imu} represents the actual measurements from the IMU with high-frequency noise, ω_n^{imu} denotes the actual angular velocity measurements, a_n^{imu} denotes the actual linear acceleration measurements, b_n represents the IMU measurement bias, η_n denotes the measurement noise, and C is the intrinsic calibration matrix of the IMU, which includes correction parameters for the effects of axis misalignment and scale factors on IMU measurements. ω_n and a_n represent the true angular velocity and linear acceleration, respectively.

According to [12,15,18], the matrix C in Equation (1) can be modeled as follows:

$$\mathrm{C} = \begin{bmatrix} \mathrm{S}_\omega \mathrm{M}_\omega & \mathrm{A} \\ 0_{3\times 3} & \mathrm{S}_\mathrm{a} \mathrm{M}_\mathrm{a} \end{bmatrix} \approx \mathrm{I}_6 \tag{2}$$

In which M is used to correct for axis misalignment, S is used to correct for scale factor errors, A represents the sensitivity to gravity and I_6 denotes the 6×6 identity matrix.

$$\mathrm{R}(n) = \mathrm{R}(n-1)\exp(\theta_n) \tag{3}$$

$$\theta_n = \omega_n^{imu} dt \tag{4}$$

$$\exp(\theta_n) = I + \frac{\sin\theta_n}{\theta_n}[\theta_n \times] + \frac{1 - \cos\theta_n}{\theta_n^2}[\theta_n \times]^2 \tag{5}$$

Equations (3)–(5) describe the attitude updates of the rigid body at each time step, where $\times$ represents the skew-symmetric operator. They calculate the angular increments based on the IMU gyroscope data and then obtain the attitude estimate through integration. Here, $R \in SO(3)$ describes the angle of rotation of the rigid body in 3D space, and θ_n represents the angular increment at each time step. Equation (5) provides a numerical solution for the matrix exponential mapping, i.e., the Rodrigues' Equation.

$$v(n) = v(n-1) + \left[\mathrm{R}(n-1)\mathrm{a}_n^{imu} - g\right]dt \tag{6}$$

$$p(n) = p(n-1) + v(n-1)dt \tag{7}$$

The velocity and position is updated as Equations (6) and (7). Next, we analyze the error propagation of the gyroscope. As the IMUs suffer from axis misalignment, scale factor errors, random walk, etc., the IMUs output can be modeled as follows:

$$\omega_n^{imu} = (I + S_\omega + \Delta C_\omega)\omega_n + b_\omega + \eta_\omega \tag{8}$$

$$\mathrm{a}_n^{imu} = (I + S_a + \Delta C_a)a_n + b_a + \eta_a \tag{9}$$

where ω_n and a_n are the true values of angular velocity and linear acceleration; S_ω and S_a are the scale factor error matrices for the gyroscope and accelerometer, respectively; C_ω and C_a are the mounting error matrices; b is the bias; and η represents the random walk. Thus, we can derive:

$$\hat{\omega}_n = (I + S_\omega + \Delta C_\omega)^{-1}\left(\omega_n^{imu} - b_\omega + \eta_\omega\right) = \hat{C}_\omega \omega_n^{imu} + \tilde{\omega}_n \tag{10}$$

where $\hat{\omega}_n$ is the corrected angular velocity, matrix C represents the correction matrix for misalignment errors, scale factor errors, and other multiplicative noises, while $\tilde{\omega}_n$ denotes the correction parameters for bias and random walk. Specifically, as misalignment errors and scale factors are constants, C is considered to be constant. In contrast, bias and random walk are time-varying, so $\tilde{\omega}_n$ is treated as time-varying noise. Additionally, due to the high-order coupling between the gyroscope data and the body's acceleration—meaning that acceleration measurements can also influence angular velocity [16,38]—this has a significant impact on $\tilde{\omega}_n$.

3.2. Simple DVL Modeling

A DVLis an instrument that measures velocity using the Doppler effect of sound waves and is widely used in underwater navigation. The Doppler effect refers to the change in frequency of a wave when the source is in relative motion to the receiver. The DVL emits sound beams at various angles into the water and receives the reflected signals from underwater objects (e.g., the seabed). If relative motion exists, the frequency of the returning sound waves changes. By analyzing the frequency shift between the transmitted and received signals, the DVL calculates its velocity relative to the water layer or seabed. Its sensor measurement model can be defined as Equation (11):

$$f_r = f_t\left(\frac{1 \mp \frac{v_{beam}}{c}}{1 \pm \frac{v_{bean}}{c}}\right) \tag{11}$$

Let f_r be the received frequency, f_t be the transmitted frequency, v_{beam} be the beam velocity, and c be the speed of sound. As the velocity of DVL is much smaller than the speed of sound, the frequency shift Δf can be simplified to:

$$\Delta f \approx \frac{2 f_t v_{beam}}{c} \tag{12}$$

The speed of beams can be defined as follows:

$$v_{beam} = \frac{c}{2 f_t} \Delta f \tag{13}$$

3.3. Network Architecture

In our method, the network model is divided into two parts: The first part consists of a single dilated convolution layer, which takes the gyroscope data as input, and it is responsible for outputting $\hat{C_\omega} \omega_n^{imu}$. The second part is composed of CNN and LSTM, which is aimed at the $\tilde{\omega}_n$. The network architecture is illustrated in Figure 2.

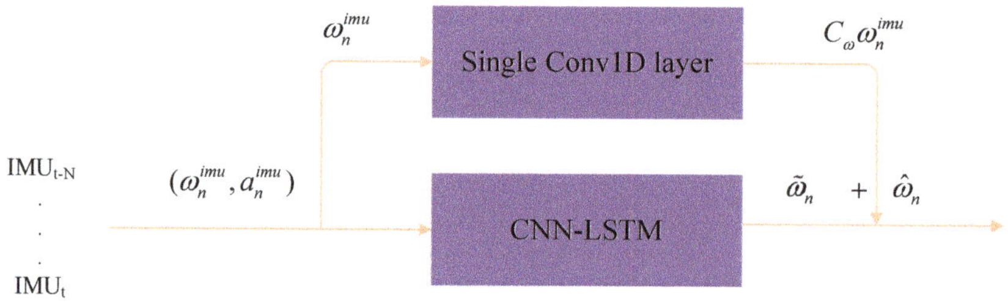

Figure 2. Network Architecture.

As described by Equation (14), the network utilizes IMU data from the past N moments to predict time-varying errors $\tilde{\omega}_n$, where $f(\cdot)$ is a nonlinear function defined by the neural network, IMU_t is the output of the IMU at time t, and N is the size of the sliding window.

$$\tilde{\omega}_n = f(\text{IMU}_{t-N}, \ldots, \text{IMU}_t) \tag{14}$$

Our network consists of 4 layers of dilated convolution and 2 layers of LSTM. As a network structure, it combines the advantages of CNN and LSTM, enabling it to capture both local spatial features and temporal relationships simultaneously. This makes it suitable for processing time series data and achieving good modeling results. It has demonstrated an excellent performance in tasks involving the processing of temporal information, such as natural language processing [39]. In our network, data are first fed into the dilated convolution layers for feature extraction, with each layer sequentially containing a Conv1d layer, a BatchNorm layer, a GELU activation function, and Dropout to accelerate convergence. Subsequently, the sequence is further transmitted to the LSTM layers to fully extract its temporal features, and finally, the prediction results are given. The network structure is shown in Figure 3 below. The table provides detailed configurations.

Table 1 provides the detailed configuration of the network. Specifically, in our method, the size of sliding window is defined as follows:

$$W_{\text{size}} = \max(\text{kernel dim} \times \text{dilation gap}) \tag{15}$$

so the window size is solved as 112, thus, the network utilizes up to 112 past sequences (i.e., data from the past 1.12 s) for learning, rather than using future data.

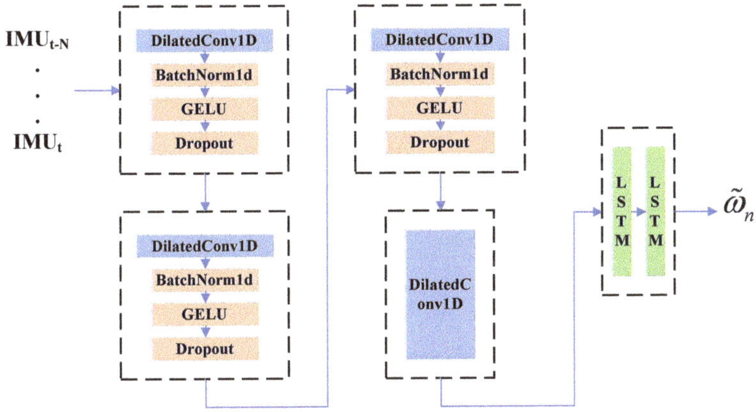

Figure 3. Proposed dilated CNN−LSTM model structure for $\tilde{\omega}_n$.

Table 1. The network configurations.

Block	In/Out Channels	Kernel Size	Dilation Factor	Num Layers
CNN layer1	(6, 16)	7	1	\
CNN layer2	(16, 64)	7	4	\
CNN layer3	(64, 128)	7	16	\
CNN layer4	(128, 3)	7	1	\
LSTM layers	(3, 3)	\	\	2

3.4. Loss Function

In our method, the relative rotation at each moment are used as the optimization variables for loss function. The training loss for each epoch is calculated by the Huber function [40,41].

$$L_{\text{Huber}}(y, f(x)) = \begin{cases} \frac{1}{2}(y - f(x))^2, & |y - f(x)| \leq \delta \\ \delta|y - f(x)| - \frac{1}{2}\delta^2, & |y - f(x)| > \delta \end{cases} \tag{16}$$

3.5. Development Environment

Our method is implemented based on PyTorch, using the ADAM optimizer [42] for network training. The learning rate is set to 0.01, with a weight decay of 0.1. The network training employs cosine warming restarts scheduler. The weight parameter for the logarithmic cosine loss is set to 1×10^6. The IMU operates at 100 Hz, so the Δt is set to 10 ms. We validated the algorithm using our own collected dataset, and it consists of 33 sequences. We select 9 sequences with diverse motion patterns as the test set. The remaining sequences are used for training, with data from 0 to 160 s of each sequence included, and validation. We trained for 1800 epochs, and it took approximately 270 s on our workstation (detailed parameters are shown in Table 2).

Table 2. Development environment.

Block	Parameters
CPU	i9-13980HX@2.20 GHz
RAM	32 G
GPU	NVIDIA GeForce RTX 4080 LAPTOP (12 GB)
Deep learning framework	Pytorch1.13.1
operating system	Windows 11

4. Experimental Evaluation

This section will focus on the collection and data processing of the dataset, and using this dataset for qualitative and quantitative analysis of our method. In addition, we will compare the performance of our method with other learning-based algorithms in attitude estimation.

4.1. Data Collection and Preprocessing

We used the motorboat (shown in Figure 4) as the data collection platform. This platform is equipped with high-precision SINS, RTK GNSS, and DVL. To ensure a diverse range of data types and a representative sample set, we collected trajectories under different times and weather conditions through various navigation tasks, including circular trajectories, irregular maneuvers, and sharp turns.

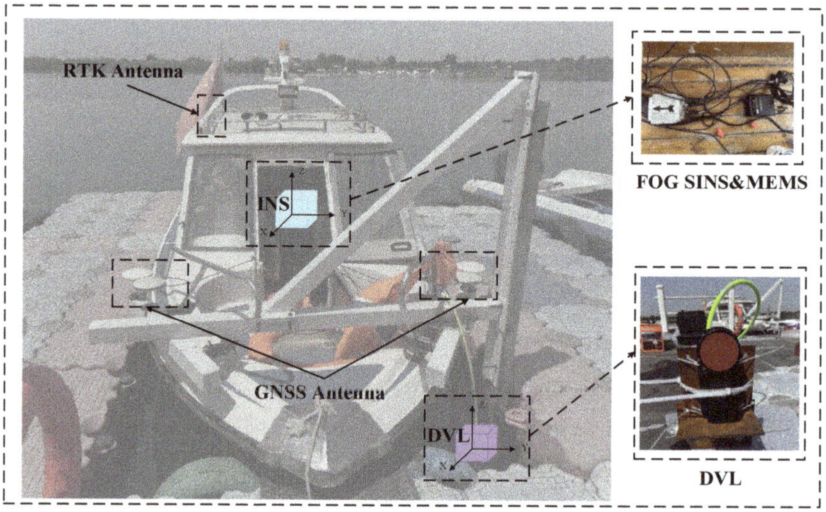

Figure 4. Data collecting platform.

Our data collection system includes the following sensors:

- A fiber optic strapdown inertial navigation system, including high-precision optical fiber gyroscopes and quartz flexible accelerometers, GNSS, and RTK positioning systems, providing high-frequency ground truth data (100 Hz) according to the NMEA-0183 protocol.
- A low-precision MEMS inertial navigation system, including MEMS gyroscopes and accelerometers, and RTK GNSS, providing training data (100 Hz).
- A Pathfinder RDI 600k DVL, providing low-frequency bottom speed information (5 Hz).

Detailed parameters can be found in Table 3.

Two INS units were installed in the cabin of the boat and synchronized using precise time provided by GPS. The DVL was mounted on the left side of the boat at a depth of 1 m and synchronized with the INS via 1pps. During the experiment, the boat traveled at a constant speed of 6 knots. Experiments were conducted in a natural canyon in northwest China (shown in Figures 5 and 6) in April 2024 and June 2024 (both experiments were conducted on sunny days, with calm lake surfaces or occasional light breezes), lasting a total of 10 h and collecting approximately 30 km of data.

Table 3. Overview of the sensor specifications and performance.

Device	Parameter	Performance	Coordinate System
FOG SINS	Heading accuracy	0.05°	Forward, left, up
	Gyro zero deviation stability	≤0.1°/h	
	Accelerometer zero offset stability	≤20 µg	
MEMS IMU	Heading accuracy	0.1°	Forward, left, up
	Gyro zero deviation stability	≤6°/h	
	Accelerometer zero offset stability	≤20 µg	
DVL	Frequency	614.4 kHz	Forward, left, up
	Velocity accuracy	±0.2% ±0.2 cm/s	
	Altitude	0.2–89 m	

(**a**) (**b**)

Figure 5. The experimental environment. (**a**) satellite photo; (**b**) natural scene.

Finally, the raw data were processed into KITTI format [43] and augmented with bottom speed and bottom height information provided by the DVL to facilitate the subsequent processing and enhance interpretability. Additionally, we visualized the trajectory by integrating the collected angular velocity, linear acceleration, and DVL bottom speed and compared it with the precise positioning provided by RTK GNSS to verify the accuracy of the data samples.

Figure 6. One sequence of our dataset, we set off from the dock, heading along the embankment deeper into the canyon.

4.2. Evaluation Metrics

To quantitatively validate the effectiveness of the method, we use the following metrics to evaluate the proposed approach, which are based on the methods proposed in the EVO toolbox [44,45] .

4.2.1. Absolute Orientation Error

The absolute orientation error (AOE) is defined as follows:

$$\text{AOE} = \sqrt{\sum_{n=1}^{M} \frac{1}{M} \|\log(\mathrm{R}_n^T \hat{\mathrm{R}}_n)\|_2^2} \tag{17}$$

AOE represents the relative rotational change between the estimated pose and the ground truth within each time interval. In our experiments, we calculate the AOE for each time interval and then average these values to obtain the AOE for the entire trajectory, where M means the length of the trajectory.

4.2.2. Absolute Positioning Error

The absolute positioning error (APE) is defined as follows:

$$\text{APE} = \sqrt{\frac{1}{M} \sum_{n=1}^{M} \|p_i - \hat{p}_i\|^2} \tag{18}$$

APE represents the difference between the estimated position and the ground truth, calculated as the RMSE (Root Mean Square Error).

4.2.3. Relative Positioning Error

The relative positioning error (RPE) is defined as follows:

$$\text{RPE} = \sqrt{\frac{1}{M} \sum_{n=1}^{M} \|p_{i+\Delta t} - p_i - (\hat{p}_{i+\Delta t} - \hat{p}_i)\|^2} \tag{19}$$

RPE represents the difference between the incremental changes in position estimates and the ground truth increments within each time interval Δt, calculated as RMSE. In our experiments, Δt is set to 20 s.

4.3. Performance

We compare the following methods:

- Raw MEMS, that is the orientation intergated from the uncalibrated IMU data;
- Gyros-net [18], which is only based on dilated convolution, set $\hat{C_\omega}$ and $\tilde{\omega}_n$ as the objects of learning, where $\tilde{\omega}_n$ is provided by the neural network, and whose training and test sets correspond to ours;
- Mahony [46], a classical nonlinear observer used for attitude estimation, which is quite effective in estimating gyroscope bias;
- Proposed method, the method demonstrated in Section 3.

Table 4 shows the AOE results. Our method achieved the best performance in predicting RPY (roll, pitch, and yaw).

To illustrate the differences more clearly, we selected several trajectories and plotted the orientation estimation and estimation error curves in Figure 7. The black line represents the ground truth (RPY), the red line represents the raw IMU output, the gold line represents our method, the turquoise line represents Gyros-Net [16], and the purple line represents the mahony. For all methods, we used integration to compute the orientation. As shown in the figure, due to uncalibrated and severe noise, the error of the raw IMU output accumulates rapidly, causing significant deviations from the ground truth. Our method has the smallest deviation from the ground truth, achieving the best results.

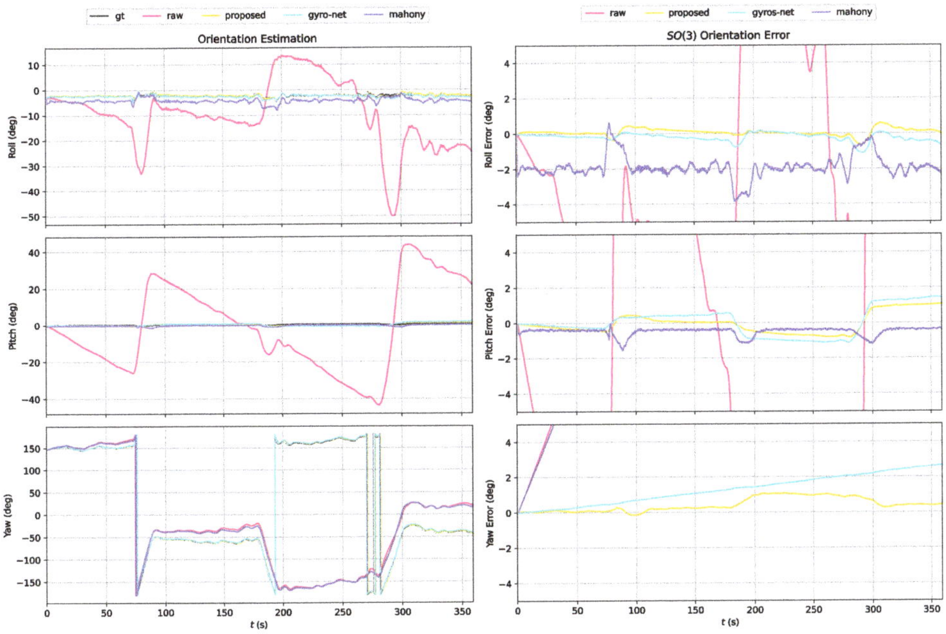

(a)

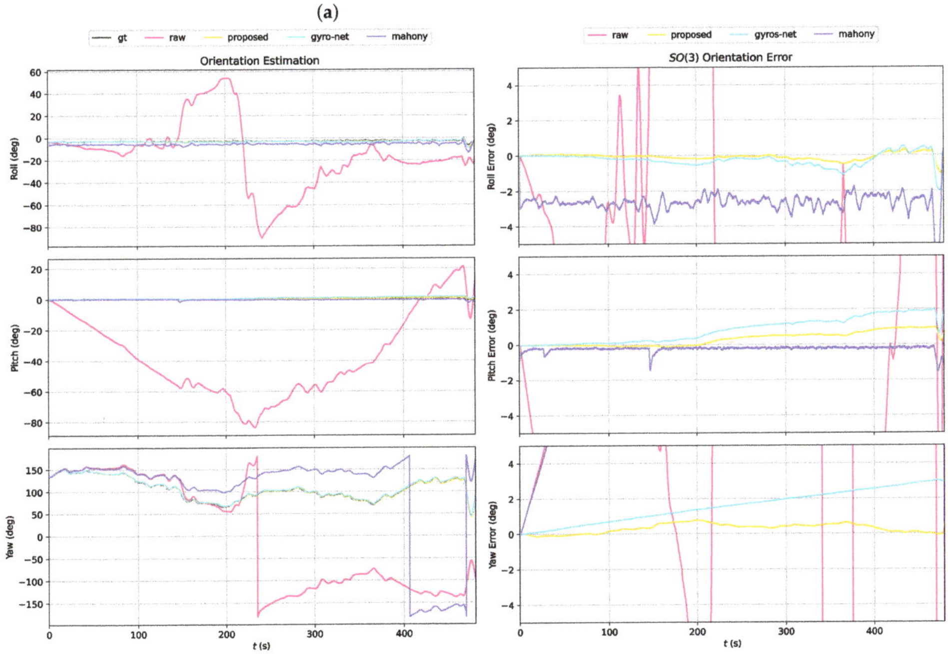

(b)

Figure 7. *Cont.*

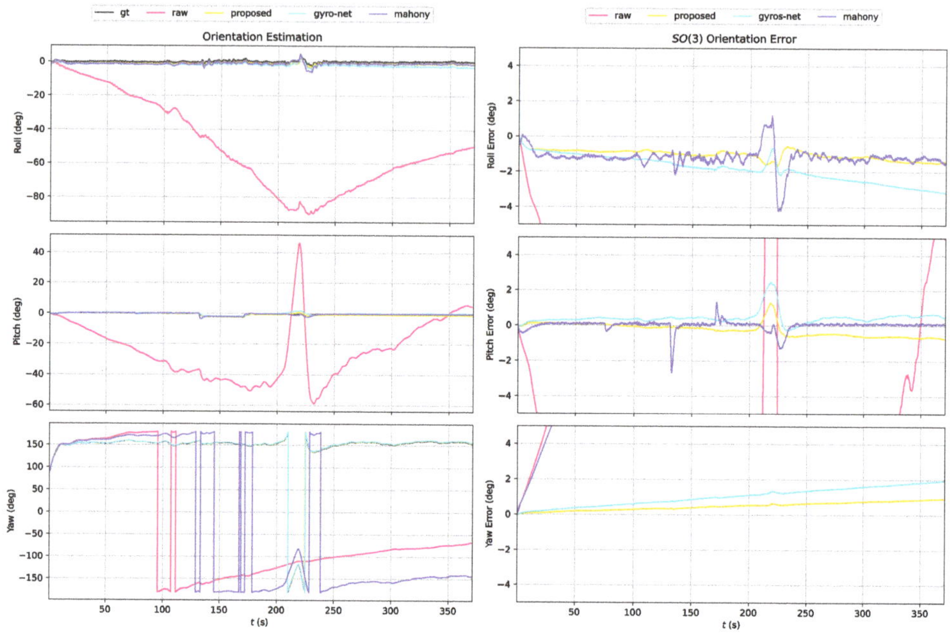

(c)

Figure 7. The overview of Orientation Estimation and SO(3) Orientation Error. In each subplot, the left part shows a comparison between the orientation estimated by these methods and the ground truth, while the right part displays the error between the estimated values and the ground truth, where (**a**) seq01; (**b**) seq03; (**c**) seq08.

Table 4. Absolute orientation error (AOE) in terms of 3D orientation(roll/pitch/yaw), in degree, on the test sequences.

	Proposed (R/P/Y)	Gyros-Net (R/P/Y)	Mahony (R/P/Y)	Raw (R/P/Y)
seq01	0.16/0.48/0.43	0.23/0.66/1.30	2.49/1.53/29.67	12.04/20.33/31.67
seq02	0.21/0.17/0.77	0.35/0.13/1.09	2.29/1.23/24.97	29.55/19.82/17.37
seq03	0.15/0.35/0.34	0.33/0.76/1.62	3.53/1.88/40.63	85.80/40.08/76.85
seq04	0.25/0.32/0.74	0.22/0.17/0.51	1.11/0.54/12.93	12.61/17.42/11.88
seq05	0.21/0.36/0.34	0.44/0.16/0.70	0.65/1.69/19.00	16.57/12.57/18.69
seq06	0.50/0.57/1.86	0.57/0.59/0.72	1.53/1.21/23.94	14.88/11.38/32.93
seq07	0.71/0.72/0.15	0.93/0.84/0.15	1.53/1.16/24.25	26.93/30.24/20.60
seq08	1.02/0.32/0.50	1.81/0.46/1.03	0.97/0.28/31.42	67.74/26.01/67.70
seq09	0.39/0.36/0.81	0.41/0.60/0.60	3.22/1.80/21.06	30.84/23.74/28.87
mean	0.40/0.41/0.66	0.59/0.49/0.86	1.92/1.26/25.32	32.99/22.40/34.06

The length of each test sequence varies.

4.4. Navigation Experiment

To demonstrate the potential of our method in low-cost underwater navigation, we conducted a series of cooperative navigation experiments, selecting four trajectories with diverse motion patterns. These trajectories cover different navigation conditions to comprehensively test the adaptability and accuracy of our method.

In the experiments, we used a DVL to collect velocity data along these trajectories. We integrated the DVL velocity data with the IMU data denoised by our method to evaluate its performance in providing accurate navigation data and compared it with the traditional SINS–DVL navigation system.

Preliminary results is shown in Figure 8, We are pleasantly surprised to find that, our method performs quite comparably to the SINS–DVL integrated navigation system. This suggests that our method has significant potential for low-cost underwater navigation while significantly reducing equipment costs.

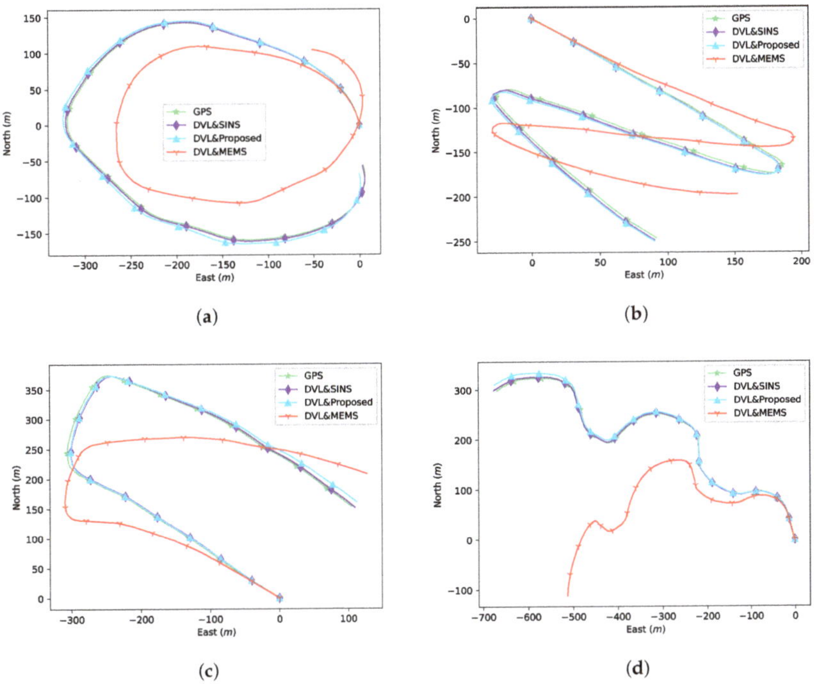

Figure 8. Comparisons of relative error between the proposed method and others, where (**a**) Seq06; (**b**) Seq05; (**c**) Seq02; (**d**) Seq09.

Tables 5 and 6 show the performance of several methods in terms of APE and RPE. We observe that our method performs comparably to the SINS–DVL integrated navigation system. The comparison indicates that our approach is effective in practical scenarios and provides new insights for reducing the cost of underwater navigation.

Table 5. APE in terms of trajectory, in meters, on the test sequences.

	SINS	Proposed	Raw
seq06	2.31	6.91	104.18
seq05	4.18	4.95	28.80
seq02	5.50	5.82	62.42
seq09	4.61	7.11	183.64
mean	4.15	6.20	94.76

Table 6. RPE in terms of trajectory, in meters, on the test sequences.

	SINS	Proposed	Raw
seq06	1.40	2.58	56.91
seq05	3.27	3.40	29.96
seq02	1.84	1.88	33.25
seq09	2.82	3.14	63.74
mean	2.33	2.75	45.97

5. Conclusions and Future Works

In this paper, we propose UGDN, a CNN–LSTM based gyroscope denoising method for underwater navigation. We design a training framework based on the gyroscope's measurement and error equations. By utilizing dilated convolution and LSTM to extract spatiotemporal features from IMU sequences, we aimed to achieve dynamic compensation for angular velocity output. We validated our method using a custom navigation dataset and compared it with other methods. The experimental results demonstrate that our method achieved a commendable calibration performance, with an AOE of 0.66 degrees in terms of yaw, an APE of 6.20 m, and an RPE of 2.75 m, approaching the performance of the SINS–DVL navigation system in DVL integrated navigation experiments. This provides an effective reference for low-cost underwater navigation. In the future, we plan to deploy the algorithm on AUVs to achieve online dynamic error compensation.

Author Contributions: Conceptualization, C.C., C.W. and F.Z.; methodology, C.C. and C.W.; validation, S.Z. and C.W.; data curation, C.C., C.W. and S.Z.; writing—original draft preparation, C.C.; writing—review and editing, C.W. and F.Z.; visualization, C.C., T.T. and L.Z.; supervision, F.Z.; project administration, F.Z.; funding acquisition, F.Z. All authors have read and agreed to the published version of the manuscript.

Funding: This research was funded by the Fundamental Research Funds for the Central Universities (G2024KY0602).

Institutional Review Board Statement: Not applicable.

Informed Consent Statement: Not applicable.

Data Availability Statement: The data presented in this study are available on request from the corresponding author.

Acknowledgments: The authors gratefully acknowledge the support provided by the Key Laboratory of Unmanned Underwater Transport Technology during the data collection process and the assistance of the research team members.

Conflicts of Interest: The authors declare no conflicts of interest.

Abbreviations

The following abbreviations are used in this manuscript:

AUV	Autonomous Underwater Vehicle
SINS	Strapdown Inertial Navigation System
DVL	Doppler Velocity Log
MEMS IMU	Micro Electro Mechanical System Inertial Measurement Unit
RTK	Real-Time Kinematic
GNSS	Global Navigation Satellite System
GPS	Global Positioning System
INS	Inertial Navigation System
USBL	Ultra Short Baseline
LBL	Long Baseline
CNN	Convolutional Neural Network
LSTM	Long Short-Term Memory
AHRS	Attitude Heading Reference System
VINS	Visual-Inertial Navigation Systems

References

1. Paull, L.; Saeedi, S.; Seto, M.; Li, H. AUV Navigation and Localization: A Review. *IEEE J. Ocean. Eng.* **2014**, *39*, 131–149. [CrossRef]
2. Yoerger, R.; Jakuba, M.; Bradley, M.; Bingham, B. Techniques for deep sea near bottom survey using an autonomous underwater vehicle. In *Robotics Research: Results of the 12th International Symposium ISRR*; Springer: Berlin/Heidelberg, Germany, 2007; pp. 416–429.
3. Bellingham, J.G.; Rajan, K. Robotics in remote and hostile environments. *Science* **2007**, *318*, 1098–1102. [CrossRef]

4. Yan, J.; Ban, H.; Luo, X.; Zhao, H.; Guan, X. Joint Localization and Tracking Design for AUV With Asynchronous Clocks and State 448 Disturbances. *IEEE Trans. Veh. Technol.* **2019**, *68*, 4707–4720. [CrossRef]
5. Su, R.; Zhang, D.; Li, C.; Gong, Z.; Venkatesan, R.; Jiang, F. Localization and Data Collection in AUV-Aided Underwater Sensor Networks: Challenges and Opportunities. *IEEE Netw.* **2019**, *33*, 86–93. [CrossRef]
6. Zhang, B.; Ji, D.; Liu, S.; Zhu, X.; Xu, W. Autonomous underwater vehicle navigation: A review. *Ocean Eng.* **2023**, *273*, 113861. [CrossRef]
7. Bresson, G.; Alsayed, Z.; Yu, L.; Glaser, S. Simultaneous localization and mapping: A survey of current trends in autonomous driving. *IEEE Trans. Intell. Veh.* **2017**, *2*, 194–220. [CrossRef]
8. Manzanilla, A.; Reyes, S.; Garcia, M.; Mercado, D.; Lozano, R. Autonomous navigation for unmanned underwater vehicles: Real-time experiments using computer vision. *IEEE Robot. Autom. Lett.* **2019**, *4*, 1351–1356. [CrossRef]
9. Mallios, A.; Ridao, P.; Ribas, D.; Hernández, E. Scan matching SLAM in underwater environments. *Auton. Robot.* **2014**, *36*, 181–198. [CrossRef]
10. Fallon, M.F.; Kaess, M.; Johannsson, H.; Leonard, J.J. Efficient AUV navigation fusing acoustic ranging and side-scan sonar. In Proceedings of the 2011 IEEE International Conference on Robotics and Automation, Shanghai, China, 9–13 May 2011.
11. Kozlov, A.; Kapralov, F. Millimeter-level calibration of IMU size effect and its compensation in navigation grade systems. In Proceedings of the DGON Inertial Sensors and Systems, Braunschweig, Germany, 30 December 2019.
12. Rohac, J.; Sipos, M.; Simanek, J. Calibration of low-cost triaxial inertial sensors. *IEEE Instrum. & Meas. Mag.* **2015**, *18*, 32–38.
13. Chen, H.; Taha,T.M.; Chodavarapu, V.P. Towards improved inertial navigation by reducing errors using deep learning methodology. *Appl. Sci.* **2022**, *12*, 3645. [CrossRef]
14. Furgale, P.; Rehder, J.; Siegwart, R. Unified temporal and spatial calibration for multi-sensor systems. In Proceedings of the IEEE/RSJ International Conference on Intelligent Robots and Systems, Tokyo, Japan, 3–7 November 2013.
15. Rehder,J.; Nikolic, J.; Schneider, T.; Hinzmann, T.; Siegwart, R. Extending kalibr: Calibrating the extrinsics of multiple IMUs and of individual axes. In Proceedings of the IEEE International Conference on Robotics and Automation, Stockholm, Sweden, 16–21 May 2016.
16. Brossard, M.; Barrau, A.; Bonnabel, S. AI-IMU Dead-Reckoning. *IEEE Trans. Intell. Veh.* **2020**, *5*, 585–595. [CrossRef]
17. Hou, X.; Bergmann, J. HINNet + HeadSLAM: Robust Inertial Navigation With Machine Learning for Long-Term Stable Tracking. *IEEE Sens. Lett.* **2023**, *7*, 1–4. [CrossRef]
18. Brossard, M.; Bonnabel, S.; Barrau, A. Denoising IMU Gyroscopes with Deep Learning for Open-Loop Attitude Estimation. *IEEE Robot. Autom. Lett.* **2020**, *5*, 4796–4803. [CrossRef]
19. Li, H.; Chang, S.; Yao, Q.; Wan, C.; Zou, G.; Zhang, D. Robust Heading and Attitude Estimation of MEMS IMU in Magnetic Anomaly Field Using a Partially Adaptive Decoupled Extended Kalman Filter and LSTM Algorithm. *IEEE Trans. Instrum. Meas.* **2024**, *73*, 9507813. [CrossRef]
20. Chen, C.; Pan, X. Deep Learning for Inertial Positioning: A Survey. *IEEE Trans. Intell. Transp. Syst.* **2024**, *25*, 10506–10523. [CrossRef]
21. Tedaldi, D.; Pretto, A.; Menegatti, E. A robust and easy to implement method for IMU calibration without external equipments. In Proceedings of the IEEE International Conference on Robotics and Automation, Hong Kong, China, 31 May–7 June 2014.
22. Cheuk, C.M.; Lau, T.K.; Lin, K.W.; Liu, Y. Automatic calibration for inertial measurement unit. In Proceedings of the International Conference on Control Automation Robotics and Vision, Guangzhou, China, 5–7 December 2012.
23. Zhang, R.; Hoflinger, F.; Reind, L.M. Calibration of an IMU Using 3-D Rotation Platform. *IEEE Sens. J.* **2014**, *14*, 1778–1787. [CrossRef]
24. Qureshi, U.; Golnaraghi, F. An Algorithm for the In-Field Calibration of a MEMS IMU. *IEEE Sens. J.* **2017**, *22*, 7479–7486. [CrossRef]
25. Qin, T.; Shen, S. Online Temporal Calibration for Monocular Visual-Inertial Systems. In Proceedings of the IEEE/RSJ International Conference on Intelligent Robots and Systems, Madrid, Spain, 1–5 October 2018.
26. Qin, T.; Li, P.; Shen, S. VINS-Mono: A Robust and Versatile Monocular Visual-Inertial State Estimator. *IEEE Trans. Robot.* **2018**, *34*, 1004–1020. [CrossRef]
27. Herath, S.; Yan, H.; Furukawa, Y. RoNIN: Robust Neural Inertial Navigation in the Wild: Benchmark, Evaluations, & New Methods. In Proceedings of the IEEE International Conference on Robotics and Automation (ICRA), Paris, France, 31 May–31 August 2020.
28. Chen, C.; Lu, X.; Markham, A.; Trigoni, N. Ionet: Learning to cure the curse of drift in inertial odometry. In Proceedings of the AAAI Conference on Artificial Intelligence, New Orleans, LA, USA, 2–7 February 2018.
29. Esfahani, M.A.; Wang, H.; Wu, K.; Yuan, S. OriNet: Robust 3-D orientation estimation with a single particular IMU. *IEEE Robot. Autom. Lett.* **2019**, *5*, 399–406. [CrossRef]
30. Liu, W.; Caruso, D.; Ilg, E.; Dong, J.; Mourikis, A.I.; Daniilidis, K.; Kumar, V.; Engel, J. TLIO: Tight Learned Inertial Odometry. *IEEE Robot. Autom. Lett.* **2020**, *5*, 5653–5660. [CrossRef]
31. Nobre, F.; Heckman, C. Learning to calibrate: Reinforcement learning for guided calibration of visual–inertial rigs. *Int. J. Robot. Res.* **2019**, *38*, 1388–1402. [CrossRef]
32. Burri, M.; Nikolic, J.; Gohl, P.; Schneider, T.; Rehder, J.; Omari, S.; Achtelik, M.W.; Siegwart, R. The EuRoC micro aerial vehicle datasets. *Int. J. Robot. Res.* **2016**, *35*, 1157–1163. [CrossRef]

33. Geneva, P.; Eckenhoff, K.; Lee, W.; Yang, Y.; Huang, G. OpenVINS: A Research Platform for Visual-Inertial Estimation. In Proceedings of the IEEE International Conference on Robotics and Automation, Paris, France, 31 May–31 August 2020.
34. Russo, P.; Di Ciaccio, F.; Troisi, S. DANAE++: A smart approach for denoising underwater attitude estimation. *Sensors* **2021**, *21*, 1526. [CrossRef] [PubMed]
35. Di Ciaccio, F.; Russo, P.; Troisi, S. DOES: A Deep Learning-Based Approach to Estimate Roll and Pitch at Sea. *IEEE Access* **2022**, *10*, 29307–29321. [CrossRef]
36. Zhang, Z.; Li, Y.; Wang, J.; Liu, Z.; Jiang, G.; Guo, H.; Zhu, W. A hybrid data-driven and learning-based method for denoising low-cost IMU to enhance ship navigation reliability. *Ocean Eng.* **2024**, *299*, 117280. [CrossRef]
37. Wang, Q.; Liu, S.; Zhang, B.; Zhang, C. FBLS-Based Fusion Method for Unmanned Surface Vessel Positioning Considering Denoising Algorithm. *J. Mar. Sci. Eng.* **2022**, *10*, 905. [CrossRef]
38. Huang, F.; Wang, Z.; Xing, L.; Gao, C. A MEMS IMU Gyroscope Calibration Method Based on Deep Learning. *IEEE Trans. Instrum. Meas.* **2022**, *71*, 1–9. [CrossRef]
39. Giezendanner, J.; Mukherjee, R.; Purri, M.; Thomas, M.; Mauerman, M.; Islam, A.K.M.l.; Tellman, B. Inferring the past: A combined CNN-LSTM deep learning framework to fuse satellites for historical inundation mapping. In Proceedings of the IEEE/CVF Conference on Computer Vision and Pattern Recognition Workshops, Vancouver, BC, Canada, 17–24 June 2023.
40. Huber, P.J. Robust Estimation of a Location Parameter. In *Breakthroughs in Statistics: Methodology and Distribution*; Kotz, S., Johnson, N.L.; Springer New York: New York, NY, USA, 1992; pp. 492–518.
41. Mur-Artal, R.; Montiel, J.M.M.; Tardós, J.D. ORB-SLAM: A Versatile and Accurate Monocular SLAM System. *IEEE Trans. Robot.* **2015**, *31*, 1147–1163. [CrossRef]
42. Kingma, DP. Adam: A method for stochastic optimization. *arXiv* **2014**, arXiv:1412.6980.
43. Geiger, A.; Lenz, P.; Stiller, C.; Urtasun, R. Vision meets robotics: The kitti dataset. *Int. J. Robot. Res.* **2013**, *32*, 1231–1237. [CrossRef]
44. evo: Python Package for the Evaluation of Odometry and SLAM. Available online: https://github.com/MichaelGrupp/evo (accessed on 12 October 2024).
45. Zhang, Z.; Davide, S. A Tutorial on Quantitative Trajectory Evaluation for Visual(-Inertial) Odometry. In Proceedings of the IEEE/RSJ International Conference on Intelligent Robots and Systems, Madrid, Spain, 1–5 October 2018.
46. Mahony, R.; Hamel, T.; Pflimlin, J.M. Nonlinear Complementary Filters on the Special Orthogonal Group. *IEEE Trans. Autom. Control* **2008**, *53*, 1203–1218. [CrossRef]

Journal of
Marine Science and Engineering

MDPI

Article

A Low-Cost Communication-Based Autonomous Underwater Vehicle Positioning System

Raphaël Garin [1], Pierre-Jean Bouvet [1,*], Beatrice Tomasi [1,2], Philippe Forjonel [1] and Charles Vanwynsberghe [1,3]

1 LabISEN, ISEN Ouest, 29200 Brest, France; beto@norceresearch.no (B.T.); philippe.forjonel@isen-ouest.yncrea.fr (P.F.); charles.vanwynsberghe@tii.ae (C.V.)
2 Norwegian Research Centre, 5008 Bergen, Norway
3 Technology Innovation Institute, Masdar City, Abu Dhabi 9639, United Arab Emirates
* Correspondence: pierre-jean.bouvet@isen-ouest.yncrea.fr

Abstract: Underwater unmanned vehicles are complementary with human presence and manned vehicles for deeper and more complex environments. An autonomous underwater vechicle (AUV) has automation and long-range capacity compared to a cable-guided remotely operated vehicle (ROV). Navigation of AUVs is challenging due to the high absorption of radio-frequency signals underwater and the absence of a global navigation satellite system (GNSS). As a result, most navigation algorithms rely on inertial and acoustic signals; precise localization is then costly in addition to being independent from acoustic data communication. The purpose of this paper is to propose and analyze the performance of a novel low-cost simultaneous communication and localization algorithm. The considered scenario consists of an AUV that acoustically sends sensor or status data to a single fixed beacon. By estimating the Doppler shift and the range from this data exchange, the algorithm can provide a location estimate of the AUV. Using a robust state estimator, we analyze the algorithm over a survey path used for AUV mission planning both in numerical simulations and at-sea experiments.

Keywords: underwater navigation; Kalman filter; particle filter; underwater acoustic communications; Doppler shift estimation; autonomous underwater vehicle; integrated sensing and communication

1. Introduction

Unmanned underwater missions can be performed by either ROVs or AUVs. Both of them integrate payloads and the mission's specific tool set, such as gripper jaws [1] or sonar detection. For the first drone category, the link between the operator and the ROV is an umbilical cable. Its length depends on the mission area, typically covering 300 m at the surface and reaching depths of up to 100 m for small ROVs. Navigation and telemetry are made through this cable, and ROVs usually have a camera. ROVs have many advantages, such as their price, handling, and live telemetry. Underwater ROVs can be used for video streaming of corals or shipwrecks for inspection, remote actuation, or real-time measurements.

In the second category of unmanned underwater vehicles, AUVs, cable communication is replaced by wireless communication, allowing the robot to extend its mission coverage as well as its movement capabilities, the only remaining constraint being the maximal battery autonomy. Compared to an ROV, an AUV can go deeper and operate in a larger radius with a given trajectory with no manual guidance except from mission planning. Usual applications are underwater surveillance, inspection, exploration, or mapping within both civil and military fields [2].

Yet, AUVs face many challenges. The underwater channel absorbs radio frequency (RF) waves after a few meters and optical waves after some tens of meters [3]. As a result, underwater acoustic (UWA) communications are privileged for long-range scenarios.

Underwater sound propagation speed is relatively low (around 1500 m/s) compared to RF or optical waves in the air (similar to light speed around 3×10^8 m/s). Low propagation speed yields an important Doppler shift, and UWA propagation is characterized by strong multipath echoes induced by boundary (i.e., sea bottom and surface) reflections. Both Doppler drift and multipath echoes need to be compensated for by the receiving system [4–6]. Moreover, in order to follow an intended trajectory, the AUV requires to know its position in real time. However, as stated before, since RF signals are absorbed by water after a few meters of depth, the GNSS localization system is not available underwater. As a result, AUV positioning and navigation are active research topics [2,7].

Despite recent research progress and industrial products, current localization and navigation technologies share two main flaws. The first consists of the use of costly inertial sensors or of an infrastructure of several acoustic beacon nodes deployed in very accurate positions to achieve the needed accuracy throughout the AUV mission. The second limitation is the lack of integration between positioning and communication systems. Even if both systems are based on acoustic waves, positioning systems used for localization use some navigation-specific transmissions that are often not suitable for data exchanges [8]. Note that in current 6G research, integrated sensing and communication is a hot topic, driving the development of theory and future standards to simultaneously provide communication and user positioning capabilities [9].

This paper proposes an algorithm where a single fixed beacon simultaneously communicates with an AUV while positioning it. In this lightweight infrastructure scenario (with only one beacon instead of four), the vehicle regularly receives data from the transmitting beacon. By assuming that these two parts are synchronized in time, range estimation between them is possible by computing the time of flight (TOF) of the transmitted signal. Moreover, the receiver has to estimate the Doppler scaling factor induced by the relative speed between the AUV and the beacon. This Doppler shift estimate will have to be compensated for when decoding the data [6]. The Doppler shift estimate is also a representation of the vehicle-beacon relative velocity. Combining the relative velocity, the bearing angle of the AUV, its heading angle, and the range to the beacon, we can compute the velocity of the vehicle projected along the linear trajectory connecting the vehicle and the beacon. While the bearing angle is estimated, the heading angle is measured by an inertial sensor such as an inertial measurement unit (IMU) that is modeled in this paper by including measurement noises and biases typical of a micro-electromechanical system (MEMS) IMU. All this information is filtered within a state estimator. Many estimators exist in the literature [10]; in this paper, we choose the extended Kalman filter (EKF). The main objective of this paper is to model and simulate this low-cost underwater localization approach and then demonstrate its feasibility in a real-world scenario.

The rest of the paper is organized as follows: we first review the state-of-the-art and previous work in Section 2. The details of our model and algorithm are provided in Section 3. In Section 4, we present the localization performance evaluation obtained in numerical simulations, by varying several system parameters and noises, and in at-sea experiments. Finally, in Section 5, a conclusion is drawn by summarizing the advantages and limitations of the proposed algorithm.

2. Existing Methodologies

2.1. Multiple-Transponder Systems

If the desired accuracy error scales at the centimeter level, multiple acoustic transponder systems are required. Combined with a powerful inertial navigation system (INS) (typically fiber optical gyroscope (FOG) technology) and Doppler velocity log (DVL) sensors, the AUV localization is precise, with a positioning error below 1% at the end of the mission. Acoustic localization/navigation techniques fall into two categories [2]:

1. Long baseline (LBL) and GPS intelligent buoy (GIB): One of the first UWA localization techniques developed in the middle of the 1970s by [11]. The long baseline (LBL) procedure consists of a set of UWA transponders precisely disposed of on the seabed

around the mission area. Each transponder has a known precise position and is synchronized with the others. Therefore, in addition to beacon deployment, the calibration of their relative position and timing becomes a key step. The AUV position is then estimated through triangulation with respect to its range to all transponders. Ranges are computed with TOF or time difference of arrival (TDOA) techniques. At least three transponders are recommended to have good accuracy. The GPS intelligent buoy (GIB) method is similar to LBL. The difference is that transponders are installed on surface buoys and not on the seafloor. This reduces calibration costs.
2. Short baseline (SBL): Beacons are deployed at opposite sides of a surface vessel (or platform). TDOA triangulation is then used to determine the AUV position. The baseline length is designed in relationship with the vessel size. The main limitation of short baseline (SBL) is its localization accuracy.

These techniques can be hybrid, such as sparse LBL. It is common to use data fusion algorithms between acoustic transponder positioning and dead-reckoning positioning systems. In that way, the inaccuracies of one technique can be compensated for by the better performance of the other one and vice versa. For instance, LBL positioning could be used without any dead-reckoning measurements, but its performance will be less precise due to the sensor's noises or the varying acoustic physical variables of the environment. Dead-reckoning (inertial navigation) can be implemented without any acoustic transponder, but the position estimates will have a considerable drift over time. For a small AUV, the currently available solution can have the following disadvantages: computational complexity, costs, infrastructures (LBL, SBL. . .), precise calibration (FOG INS), and heavy and costly DVL sensors. These solutions are better suited for offshore industry or bathymetric applications where precision is crucial.

2.2. Single-Transponder Systems

For systems using a single beacon at a known position, it is common to use ultra-short baseline (USBL) [7] or range-only single beacon (ROSB), as the methods proposed in [10,12,13]. These methods are easier to deploy for low-cost missions than those described in Section 2.1. The USBL system estimates the AUV's position using multiple transmissions done by the UWA communication system feeding an EKF with TDOA and bearing angles. Depending on the user's choice, it is possible to only carry out live tracking of the drone rather than helping the drone to reposition itself. The architecture of the anchor is more sophisticated than a simple underwater transceiver. More than three receiving hydrophones are needed to compute the bearing angle estimation. But, it is common to use four receivers at a short distance to optimize this estimation (usually below 10 cm). Kongsberg, EXail, Blueprint, and Sonardyne propose these systems [14]. Range-based navigation (ROSB) is the fastest to implement and the cheapest. Using a series of TOF or TDOA measurements, the only estimation made is the range between the fixed beacon at a known position and the moving AUV. This solution and USBL require information about sound speed and depth. USBL also requires either a differential global positioning system (DGPS) and/or an attitude and heading reference system (AHRS) in order to position the bearing angle in relation to the north. The physical implementation thus requires physical stability in rotation, whereas ROSB is rotationally independent.

One can also cite research subjects such as virtual LBL [15]. The main idea is to add virtual transponders with range measurements and predicted dead-reckoning positions. But, dead-reckoning navigation precision is greatly affected by the presence of a high-cost DVL sensor.

Lastly, industrial products from companies such as Exail and Sonardyne developed sparse LBL [16]. This is a positioning system that uses a single transponder. Range measurements from acoustics are coupled with a mandatory INS (or INS/DVL) to enhance AUV positioning. It has the advantage of being usable with more transponders according to the desired accuracy. It consists of a series of acoustical pings.

It is important to note that it is possible to use dead-reckoning/inertial navigation INS without any acoustic system. It can be implemented through three technologies: MEMS for the cheapest and smallest solution, FOG for strong stability over time, and ring laser gyroscope (RLG) as the historically strong stability solution, but its space requirement is not suitable for a micro-AUVs. Depending on the sensor performance, the drift bias makes the positioning error grow even more. This is usually called the bias instability. The instability range is from 0.01°/h for the high-end FOG to 2°/h for the cheapest FOG. As for MEMS technology, the range is from 2°/h for high-end systems to 10°/h for the smallest systems (such as smartphones, for example). The DVL can reduce this error, but its technology range is limited by the depth level of the AUV. A DVL works by emitting acoustic pulses downward and measuring the frequency shift of the returned signals reflected from the seabed or particles in the water. This frequency shift, caused by the Doppler effect, is used to calculate the vehicle's velocity relative to the bottom or water column. Equipped with multiple beams and an IMU, a DVL can determine not only the speed in three dimensions but also the heading and pitch, making it essential for precise navigation of underwater vehicles like ROVs and AUVs. Depending on the DVL, the distance between the seafloor and AUV must be 300 m to 6000 m at most for the smallest ones.

As shown previously, ROSB is a good solution to perform simultaneous navigation and communication with a minimum number of sensors. However, such a system provides large localization errors [10] and is demonstrated to be current dependent without a DVL [13]. In practice, the ROSB method needs to be combined with an acoustic Doppler current profiler (ADCP) or DVL to achieve good accuracy. The idea of combining range and Doppler-shift measurements to perform underwater target tracking was first introduced by Diamant et al. in [17]. However, the authors used several anchor nodes to provide the TOF, bearing angle, and Doppler information. In [18], the authors proposed a ROSB method combined with Doppler shift measurements as a navigation algorithm for an AUV performing a lemniscate trajectory. Simulation results were promising: for communication made every 10 s, the positioning accuracy error was contained below 10 m for a 6000 m^2 mission area. Later on [19], the authors conducted an experiment using the same trajectory in a 3 m^2 pool. Despite numerous reflections, sensor noises, and a small induced Doppler shift, the algorithm kept a low positioning error. Finally, the algorithm was evaluated during at-sea experiments in [20].

In the current paper, we extend the work in [20] by conducting extensive simulations with realistic estimation noises and biases. The impact of time between successive communications on the positioning accuracy is also analyzed. Performance comparison against the conventional ROSB method is carried out. Finally, numerical simulation results are compared against experimental results.

3. Materials and Methods

3.1. Problem Formulation

We focus on a two-dimensional problem by omitting the depth axis. In fact, it is straightforward to implement the last axis by adding a low-cost pressure sensor. The accuracy of this sensor category is in centimeters and is easily implementable in a small AUV.

We consider a mission where an AUV performs a survey path (as typically used in side scan sonar missions) and is equipped with an IMU, a global positioning system (GPS), and an acoustic modem. While navigating, the AUV receives acoustic data from a fixed surface beacon with a submerged transponder. The chosen trajectory for the mission is a classical boustrophedon trajectory, also called a survey path, and has a mission duration named T_{mission}. The reference beacon position is denoted as $\begin{pmatrix} x_{\text{ref}} & y_{\text{ref}} \end{pmatrix}^T$, and is obtained in practice with a GNSS receiver that provides typically metric accuracy. To enhance this value, a DGPS system can be mounted to the station. DGPS works with two spaced GNSS receivers. Compared to one GNSS antenna system, it also provides a precise azimuth angle. To obtain centimetric accuracy, it is possible to add a RTK (real-time kinematic) antenna to the DGPS system.

Table 1 summarizes the sensor's public price available. As the AHRS is a critical component, this sensor should not be subject to a cost-reduction target. Propeller parts and embedded electronics are not taken into account here. The costly component of the proposed system is the UWA modem. Furthermore, there is no ultra-low-cost mass-market solution (below EUR 2000). The final solution embedded into a micro-AUV offers real-time communication between surfaces while positioning the AUV. Using the less expensive mass-market UWA modem, the proposed navigation algorithm cost is below EUR 4000, which gives the proposed system a relatively low-cost label compared to conventional positioning systems based on USBL or DVL.

Table 1. Navigation algorithm sensors standard price.

Sensor	Standard Public Price	Measurement(s)
AHRS	EUR 1200	Euler angles
Pressure	EUR 90	Depth and temperature
GPS	EUR 20	Surface position
UWA modem	from EUR 500 [21] to 2500 € [22]	Communication, Doppler, and range

The proposed positioning algorithm is depicted in Figure 1. Parameters and noise initialization correspond to the initialization of mission parameters and filter input parameters (observation and state functions, covariance matrices, and command). The command update refers to the linearization feedback method leading to command vector **u** described in Section 3.2. If no acoustic communication frame is available, the measurements rely only on IMU data (dead reckoning); if yes, acoustic communication decoding brings additional measurements (distance and relative speed), as described in Section 3.5. The time between two received communication frames is denoted as T_{com}. Measurements are then fed to the filter estimator (see Section 3.6).

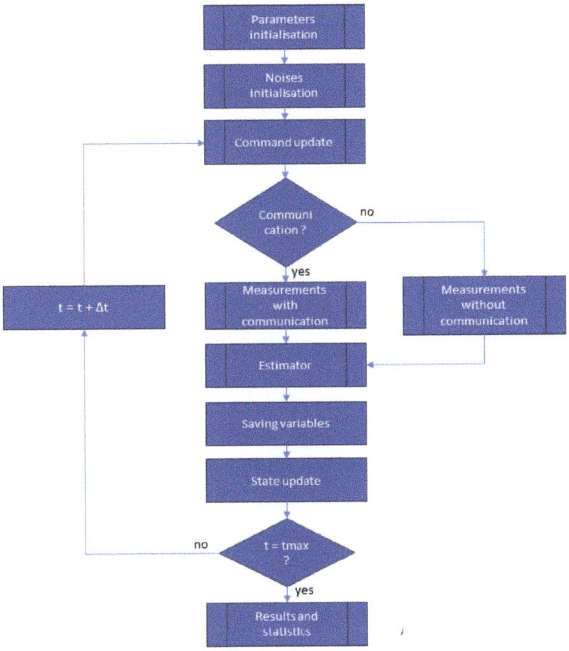

Figure 1. Positioning algorithm.

3.2. Drone Scheduling

The dynamic of the AUV in a 2D plan is depicted in Figure 2. We define the ground truth vector $\mathbf{X} = \begin{pmatrix} x & y & \psi & v \end{pmatrix}^T$ with $\begin{pmatrix} x & y \end{pmatrix}^T$ as the actual position of the drone; ψ is the heading angle of the drone, and v is its speed. We note ω_b as the bearing angle between the drone and the beacon, and v_r is the relative speed of the drone projected in the direction of the reference beacon. The 2D trajectory is defined by $\begin{pmatrix} x_d & y_d \end{pmatrix}^T$ with x_d and y_d, respectively, and the x-axis and the y-axis coordinates of a point on the trajectory at sampling time t.

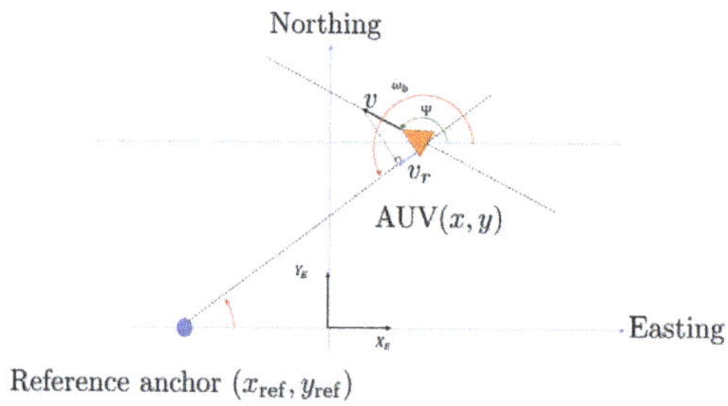

Figure 2. Dynamic of the AUV.

We assume that the drone command is made through the $\mathbf{u}$ vector:

$$\mathbf{u} = \begin{pmatrix} \Omega \\ a \end{pmatrix} = \begin{pmatrix} \dot{\psi} \\ \dot{v} \end{pmatrix}, \tag{1}$$

where Ω is the rotational speed of the AUV, and a is its acceleration.

Using the linearization feedback method, the drone command vector can be expressed as follows:

$$\begin{pmatrix} \Omega \\ a \end{pmatrix} = \begin{pmatrix} -v \cdot \sin(\psi) & \cos(\psi) \\ v \cdot \cos(\psi) & \sin(\psi) \end{pmatrix}^{-1} \begin{pmatrix} (x_d - x) + 2(\dot{x}_d - v \cdot \cos(\psi)) + \ddot{x}_d \\ (y_d - y) + 2(\dot{y}_d - v \cdot \sin(\psi)) + \ddot{y}_d \end{pmatrix}. \tag{2}$$

The model is updated at each time Δt by the following equation :

$$\mathbf{X}(t) = \mathbf{X}(t - \Delta t) + f_t(\mathbf{X}, \mathbf{u}) \cdot \Delta t, \tag{3}$$

with

$$f_t(\mathbf{X}, \mathbf{u}) = \begin{pmatrix} v\cos(\psi) \\ v\sin(\psi) \\ \Omega \\ a \end{pmatrix} = \dot{\mathbf{X}}. \tag{4}$$

In practice, the open-source software "ArduPilot" [23] can carry out the planning part of the drone (trajectory/linearization/control) alone from way-points given by positioning estimation.

3.3. AUV Model

The use of Kalman filtering for motion estimation requires a discrete time representation of the state space of the drone behavioral model. For the model, we define the following state equations:

$$\begin{cases} \mathbf{z}_{k+1} = f(\mathbf{z}_k, \mathbf{u}_k) + \mathbf{w}_k \\ \mathbf{y}_k = g(\mathbf{z}_k) + \mathbf{v}_k \end{cases}, \tag{5}$$

where $\mathbf{z}_k$ is the state vector of the AUV at sampling time k, $\mathbf{u}_k$ is the command vector $\mathbf{u}$ at sampling time k, $f(\mathbf{z}_k, \mathbf{u}_k)$ is the state equation, $g(\mathbf{z}_k)$ is the observation function, and finally, $\mathbf{w}_k$ and $\mathbf{v}$ are, respectively, the state and observation noises. The state vector of the AUV is defined as follows:

$$\mathbf{z} = \begin{pmatrix} x & y & \psi & v & \Omega & a \end{pmatrix}^T. \tag{6}$$

The state equation becomes the following:

$$\mathbf{f}(\mathbf{z}, \mathbf{u}) = \begin{pmatrix} x + v\,\Delta_t \cos\psi \\ y + v\,\Delta_t \sin\psi \\ \psi + \Delta_t \Omega \\ v + \Delta_t a \\ \Omega \\ a \end{pmatrix}. \tag{7}$$

3.4. Underwater Acoustic Communication System

We assume that the beacon and the AUV regularly exchange data with each other using underwater acoustic communications. At each time interval T_{com}, the beacon transmits the acoustic frame depicted in Figure 3, which is received by the AUV. On the one hand, the acoustic frame includes a main signal centered on f_0 with bandwidth B carrying N_p pilot symbols (for synchronization and channel estimation) and N_d useful data symbols. On the other hand a pure-tone signal with frequency f_{pt} is used for Doppler shift estimation. The frame starts and ends with N_d null symbols to avoid interference between successive frames. The passband transmit signal can be written as follows:

$$s(t) = \Re\left[e^{j2\pi f_{\text{pt}} t} + \sum_{k=0}^{N_p+N_d-1} x[k] g_T(t - kT) e^{j2\pi f_0 t} \right], \tag{8}$$

where $x[k]$ are phase shift keying (PSK) symbols, T is the modulation symbol duration, and $g_T(t)$ is a pulse-shaping waveform chosen as square root raised cosine (SRRC).

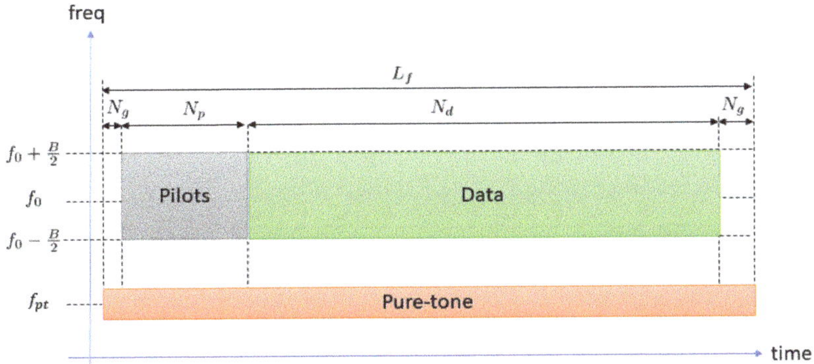

Figure 3. Architecture of the communication frame.

On the receiver side, underwater acoustic frames are decoded using the algorithm described in Figure 4. After analog-to-digital conversion, the main signal is base–band translated and match-filtered with an SRRC filter. Frame synchronization is then performed using cross-correlation between matched filtered samples and the pilot sequence. On the other hand, the pure-tone signal is base–band converted and then low pass filtered. Since the motion-induced Doppler shift f_d is linked to relative speed v_r with $f_d = v_r f_{\text{pt}}/c_w$, the low pass filter cutting frequency is set to $f_{\text{max}} = \frac{v_{\text{max}} f_{\text{pt}}}{c_w}$ with v_{max} as the maximum AUV speed and c_w as the sound speed in water. The chosen Doppler estimation algorithm consists of extracting the Doppler shift from the phase derivative of the pure-tone base-band signal $\tilde{r}_{pt}(t)$ by computing the phase angle from two successive samples. As a result, an estimation of the relative speed at sampling time k is formed as follows:

$$\hat{v}_r(kT) = \frac{c_w}{2\pi T f_{pt}} \arg(\tilde{r}_{pt}(kT)\tilde{r^*}_{pt}((k-1)T)). \tag{9}$$

The relative speed estimation is then fed to the SRRC filter in order to match the SRRC response to the Doppler shift effect and to the Doppler compensation stage. In fact, the motion-induced Doppler shift is usually removed from the received useful base-band signal $\tilde{r}(t)$ via resampling and phase compensation [6]:

$$y[k] = \tilde{r}(\hat{t}_k)e^{-j\hat{\phi}_k}, \tag{10}$$

where $\hat{t}_k$ and $\hat{\phi}_k$ are updated dynamically from the instantaneous relative velocity estimation:

$$\begin{cases} \hat{t}_k = \hat{t}_{k-1} + T(1 - \frac{\hat{v}_r(kT)}{c_w}) \\ \hat{\phi}_k = \hat{\phi}_{k-1} + 2\pi f_0 T \frac{\hat{v}_r(kT)}{c_w} \end{cases}. \tag{11}$$

Data decoding is finally performed using the received signal $y[k]$ by advanced equalization processing and channel decoding. In our case, we consider a turbo-equalizer with residual phase compensation as described in [5].

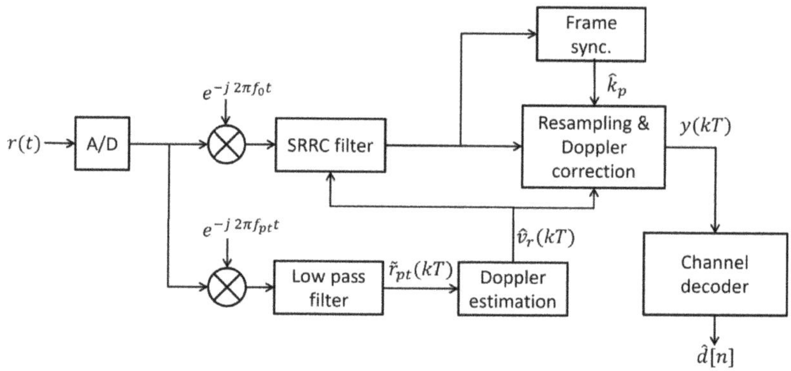

Figure 4. Architecture of the communication decoder.

3.5. Measurements

3.5.1. Distance Estimation

The distance estimation $\hat{d}$ is provided by using the frame synchronization process described in Figure 4. By assuming perfect clock synchronization between the beacon and AUV, if a UWA frame is sent at time 0, we can estimate the distance with the TOF of the acoustic signal as follows:

$$\hat{d} = \frac{\hat{t}_0 - N_g T}{c_w}, \tag{12}$$

where $\hat{t}_0$ denotes the estimated time of arrival of the first pilot symbol of the UWA frame. For simplification, this approach is considered in this paper. However, in reality, achieving perfect synchronization between the transmitter and receiver clocks is challenging. One potential solution is to use the pulse per second (PPS) signal from a surface buoy GNSS to aid synchronization for the AUV. The AUV would reproduce this PPS synthetically. But, it drifts over time and is not a long-term solution. This is, for instance, suitable for a micro-AUV. It is also possible to use an atomic clock, but it is not suitable for a micro-AUV because of its size and cost.

Another option is to use bidirectional communication. Using time of emission and communication transit time, it is possible to compute distance estimation. No synchronization between the clocks of the transmitter and receiver is needed. Precise time computation is made through stable electronics. Multiple acoustic sources are commonly used to get the most precise estimation possible. To remove the need for precise emission time knowledge, TDOA is another popular ranging technique. This method only requires the time signal that was received and the speed that the signal travels. Once the signal is received, the difference in arrival time with the last communication is used to calculate the difference in distances.

No matter which technique is used, at least three beacons are necessary to carry out a triangulation and, hence, estimate the position of the AUV. However, in this paper, time of arrival (TOA) and TDOA could be chosen for our unique beacon solution as we only need to know the distance between the beacon and the AUV.

3.5.2. Speed Estimation

Obtaining the projected speed from $\hat{v}_r$ requires two angles, ω_b and ψ, as illustrated in Figure 2, and this is expressed as follows:

$$v_r = v \cdot \cos(\omega_b - \psi). \tag{13}$$

The bearing angle ω_b is defined as follows:

$$\omega_b = \arctan 2\left(\frac{y - y_{\text{ref}}}{x - x_{\text{ref}}}\right) + \pi. \tag{14}$$

As a result, an estimation of the AUV speed can be formed as follows:

$$\hat{v} = \frac{\hat{v}_r}{\cos(\arctan 2(\hat{y} - y_{\text{ref}}, \hat{x} - x_{\text{ref}}) - \hat{\psi})}. \tag{15}$$

If the relative speed estimation $\hat{v}_r$ is provided by the Doppler shift estimation and the heading angle $\hat{\psi}$ comes from the AHRS, Equation (15) requires the knowledge of the AUV position $(\hat{x}, \hat{y})$, which is provided by the prediction state $\mathbf{z}_k$ of the Kalman filter.

3.5.3. Proprioceptive Sensors

Estimates of the heading angle of the drone (yaw angle), labelled $\hat{\psi}$, can be measured either by a low-cost AHRS, which delivers a heading directly (e.g. MTI-630, Xsens, Enschede, Netherlands or ELLIPSE series, SBG systems, Carrières-sur-Seine, France), or by an IMU, which delivers the accelerations, gyroscope (angular rates), and magnetic fields on the x, y and z axes. Magnetic fields are, respectively, H_x, H_y and H_z. The magnetic heading is then calculated according to the following method for a 2D plane :

$$\begin{aligned}
\hat{\psi}(y > 0) &= 90 - \arctan(H_x/H_y) \cdot 180/\pi, \\
\hat{\psi}(y < 0) &= 270 - \arctan(H_x/H_y) \cdot 180/\pi, \\
\hat{\psi}(y = 0, x < 0) &= 180, \\
\hat{\psi}(y = 0, x > 0) &= 0.
\end{aligned} \tag{16}$$

The magnetic heading is, however, subject to strong errors from the environment. It is then usual to fuse it to an EKF with acceleration and a gyroscope labelled $\hat{\Omega}$ and $\hat{a}$. The usual sensor noise for a low-cost sensor in static is given for $b_\theta \in [0.1; 1]^\circ$. In dynamics, MEMS inertial unit noises are kept below $b_\theta = 2^\circ$ but drift over time with a bias instability around $5^\circ/\text{h}$.

3.5.4. Equations

The measurement equation is defined as follows:

$$\begin{cases} \mathbf{y}_k = g(\mathbf{z}_k) + \mathbf{v_k} & \text{if no UWA communication} \\ \mathbf{y}_k^{\text{com}} = g^{\text{com}}(\mathbf{z}_k) + \mathbf{v}_k^{\text{com}} & \text{else} \end{cases}, \tag{17}$$

where $\mathbf{y}_k^{\text{com}}$ is the observation vector, $g^{\text{com}}(.)$ is the observation function when a UWA communication frame is received and then decoded, and $\mathbf{y}_k$ and $g(.)$ are observed when the algorithm is updated in dead-reckoning mode. Observation vectors are defined as follows:

$$\mathbf{y}_k^{\text{com}} = \begin{pmatrix} \hat{d} \\ \hat{v}_r \\ \hat{\psi} \\ \hat{\Omega} \\ \hat{a} \end{pmatrix} \quad \mathbf{y_k} = \begin{pmatrix} \hat{\psi} \\ \hat{\Omega} \\ \hat{a} \end{pmatrix}. \tag{18}$$

The distance between the drone and the reference point can be computed as a function of the drone position:

$$d = \sqrt{(x - x_{\text{ref}})^2 + (y - y_{\text{ref}})^2}. \tag{19}$$

By combining (15), (18), and (19), we obtain the following observation functions:

$$\begin{aligned} g^{\text{com}}(\mathbf{z}) &= \left(\sqrt{(x - x_{\text{ref}})^2 + (y - y_{\text{ref}})^2} \quad -v\cos\left(\arctan 2(y - y_{\text{ref}}, x - x_{\text{ref}}) - \psi\right) \quad \psi \quad \Omega \quad a \right)^T \\ g(\mathbf{z}) &= \left(\psi \quad \Omega \quad a \right)^T \end{aligned}. \tag{20}$$

3.6. *Estimator Filter*

In the proposed positioning system, we consider the EKF estimator [24,25] described in Algorithm 1. The filter takes as input the state vector and the estimator covariance matrix at sampling step $k-1$, i.e., $\mathbf{z}_{k-1}$ and $\mathbf{P}_{k-1}$, input command vector $\mathbf{u}_k$, and measurement vector $\mathbf{y}_k$. The estimator produces state vector $\hat{\mathbf{z}}_k$ and an updated covariance matrix $\mathbf{P}_{k-1}$. The algorithm requires the state transition matrix $\mathbf{F}$, which is computed as follows.

Algorithm 1: Extended Kalman Filter

Data: $\mathbf{z}_{k-1}, \mathbf{P}_{k-1}, \mathbf{u}_k, \mathbf{u}_k$
Result: $\hat{\mathbf{z}}_k, \mathbf{P}_k$

$\mathbf{F} = \nabla_f$;
$\mathbf{H} = \nabla_h$;
$\hat{\mathbf{z}}_k^- = f(\mathbf{z}_{k-1}, \mathbf{u}_k)$;
$\mathbf{P}_k^- = \mathbf{F}\mathbf{P}_{k-1}\mathbf{F}^T + \mathbf{Q}$;
$\mathbf{K}_k = \mathbf{P}_k^- \mathbf{H}^T (\mathbf{H}\mathbf{P}_k^- \mathbf{H}^T + \mathbf{R})^{-1}$;
$\mathbf{y} = (\mathbf{y}_k - h(\hat{\mathbf{x}}_k^-))$;
$\hat{\mathbf{z}}_k = \hat{\mathbf{z}}_k^- + \mathbf{y}\mathbf{K}_k$;
$\mathbf{P}_k = \mathbf{P}_k^- - \mathbf{K}_k\mathbf{H}\mathbf{P}_k^-$.

$$\mathbf{F} = \frac{\partial f}{\partial \mathbf{z}} = \begin{pmatrix} 1 & 0 & -v\cdot\Delta_t\cdot\sin(\psi) & \Delta_t\cdot\cos(\psi) & 0 & 0 \\ 0 & 1 & v\cdot\Delta_t\cdot\cos(\psi) & \Delta_t\cdot\sin(\psi) & 0 & 0 \\ 0 & 0 & 1 & 0 & \Delta_t & 0 \\ 0 & 0 & 0 & 1 & 0 & \Delta_t \\ 0 & 0 & 0 & 0 & 1 & 0 \\ 0 & 0 & 0 & 0 & 0 & 1 \end{pmatrix}, \tag{21}$$

The algorithm also requires the so called observation matrix, which is computed as follows when an UWA communication frame is available:

$$\mathbf{H} = \frac{\partial g_{\text{com}}}{\partial \mathbf{z}} = \begin{pmatrix} 0 & 0 & 1 & 0 & 0 & 0 \\ 0 & 0 & 0 & 0 & 1 & 0 \\ 0 & 0 & 0 & 0 & 0 & 1 \\ A \cdot (x - x_{\text{ref}}) & A \cdot (y - y_{\text{ref}}) & 0 & 0 & 0 & 0 \\ -B \cdot (y - y_{\text{ref}}) & B \cdot (x - x_{\text{ref}}) & -v \cdot \sin(C) & -\cos(C) & 0 & 0 \end{pmatrix}, \quad (22)$$

with

$$\begin{aligned} A &= \frac{1}{\sqrt{(x - x_{\text{ref}})^2 + (y - y_{\text{ref}})^2}} \\ B &= \frac{v \cdot \sin(\arctan 2(y - y_{\text{ref}}, x - x_{\text{ref}}) - \psi)}{(x - x_{\text{ref}})^2 + (y - y_{\text{ref}})^2}. \\ C &= \arctan(y - y_{\text{ref}}, x - x_{\text{ref}}) - \psi \end{aligned} \quad (23)$$

In the case of a lack of an UWA communication frame, the observation matrix becomes the following:

$$\mathbf{H} = \frac{\partial g}{\partial \mathbf{z}} = \begin{pmatrix} 0 & 0 & 1 & 0 & 0 & 0 \\ 0 & 0 & 0 & 0 & 1 & 0 \\ 0 & 0 & 0 & 0 & 0 & 1 \end{pmatrix}. \quad (24)$$

Finally, the algorithm requires the covariance matrix of the process noise noted as **Q** and the covariance matrix of measurement noise labelled **R**.

4. Results

4.1. Simulation

In this section, the performance of the positioning system is studied using simulation. The global architecture of the simulation system is depicted in Figure 5, which includes two main parts exchanging data in an iterative manner: the positioning algorithm (see Figure 1) and the simulation algorithm computing AUV trajectory, generating noises and collecting results. The simulation script is implemented in MATLAB™ 2024 software, and the algorithm is based on the Monte-Carlo method, i.e., each noise datum is random, and the simulation is performed N_{runs} times in order to average the results.

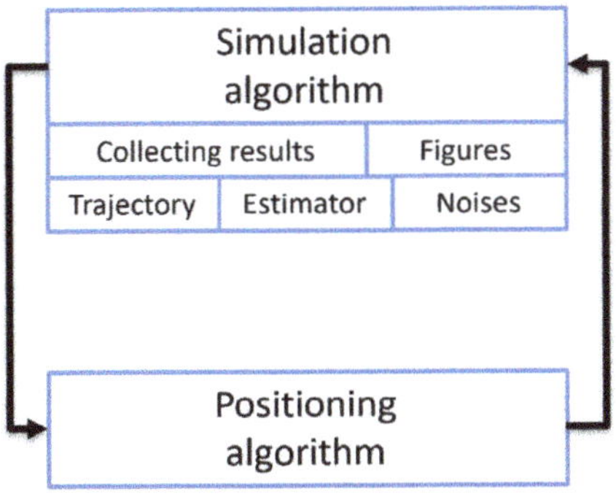

Figure 5. Global architecture of the simulation script.

4.1.1. Parameters

The main simulation parameters are summarized in Table 2. The simulated trajectory is a boustrophedon with a traveled distance of 1567 m during around 26 min, representing an average speed of 1 m/s. The algorithm is updated every $\Delta t = 0.25$ s, which is aligned with a frequency update of 4 Hz. The area of the mission is 100 m per 250 m, and the reference beacon is located in the center of this area.

Table 2. Simulation parameters.

Parameters	Description	Value
$T_{mission}$	Mission duration	1550 s
Δt	Algorithm time step	250 ms
T_{com}	UWA communication period	$[3, 20]$ s
N_{runs}	Number of simulation runs	300
$\begin{pmatrix} x_{ref} & y_{ref} \end{pmatrix}^T$	Beacon position	$\begin{pmatrix} 50 & 100 \end{pmatrix}^T$
v	drone velocity	$[-3, -1] \cup [1, 3]$ m/s

To obtain accurate results, we conduct $N_{runs} = 300$ iterations by generating new noise samples for each parameters at each iteration where one iteration gives the following:

- Mean positioning error;
- Variance of the positioning error;
- Root mean square error and positioning error curves;
- Noises for each measurement;
- Bearing angle estimation errors.

At the end of N_{runs} iterations, each of the values is averaged. The average of the mean positioning error is used to compute outlier results. The outliers are detected if the number is greater than $q_3 + 1.5 \times (q_3 - q_1)$ or less than $q_3 - 1.5 \times (q_3 - q_1)$. These outlier results are not considered, as it is assumed that the algorithm has diverged or is too far from the true position. For the remainder of this article, we will note the number of simulations that diverged. In the simulation script, the UWA communication process, as described in Section 3.4, is not simulated: estimated range and speed values are only emulated by adding noise representative of estimation errors. Noises from the measurements are shown in three box plots in Figure 6. White Gaussian noises centered around 0 from the IMU (acceleration and angular velocity) are also added, with values contained in ± 0.1 m/s^2 and ± 0.1 s^{-1}. For the EKF estimator, matrices **R** and **Q** are defined as follows:

$$\begin{aligned} \mathbf{Q} &= \mathrm{diag}(\Delta_t^8, \cdots, \Delta_t^8) = \mathrm{diag}(0.25^8, \cdots, 0.25^8) \\ \mathbf{R} &= \mathrm{diag}(1 \;\; 0.005 \;\; 0.005 \;\; 12 \;\; 0.2) \end{aligned}, \quad (25)$$

with $\mathbf{Q} \in \mathbb{R}^{6\times6}$ and $\mathbf{R} \in \mathbb{R}^{5\times5}$. These covariance matrices are fixed during all simulations.

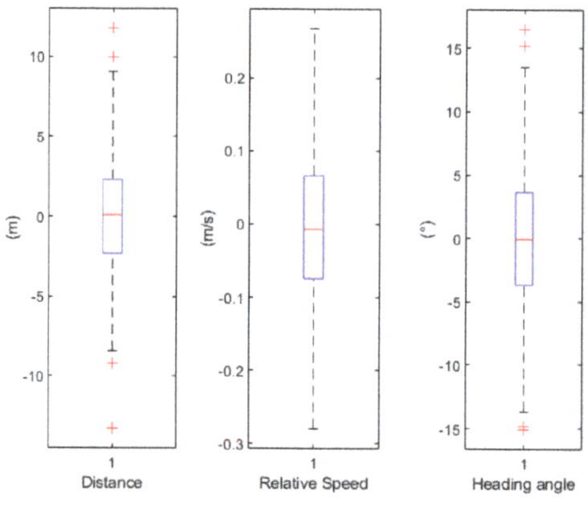

Figure 6. Noise statistics added for each mesurements. The central mark indicates the median, and the bottom and top edges of the box indicate the 25th and 75th percentiles, respectively. The outliers are plotted individually using a red cross marker symbol.

4.1.2. Simulation Results

A two-dimensional top-view of the mission is depicted in Figure 7, and the associated statistical results are carried out in Figure 8. To keep the top view as clear as possible, only the four lowest communication periods are exploited. The acoustic beacon, at the same depth of the AUV for 2D simplification, is represented by a blue star marker.

Results are displayed in their entirety in Table 3. Numbers displayed in parenthesis are the number of outliers in the N_{runs} iterations, whereas %etd (error over traveled distance) denotes the positioning error in percentage of the traveled distance. Whatever the value of T_{com} is, any divergence of the algorithm is observed; however, as T_{com} increases, both positioning error and its variance increases. This phenomenon is easily explained by the fact that between two UWA communications, the positioning algorithm works only in dead-reckoning navigation, which results in higher drift errors.

The Boustrophedon-type trajectory is one of the most commonly used trajectories in the AUV and micro-AUV fields, among other trajectories such as pipeline tracking or depth saw-tooth. The objective of a Boustrophedon trajectory is to collect bathymetry and side-scan sonar data. The overlap of scan images has not been taken into account here. Therefore, the positioning results must be linked to the position of the acquired data. Traditional side-scan sonar embedded on ROVs or micro-AUVs has a maximum beam of around 50 m. Results shown in Figure 8 highlight the maximum positioning error at around 18 m after 1500 s of navigation. Furthermore, the peak of these errors occurs during turns (curvilinear sections) where scan and bathymetry data are not necessarily required.

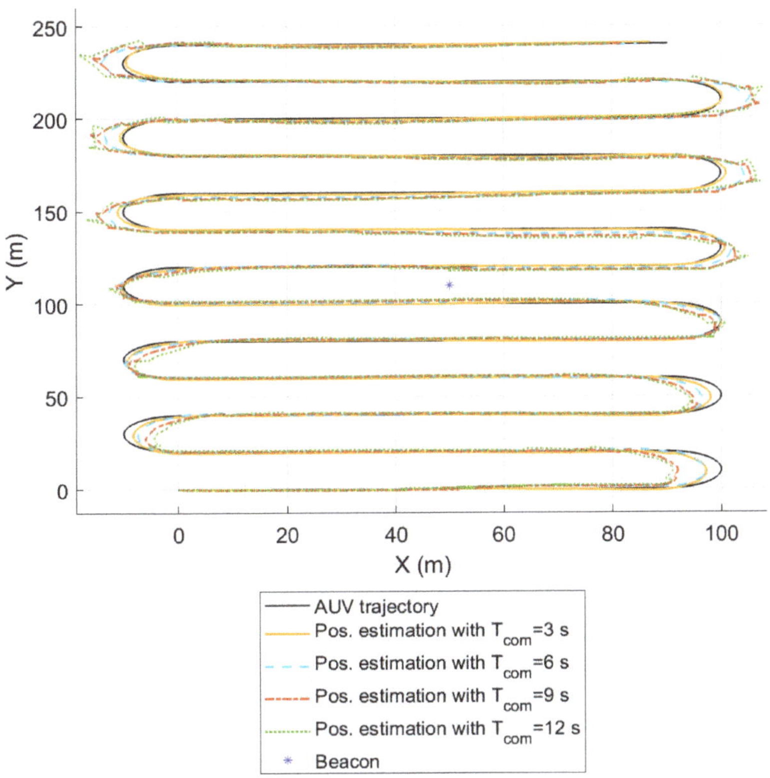

Figure 7. Two-dimensional top view of the averaged positioning estimation in comparison to the actual AUV position in black.

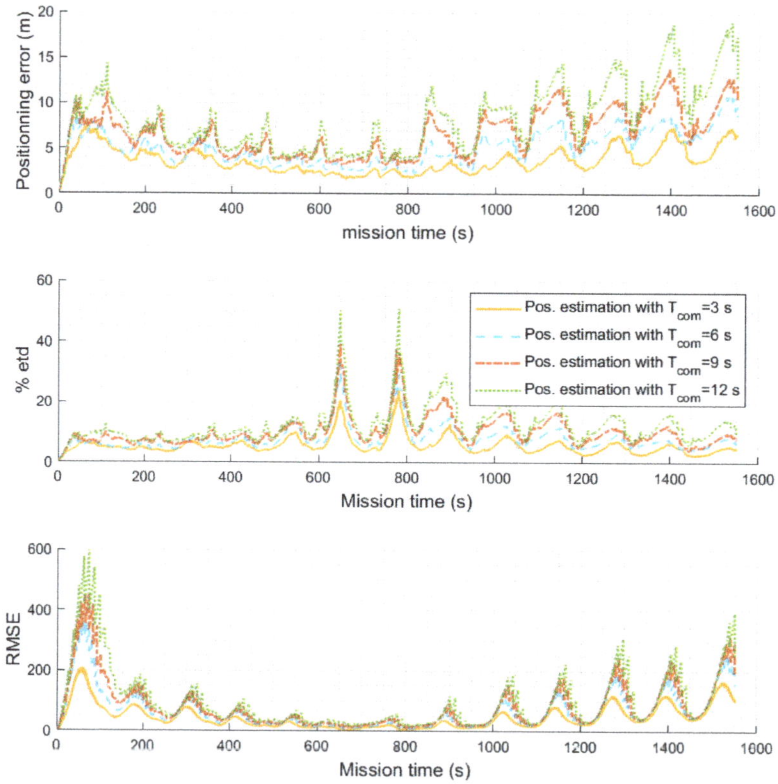

Figure 8. Statistical performance results of the simulated positioning system as a function of mission time and the UWA communication period T_{com}. **Up**: averaged positioning error (% etd), **middle**: error as a percentage of the traveled distance, and **down**: root mean square error (RMSE).

The lack of linear speed measured by the system (which could be measured through a DVL) induces growing errors during curvilinear sections of the trajectory. This phenomenon can also be explained by the absence of the bearing angle estimation of the UWA communication, which causes a delay in the projection of the relative speed of the AUV.

Table 3. Algorithm results for 300 simulated missions.

T_{com}	Mean Positioning Error (Outliers) in m	Variance of the Positioning Error	%etd Max	%etd Mean
3 s	3.6054 (1)	1.7549	0.5268	0.2281
6 s	5.0440 (5)	3.2143	0.5850	0.3152
9 s	6.4518 (1)	6.2173	0.7662	0.4032
12 s	8.0853 (6)	12.7455	1.0938	0.5053
15 s	9.5457 (6)	18.3823	1.3950	0.5966
20 s	12.1322 (6)	33.9161	1.7825	0.7583

A comparison between the proposed algorithm and the well-known ROSB technique [10] is provided in Table 4. We can note that the ROSB method is both less accurate and less precise than the proposed method. As T_{com} increases, the difference between the two methods is reduced; however, the number of outliers with the ROSB method increases more and more often.

Table 4. Comparison between proposed method and state-of-the-art method, i.e., the ROSB method.

T_{com}	Mean Positioning Error (Outliers) in m with Proposed Method	Mean Positioning Error (Outliers) in m with ROSB
3 s	3.6054 (1)	4.5870 (4)
6 s	5.0440 (5)	6.2255 (6)
9 s	6.4518 (1)	7.7197 (5)
12 s	8.0853 (6)	9.2462 (15)
15 s	9.5457 (6)	10.6684 (10)
20 s	12.1322 (6)	12.9695 (29)

The results of these simulations show that for bathymetric or scan data collection, where the position of the data is a critical value, a low-cost positioning system communicating in real time simultaneously with a beacon is effective. In the case of analysis of a large underwater zone, where the accuracy of the data position is important, it is perfectly feasible to make surface ascents to recover the GPS before diving again to continue the mission.

4.2. Experiments

4.2.1. Description

The experiment took place in the Bay of Brest between the 5 and 6 of July 2022. Since the positioning is post-processed, everything was set up so that measurements were data-logged and synchronized using PPS. While moving with a rigid-hull inflatable boat with one hydrophone submerged, data from two GNSS signals (one from the boat and one from the beacon), the hydrophone signal, and measurements from an IMU were recorded. Sound speed was measured hourly using a sound velocity probe (SVP) instrument. The measured sound speed remained constant between 1512 and 1513 m/s over the two days. For simplicity, in the following, we set c_w in the positioning algorithm as an average value during the 2 days of the experiment. In practice, the algorithm should use the instantaneous value of c_w provided by the SVP. The water depth was between 3 and 5 m and the boat speed was between 1 and 5 knots. Other parameters are detailed in Table 5.

From the experiment session, four trajectories were selected in this paper. Due to operational reasons, it was not possible to reproduce exactly the Boustrophedon trajectory detailed in the simulation section. Trajectory #1 is a 10 min long, complex route with many U-turns, but it is close to the beacon. #2 is a 5 min simple and close route. #3 is a 10 min long survey path going further and further away from the beacon. #4 is a 450 m away simple path for 3 min and 20 s. Figure 9 represents the GNSS positioning values of the first trajectory. Acoustic data are processed as described in Section 3.5 in order to estimate both distance $\hat{d}$ and $\hat{v}$, and then estimates are fed to the position algorithm. All the processing scripts are implemented in MATLAB™.

Table 5. Experimental parameters.

Parameters	Description	Value
Δt	Algorithm time step	1 s
T_{com}	UWA communication period	3 s
$(x_{ref} \quad y_{ref})^T$	Beacon position	$(48.38243 \text{ lat.}, -4.407295 \text{ long.})^T$
T_F	UWA frame duration	300 ms
f_{pt}	Pure-tone signal frequency	20 kHz
f_0	Data signal center frequency	28 kHz
$1/T$	Modulation speed	6.4 kHz
B	Signal bandwidth	8.96 kHz
f_s	Sampling frequency	96.153 kHz

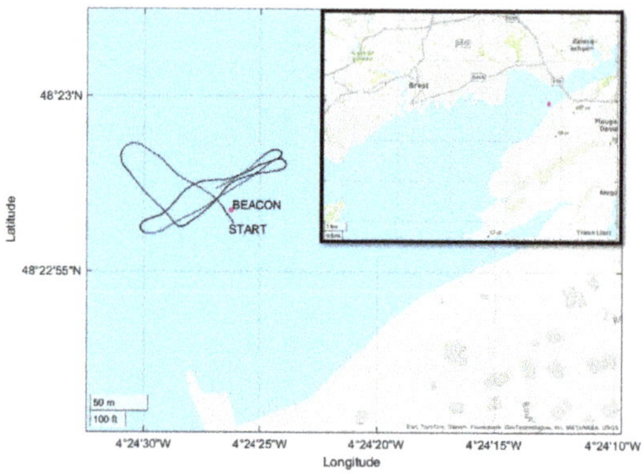

Figure 9. Map of trajectory #1 made by the boat.

4.2.2. Experimental Results

Figures 10 and 11 show the four trajectories chosen for this paper. The blue dashed line represents the ROSB state-of-art positioning estimation, whereas the red dotted line represents the positioning algorithm of this article. Every route starts at $(0,0)$, and the known position of the buoy is represented as a red star. The real position is the GNSS measurement.

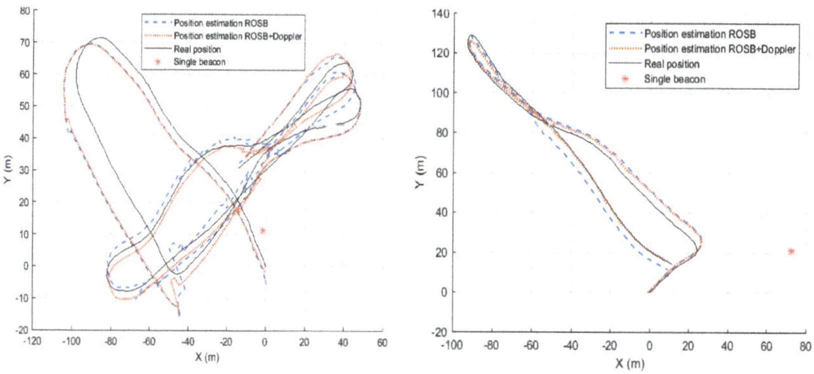

Figure 10. Trajectory #1 (**left**), trajectory #2 (**right**).

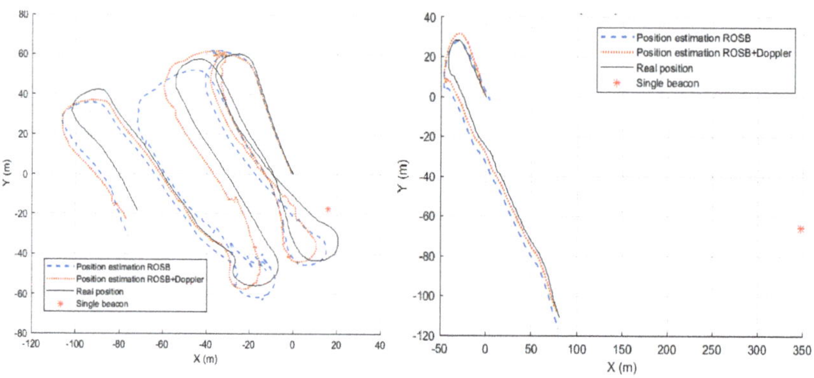

Figure 11. Trajectory #3 (**left**), trajectory #4 (**right**).

Table 6 exhibits the mean positioning error for the 4 studied trajectories and a comparison with the state-of-the-art method, i.e., ROSB. Each one of the shown trajectories has a mean positioning error, compared to the GNSS, below 6 m. This is an affirmation that can be extended to all sessions of experimentation regardless of the distance or the complexity of the followed route. However, we can observe that a simpler path is obviously easier to estimate (such as #2) and better closer than further (#2 is better than #4). In all cases and in a similar manner to simulation, the proposed method outperforms the ROSB, demonstrating the interest in estimating the AUV speed through Doppler estimation.

Trajectories #1 and #3 highlight the fact that estimating the relative speed of the AUV using the Doppler of the acoustic communication improves positioning over time compared to the ROSB solution, especially when the drift on heading is important. There are two striking examples in the trajectories cited above. For the first trajectory, around $x = -45$ and $y = -5$, we observe a shift linked to the acoustic communication. In the case of the ROSB, the realignment is too far and degrades positioning accuracy, whereas, in the case of the system proposed in this article, the realignment is correct, allowing the straight line that follows to be positioned accurately. This is important for scan recording. For trajectory #3, the authors would like to draw the reader's attention to the line located after the turn at $x = -50$ and $y = 55$. Of all the trajectories, this is the area where the difference between ROSB and the proposed system is the largest. This can be explained by the start of the turn, where the trajectory estimated by ROSB does not take speed into account and relies on its heading, which leads to errors throughout the rest of the trajectory. The acoustic range observation does not seem to correct this initial error. Unlike a USBL, which provides latitude and longitude measurements, the range information alone does not allow the AUV to be located precisely in the area. In contrast, the trajectory estimated by the proposed system is much better. It is even possible to visually observe the acoustic corrections, although this is still insufficient compared with a USBL/DVL positioning system. Finally, one can notice that, on trajectory #4, during the turn, the ROSB system performs better than the ROSB+Doppler proposed solution. Since the estimated heading at this moment is correct, the addition of the relative speed by Doppler estimation without the bearing angle of the UWA communication degrades the positioning compared with the simpler system. However, the algorithm does not diverge and still estimates the correct position at the end of the turn. This is in line with the simulations carried out with varying degrees of heading error during boustrophedon scan trajectories. At a certain distance from the beacon, the estimated position in turns exceeds the actual position but is correctly aligned in straights.

Figure 12 exhibits the positioning error and RMSE over time for each trajectory, which is the image of the confidence of the EKF estimation. We can see that after 10 min of localization, the error does not grow; thus, the algorithm does not diverge. The error mainly increases during turns for a few reasons:

- The sea current implies a boat drift, which impacts the heading angle, bearing angle and then relative speed v_r. In fact, due to (13), we estimate only the forward speed and not the transverse speed.
- The bearing angle is not measured or estimated through UWA but through EKF linearity.
- Between each communication, the EKF only measures the heading angle ψ.

Table 7 provides results in the form of percentage error on traveled distance that can be compared with values obtained in simulation (see Table 3). For the first trajectories, the results are very close to the values provided in Table 3. The results of the 4-th trajectory can be explained by the low distance traveled in 200 s; in fact, the boat speed was lower than other trajectories. Moreover, this trajectory has the highest range between beacon and boat, although a less complex y trajectory than other ones. Nevertheless, the proposed method still outperforms ROSB for this trajectory.

Finally, Table 8 provides the mean positioning error as a function of T_{com}. As shown in the simulation, the positioning increases gradually as T_{com} increases. One can note that for trajectory #1, a divergence of the algorithm is observed for $T_{\text{com}} = 15$ s.

Table 6. Experimental results and comparison with state-of-the-art methods.

Trajectory	Mean Positioning Error in m with Proposed Method	Mean Positioning Error in m with ROSB Method
#1 (t_{max} = 600 s)	4.6225 m	5.4287 m
#2 (t_{max} = 300 s)	2.1537 m	2.9931 m
#3 (t_{max} = 600 s)	5.0740 m	8.6719 m
#4 (t_{max} = 200 s)	4.2069 m	6.2746 m

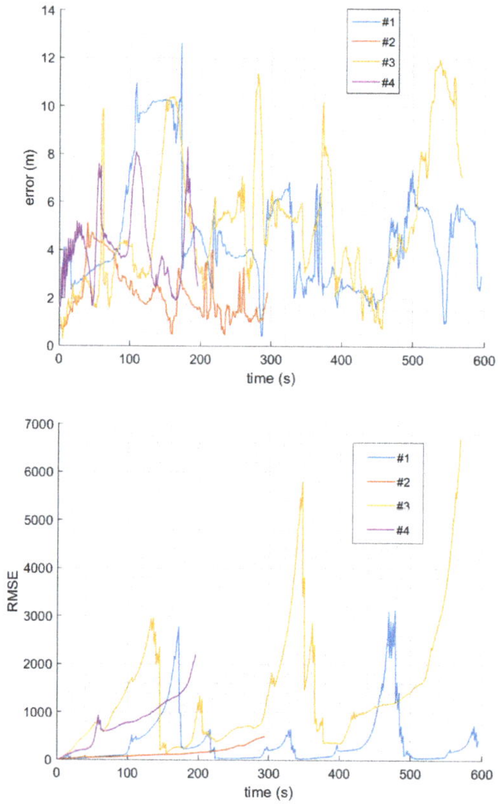

Figure 12. Experimental performance of positioning system for the 4-th selected trajectory as function of mission time. **Up**: averaged positioning error, and **down**: error dispersion (RMSE).

Table 7. Experimental results in the form of a positioning error as a percentage of the traveled distance for the 4 studied trajectories.

Trajectory	Traveled Distance in m	%etd Max	%etd Mean
#1 (t_{max} = 600 s)	698.4989	1.7179	0.6617
#2 (t_{max} = 300 s)	353.3752	1.4149	0.6094
#3 (t_{max} = 600 s)	681.5850	1.7606	0.7444
#4 (t_{max} = 200 s)	190.5176	4.1990	2.2081

Table 8. Experimental results of the 4 studied trajectories as function T_{com}. The DIV label refers to a divergence of the algorithm.

T_{com}	Mean Positioning Error for #1 in m	Mean Positioning Error for #2 in m	Mean Positioning Error for #3 in m	Mean Positioning Error for #4 in m
3 s	4.6225	2.1537	5.0740	4.2069
6 s	5.0982	3.3472	7.1929	4.4613
9 s	5.6782	3.3813	7.4395	5.0431
12 s	6.0585	3.4354	8.4071	5.6932
15 s	DIV	2.5071	8.2798	5.3535
18 s	13.71	4.4618	9.3593	7.5195

5. Conclusions

The objective of this paper was to describe an innovative localization approach using cost-effective sensors for AUV navigation. In the proposed method, the positioning algorithm relies only on an IMU, a pressure sensor and a UWA communication modem, also used to exchange data between the AUV and the surface. By decoding a UWA frame, the AUV can determine, on the one hand, the TOF of the acoustic link (and, thus, the range) via the frame synchronization algorithm and, on the other hand, its relative speed with respect to the beacon owing to the Doppler shift estimation required for UWA channel equalization.

In comparison to the traditional method of beacon range localization, i.e., ROSB, we performed both simulations and conducted experiments to analyze the performance of the proposed approach in depth. Via simulation over a boustrophedon trajectory and using noise representing error measurements, we show that the proposed method outperforms the ROSB method with lower mean positioning error and fewer outliers. The Boustrophedon type of trajectory is one of the most commonly used trajectories in the field of AUVs, along with others such as depth saw-tooth and pipeline tracking. In both simulation and experimentation, the proposed system approaches the planned mission with errors maintained below 20 m. Because the sensors used are cost-effective and low energy (which is critical for a micro-AUV, where autonomy is limited due to the small size of the battery), the acquired data are positioned satisfactorily. The performance of both algorithms is also closely linked to the time between two communication frames; as this time increases, the positioning error increases gradually. The lack of DVL on both solutions impacts the velocity estimation of the AUV and, thus, implies positioning errors and deviation during turns. It is valuable to note that the main error comes from the bearing angle being unknown rather than the lack of the DVL due to the relative speed estimation from the Doppler shift. Nevertheless, it may be considered that in the context of a sonar scan or bathymetry pattern, positioning accuracy during turns is less critical and may be considered by the mission planning operator. The important point, however, is not to cause the navigation algorithm to diverge. It is possible to overcome this problem by making regular ascent to the surface to obtain a GPS fix, which is a subject for future research and implementation. However, a mission at deep depth (e.g., beyond 200 m) is less prone to regular returns to the surface.

On the other hand, INS, in combination with a USBL as used in the state-of-the-art methods, is allegedly more accurate, even if it is difficult to obtain figures as it differs greatly depending on the hardware used for the system. However, it is also true to say that the proposed solution uses less energy (thanks to fewer hydrophones) and is not affected by the error in computing the bearing angle due to the rotation of the USBL head by the waves but only the estimation error by the filter. Due to the simplification of the acoustic computing compared to the USBL, we can also argue that the positioning and data exchange through the UWA channel is more robust. Adding a pure tone before the UWA communication allows an instant Doppler estimation [20].

In the sea environment, the whole positioning system (including the UWA communication link) was evaluated over several trajectories. Experimental results confirmed simulation results, and we showed that using Doppler shift estimation improves the localization estimation of the AUV by increasing overall accuracy compared to state-of-the-art ROSB. The accuracy of the system is fairly independent from the range, but the low-speed propagation of the sound underwater will reduce the frequency of UWA communications at long range, which then implies an increase in the positioning error. However, in our algorithm, only forward speed in estimated; thus, transverse speed due to current is not estimated. Therefore, the performance of the proposed system will be substantially affected in case of a large current w.r.t. to forward speed of the AUV. Moreover, we assume perfect timing synchronization between the beacon and AUV; a large timing error between beacon and AUV will result in a Doppler-shift effect at the receiving end, yielding to an erroneous speed estimation error. This phenomenon can be mitigated by measuring this clock drift at the beginning of the mission and by compensating it with Doppler-shift estimation.

In addition, we demonstrated that with a very simple architecture, positioning approximation is possible, and then small AUV navigation can be considered using cost-efficient transducers. Having real-time communication between the AUV and the base of operations is a critical issue, whether for real-time data acquisition without waiting for the end of the mission or a return to the surface or for applications where time is critical, such as mine detection or the detection of a vessel in distress. This simple architecture with an acoustic transponder also opens up a whole range of possibilities. It is feasible to communicate with multiple AUVs while repositioning them from a single beacon. This would make it possible to carry out several short missions to cover a large area with small AUVs, reducing problems of autonomy or positioning error over the long term.

Future work will focus on three-dimensional positioning, adding inertial unit measurements, and pressure sensor implementation. In addition, emphasis will be placed on the measurement covariance matrix as it is possible to obtain the matrix variable, depending on the signal-to-noise ratio (SNR) of UWA acoustic channel, for example.

Author Contributions: Conceptualization, R.G., P.F. and P.-J.B.; methodology, R.G., B.T. and C.V.; software, R.G.; writing—original draft preparation, R.G. and P.-J.B.;writing—review and editing, P.F., C.V. and B.T. All authors have read and agreed to the published version of the manuscript.

Funding: This research received no external funding.

Informed Consent Statement: Not applicable.

Data Availability Statement: The raw data supporting the conclusions of this article will be made available by the authors on request.

Conflicts of Interest: The authors declare no conflicts of interest.

References

1. Ridao, P.; Carreras, M.; Ribas, D.; Sanz, P.J.; Oliver, G. Intervention AUVs: The Next Challenge. *IFAC Proc. Vol.* **2014**, *47*, 12146–12159. [CrossRef]
2. González-García, J.; Gómez-Espinosa, A.; Cuan-Urquizo, E.; García-Valdovinos, L.G.; Salgado-Jiménez, T.; Cabello, J.A.E. Autonomous Underwater Vehicles: Localization, Navigation, and Communication for Collaborative Missions. *Appl. Sci.* **2020**, *10*, 1256. [CrossRef]
3. Gussen, C.M.G.; Diniz, P.S.R.; Campos, M.L.R.; Martins, W.A.; Costa, F.M.; Gois, J.N. A Survey of Underwater Wireless Communication Technologies. *J. Commun. Inf. Syst.* **2016**, *31*, 242–255. [CrossRef]
4. Stojanovic, M.; Beaujean, P.P.J. Acoustic Communication. In *Springer Handbook of Ocean Engineering*; Dhanak, M.R., Xiros, N.I., Eds.; Springer International Publishing: Cham, Switzerland, 2016; pp. 359–386. [CrossRef]
5. Bouvet, P.; Auffret, Y.; Munck, D.; Pottier, A.; Janvresse, G.; Eustache, Y.; Tessot, P.; Bourdon, R. Experimentation of MIMO underwater acoustic communication in shallow water channel. In Proceedings of the OCEANS 2017—Aberdeen, Aberdeen, UK, 19–22 June 2017; pp. 1–6. [CrossRef]
6. Eggen, T.H.; Baggeroer, A.B.; Preisig, J.C. Communication over Doppler spread channels. Part I: Channel and receiver presentation. *IEEE J. Ocean. Eng.* **2000** *25*, 62–71. [CrossRef]
7. Paull, L.; Saeedi, S.; Seto, M.; Li, H. AUV Navigation and Localization: A Review. *IEEE J. Ocean. Eng.* **2014**, *39*, 131–149. [CrossRef]
8. Thomson, D. Acoustic Positioning Systems—The Hydrographic Society UK. Available online: https://www.yumpu.com/en/document/view/10350341/acoustic-positioning-systems-the-hydrographic-society-uk (accessed on 17 May 2022).

9. Kaushik, A.; Singh, R.; Dayarathna, S.; Senanayake, R.; Di Renzo, M.; Dajer, M.; Ji, H.; Kim, Y.; Sciancalepore, V.; Zappone, A.; et al. Toward Integrated Sensing and Communications for 6G: Key Enabling Technologies, Standardization, and Challenges. *IEEE Commun. Stand. Mag.* **2024**, *8*, 52–59. [CrossRef]
10. Masmitjà Rusiñol, I. Acoustic Underwater Target Tracking Methods Using Autonomous Vehicles. Ph.D. Thesis, Universitat Politècnica de Catalunya, Barcelona, Spain, 2020.
11. Hunt, M.M.; Marquet, W.M.; Moller, D.A.; Peal, K.R.; Smith, W.; Spindel, R.C. *An Acoustic Navigation System*; Woods Hole Oceanographic Institution: Woods Hole, MA, USA, 1974.
12. Yuan, M.; Li, Y.; Li, Y.; Pang, S.; Zhang, J. A fast way of single-beacon localization for AUVs. *Appl. Ocean. Res.* **2022**, *119*, 103037. [CrossRef]
13. Casey, T.; Guimond, B.; Hu, J. Underwater Vehicle Positioning Based on Time of Arrival Measurements from a Single Beacon. In Proceedings of the OCEANS, Vancouver, BC, Canada, 29 September–4 October 2007; pp. 1–8. . ISSN: 0197-7385. [CrossRef]
14. Vickery, K. Acoustic positioning systems. A practical overview of current systems. In Proceedings of the 1998 Workshop on Autonomous Underwater Vehicles (Cat. No.98CH36290), Cambridge, MA, USA, 21 August 1998; pp. 5–17. [CrossRef]
15. LaPointe, C.E.G. Virtual Long Baseline (VLBL) Autonomous Underwater Vehicle Navigation Using a Single Transponder. Mater's Thesis, Massachusetts Institute of Technology, Cambridge, MA, USA, 2006; p. 94.
16. iXblue. RAMSES Sparse-LBL Positioning System Datasheet. Available online: https://www.ixblue.com/wp-content/uploads/2021/12/ramses-lbl-transceiver-datasheet.pdf (accessed on 17 May 2022).
17. Diamant, R.; Wolff, L.M.; Lampe, L. Location Tracking of Ocean-Current-Related Underwater Drifting Nodes Using Doppler Shift Measurements. *IEEE J. Ocean. Eng.* **2014**, *40*, 887–902. [CrossRef]
18. Aubry, C.; Forjonel, P.; Bouvet, P.; Pottier, A.; Auffret, Y. On the use of Doppler-shift estimation for simultaneous underwater acoustic localization and communication. In Proceedings of the OCEANS 2019—Marseille, Marseille, France, 17–20 June 2019; pp. 1–5. [CrossRef]
19. Garin, R.; Vanwynsberghe, C.; Forjonel, P.; Tomasi, B.; Bouvet, P.J. Simultaneous underwater acoustic localization and communication: An experimental study. In Proceedings of the OCEANS 2021: San Diego—Porto, San Diego, CA, USA, 20–23 September 2021; pp. 1–6. [CrossRef]
20. Garin, R.; Bouvet, P.J.; Forjonel, P.; Tomasi, B. Sea experimentation of single beacon simultaneous localization and communication for AUV navigation. In Proceedings of the OCEANS 2023, Limerick, Ireland, 5–8 June 2023.
21. UCNL Website. Available online: https://unavlab.com/en/o-nas/articles/myi-sdelali-samyiy-malenkiy-i-samyiy-deshevyiy-v-mire-gidroakusticheskiy-modem-uwave/ (accessed on 12 October 2024).
22. Waterlinked Website. Available online: https://waterlinked.com/shop/modem-m16-186 (accessed on 12 October 2024).
23. ArduPilot Website. Available online: https://ardupilot.org/ (accessed on 30 August 2024).
24. Julier, S.; Uhlmann, J.; Durrant-Whyte, H. A new approach for filtering nonlinear systems. In Proceedings of the 1995 American Control Conference—ACC'95, Seattle, WA, USA, 21–23 June 1995; Volume 3; pp. 1628–1632. [CrossRef]
25. Julier, S.; Uhlmann, J.; Durrant-Whyte, H. A new method for the nonlinear transformation of means and covariances in filters and estimators. *IEEE Trans. Autom. Control.* **2000**, *45*, 477–482. [CrossRef]

Journal of
Marine Science and Engineering

MDPI

Article

Autonomous Underwater Pipe Damage Detection Positioning and Pipe Line Tracking Experiment with Unmanned Underwater Vehicle

Seda Karadeniz Kartal * and Recep Fatih Cantekin

Electrical and Electronic Engineering Department, Zonguldak Bulent Ecevit University, 67100 Zonguldak, Turkey; recep_cantekin@hotmail.com

* Correspondence: sedakaradeniz@gmail.com

Abstract: Underwater natural gas pipelines constitute critical infrastructure for energy transportation. Any damage or leakage in these pipelines poses serious security risks, directly threatening marine and lake ecosystems, and potentially causing operational issues and economic losses in the energy supply chain. However, current methods for detecting deterioration and regularly inspecting these submerged pipelines remain limited, as they rely heavily on divers, which is both costly and inefficient. Due to these challenges, the use of unmanned underwater vehicles (UUVs) becomes crucial in this field, offering a more effective and reliable solution for pipeline monitoring and maintenance. In this study, we conducted an underwater pipeline tracking and damage detection experiment using a remote-controlled unmanned underwater vehicle (UUV) with autonomous features. The primary objective of this research is to demonstrate that UUV systems provide a more cost-effective, efficient, and practical alternative to traditional, more expensive methods for inspecting submerged natural gas pipelines. The experimental method included vehicle (UUV) setup, pre-test calibration, pipeline tracking mechanism, 3D navigation control, damage detection, data processing, and analysis. During the tracking of the underwater pipeline, damages were identified, and their locations were determined. The navigation information of the underwater vehicle, including orientation in the x, y, and z axes (roll, pitch, yaw) from a gyroscope integrated with a magnetic compass, speed and position information in three axes from an accelerometer, and the distance to the water surface from a pressure sensor, was integrated into the vehicle. Pre-tests determined the necessary pulse width modulation values for the vehicle's thrusters, enabling autonomous operation by providing these values as input to the thruster motors. In this study, 3D movement was achieved by activating the vehicle's vertical thruster to maintain a specific depth and applying equal force to the right and left thrusters for forward movement, while differential force was used to induce deviation angles. In pool experiments, the unmanned underwater vehicle autonomously tracked the pipeline as intended, identifying damages on the pipeline using images captured by the vehicle's camera. The images for damage assessment were processed using a convolutional neural network (CNN) algorithm, a deep learning method. The position of the damage relative to the vehicle was estimated from the pixel dimensions of the identified damage. The location of the damage relative to its starting point was obtained by combining these two positional pieces of information from the vehicle's navigation system. The damages in the underwater pipeline were successfully detected using the CNN algorithm. The training accuracy and validation accuracy of the CNN algorithm in detecting underwater pipeline damages were 94.4% and 92.87%, respectively. The autonomous underwater vehicle also followed the designated underwater pipeline route with high precision. The experiments showed that the underwater vehicle followed the pipeline path with an error of 0.072 m on the x-axis and 0.037 m on the y-axis. Object recognition and the automation of the unmanned underwater vehicle were implemented in the Python environment.

Keywords: unmanned underwater vehicle; convolution neural network; underwater pipe line tracking; underwater pipe damage detection; navigation of unmanned underwater vehicle

1. Introduction

1.1. Motivation

Unmanned underwater observation vehicles are critically important for various military and civilian applications. These unmanned vehicles are used in civilian fields such as underwater mapping, port security, geological geophysics, and fisheries, and in military areas for mine detection, enemy ship detection, ship safety, coastal security, and human detection, as well as in underwater cable and pipeline laying operations [1–3]. The highly variable nature of the underwater environment makes these operations challenging. Using unmanned underwater vehicles instead of human divers in long-term operations in dark and deep waters is both safer and more cost-effective due to the potential risks to human life [4,5]. Divers can only remain submerged for a limited time during any underwater operation due to the risk of hypothermia from prolonged exposure. For this reason, the duration of underwater operations tends to be extended. To eliminate the negative aspects arising in similar underwater operations, unmanned underwater vehicles have started to be preferred for underwater tasks.

Underwater natural gas pipelines form a critical infrastructure for energy transportation. Any damage or leakage occurring in these pipelines can create serious security risks. Proper monitoring and damage detection facilitate the early identification and prevention of potential hazards. Additionally, underwater natural gas pipelines can pose a direct threat to marine and lake ecosystems. Any leakage or damage can lead to environmental pollution and harm aquatic life. Timely damage detection can help minimize environmental impacts. Damages to underwater natural gas pipelines can also cause interruptions in the transmission of natural gas, leading to serious operational problems in the energy supply chain. Continuous monitoring and damage detection are crucial for maintaining operational continuity. Furthermore, damages, repairs, and interruptions in the pipeline can lead to significant economic losses for energy companies. Timely damage detection can reduce costly emergency interventions and increase operational efficiency. There are legal and regulatory requirements for underwater natural gas pipelines in many systems. These requirements mandate regular monitoring of the pipelines and the performance of damage detection.

In this study, damages on an underwater pipeline were detected using an unmanned underwater vehicle. As important as detecting damages on the pipeline is knowing the locations of these damages. If the location of the damage is unknown after it has been detected, finding the locations of the damages would require additional time due to the extensive length of the pipelines. In this study, autonomous features were added to the unmanned underwater vehicle, enabling autonomous tracking of the damaged pipeline underwater, and while tracking the pipeline, the damages were diagnosed using artificial learning and their locations were determined. This study involves the autonomous tracking of an underwater pipeline of a certain length by an unmanned underwater vehicle (UUV), focusing on detecting and localizing damages on the pipeline during this tracking. The key focus areas and objectives of the study are as follows:

- Autonomous Pipeline Tracking: The remote-controlled UUV autonomously follows the underwater pipeline using its navigation system.
- Damage Detection: Damages on the pipeline are identified using images captured by the UUV's camera, processed through a Convolutional Neural Network (CNN) algorithm.
- Damage Localization: The location of the detected damage is determined by correlating it with the vehicle's INS (Inertial Navigastion System) based navigation data, providing an accurate position.
- Real-World Application: The system is tested and validated for the usability of underwater pipeline maintenance and monitoring in real-world environments.

1.2. Related Works

For successful execution of unmanned underwater operations, both a well-defined underwater environment and accurate knowledge of the underwater vehicle's position are necessary [6–8]. Therefore, recognizing underwater objects and navigation of underwater vehicles are important. Various types of identification algorithms are available to identify an object. In this study, the convolutional neural network (CNN) training algorithm, a deep learning method, was used. CNNs are used in various fields including image classification, object tracking, object recognition, exposure estimation, text detection and recognition, visual projection detection, action recognition, scene labeling, speech processing, and natural language processing [9]. Other training models require a large amount of prior knowledge at the end of training to achieve high accuracy in object recognition. However, in the CNN model, input data is provided to the model without the need for feature extraction or creation processes. Since the CNN model trains by altering the depth and width of the input image, it determines the features of the image and makes accurate assumptions [9,10]. CNN training requires significant computational resources, but several methods have been developed to address this issue [11–27]. The most important of these developments is the development of a CNN model trained using the ImageNet dataset in 2012. This model produced more accurate image classifications than previous methods [9]. He and Zhang in 2018 suggested predicting movements from an image with CNN [28]. Pertusa and Gallego in 2018 used CNN for common object identification on smartphones [29]. In 2015, Li and Shang used the fast region-based CNN (R-CNN) algorithm for underwater fish detection [30]. In 2017, Gomez Chavez and Mueller predicted body posture using the Long Short-Term Memory Recurrent Neural Network (LSTM-RNN) method [31]. The CNN algorithm continues to be used in classification studies today [32–34]. Another current machine learning algorithm is the Support Vector Machine (SVM). The SVM algorithm emerged in 1995 [35]. It is a high-performance algorithm frequently chosen for regression and prediction problems [36]. To date, SVM has been used in various contexts including battery life prediction, housing price forecasting, and predicting potential inflation [37–42]. Although SVM is more commonly chosen for classification problems, Smola and others have shown it can also be used for regression problems [43]. The algorithm used for regression problems is named Support Vector Regression (SVR). SVR has been applied in various regression problems such as motion prediction [44], electric load forecasting [45], and enhancing the performance of filters [46–48] . In this study, CNN and SVM were used in a hybrid manner. CNN served as a feature extractor and SVM as a classifier to identify the target object.

Various methods are employed in the literature for detecting damage to underwater pipelines. While image processing and deep learning algorithms (CNN) can detect damages on pipelines with high accuracy, methods such as sonar and acoustic imaging offer the advantage of scanning larger areas. Techniques like magnetic flux leakage (MFL) and ultrasonic testing (UT) detect corrosion and cracks within the internal structure of the pipes, while fiber optic sensors and vibration-based monitoring methods provide continuous monitoring and early detection capabilities. However, each method has its advantages and limitations depending on environmental conditions, pipeline material, and the size of the damage. While modern image processing techniques such as CNN analyze visual data, acoustic and magnetic methods like sonar and MFL can scan broader areas regardless of environmental conditions. The choice of method depends on the characteristics of the pipeline and environmental factors. Unmanned underwater vehicles (UUVs) are effective tools for detecting underwater pipeline damage when integrated with various detection technologies. For visual-based detection, deep learning algorithms such as CNN can offer higher accuracy compared to traditional methods like sonar and magnetic techniques. The methods used for underwater pipeline detection and the advantages and limitations of these methods are presented in the comparison table in Table 1. As seen from the comparison, the use of both the unmanned underwater vehicle and the CNN algorithm in this study has significantly enhanced performance. In addition, to facilitate fault detection in subsea pipelines, transient test-based techniques were used to generate safe and small overpressure waves, allowing the detection of anomalies such as leaks and wall deterioration [39,40].

Table 1. Performance comparision of pipeline damage detection methods.

Method	Key Strengths	Key Limitations	Paper Citation
CNNs (Optical Image)	High accuracy with image data Automatic Feature xtraction Handles compex surface defects Adaptable and scalable with large datasets	Require large training datasets Computationally intensive	[14–16]
Acoustic Imaging	Effective in deep or turbid water Covers large areas	Lower resolution, noisy data Difficult to detect small defects like cracks or corrosion Complex data interpretation and prone to noise	[17]
Magnetic Flux Leakage	Good for internal defects and corrosion Real time monitoring	Only works for metallic pipelines Ability to detect external surface damage like cracks or dents Limited compared to CNN-based visual inspection	[18]
Ultrasonic Testing	High precision for internal defect, non-invasive	Requires physical contact or close proximity to the pipeline surface, which can be difficult in certain underwater conditions. Doesn't provide as detailed information on surface defects as CNNs do with optical imagery	[19]
Fiber Optic Sensing	Real-time, continuous monitoring, detects strain and deformation	Installation is complex and expensive, as it requires physically embedding the sensors along the pipeline. It doesn't provide detailed visual feedback, unlike CNN-based methods.	[20]
Vibration based Monitoring	Sensitive to structural changes, covers long distances	Influenced by external factors such as sea currents, environmental noise, or nearby operations, making data interpretation challenging. Unlike CNNs, it doesn't provide a direct visual assessment of the damage.	[21]
Visual Inspection Using Underwater Vehicle	Real-time inspection High-resolution data Can reach inaccessible areas Lower cost Continuous monitoring Long-term stability	High cost Limited battery life	[22–26]

Another important factor for the successful execution of unmanned underwater operations, such as underwater pipeline tracking and underwater pipeline damage detection, is the localization of the underwater vehicle [49]. Due to the attenuation/damping of electromagnetic waves underwater, high-accuracy global positioning systems cannot be used. With inertial measurement systems, linear and angular position information is derived from the measured acceleration and angular velocity of the underwater vehicle [50]. Some studies use integrated navigation systems for high accuracy and continuous data transmission. If an INS-GPS integration system is to be used, a surface platform synchronized with the underwater vehicle is essential [51]. In this study, the navigation information of the underwater vehicle was obtained from integrated gyroscopes, magnetic compasses, accelerometers, and pressure sensors.

Despite extensive work on aerial and terrestrial object tracking, there is much less research on underwater object tracking. This is due to the various challenges of working underwater and the degradation of underwater visual data quality, which varies depending on light refraction, water depth, color, and nature. In 2013, Min Li and colleagues presented a method for underwater object identification and tracking based on multi-beam sonar imaging [52]. In 2016, Filip Mandic and colleagues combined sonar and USBL (Ultra Short Baseline) measurements to develop an autonomous surface vehicle and perform underwater object tracking [53]. They developed a filter that combines USBL and sonar image measurements to obtain reliable object tracking predictions even when sonar or USBL measurements are unavailable or erroneous. In addition to object tracking, they focused on adapting only the desired region within the sonar image using the tracking filter's covariance transformation to improve object identification and filter out erroneous sonar measurements. In 2016, Xianbo Xiang and colleagues proposed a method using magnetic sensing to autonomously track underwater buried cables with a three degrees of freedom (3-DOF) autonomous underwater vehicle [54]. They used feedback linearization technique to design a simplified cable tracking controller based on the geometric relationship between the vehicle and the cable by creating a specialized magnetic line of sight guide. In 2020, Caterina Bigoni and Jan S. Hesthaven suggested a simulation-based decision strategy with machine learning techniques for anomaly detection and damage localization [55]. In 2021, Kakani Katija and colleagues proposed using an underwater vehicle for the visual tracking of deep-sea animals controlled by machine learning [56]. In their study, they presented an integrated tracking algorithm using machine learning that includes multi-class detectors and 3D stereo imaging to track underwater animals over extended periods. There are many studies like these focused on underwater object tracking, and the research continues. Studies such as continuous autonomous tracking and imaging of great white sharks with an autonomous underwater vehicle, and performance analysis of existing underwater object tracking algorithms and dataset creation are available [57,58].

1.3. Contribution

In this study, an autonomous system was developed to monitor underwater pipelines using an unmanned underwater vehicle (UUV). Damages to underwater pipelines were detected with high accuracy through the CNN algorithm. The precise location of the detected damage was determined using the UUV's navigation data, ensuring precise positioning. The system's usability for the maintenance and monitoring of underwater pipelines was tested and validated. There is no experimental study in the literature that autonomously tracks underwater pipelines with a UUV while simultaneously detecting pipeline damage using deep learning algorithms and pinpointing the damage's location. The research gap, contribution, and innovation of this study can be summarized in the following key points:

- Underwater Damage Detection with Deep Learning: Autonomous systems for detecting damages in underwater pipelines are limited in the literature, and the use of deep learning algorithms such as CNN in this area is rare. This study aims to automatically detect damages in underwater pipelines using deep learning algorithms, offering a faster and more accurate approach compared to manual methods.
- Autonomous Damage Detection and Navigation Relationship: The ability to correlate the damage location with the vehicle's INS-based navigation data is an innovation in terms of damage detection and accurate positioning of underwater vehicles. There is no example in the literature where these two capabilities are combined in real-time and autonomously.
- Experimental Application: Another strength of this study is the experimental application of the CNN algorithm. Unlike simulation or theoretical studies, this experimental part, which demonstrates real-world applicability, distinguishes this research from others in the literature.
- Potential for Industrial Application: This technology could be used for the monitoring and maintenance of underwater infrastructures, such as oil, gas, and other underwater pipelines, with the potential to improve the management of these infrastructures.

In conclusion, this study aims to address the gap in the literature by offering an innovative approach to real-time and reliable underwater pipeline damage detection using autonomous underwater vehicles and deep learning algorithms.

1.4. Organization

The paper is organized as follows. The unmanned underwater vehicle used in our underwater experiments is introduced in Section 2. The method used for the underwater pipe damage detection experiment, involving a convolution neural network, and the experimental results of the damage diagnosis are explained in Section 3. The experiment on underwater autonomous pipe tracking and damage location detection is detailed in Section 4. In Section 4, the unmanned underwater vehicle navigation, autopilot, underwater damage location detection, and experimental results are also presented sequentially. In final, the paper is concluded in Section 5.

2. Unmanned Underwater Vehicle Used in Experiments

In this study, the remotely operated underwater vehicle (ROV) that will be endowed with autonomous features consists of a user computer, an operator console, and cable section. The unmanned underwater vehicle is equipped with two forward thrusters, one on the left and one on the right, a vertical thruster, and a camera that can rotate 180 degrees. The forward thrusters provide forward movement and yaw orientation, while the vertical thruster provides diving movement. The operator console is used for controlling the vehicle and transferring the data obtained from the vehicle to the computer. The cable ensures data and power transmission between the underwater vehicle and the operator console. In this study, the experimental equipment used in the pool experiment is presented in Figure 1 [59].

The unmanned underwater vehicle used in this study measures 42 cm in length, 33 cm in width, and 27 cm in height, with a total mass of 10 kg. The vehicle is designed to operate at depths of up to 200 m. Its moments of inertia are 0.223 kg/m^2 along the x-axis, 0.225 kg/m^2 along the y-axis, and 0.06 kg/m^2 along the z-axis. The maximum surge speed of the vehicle is 1.5 m per second. The vehicle's sensor suite includes an Inertial Navigation System (INS) from the MPU-6000 series, featuring accelerometers, gyroscopes, and a depth sensor (MS5837-30BA), as well as a camera.

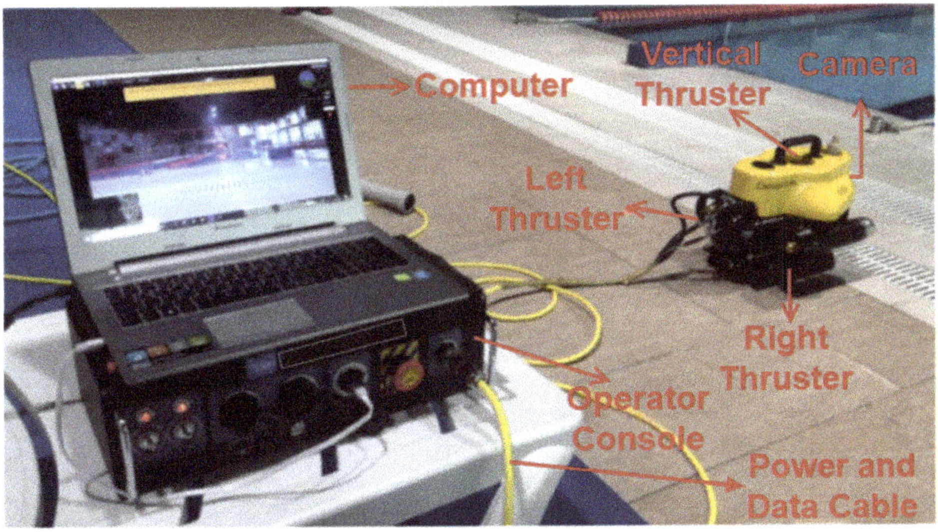

Figure 1. Unmanned underwater vehicle used in the experiment.

The experimental data related to underwater object detection was obtained from a camera integrated into the remotely operated underwater vehicle shown in Figure 1. The camera, with vertical and horizontal fields of view of 128 and 96 degrees respectively and a resolution of 700 TVL, is placed in a waterproof compartment at the front of the vehicle. This vehicle can reach depths of 200 m to perform tasks such as underwater observation, real-time high-resolution video and photo capture, data collection, and underwater mapping.

3. Underwater Pipeline Damage Detection Using Convolutional Neural Networks

In this study, damage assessment on the pipeline was conducted using pool images taken from a camera integrated into the unmanned underwater vehicle. The camera, placed in a waterproof compartment at the front of the vehicle, has a vertical field of view of 128 degrees, a horizontal field of view of 96 degrees, and a resolution of 700 TVL [51]. The damage assessment study was carried out in a pool environment based on experimental data. The video footage from the vehicle's camera was transmitted via fiber optic cable to the operator console and then to a computer via Ethernet cable. To diagnose the damage, the images were processed using a deep learning method, the Convolutional Neural Network (CNN) algorithm. The all study was conducted in a Python environment. The pipes and damages used for the pipeline damage assessment experiment are presented in Figure 2.

In our study, the type of pipe used is polyvinyl chloride pipes, which are commonly used in water transportation systems. These pipes are frequently preferred due to their strength and resistance to corrosion. In this study, polyvinyl chloride pipes were used for testing purposes. In real-world aquatic environments, biofouling, mineral deposits, wear, scratches, and UV light cause damage that manifests as color changes on the surface of underwater pipes. In marine environments, the accumulation of algae, bacteria, shellfish, and certain minerals (especially iron) results in biofouling, creating reddish-brown discolorations on the pipe surface. This buildup can damage the pipe's outer surface and negatively affect the functionality of the pipeline. Underwater PVC pipes may also experience surface scratches and abrasions due to the friction of particles carried by the water flow. These physical deformations lead to color changes and reduced performance. Additionally, chemicals present in seawater or maintenance chemicals used around the pipe can cause discoloration, impacting the material's long-term durability [60]. In this study, as seen in Figure 2, the damages identified are associated with these types of color changes observed on pipe surfaces in real-world marine environments. Specifically, damage types such as surface scratches, dents, abrasions and color changes were identified, and a Convolutional Neural Network (CNN) algorithm was trained using this visual data. The pipe deformations shown in Figures 2 and 3 focus on how our developed algorithm detects these types of damages. The damage types addressed in our study include physical deterioration that could have potential negative effects on the structural integrity and safety of the pipes.

In this study, the Faster R-CNN Inception v.2 object detection algorithm, widely used in various object detection tasks, was utilized as the object detection algorithm [61,62]. The hybrid use of the CNN model with SVM was addressed similarly to the technique mentioned in the article "A hybrid CNN–SVM classifier for weed recognition in winter rape field." In this study, the Inception v.2 part of the model was responsible for extracting feature vectors from the images. These features were then used to train the SVM classifier, yielding good results. The hardware specifications used for the training are: Windows 10 operating system, Nvidia GTX 1060 GPU, Core i5-7700 CPU, Python 3.6.5, and 16 GB RAM. With these specifications, the training process took approximately 16 h.

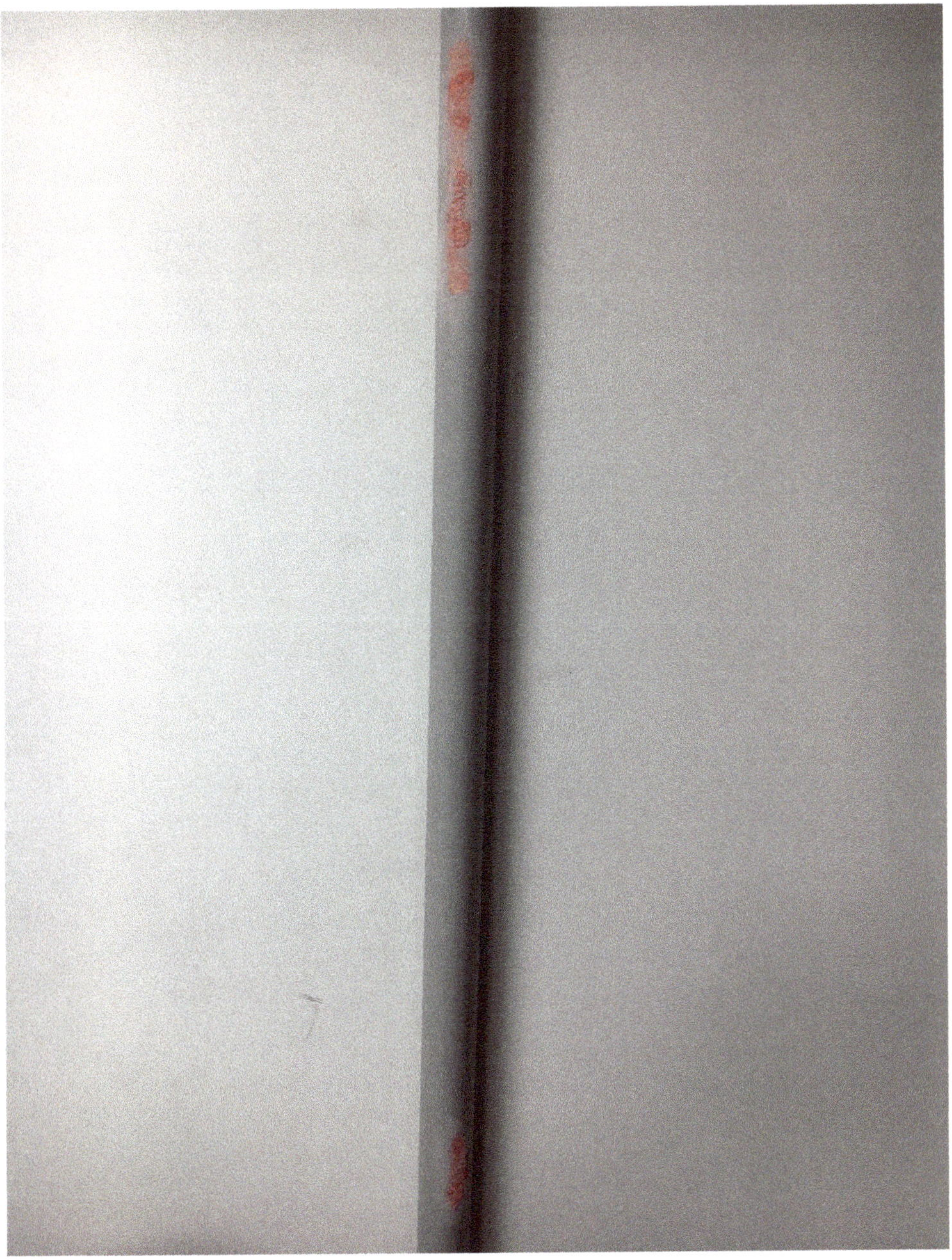

Figure 2. Damaged pipe used in the experiment (red-colored region is defined as "damage").

Figure 3. Damaged pipe used in the experiment (red-colored region is defined as "damage").

3.1. Convolutional Neural Network

3.1.1. Input Layer

The size of the data provided to this layer is crucial for the success of the model. If the amount of data is too large, the training can yield very successful results but will take a long time. Conversely, if the amount of data is too small, the success rate of the training will significantly decrease. While data size in the input layer plays an important role, it is not sufficient on its own for the overall performance of the model. The number of filters, layer depth, activation functions, data variety, data preprocessing, and similar factors also play significant roles in all layers of the model and affect the model's overall performance.

3.1.2. Convolutional Layer

This layer is the first layer that extracts features from the input data. Different filters are applied to the input data. The applied filters are passed over the entire image to produce an output. The output after applying the filter is known as a feature map [63].

3.1.3. Rectified Linear Unit Layer

The Rectified Linear Unit (ReLU) is a commonly used activation function in CNN [64]. ReLU is defined as follows:

$$g(y) = max(0, y) \quad (1)$$

Here, $g(y)$ is a function corresponding to the input y. ReLU sets the negative values of the data applied to its input to zero. This reduces the computational load and training time.

3.1.4. Pooling Layer

The pooling layer is used between successive convolutional layers. The primary purpose of using this layer is to reduce the computational intensity in subsequent layers. There are various pooling methods; one commonly used method is max pooling [65]. In max pooling, the maximum values from the values corresponding to the filter are selected, and the filter moves two steps after each application area. The method used in this study is the max pooling method.

3.1.5. Fully Connected Layer

A fully connected layer connects to all nodes in all layers before and after it. The fully connected layer adds weights to the data to enable accurate classification of the data received from previous layers. After this process, the network provides predictions. Predictions are obtained by calculating probabilities between feature classes detected in previous layers. If the weighting is incorrect before producing a prediction, the predictions are incorrect and a cost function is calculated. The cost function serves as a guide to optimize our model. The cost function between the actual and predicted networks has been minimized using the backpropagation algorithm [66]. Additionally, overfitting is an undesirable condition in CNNs. The dropout method has been developed in this layer to prevent overfitting [9].

3.1.6. DropOut Layer

In CNNs, excessive training can lead to overfitting—memorization, which reduces the training error at each step during the training of the CNN model, but the test error may not decrease in the same direction. The reliability of a training with overfitting is low, and a training model that starts memorizing is formed. A training model that has overfitted will perform poorly when presented with an image outside of the training dataset. Large datasets like ImageNet have labeled data samples to prevent overfitting [9]. In this study, since the dataset created is not as large as ImageNet, dropout has been used to prevent overfitting.

3.1.7. Classifier Layer

The output value of the classification layer should be equal to the number of objects to be classified. For instance, if five classifications are to be made, the output of the layer should be five. Model predictions are assigned as values in the range of 0–1. A classification can be added to the CNN architecture or created as a separate model. There are different classifier models. In this study, CNN and SVM were used as a hybrid. CNN is a feature extractor, and SVM has been used as a classifier to recognize the object being sought [11].

SVMs choose from an infinite number of decision boundaries that minimize error with the greatest distance between two classes [67]. SVM uses the output of CNN as input and determines the classes by extracting features obtained by CNN. Thus, classification is achieved.

3.2. *Convolutional Neural Network Training*

To train the CNN architecture, a dataset was created using deep learning and image augmentation methods from photographs of damage taken on the pipeline. For the damage diagnosis study, the dataset from damages on the underwater pipeline was created through data augmentation without altering the characteristic features of the images. The data augmentation methods used for this study include rotating the image horizontally and vertically, rotating the image at specific angles, shifting the image horizontally and vertically, zooming, darkening, lightening, and changing the color. In CNN, each photo collected with the LabelImg program was individually labeled. The labeling process generated files with a .xml extension. After completing the labeling process, the photos were divided into two separate folders to run the training algorithm. These folders are the training and testing folders. Eighty percent of the photos of the object to be detected were placed in the training folder, and twenty percent were placed in the testing folder, and the training of the CNN was initiated. A total of 2000 data were collected, with 1500 used for training and 500 for testing purposes. Each image in the training set contains at least one damage label.

3.3. *Underwater Damage Detection Performance of the CNN Algorithm*

Using the CNN algorithm, damages on the underwater pipeline in a pool environment were diagnosed online using an unmanned underwater vehicle. The damage diagnosis results are presented in Figure 4. Although the damaged section on the underwater pipeline is very small, it was successfully detected using the CNN algorithm. The accuracy of detection in the pool environment can vary depending on lighting conditions, light refraction, reflections, and glare. In the pool, light usually comes from artificial sources and provides more homogeneous illumination. The reflections created by artificial lights on the water surface or direct entry of light into the water can cause glare or shadows in the image, making it difficult for the CNN algorithm to accurately recognize objects. Due to light refraction, objects may appear different from their actual form, further complicating the algorithm's ability to identify them correctly. As seen from in the Figure 4, the unmanned underwater vehicle has been endowed with the ability to detect objects through the underwater damage detection test.

Figure 4. Pipeline damage detection experiment results.

In Figure 5, the accuracy and loss curves of the CNN training are visible. The ESA training for damage diagnosis was carried out over 50 Epochs. An epoch means that the model sees each data in the dataset once. As seen in Figure 5, both the training loss and the validation/consistency loss have been observed to decrease, which is a desired outcome.

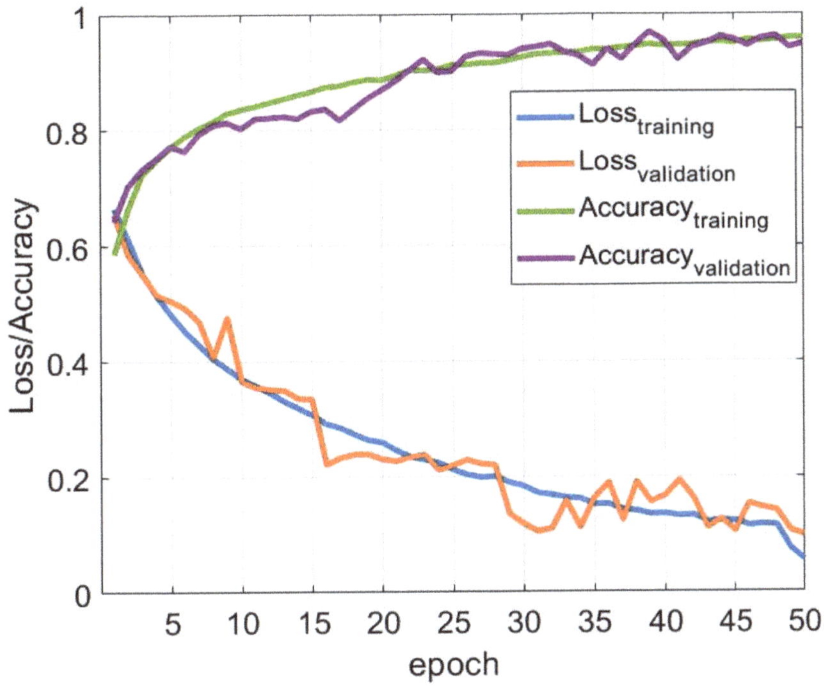

Figure 5. CNN training consistency and loss curves.

Light refraction in water, especially at different water depths, can cause image distortion. This issue can be minimized by positioning the cameras on the vehicle at appropriate angles and using multiple sensors. Additionally, adjusting the artificial light sources used in the underwater environment to the correct spectrum can also improve image quality. As depth increases, the amount of light decreases, making detection more difficult. Therefore, in deep waters, especially in deep-sea environments, more sensitive camera systems and lighting devices should be used. By using deep learning algorithms and image enhancement techniques, the quality of images obtained in low-light conditions can be improved through pre-processing. The depth and clarity of the water can cause changes in colors, making it difficult to detect real damage or anomalies in the image. To mitigate this issue, color balancing algorithms can be used. Furthermore, multi-spectral camera systems are less affected by color changes underwater, and the use of such systems can improve image quality. As in this study, deep learning-based image enhancement algorithms can make distorted underwater images clearer.

4. Autonomous Pipeline Tracking and Damage Localization with an Unmanned Underwater Vehicle

In this study, autonomous features were endowed to a remotely operated underwater vehicle, which was then used for underwater pipeline tracking in a pool environment, and damage detection on the pipe as well as the location of this damage were achieved. In the experiment, a schematic representation of the scenario created for the underwater pipeline and images from the pool experiment are given in Figures 6 and 7. As seen in Figure 6, during the experiment, an underwater pipeline was created inside the pool using one 3-m and two 2-m pipes.

Figure 6. Pipeline tracking experiment scenario schematic representation.

Figure 7. Pipeline tracking pool experiment.

Preliminary tests were conducted in a pool environment to enable the unmanned underwater vehicle to autonomously follow the desired pipeline. In these tests, the PWM values that need to be applied to the vehicle's thrusters were determined to enable the vehicle to autonomously perform the following movements: moving forward 3 m, turning right, moving 2 m, turning right, and then moving another 2 m in a linear and angular motions.

In these tests, the vehicle's speed information corresponding to the PWM values of the thruster motors was obtained by remotely controlling the vehicle to move along different routes. The speed information corresponding to these PWM values and, consequently, the distance information obtained were observed. The necessary PWM values for the vehicle to follow a pipeline of known length, which will be used in the experiment, were determined as a result of these tests and sent to the vehicle's thruster motors as input information. Consequently, an object tracking algorithm was developed using the data obtained from these preliminary tests to enable the vehicle to autonomously follow the desired object.

During the vehicle's tracking of the pipeline, the damages on the pipe were detected using the CNN method detailed above. The damage location detection information has been supported with the vehicle navigation and autopilot described in the next section.

4.1. Navigation of Unmanned Underwater Vehicle

The navigation information of unmanned underwater vehicle comes from IMU, depth sensor. IMU with MPU-6000 series used in the experiment combines 3-axis gyroscopes integrated with magnetic compass, 3-axis accelerometer. The depth sensor used in the experiment is the measurement specialties MS5837-30BA, which can measure up to 30 bar (300 m/1000 ft depth). In the pool experiments, the speed and orientation information of the unmanned underwater vehicle was obtained from the Inertial Measurement Unit (IMU) sensor integrated into the vehicle. The linear speed of the vehicle used in the pool experiments was obtained by integrating the acceleration. Linear position information was obtained by taking the double integral of the measurement from the accelerometer sensor. The yaw angle (rotation around the z-axis) information was obtained by taking the integral of the gyroscope measurement data once. The depth information of the vehicle from the pool surface was obtained from the pressure sensor integrated into the vehicle [68]. The equations for the vehicle's linear motion along the x and y axes and angular motion around the z-axis are given in (2)–(4) [69].

$$\dot{x} = u_r * \cos_{\varphi} - v_r * \sin_{\varphi} + \dot{x}_c \quad (2)$$

$$\dot{y} = u_r * \sin_{\varphi} + v_r * \cos_{\varphi} + \dot{y}_c \quad (3)$$

$$\dot{\varphi} = r \quad (4)$$

In this study, since the vehicle used only has forward speed, the relative velocity, v_r has been neglected, and external disturbances were disregarded in the pool experiments. Thus, the relationship between the vehicle speed and the absolute speed is reduced as in (5) and (6) [69].

$$\dot{x} = u_r * \cos_{\varphi} \quad (5)$$

$$\dot{y} = u_r * \sin_{\varphi} \quad (6)$$

4.2. Autopilot of Unmanned Underwater Vehicle

In this study, the Pixhawk control board was used for the autopilot of the unmanned underwater vehicle as seen Figure 8. Pixhawk is frequently used as a control board for unmanned aerial vehicles, surface, and underwater vehicles due to its low cost and high-performance advantages. In this study, communication between the user computer and the unmanned underwater vehicle, as well as sending inputs to the vehicle via Python software, (3.9.4 version) were carried out using the MAVLink protocol. The pymavlink library was used to establish the connection between the vehicle and Python.

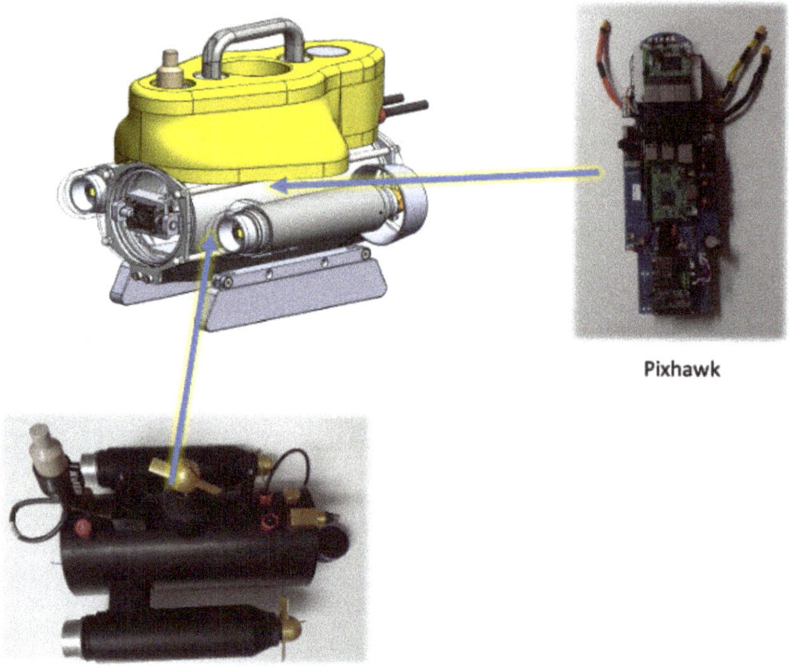

Figure 8. Vehicle's pixhawk and thrusters used in the experiment.

In the experiment, to automate the pipeline tracking, a series of preliminary pool tests were initially conducted to observe the vehicle's responses to PWM (pulse-width modulation) signals sent to the motors by the Pixhawk. Subsequently, the necessary input value of each thrusters' values to be sent to the motors for the created pipeline tracking scenarios were determined. Before the experiment for the established pipeline scenario, the Speed-PWM and Depth-PWM relationships were observed, and the necessary PWM information to enable the vehicle to follow the pipeline was sent to the vehicle.

The vehicle has successfully followed the designated route. The reference and measured forward speed information, yaw angle, and depth information while the vehicle was following the designated pipeline route are shown sequentially in Figures 9–11.

Figure 9 shows the necessary reference forward speed information for following the specified pipeline, along with the forward speed information obtained from the accelerometer during the pipeline tracking. Figure 10 presents the necessary yaw angle for the vehicle to follow the specified pipeline in the experiment, along with the yaw angle measured from the gyroscope during the experiment.

It can be seen from Figures 9–11 that the vehicle successfully followed the necessary reference forward speed, reference depth, and reference yaw angle required for autonomously tracking the specified pipeline.

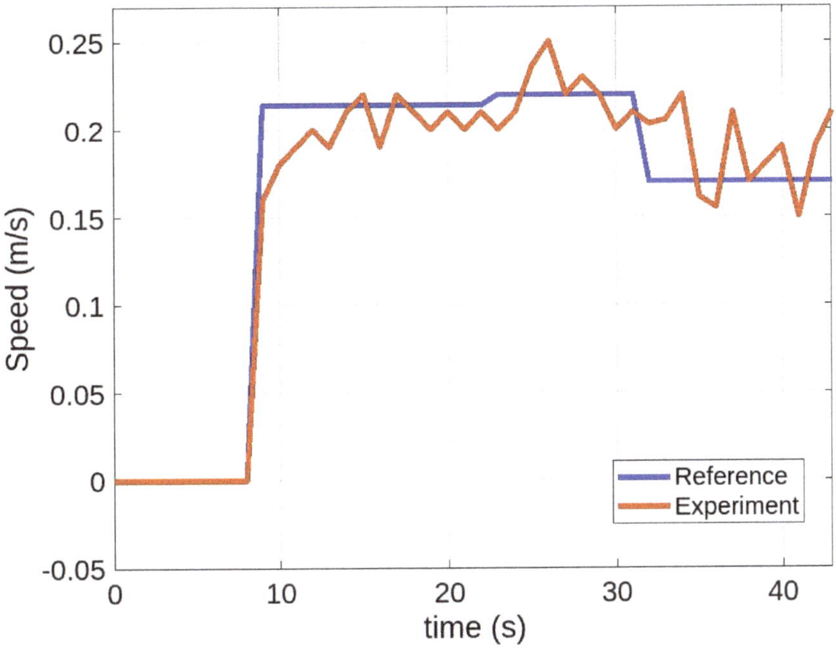

Figure 9. Reference surge speed (blue line) and measured (red line) surge speed.

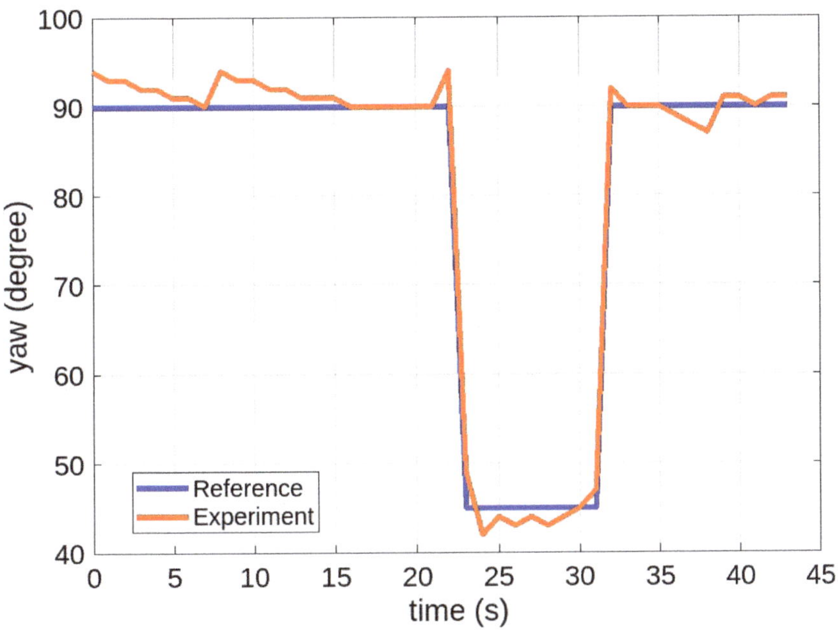

Figure 10. Reference yaw angle (blue line) and measured yaw angle (red line).

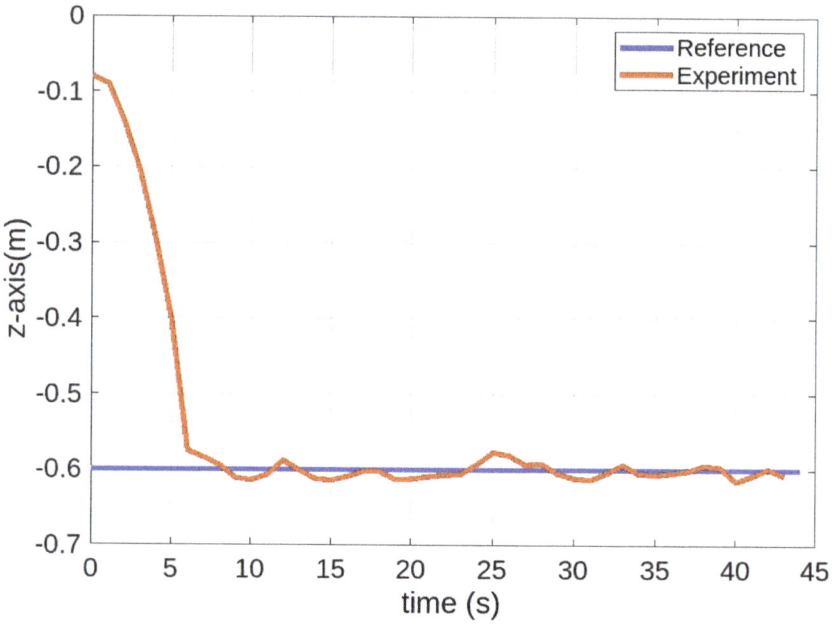

Figure 11. Reference depth value (blue line) and measured depth value (red line).

4.3. Damage Localization Using Navigation Data and Image Processing

In the pool experiment, while the unmanned underwater vehicle autonomously followed the pipeline as desired, damages on the pipeline were diagnosed using images obtained from the vehicle's camera. The position of the diagnosed damage relative to the vehicle was identified using pixel data obtained from images captured by the underwater vehicle's camera, and the real-world position was estimated based on the pixel dimensions. A dataset was created by calculating pixel values from images of damages taken from different distances. With this dataset, a support vector machine (SVM) was trained to estimate the distance corresponding to the pixel sizes, and consequently, a model was developed to predict the position of the damage. Additionally, the vehicle's position at the moment the damage was diagnosed was determined by taking the double integral of the accelerometer data, and the yaw angle was known from gyroscope measurements. By combining these two pieces of location information, the position of the damage relative to the starting point was determined.

The underwater vehicle is equipped with a specialized cable that provides both power and data transmission. This cable allows the collected data to be transmitted to a computer. The image data received by the computer is processed in the Python environment and passed through a pre-trained deep learning model. Since the pipeline tracking experiment was conducted in a pool environment, which is smaller compared to real-world scenarios, data transmission delays were neglected.

EXPERIMENT RESULTS

For this experiment, one 3-m and two 2-m pipes were placed in the pool at 45-degree angles to each other. Prior to the experiment, three separate points on the pipeline positioned in the pool were damaged. These damage images were provided in Section 1, Figure 2. The results of the damage diagnosis with CNN for the pool experiment and the location of this damage are given in Figures 12–14.

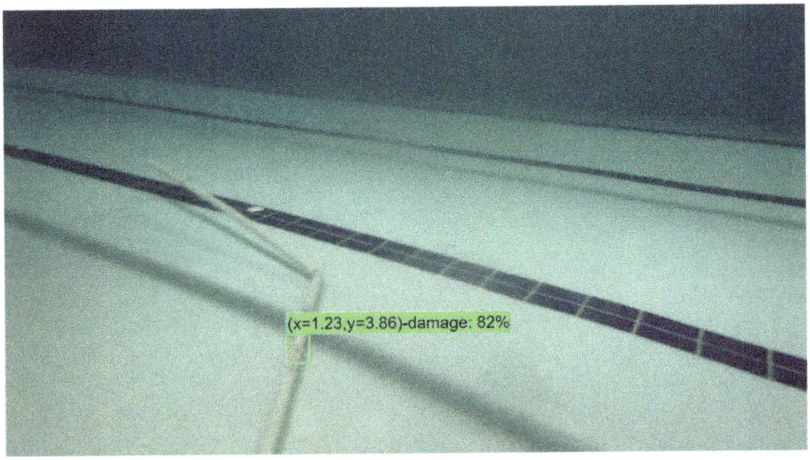

Figure 12. Location 1 of pipe line damage detection.

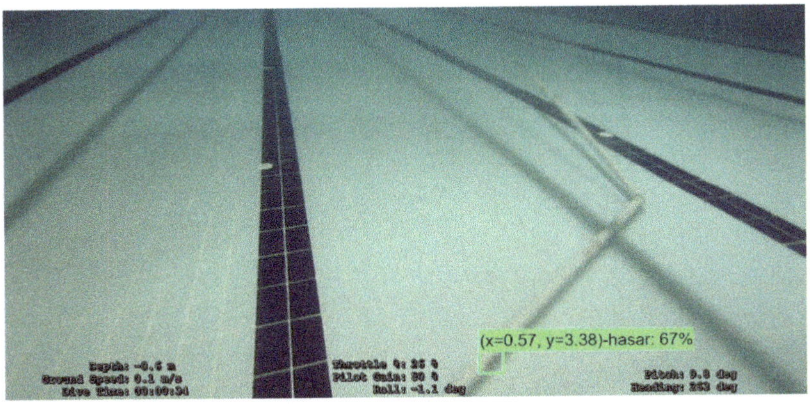

Figure 13. Location 2 of pipe line damage detection.

Figure 14. Location 3 of pipe line damage detection.

In Figure 15, the 3D movement of the unmanned underwater vehicle during the autonomous underwater pipeline tracking experiment are presented, respectively. The pipeline that was intended to be followed is shown as a blue line as the reference path (the known pipeline length is the actual data), and the path the vehicle followed during the experiment is shown as a red line (measured comes from navigation of the vehicle). As can be seen from Figure 15, the vehicle successfully followed the pipeline placed in the pool with high performance.

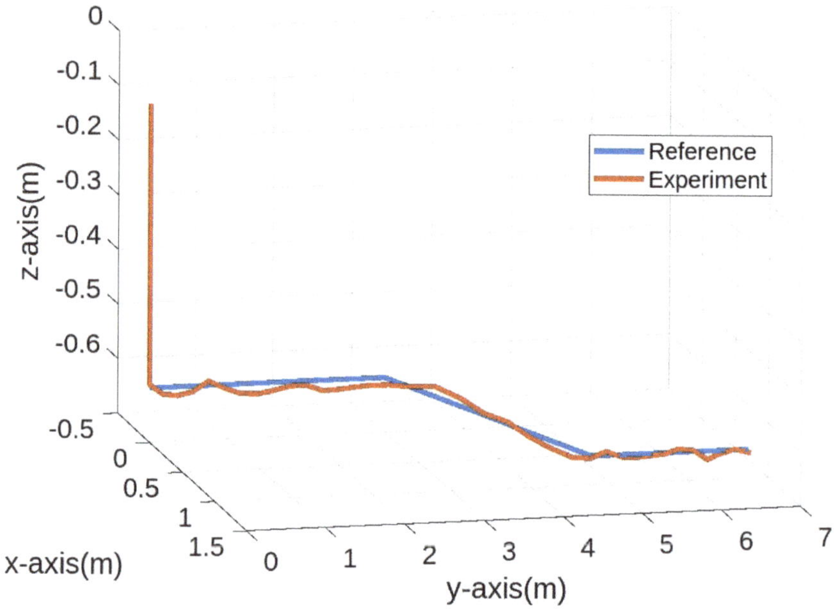

Figure 15. Reference path (blue line) and followed path (red line).

Figure 11 presents the depth information of the vehicle. In this experiment, the pipeline was placed at a depth of 1.8 m in the pool. To follow the pipeline from above at this depth, the unmanned underwater vehicle was submerged to a depth of 60 cm in the pool. Figure 10 presents the changes in the yaw angle made by the vehicle while following the pipeline. Initially, the vehicle followed the 3-m section of the pipeline at a 90-degree deviation angle in about 22 s (including 6 s of submersion time), the 2-m section at a 45-degree angle in 10 s, and the final 2-m section again at a 90-degree angle in 10 s.

Table 2 presents the root mean square error (RMSE) values for the study of underwater pipeline tracking with an autonomous unmanned underwater observation vehicle. It is the difference between the reference position value and the tracked position value. These RMSE values are known to the user as the path where the pipeline was placed, and were obtained from the path information recorded by the vehicle after completing the pipeline tracking.

Table 2. RMSE values between reference and tracked position value.

Position	RMSE
x	0.072 m
y	0.037 m
z	0.161 m
yaw	1.9 deg

5. Conclusions

The remotely controlled unmanned underwater vehicle, equipped with autonomous capabilities, successfully followed the pipeline placed in the pool. During the tracking process, damage to the pipeline was identified using a deep learning algorithm, CNN, from the images taken by the camera integrated into the vehicle. From our pool experiments, it was observed that factors such as light refraction, reflection, glare negatively affected the quality of the images obtained from the camera in underwater environments. Although the detection performance of the CNN algorithm decreased in a pool environment where light is refracted, reflected and shined environments, its object detection performance was still observed to be successful in experimentally. After detecting the damages on the autonomously tracked pipeline in the pool environment with CNN, the location of the damage was identified by correlating it with the vehicle's navigation data. The navigation data of the vehicle comes from the accelerometer, gyroscope, and pressure sensor. When comparing the actual path (reference path) values of the pipeline, which has a known length, with the measured position and deviation angle (experimental data) values during the vehicle's pipeline tracking, it was observed that the vehicle followed the pipeline with high accuracy. The results obtained in this study demonstrate that the detection and regular monitoring of defects in underwater pipelines can be carried out autonomously, safely, and continuously using unmanned underwater vehicles and deep learning algorithms.

Despite this achievement, the study has some limitations. The system in this study operates based on the images provided by the camera integrated into the unmanned underwater vehicle. Since the camera can only focus on the upper and side parts of the pipeline, it may not be possible to detect damage occurring on the underside of the pipeline. This is an important limitation, especially in cases where damage such as corrosion or cracks may occur on the lower surface of the pipeline. To overcome this limitation, multi-angle camera systems capable of imaging the lower part of the vehicle, or additional sensors capable of providing a full circular view around the pipeline, could be used. Another limitation is energy management for long-term underwater operations. Extending the battery capacity, turning off unused sensors, or processing only critical data could extend the operational duration of the vehicle. In the future work, autonomous tracking, damage detection, and damage localization of a pipeline laid in a marine environment will be tested using additional sensors such as sonar, under different scenarios and for various pipeline types.

Author Contributions: Writing, Data curation, Formal analysis, Investigation, Methodology, Visualization, S.K.K.; Software, Research, Data curation, R.F.C. All authors have read and agreed to the published version of the manuscript.

Funding: This work is supported by the Scientific and Technological Research Council of Turkey (grant 119E037) and grant 2210-C.

Institutional Review Board Statement: Not applicable.

Informed Consent Statement: Not applicable.

Data Availability Statement: Data are contained within the article.

Acknowledgments: This work is supported by the Scientific and Technological Research Council of Turkey (grant 119E037) and 2210-C. The authors are grateful for the support of the Scientific and Technological Research Council of Turkey and project of 119E037 team members and Berna Erol Cantekin, Ş. Hakan Kutoğlu , M. Kemal Leblebicioğlu, Rıfat Hacıoğlu, K. Sedar Görmüş.

Conflicts of Interest: The authors declare no conflict of interest.

References

1. Alvarez, A. Redesigning the SLOCUM Glider for Torpedo Tube Launching. *IEEE J. Ocean. Eng.* **2010**, *35*, 984–991. [CrossRef]
2. Bishop, G.C. Gravitational Field Maps and Navigational Errors. In Proceedings of the 2000 International Symposium on Underwater Technology, Tokyo, Japan, 23–26 May 2000; pp. 149–154.

3. Fattah, S.A.; Abedin, F. R3Diver: Remote Robotic Rescue Diver for Rapid Underwater Search and Rescue Operation. In Proceedings of the 2016 IEEE Region 10 Conference (TENCON), Singapore, 22–25 November 2016; pp, 3280–3283.
4. Wynn, R.B.; Huvenne, V.A.; Le Bas, T.P.; Murton, B.J.; Connelly, D.P.; Bett, B.J.; Ruhl, H.A.; Morris, K.J.; Peakall, J.; Parsons, D.R.; et al. Autonomous Underwater Vehicles (AUVs): Their Past, Present and Future Contributions to the Advancement of Marine Geoscience. *Mar. Geol.* **2014**, *352*, 451–468. [CrossRef]
5. Dinc, M.; Chingiz, H. Autonomous Underwater Vehicles. *J. Mar. Eng. Technol.* **2015**, *14*, 32–43. [CrossRef]
6. Lee, J.; Park, J.-H.; Hwang, J.-H.; Noh, K.; Choi, Y.; Suh, J. Artificial Neural Network for Glider Detection in a Marine Environment by Improving a CNN Vision Encoder. *J. Mar. Sci. Eng.* **2024**, *12*, 1106. [CrossRef]
7. Xia, T.; Cui, D.; Chu, Z.; Yu, X. Autonomous Heading Planning and Control Method of Unmanned Underwater Vehicles for Tunnel Detection. *J. Mar. Sci. Eng.* **2023**, *11*, 740. [CrossRef]
8. Kartal, S.K.; Hacıoğlu, R.; Görmüş, S.K.; Kutoğlu, Ş.H.; Leblebicioğlu, M.K. Modeling and Analysis of Sea-Surface Vehicle System for Underwater Mapping Using Single-Beam Echosounder. *J. Mar. Sci. Eng.* **2022**, *10*, 1349. [CrossRef]
9. Gu, J.; Wang, Z. Recent Advances in Convolutional Neural Networks. *Pattern Recognit.* **2018**, *77*, 354–377. [CrossRef]
10. Kaya Ustabaş, G.; Kocabaş, S.; Kartal, S.; Kaya, H.; Tekin, I.Ö.; Tığlı Aydin, R.S.; Kutoğlu, Ş.H. Detection of airborne nanoparticles with lateral shearing digital holographic microscopy. *Opt. Lasers Eng.* **2022**, *151*, 106934. . [CrossRef]
11. Niu, X.; Ching, Y.S. A Novel Hybrid CNN-SVM Classifier For Recognizing Handwritten Digits | Pattern Recognition. *Pattern Recognit.* **2011**, *45*, 131825.
12. Russakovsky, O.; Deng, J. ImageNet Large Scale Visual Recognition Challenge. *Int. J. Comput. Vis.* **2015**, *115*, 211–252. [CrossRef]
13. Simonyan, K.; Andrew, Z. Very Deep Convolutional Networks for Large-Scale Image Recognition. *arXiv* **2015**, arXiv:1409.1556.
14. Spahic, R.; Poolla, K. Image-based and Risk-informed Detection of Subsea Pipeline Damage. *Discov. Artif. Intell.* **2023**, *3*, 23. [CrossRef]
15. Bastian, B.; Jaspreeth, N. Visual Inspection and Characterization of External Corrosion in Pipelines Using Deep Neural Network. *NDT E Int.* **2019**, *107*, 102134. [CrossRef]
16. Avci, A.; Karadeniz Kartal, S. Leakage Detection in Underwater Oil and Natural Gas Pipelines Using Convolutional Neural Networks *Int. J. Energy Appl. Technol.* **2021**, *8*, 197–202.
17. Nadimi, N.; Javidan, R. Efficient Detection of Underwater Natural Gas Pipeline Leak Based on Synthetic Aperture Sonar (SAS) Systems. *Mar. Sci. Eng.* **2021**, *9*, 1273. [CrossRef]
18. Shi, Y.; Zhang, C. Theory and Application of Magnetic Flux Leakage Pipeline Detection. *Sensors* **2015**, *15*, 31036–31055. [CrossRef]
19. Hong, X.; Huang, L. Shedding Damage Detection of Metal Underwater Pipeline External Anticorrosive Coating by Ultrasonic Imaging Based on HOG + SVM. *Mar. Sci. Eng.* **2021**, *9*, 364. [CrossRef]
20. Bertulessi, M.; Bignami, D. Experimental Investigations of Distributed Fiber Optic Sensors for Water Pipeline Monitoring. *Sensors* **2023**, *23*, 6205. [CrossRef]
21. Peng, X.; Hao, H. A Numerical Study Of Damage Detection Of Underwater Pipeline Using Vibration-Based Method. *Int. J. Struct. Stab. Dyn.* **2012**, *12*, 1250021. [CrossRef]
22. Shitong, Z.; Yanhui, W. Subsea Pipeline Leak Inspection by Autonomous Underwater Vehicle. *Appl. Ocean. Res.* **2020**, *107*, 102321.
23. Khan, A.; Azhar, S. Visual Feedback–based Heading Control of Autonomous Underwater Vehicle for Pipeline Corrosion Inspection. *Int. J. Adv. Robot. Syst.* **2017**, *14*, 1729881416658171. [CrossRef]
24. Foresti, G.L. Visual Inspection of Sea Bottom Structures by an Autonomous Underwater Vehicle. *Int. J. Adv. Robot. Syst.* **2001**, *31*, 691–705. [CrossRef] [PubMed]
25. Cang, X.; Jouvencel, B. Coordinated Formation Control of Multiple Autonomous Underwater Vehicle for Pipeline Inspection. *Int. J. Adv. Robot. Syst.* **2010**, *7*, 3.
26. Cang, X.; Jouvencel, B. An Autonomous Underwater Vehicle Simulation With Fuzzy Sensor Fusion for Pipeline Inspection. *IEEE Sensors J.* **2023**, *23*, 8941–8951.
27. Szegedy, C.; Liu, W. Going Deeper With Convolutions. In Proceedings of the IEEE Conference on Computer Vision and Pattern Recognition (CVPR), Boston, MA, USA, 7–12 June 2015; pp.1–9.
28. He, X.; Wei, Z. Emotion Recognition By Assisted Learning With Convolutional Neural Networks. *Neurocomputing* **2018**, *291*, 187–194. [CrossRef]
29. Pertusa, A.; Gallego, A. MirBot: A Collaborative Object Recognition System for Smartphones Using Convolutional Neural Networks. *Neurocomputing* **2018**, *293*, 87–99. [CrossRef]
30. Li, X.; Min, S. Fast Accurate Fish Detection and Recognition Of Underwater İmages with Fast R-CNN. In *OCEANS MTS/IEEE Washington*; IEEE: Piscataway, NJ, USA, 2015.
31. Chavez, A.G.; Birk, A. Stereo-Vision Based Diver Pose Estimation Using LSTM Recurrent Neural Networks For AUV Navigation Guidance. In *OCEANS-Aberdeen*; IEEE: Piscataway, NJ, USA, 2017; pp.1–7.
32. Buß, M.; Steiniger, Y. Hand-Crafted Feature Based Classification against Convolutional Neural Networks for False Alarm Reduction on Active Diver Detection Sonar Data. In *OCEANS MTS/IEEE Charleston*; IEEE: Piscataway, NJ, USA, 2018; pp.1–7.
33. Williams, D.P. *Demystifying Deep Convolutional Neural Networks For Sonar Image Classification*; NATO STO Centre for Maritime Research and Experimentation (CMRE) Viale San Bartolomeo: La Spezia, Italy, 2019; Volume 400, pp. 513–520.
34. Williams, D.P. On the Use of Tiny Convolutional Neural Networks for Human-Expert-Level Classification Performance in Sonar Imagery. *IEEE J. Ocean. Eng.* **2020**, *4*, 236–260. [CrossRef]

35. Cortes, C.; Vapnik, V. *Support-Vector Networks Machine Learning*; Springer: Cham, Switzerland, 1995; Volume 20, pp. 273–297.
36. Zhang Z., Ding S. MBSVR: Multiple birth support vector regressions. *Inf. Sci.* **2021**, *552*, 65–79. [CrossRef]
37. Zhao, Q.; Qin, X. A novel prediction method based on the support vector regression for the remaining useful life of lithium-ion batteries. *Microelectron. Reliab.* **2018**, *85*, 99–108. [CrossRef]
38. Li, X.; Shu, X. An On-Board Remaining Useful Life Estimation Algorithm for Lithium-Ion Batteries of Electric Vehicles. *Energies* **2017**, *10*, 691. [CrossRef]
39. Meniconi, S.; Brunone, B.; Tirello, L.; Rubin, A.; Cifrodelli, M.; Capponi, C. Transient tests for checking the Trieste subsea pipeline: Towards the field tests. *J. Mar. Sci. Eng.* **2024**, *12*, 374. [CrossRef]
40. Meniconi, S.; Brunone, B.; Tirello, L.; Rubin, A.; Cifrodelli, M.; Capponi, C. Transient tests for checking the Trieste subsea pipeline: Diving into fault detection. *J. Mar. Sci. Eng.* **2024**, *12*, 391. [CrossRef]
41. Oktanisa, I.; Mahmudy, W.F. Inflation Rate Prediction in Indonesia using Optimized Support Vector Regression Model. *J. Inf. Technol. Comput. Sci.* **2020**, *5*, 104–114. [CrossRef]
42. Manasa, J.; Grupta, R. Machine Learning based Predicting House Prices using Regression Techniques. In Proceedings of the 2nd International Conference on Innovative Mechanisms for Industry Applications, (ICIMIA), Bangalore, India, 5–7 March 2020; pp. 624–630.
43. Smola, A.J.; Schölkopf, B. A Tutorial On Support Vector Regression. *Stat. Comput.* **2004**, *14*, 199–222. [CrossRef]
44. Li, M.W.; Geng, J. Periodogram Estimation Based On LSSVR-CCPSO Compensation For Forecasting Ship Motion. *Nonlinear Dyn.* **2019**, *97*, 2579–2594. [CrossRef]
45. Dong, Y.; Zhang, Z. A Hybrid Seasonal Mechanism With A Chaotic Cuckoo Search Algorithm with A Support Vector Regression Model For Electric Load Forecasting. *Energies* **2018**, *11*, 1–21. [CrossRef]
46. Cheng, K.; Lu, Z. Active Learning Bayesian Support Vector Regression Model For Global Approximation. *Inf. Sci.* **2021**, *544*, 549–563. [CrossRef]
47. Zhang, Z.; Ding, S. A Support Vector Regression Model Hybridized With Chaotic Krill Herd Algorithm And Empirical Mode Decomposition For Regression Task. *Neurocomputing* **2020**, *410*, 185–201. [CrossRef]
48. Erol, B.; Cantekin, R.; Kartal, S.K.; Hacioglu, R.; Gormus, S.; Kutoglu, H.; Leblebicioglu, K. Estimation of Unmanned Underwater Vehicle Motion with Kalman Filter and Improvement by Machine Learning. *Int. J. Adv. Eng. Pure Sci.* **2021**, *33*, 67–77.
49. Liang, Z.; Wang, K. Multisensor Fusion Simultaneous Localization and Mapping System Based on Image Enhancement. *J. Mar. Sci. Eng.* **2024**, *17*, 1170. [CrossRef]
50. Wang, C.; Cheng, C.; Cao, C.; Guo, X.; Pan, G.; Zhang, F. An Invariant Filtering Method Based on Frame Transformed for Underwater INS/DVL/PS Navigation. *J. Mar. Sci. Eng.* **2024**, *12*, 1178. [CrossRef]
51. Kartal, S.; Leblebicioğlu, M.K. Experimental Test of the Acoustic-Based Navigation and System Detection of an Unmanned Underwater Survey Vehicle (SAGA). *Trans. Inst. Meas. Control* **2018**, *40*, 247–687.
52. Li, M.; Ji, H. Underwater Object Detection And Tracking Based On Multi-Beam Sonar İmage Processing. In Proceedings of the IEEE International Conference on Robotics and Biomimetics (ROBIO), Guilin, China, 19–13 December 2013; pp. 1071–1076.
53. Mandic, F.; Ivor, R. Underwater Object Tracking Using Sonar and USBL Measurements. *J. Sensors* **2016**, *2016*, 8070286. [CrossRef]
54. Xiang, X.; Caoyang, Y. Autonomous Underwater Vehicle with Magnetic Sensing Guidance. *Sensors* **2016**, *16*, 1335. [CrossRef]
55. Bigoni, C.; Hesthaven, J. Simulation-Based Anomaly Detection and Damage Localization: An Application to Structural Health Monitoring. *Comput. Methods Appl. Mech. Eng.* **2020**, *363*, 112896. [CrossRef]
56. Katija, K.; Roberts, P. Visual Tracking Of Deepwater Animals Using Machine Learning-Controlled Robotic Underwater Vehicles. In Proceedings of the IEEE Winter Conference on Applications of Computer Vision (WACV), Waikoloa, HI, USA, 3–7 January 2021; pp. 859–868. [CrossRef]
57. Packard, G.E.; Kukulya, A. Continuous Autonomous Tracking And İmaging Of White Sharks And Basking Sharks Using a REMUS-100 AUV. In *OCEANS*; IEEE: Piscataway, NJ, USA, 2013; pp.1–5.
58. Kezebou, L.; Oludare, V. Underwater Object Tracking Benchmark and Dataset. In Proceedings of the IEEE International Symposium on Technologies for Homeland Security (HST), Woburn, MA, USA, 23–24 October 2019; pp.1–6.
59. Kartal, S.; Leblebicioğlu, M.K. Experimental test of vision-based navigation and system identification of an unmanned underwater survey vehicle (SAGA) for the yaw motion. *Trans. Inst. Meas. Control* **2019**, *41*, 2160–2170. [CrossRef]
60. Khan, A.; Ali, S.S.A.; Anwer, A.; Adil, S.H.; Mériaudeau, F. Subsea Pipeline Corrosion Estimation by Restoring and Enhancing Degraded Underwater Images. *IEEE Access* **2018**, *6*, 40585–40601. [CrossRef]
61. Hong, S.-J.; Kim, S.-Y. Moth Detection from Pheromone Trap Images Using Deep Learning Object Detectors. *Agriculture* **2020**, *10*, 170. [CrossRef]
62. Ning, W.; Yuanyuan, W. Marine vessel detection dataset and benchmark for unmanned surface vehicles. *Appl. Ocean. Res.* **2024**, *142*, 103835.
63. He, K.; Xiangyu, Z. Spatial Pyramid Pooling in Deep Convolutional Networks for Visual Recognition. *IEEE Trans. Pattern Anal. Mach. Intell.* **2014**, *37*, 346–361.
64. Nair, V.; Geoffrey, E.H. Rectified Linear Units Improve Restricted Boltzmann Machines. In Proceedings of the International Conference on Machine Learning, ICML10, Omnipress, Haifa, Israel, 21–24 June 2010; Volume 27, pp. 807–814.
65. Albawi, S.; Tareq, A.M. Understanding of a Convolutional Neural Network. In Proceedings of the 2017 International Conference on Engineering and Technology (ICET), Antalya, Turkey, 21–23 August 2017; pp. 16–21.

66. Albawi, S.; Tareq, A.M. Artificial Convolution Neural Network for Medical Image Pattern Recognition. *Neural Netw.* **1995**, *8*, 1201–1214.
67. De Oliveira, D.C.; Wehrmeister, M.A. Using Deep Learning and Low-Cost RGB and Thermal Cameras to Detect Pedestrians in Aerial Images Captured by Multirotor UAV. *Sensors* **2018**, *18*, 2244. [CrossRef] [PubMed]
68. Caccia, M.; Bibuli, R. Basic Navigation, Guidance And Control Of An Unmanned Surface Vehicle. *Auton. Robot.* **2008**, *25*, 349–365. [CrossRef]
69. Fossen, T.I. *Guidance and Control of Ocean Vehicles*; Wiley: Hoboken, NJ, USA, 1999.

MDPI AG
Grosspeteranlage 5
4052 Basel
Switzerland
Tel.: +41 61 683 77 34

Journal of Marine Science and Engineering Editorial Office
E-mail: jmse@mdpi.com
www.mdpi.com/journal/jmse